MOUNTAIN METEOROLOGY

MOUNTAIN METEOROLOGY

FUNDAMENTALS AND APPLICATIONS

C. David Whiteman

New York Oxford

Oxford University Press

2000

Oxford University Press

Oxford New York
Athens Auckland Bangkok Bogotá Buenos Aires Calcutta
Cape Town Chennai Dar es Salaam Delhi Florence Hong Kong Istanbul
Karachi Kuala Lumpur Madrid Melbourne Mexico City Mumbai
Nairobi Paria São Paulo Singapore Taipei Tokyo Toronto Warsaw

and associated companies in
Berlin Ibadan

Published by Oxford University Press, Inc.
198 Madison Avenue, New York, New York 10016

Library of Congress Cataloging-in-Publication Data
Whiteman, C. D. (Charles David), 1946–
Mountain meteorology : fundamentals and
applications / C. David Whiteman.
p. cm.
Includes bibliographical references and index.
ISBN 0-19-513271-8
1. Mountain climate. I. Title.
QC993.6.W48 2000
551.6914′3—dc21 99-24940

9 8 7 6 5 4 3 2 1
Printed in Hong Kong
on acid-free paper

To Johanna,
love of my life.

Preface

Mountain Meteorology: Fundamentals and Applications aims to heighten awareness and appreciation of the weather in mountainous areas by introducing the reader to the basic principles and concepts of mountain meteorology and by discussing applications of these principles and concepts in natural resource management. The reader will learn to recognize characteristic mountain weather patterns and events, to anticipate their evolution, and to evaluate their impact on planned activities. Two hundred and seventy-four figures, diagrams, and photographs, most in full color, support the text and promote a conceptual understanding of mountain meteorology. In the figures and in the text, observable indicators (winds, temperature, clouds) of atmospheric processes are emphasized to facilitate the recognition of weather systems and events.

Mountain Meteorology will interest anyone who spends time in or near mountains and whose daily life, work, or recreational activities are affected by the weather. It was written, however, to meet the specific needs of three U.S. government agencies. Work on the book was initiated with the support of the USDA Forest Service to address the need for a training manual for aerial spraying operations in national forests. Support was also provided by the National Weather Service, which needed a reference and training book for their meteorologists, who forecast mountain weather for the general public, as well as for natural resource agencies responsible for air pollution investigations, forest fire and smoke management, aerial spraying, and other land management activities. The U.S. Army also supported the project because of its interest in aerosol dispersion in mountainous terrain and in training personnel to meet land management responsibilities at army facilities in complex terrain.

The mountains of North America provide most of the examples included in this text, although the principles behind the examples apply to mountainous regions around the world. Examples from the European

Alps, the Southern Alps of New Zealand, and the Himalayas are included when the phenomena being described are generally associated with that particular region or are particularly well developed there.

Both English and metric units are usually used. The order in which the units are given varies and depends on the specific context. Mathematical equations are provided in an appendix rather than in the main text. Throughout the book, "Points of Interest" expand on mountain weather topics, provide specific examples, or explore a closely related subject. These sections are identified by a colored background. "Key Points" appear in the margins and highlight basic concepts. Technical terms that may be new to readers are italicized and usually defined on first use. These terms, and many others, are included in a glossary at the end of the book. An index assists the reader in quickly finding topics or geographical place names of interest.

Mountain Meteorology is divided into four parts. Part I (chapters 1 and 2) discusses four factors that influence climate (chapter 1) and describes the characteristic climates of the mountain areas of North America (chapter 2). Part II (chapters 3–9) sets the stage for the discussions of mountain wind systems and applications in parts III and IV by describing basic weather elements and processes. These chapters also highlight several mountain meteorology topics, including mountain clouds, mountain thunderstorms, and lightning safety. Part III (chapters 10 and 11) focuses on mountain wind systems. The wind systems are key to understanding all types of mountain weather. They affect the movement of fronts and air masses, the development of clouds and precipitation, and the daily and seasonal cycles of temperature and humidity in mountain areas. Chapter 10 focuses on the terrain-forced flows produced when air currents approach a mountain barrier and are forced to flow over or around the barrier or through gaps in the barrier. These flows connect the atmosphere within a mountain region to the larger scale winds aloft and can impact wildfires and atmospheric dispersion within the mountain massif. Chapter 11 discusses diurnal circulations that develop within mountain areas. These circulations, caused by temperature contrasts within the mountain massif or between the mountain massif and the plains, occur regularly on fair weather days and are characteristic of the mountain environment. Part IV (chapters 12–14) applies the meteorological principles explained in the previous chapters to selected forest and land management practices and operations. Chapter 12 discusses air pollution dispersion. Chapter 13, written by Carl J. Gorski and Allen Farnsworth, discusses fire weather and the management of smoke from prescribed fires and wildfires. Chapter 14, written by Harold W. Thistle and John W. Barry, discusses aerial spraying of pest control agents, seeds, and fertilizers.

Appendices provide key equations (appendix A), tables for computation of relative humidity (appendix B), a compilation of source materials on meteorological monitoring and instrumentation (appendix C), units conversion tables (appendix D), computer programs for calculating theoretical solar radiation on slopes (appendix E), a list of additional reading materials (appendix F), and a list of abbreviations used in meteorological codes (appendix G).

Acknowledgments

I am deeply grateful to my wife, Johanna, whose encouragement, support, and enthusiasm helped me reach a long-term goal that was set when we were exploring the Rocky Mountains together with our Colorado Mountain Club and ski patrol friends in high school. She was a true partner in completing this book. She edited and rewrote the manuscript numerous times, answering many organizational and formatting questions. Her perspective as a nonmeteorologist and her training in grammar and language helped identify jargon and clarify the text.

Kathy Kachele of Lockheed Martin Services, Inc., in Richland, Washzington, produced most of the figures for the manual. Assistance on special figures came from Xindi Bian, Harlan Foote, and Jerome Fast of Battelle Pacific Northwest Laboratories in Richland, Washington.

Helpful review comments came from many individuals, including W. R. Barchet, John W. Barry, Stephan de Wekker, Bob Hammer, Harold Thistle, Milt Teske, Paul Stokols, Sue Ferguson, Mike Johnson, and James F. Bowers.

Permission to reproduce copyrighted or original source materials was provided by

Allen Farnsworth, USDA Forest Service, Flagstaff, Arizona
Jerome D. Fast, Pacific Northwest National Laboratory, Richland, Washington
Harlan P. Foote, Pacific Northwest National Laboratory, Richland, Washington
Carl Gorski, National Weather Service, Salt Lake City, Utah
Edward E. Hindman, City University of New York, New York
Ronald L. Holle, National Severe Storms Laboratory, Norman, Oklahoma
Peter F. Lester, San Jose State University, San Jose, California

Brooks Martner, NOAA Environmental Technology Laboratory, Boulder, Colorado
John Thorp, Pacific Northwest National Laboratory, Richland, Washington
Mike Ziolko, Oregon Department of Forestry, Salem, Oregon
The American Alpine Club, Golden, Colorado
The American Meteorological Society, Boston, Massachusetts
The European Center for Medium-Range Forecasting, Reading, United Kingdom
The Montana Stockgrowers Association, Helena, Montana
The Mount Washington Observatory, North Conway, New Hampshire
The National Aeronautics and Space Administration, Washington, DC
The National Oceanic and Atmospheric Administration, Washington, DC
The National Park Service, Washington, DC
The National Weather Service, Washington, DC
The United States Air Force, Washington, DC
The United States Department of Agriculture Forest Service, Washington, DC

An Editorial Board provided technical resources and guidance on the technical content and arranged for the review and publication of the manual. Board members were Harold Thistle, John W. Barry, Carl Gorski, Bruce Grim, and Rusty Billingsley.

Funding for this book was provided by the USDA Forest Service, the National Weather Service, and the U.S. Army. I thank John W. Barry and Harold Thistle of the USDA Forest Service, Jeanne Hoadley, Paul Stokols, and Andy Edman of the National Weather Service, and Bruce Grim of the U.S. Army for arranging the necessary funding. I also thank the U.S. Department of Energy for their support of my research programs in mountain meteorology over the last decade and the opportunity to collaborate with other federal agencies on this project.

This work was accomplished at Pacific Northwest National Laboratory, which is operated for the U.S. Department of Energy under contract DE-AC06-76RLO 1830 by Battelle Memorial Institute.

Contents

Part I Mountain Climates

1. Four Factors that Determine Climate 3
    - 1.1. Latitude 3
    - 1.2. Altitude 4
    - 1.3. Continentality 7
    - 1.4. Regional Circulations 7
2. Mountain Climates of North America 11
    - 2.1. The Appalachians 13
    - 2.2. The Coast Range, the Alaska Range, the Cascade Range, and the Sierra Nevada 15
    - 2.3. The Rocky Mountains 18
    - 2.4. Between the Mountains 19

Part II An Introduction to the Atmosphere

3. Atmospheric Scales of Motion and Atmospheric Composition 25
    - 3.1. Atmospheric Scales of Motion 25
    - 3.2. Atmospheric Composition 26
4. Atmospheric Structure and the Earth's Boundary Layer 31
    - 4.1. Vertical Structure of the Atmosphere 31
    - 4.2. Temperature 33
    - 4.3. Atmospheric Stability 38
    - 4.4. The Atmospheric Boundary Layer and the Surface Energy Budget 42

5. Pressure and Winds 49
5.1. Atmospheric Pressure 49
5.2. Winds 60
6. Air Masses and Fronts 73
6.1. Air Mass Source Regions and Trajectories 73
6.2. Fronts 74
7. Clouds and Fogs 81
7.1. Clouds 81
7.2. Fogs 94
8. Precipitation 99
8.1. Types of Precipitation 99
8.2. Intensity of Precipitation 101
8.3. Measuring Precipitation 101
8.4. Formation of Precipitation 102
8.5. Spatial and Temporal Distribution of Precipitation 105
8.6. Icing 111
8.7. Mountain Thunderstorms 112
9. Weather Maps, Forecasts, and Data 127
9.1. Weather Maps 127
9.2. Forecasting Guidelines 128
9.3. Weather Information: Data Collection and Dissemination 129
9.4. Obtaining Professional Forecasts for Major Federal Projects 138

Part III Mountain Winds

10. Terrain-Forced Flows 141
10.1. Three Factors that Affect Terrain-Forced Flows 141
10.2. Flow over Mountains 146
10.3. Flow around Mountains 158
10.4. Flows through Gaps, Channels, and Passes 161
10.5. Blocking, Cold Air Damming, and Obstruction of Air Masses 165
10.6. On the High Plains: The Low-Level Jet 168
11. Diurnal Mountain Winds 171
11.1. The Daily Cycle of Slope and Along-Valley Winds and Temperature Structure 172
11.2. Modification of Diurnal Mountain Winds by Variations in the Surface Energy Budget 174
11.3. Disturbances of the Daily Cycle by Larger Scale Flows 182
11.4. The Four Components of the Mountain Wind System 186
11.5. Diurnal Mountain Winds in Basins 197
11.6. Diurnal Mountain Winds over Plateaus 198
11.7. Other Local Thermally Driven Wind Systems 199

Part IV Selected Applications of Mountain Meteorology

12. Air Pollution Dispersion 205
    - 12.1. Classification and Regulation of Air Pollutants 205
    - 12.2. Air Quality Studies and Air Pollution Models 209
    - 12.3. Wind Speed and Air Pollution Concentrations 212
    - 12.4. Stability, Inversions, and Mixing Depth 213
    - 12.5. Synoptic Weather Categories and Air Pollution Dispersion 218
    - 12.6. Mountainous Terrain and Atmospheric Dispersion 221
    - 12.7. Assessing Air Pollution Potential in Mountain Terrain 235

13. Fire Weather and Smoke Management (by Carl J. Gorski and Allen Farnsworth) 239
    - 13.1. The Fire Environment 239
    - 13.2. Fuel Moisture Content 241
    - 13.3. Fire Weather in Complex Terrain 242
    - 13.4. Critical Fire Weather 254
    - 13.5. Prescribed Fire and Smoke Management 259
    - 13.6. Monitoring Fire Weather and Smoke Dispersion Parameters 265

14. Aerial Spraying (by Harold W. Thistle and John W. Barry) 273
    - 14.1. Overview of Aerial Spraying 275
    - 14.2. Meteorological Factors that Affect Aerial Spraying Operations 279
    - 14.3. Spray Deposition 285
    - 14.4. Additional Considerations in Complex Terrain 286
    - 14.5. Collection of Meteorological Data 288
    - 14.6. Computer Modeling 293
    - 14.7. Integration of Meteorological Information into Operations 295

References 299

Appendixes 303
- A. Formulas 303
- B. Psychrometric Tables 307
- C. Sources of Information on Weather Monitoring and Instrumentation 308
- D. Units, Unit Conversion Factors, and Time Conversions 309
- E. Solar Radiation on Slopes 312
- F. Additional Reading 319
- G. METAR and TAF Code Abbreviations 321

Glossary 325

Abbreviations and Acronyms 341

Index 343

Meteorology is the study of the earth's atmosphere and the atmospheric processes that produce *weather* and *climate*. Weather is the state of the atmosphere during a short period of time (e.g., days or weeks), as measured by variables such as temperature, humidity, wind speed and direction, cloudiness, precipitation, and pressure. Climate is the average or generally prevailing weather of a given region over a long period of time (e.g., months, years, or centuries). *Mountain meteorology* focuses on the weather and climate of mountainous regions.

Part I

MOUNTAIN CLIMATES

Four Factors That Determine Climate

1

Climate differs from one location to another because of differences in

- latitude, the angular distance north or south from the equator
- altitude, the height above sea level
- *continentality*, the distance from the sea
- exposure to regional circulations, including winds and ocean currents

1.1. Latitude

The latitude of a given site determines the length of the day and the angle of incoming sunlight and therefore the amount of solar *radiation* received at that site. Seasonal and *diurnal* (day–night) variations in the amount of solar radiation received cause seasonal and diurnal variations in the weather. Near the equator, the days of the year are all about the same length, and the noon sun is nearly overhead year-round. Because *day length* and solar angle change little with the season, there is little seasonal variability in the weather. In the polar regions, on the other hand, the sun does not rise at all in the winter, and in the summer it never sets, although it remains low in the sky. Thus, polar weather has a high seasonal variability, but a low diurnal variability. In the midlatitudes, the climate is characterized by both seasonal and diurnal changes. Except at the equator, day length varies throughout the year (figure 1.1). In the Northern Hemisphere, the longest day of the year is at the summer *solstice* (June 21), the shortest day of the year is at the winter solstice (December 21), and the day is 12 hours long on the vernal and autumnal *equinoxes* (March 20 and September 22). The altitude angle of the sun also varies throughout the year, with an increase of about 47° from winter to

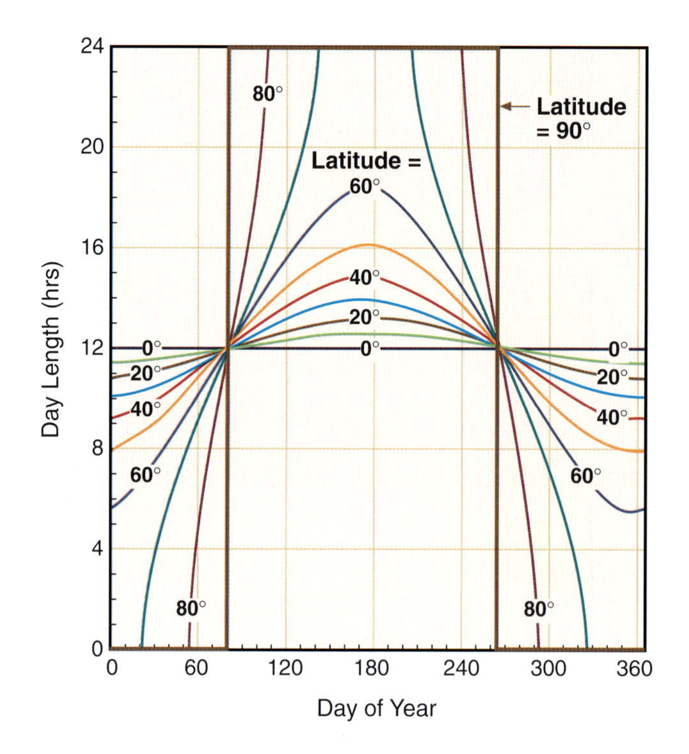

Figure 1.1 Day length (the period of time during which the sun is above the horizon) depends on time of year and latitude. The sun shines for 12 hours per day at the equator year-round. At the North Pole, the sun is above the horizon for 24 hours per day in the summer, but it does not rise at all in winter. In the midlatitudes, days are longer than 12 hours in summer and shorter than 12 hours in winter.

summer (figure 1.2). The more direct summer sunlight produces more heating than the slanted rays of the winter sun.

The latitude of a given site affects its climate not only because it determines the angle of solar radiation and the length of a day, but also because it determines the site's exposure to latitudinal belts of surface high and low *pressure* that encircle the earth (figure 1.3). High pressure belts (i.e., zones where high pressure centers are often found) are associated with sinking motions, or *subsidence,* in the atmosphere, clear skies, dry air, and light winds. Low pressure belts are associated with rising motions, or *convection,* in the atmosphere, cloudiness, precipitation, and strong winds. Belts of low pressure occur in the equatorial (0–20° latitude) and subpolar (40–70°) regions and alternate with belts of high pressure that form in the subtropical (20–40°) and polar (70–90°) regions. The United States's contiguous 48 states are influenced primarily by the subtropical high pressure belt and the subpolar low pressure belt. Alaska and Canada are affected by the belts of polar high pressure and subpolar low pressure. Pressure belts shift north in the summer and south in the winter, causing significant seasonal changes in weather and climate. *High* and *low pressure centers* that are characteristic of a given area are named for that area, such as the Bermuda–Azores High or Bermuda High, the Pacific High, and the Aleutian Low.

1.2. Altitude

Temperature, atmospheric moisture, precipitation, winds, incoming solar radiation, and air density all vary with altitude. Up to an altitude of

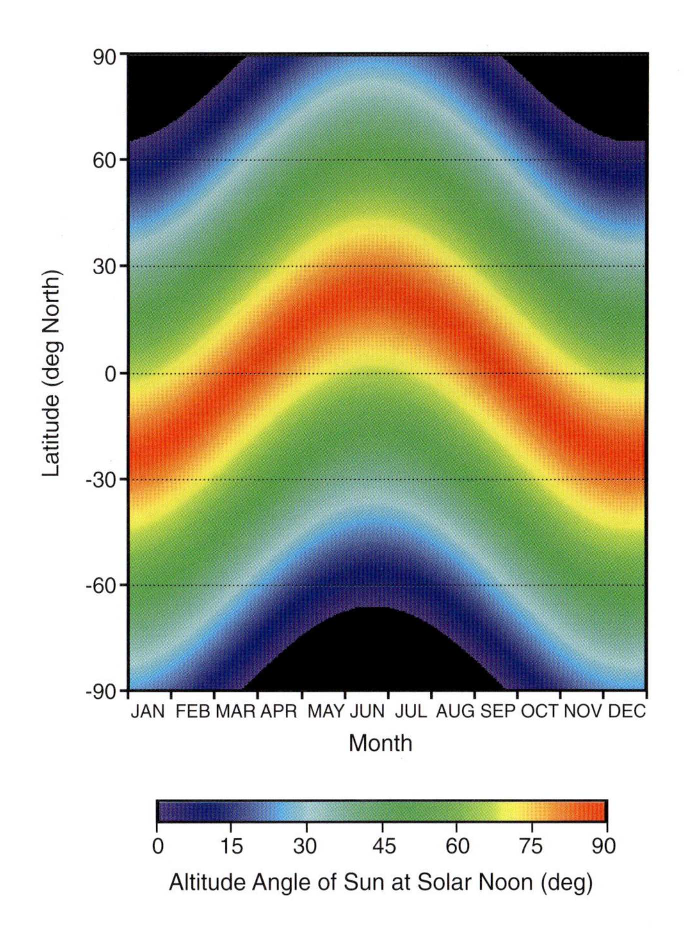

Figure 1.2 The altitude angle of the sun at noon depends on the latitude and the day of the year. At latitudes between the polar circles (67 ½°N and S), the altitude angle of the sun shifts about 47° from winter to summer. At the poles, the sun does not rise above the horizon during the 6-month-long polar night (black shading).

7 miles (11 km), temperature generally decreases with altitude. The rate of decrease is typically 3.5°F per 1000 ft or 6.5°C/km. Thus, locations at high elevations generally have a cooler climate than locations at lower elevations.

Incoming solar radiation increases with altitude. As radiation passes through the earth's atmosphere, a small fraction is absorbed by the atmosphere, resulting in a minor increase in the air temperature. Another small fraction is scattered by atmospheric constituents and redirected into space. Because solar radiation reaches higher elevation land surfaces before lower elevation land surfaces, there is less depletion of the solar beam through *absorption* and *scattering* at higher elevations than at lower elevations.

Although more incoming solar radiation reaches the ground at higher elevations, the effect on air temperature is minimal, and changes in air temperature from day to night on exposed mountainsides and peaks are smaller than the diurnal changes at lower altitudes. This is explained by the way the earth's atmosphere is heated and cooled and by the decrease of land surface area with elevation. Most of the radiation received from the sun does not heat the earth's atmosphere directly but rather passes largely unimpeded through the atmosphere, is received at the earth's surface, and heats the ground. The ground, in turn, heats the atmosphere from below. Because there is less land surface area at higher elevations, less heat is transferred to the atmosphere during the day.

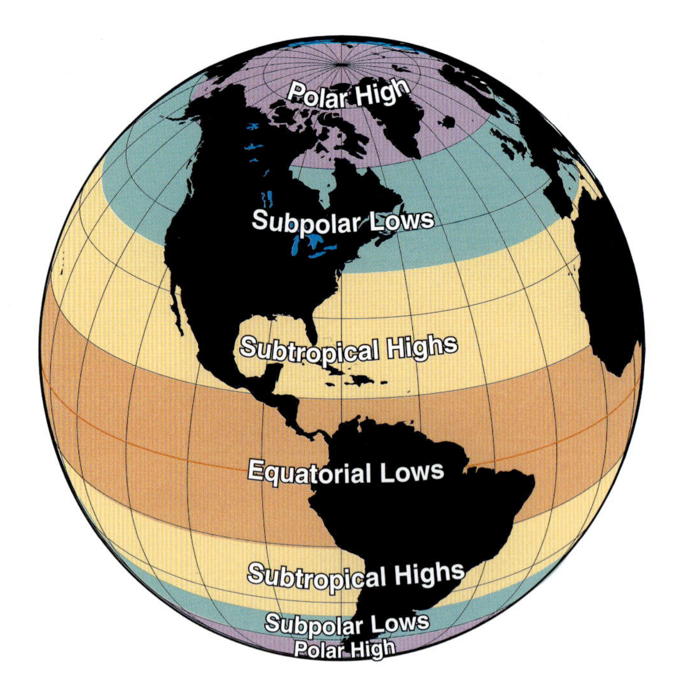

Figure 1.3 Latitudinal belts of high and low pressure. Rising motions and high rainfall amounts are usually concentrated in the low pressure belts, whereas sinking motions and dry conditions are found in the high pressure belts.

At night, radiation loss from the ground cools the earth's surface, which then cools the air above it. Air near the surface, for example, on mountain slopes, thus cools more than air at the same elevation in the *free atmosphere*. Air that cools over mountain slopes flows down the slopes and collects in valleys and basins, resulting in the formation of *temperature inversions*, layers in which temperature increases with height instead of decreasing. Temperature inversions are commonly seen both in mountainous areas and over flat terrain at night or at all times of day in winter; they are an important exception to the rule that temperature decreases with altitude.

The temperature at a given site within a mountain *massif* depends not only on the site's altitude, but also on its exposure to incoming solar radiation. South-facing slopes receive more solar radiation than north-facing slopes, for example. Temperature differences within a mountain massif are important because they drive local winds (see chapter 11).

Atmospheric moisture also generally decreases with altitude. The earth's surface supplies the atmosphere not only with heat, but also with water vapor through evaporation of water (the oceans are the primary source of atmospheric moisture) and transpiration from plants. As altitude increases, the distance from the source of moisture increases, and, therefore, the amount of moisture in the atmosphere decreases.

Although moisture decreases with altitude, precipitation usually increases. The cooler air at higher altitudes can hold less moisture than the warmer air at sea level. Thus, when warm, moist air from sea level is lifted over a mountain range, the air cools, its capacity to hold moisture decreases, and much of the moisture is released as precipitation.

Wind speeds generally increase with altitude. Wind speeds are lowest at the earth's surface because of friction. Winds measured on mountain peaks tend to be stronger than winds at lower elevations because the peaks extend high into the atmosphere where wind speeds are higher. The limited surface area of the peaks themselves produces little friction to slow the winds. Winds that are carried over mountains or through mountain passes may even speed up because of the influence of the *complex terrain* (chapter 10). However, valleys, basins, and *lee* slopes within a mountain area are often sheltered from the generally stronger winds at high altitudes by the surrounding topography.

Air *density*, the mass of a unit volume of air, decreases exponentially with height. The same amount of heat input results in a greater change in temperature at higher elevations than at lower elevations. The less dense mountain atmosphere also responds more quickly to the input of heat than the denser atmosphere at lower elevations. This quick response to heat input combined with the rapid transport of warm and cold air, clouds, and storms by the strong winds at high elevations contributes to the perception of the high "changeability" of mountain weather. Lower air density also affects the perception of wind velocity. The same wind velocity is perceived as somewhat weaker at higher elevations than at lower elevations because it is transporting less mass and thus less momentum.

1.3. Continentality

Locations at the center of a continent experience larger diurnal and seasonal temperature changes than locations on or near large bodies of water because land surfaces heat and cool more quickly than oceans. Interior locations also experience more sunshine, less cloudiness, less moisture, and less precipitation than coastal areas, where the maritime influence produces more cloudiness and precipitation and moderates temperatures. Precipitation is especially heavy on the *windward* side of coastal mountain ranges oriented perpendicular to prevailing winds from the ocean. As mentioned previously (section 1.2), marine air that is lifted up a mountain range releases much of its moisture as precipitation. As a result, far less precipitation is received on the *leeward* side of a mountain range.

1.4. Regional Circulations

Although latitude, altitude, and continentality are the primary determinants of climate in a mountainous region, exposure to regional winds and ocean currents is also a factor. Regional winds are associated with the semipermanent atmospheric *high* and *low pressure systems* that form in different latitude belts and directly affect climate (figure 1.4). In the Northern Hemisphere, winds blow clockwise around high pressure centers and counterclockwise around low pressure centers.

In summer (figure 1.4a), two large high pressure centers influence the weather over much of North America. The Bermuda–Azores High is located over the Atlantic Ocean in the subtropical belt of high pressure. The large clockwise circulation around the high pressure center carries warm tropical air northward into the central United States. This moist air contributes to the development of thunderstorms. In midsummer, the influ-

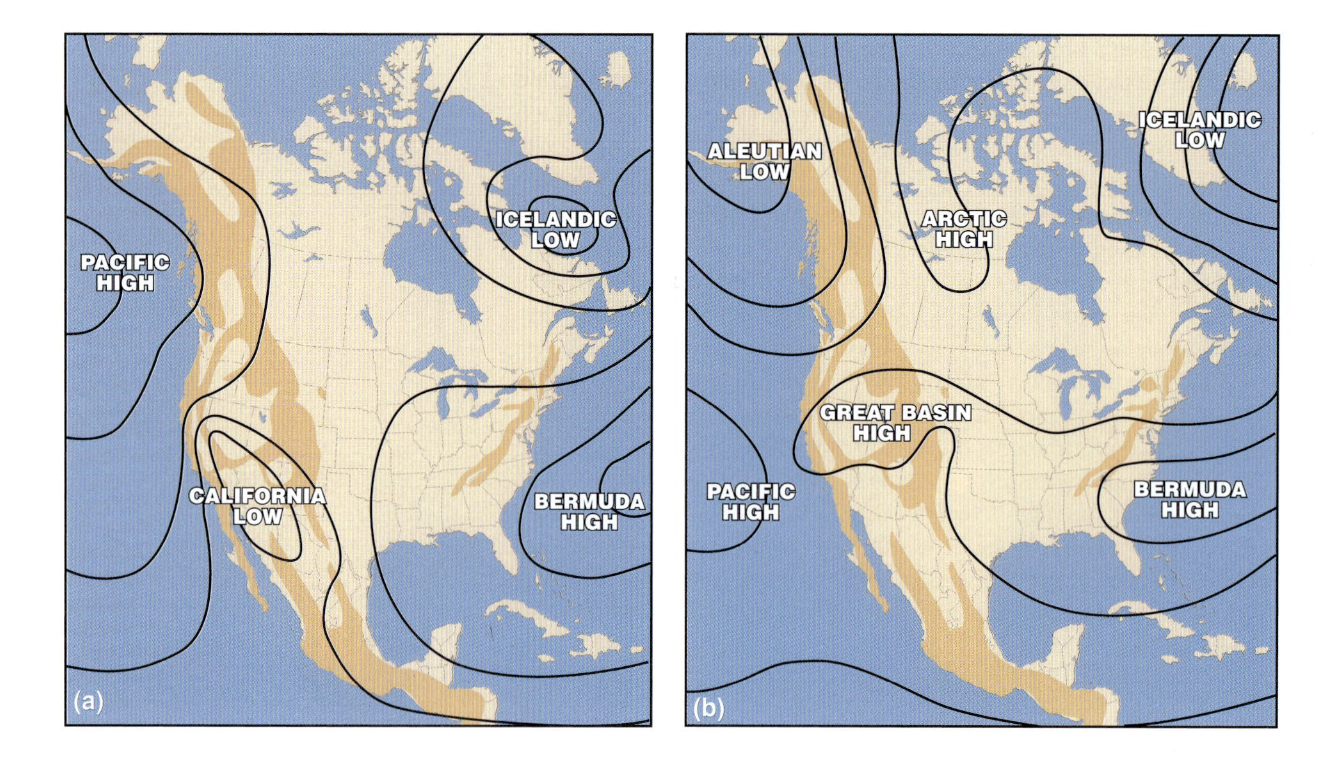

Figure 1.4 Mean surface pressure patterns over the United States during (a) summer and (b) winter.

ence of the Bermuda–Azores High can extend as far west as Nevada. The Pacific High extends over most of the eastern Pacific, causing weak winds to blow southward along the West Coast of North America. Air over the high pressure center sinks and warms, producing cloudless skies, low humidity, and high temperatures. Between these two high pressure centers, a shallow and weak low pressure center, the California Low, forms over the hot desert and semiarid regions of the southwestern United States and northern Mexico.

In winter, the Bermuda–Azores High continues to dominate the weather of the eastern part of the North American continent. The Pacific High also retreats southward and weakens (figure 1.4b), exposing much of the coast to strong westerly winds associated with the Aleutian Low, a low pressure center that forms near the Aleutian Islands where cold air moving southward off the Bering Sea meets warmer air moving northward from Japan. The strong westerly winds produce extensive wintertime precipitation as the low and related *fronts* move south and east and as maritime air is lifted over the coastal mountain barriers. A weak high pressure center, the Great Basin High, forms in the intermountain region between the Rockies to the east and the Sierra Nevada and Cascade ranges to the west as the surfaces of the mountains cool (especially during winter nights), causing cold air to drain into the Great Basin.

The climates of the east and the west coasts of North America are affected by ocean currents. The Gulf Stream in the Atlantic and the Japanese Current in the Pacific moderate the climate of nearby coastal regions. The Gulf Stream, a narrow, intense, northward flow of warm water that follows the east coast to Cape Hatteras, North Carolina, and then turns seaward, is partly driven by the drag on the ocean's surface of winds flowing clockwise around the Bermuda–Azores High. The eastward-flowing Japanese Current in the northern Pacific, partly driven by the clockwise air flow around the Pacific High, reaches North America at latitudes between Puget Sound and Juneau, Alaska. The current diverges into northward and southward currents that parallel the Pacific coastline. Because of the rotation of the earth, ocean currents veer somewhat to the right of the winds that drive them. Thus, the southern branch of the Japanese Current draws surface water away from the Pacific coast, allowing much colder water from below to rise to the surface, thereby producing coastal fog and stratus clouds.

Mountain Climates of North America

2

The basic climatic characteristics of the major mountain ranges in the United States—the Appalachians, the Coast Range, the Alaska Range, the Cascade Range, the Sierra Nevada, and the Rocky Mountains—can be described in terms of the four factors discussed in chapter 1.

The mountains of North America extend latitudinally all the way from the Arctic Circle (66.5°N) to the tropic of Cancer (23.5°N) (figure 2.1). There are significant differences in day length and angle of solar radiation over this latitude belt that result in large seasonal and diurnal differences in the weather from north to south.

Elevations in the contiguous United States extend from below sea level at Death Valley to over 14,000 ft (4270 m) in the Cascade Range, the Sierra Nevada, and the Rocky Mountains. Several prominent peaks along the Coast Range in Alaska and Canada (e.g., Mount St. Elias and Mount Logan) reach elevations above 18,000 ft (5486 m). Denali (20,320 ft or 6194 m) in the Alaska Range is the highest peak in North America. The highest peak in the Canadian Rockies is Mt. Robson, with an elevation of 12,972 ft (3954 m).

The climates of the Coast Range, the Cascade Range, and the Sierra Nevada, all near the Pacific Ocean, are primarily maritime. The Appalachian Mountains of the eastern United States are subject to a maritime influence from the Atlantic Ocean and the Gulf of Mexico, but they are also affected by the prevailing westerly winds that bring continental climatic conditions. Only the climate of the Rocky Mountains, far from both the Pacific and Atlantic Oceans, is primarily continental.

Each of the mountain ranges is influenced by regional circulations. For example, the Appalachians are exposed to the warm, moist winds brought northward by the Bermuda–Azores High and to the influence of the Gulf Stream. Similarly, the Coast Range feels the impact of the Pacific High, the Aleutian low, and the Japanese Current.

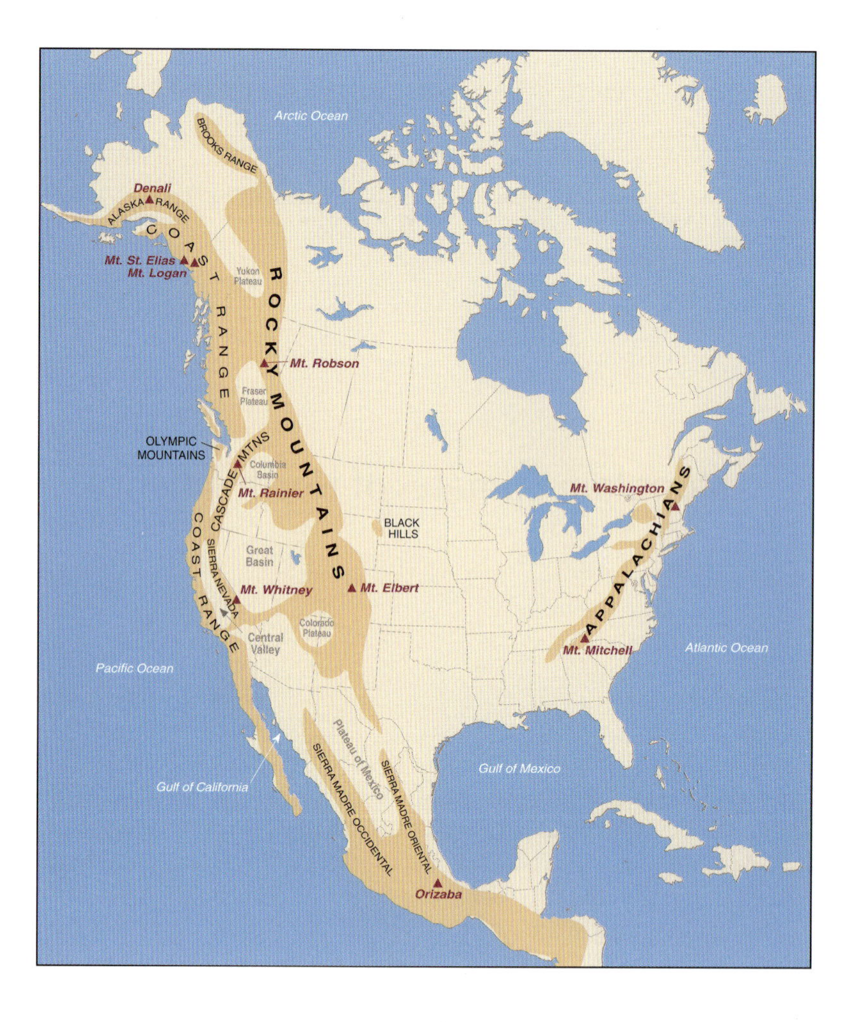

Figure 2.1 Map of North America showing the locations of the major mountain ranges and peaks.

A mountain range, depending on its size, shape, orientation, and location relative to *air mass* source regions, can itself affect the regional climate by acting as a barrier to regional flows. For example, the Brooks Range acts as a barrier to the extremely cold air masses that develop over the Arctic Ocean in winter and occasionally blocks the cold air, protecting central Alaska from many extremely cold air outbreaks. Similarly, polar and arctic air masses that descend from northern Canada in the winter are often trapped between the Rockies and the Appalachians (or less commonly, between the Rockies and the Coast Range or Cascades).

Although latitude, altitude, continentality, and exposure to regional circulations provide a general description of the climate of a mountainous region, two factors make a more detailed description difficult. First, the varied topography within a mountain range produces microclimatic differences even over very short distances. Temperatures on sunny slopes are higher than on a shaded valley floor, and winds can be affected by vegetation and terrain projections. *Microclimates* can be identified by observing differences in plant and animal species from one area to another. Second, there are few high-altitude climate measurement stations. Measurements taken at these stations, especially precipitation measurements, are often inaccurate as a result of the severe high-altitude environment. Further, the measurements may not be representative of the region surrounding the climate station because of microclimatic variations.

Detailed climate descriptions for selected sites in each of the U.S. mountain ranges can be found in Reifsnyder (1980). Another source for detailed climate summaries and long-term weather data for many climate stations in the United States is the National Climatic Data Center, Federal Center, Asheville, North Carolina 28801.)

2.1. The Appalachians

The Appalachian Mountains (figure 2.2), an ancient mountain range characterized by low ridges and smooth rounded mountain summits, parallel the coastline of the eastern United States, extending 1500 miles from southwest to northeast. Throughout most of the range, elevations are be-

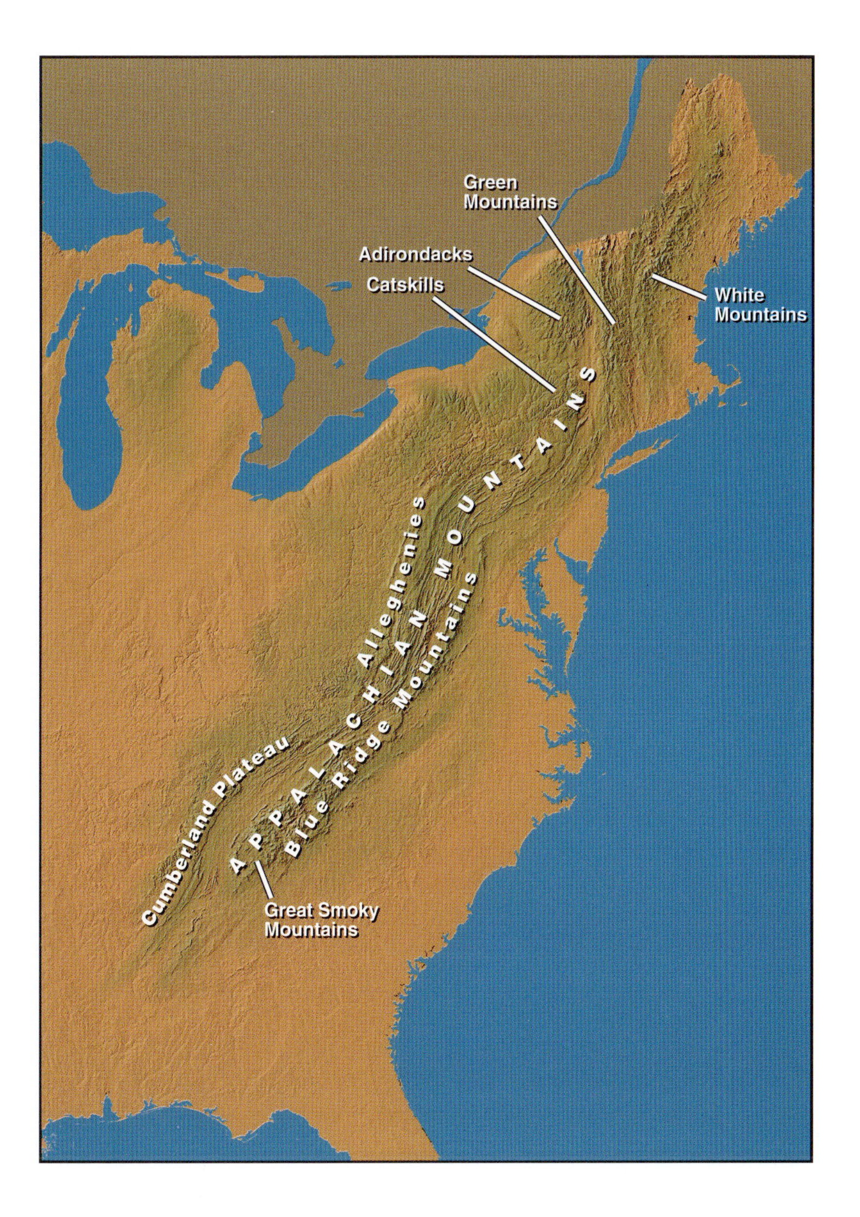

Figure 2.2 Shaded relief map of the eastern United States, showing the Appalachian Mountains and its subranges.

low that of Denver, Colorado (5280 ft or 1609 m), at the foot of the Rocky Mountains. The highest peak in the Appalachians is Mt. Mitchell (6684 ft or 2037 m) in North Carolina. The Appalachians are heavily forested, with hardwood and pine forests in the south and spruce and fir forests in the north.

The southern Appalachians consist of long ridges and intervening valleys extending northeastward from central Alabama. Major ridges include the Cumberland Mountains to the west, the Alleghenies, which extend from Virginia northward into Pennsylvania, and the Blue Ridge Mountains, including the Great Smoky Mountains, on the eastern edge of the Appalachians.

The northern Appalachians, in contrast, are composed of less extensive clusters of mountains, including the Green Mountains of Vermont, the White Mountains of New Hampshire, and the Catskills and Adirondacks of New York. New Hampshire's Mt. Washington (6288 ft or 1917 m), the highest peak in the northern Appalachians, experiences some of the worst weather on the continent (Putnam, 1991). It averages 310 foggy days per year and has an annual average wind speed of 35 miles per hour or mph (16 meters per second or m/s). The highest wind ever recorded at the earth's surface, 231 mph (103 m/s), was recorded on its summit. The average temperature for the year is only 26.5°F (−3.1°C), and winter temperatures can fall below −30°F (−34°C). The lowest temperature recorded on Mt. Washington is −47°F (−44°C).

Because the Appalachians are a relatively low mountain range, altitude is less important as a climate determinant than is latitude. The northern Appalachians have a cool climate; the southern Appalachians, which extend into the subtropics, have a temperate and rainy climate. The northern Appalachians receive most of their winter precipitation as snow and can maintain a snow cover for most of the winter. In the southern Appalachians, winter precipitation usually falls as rain. Only the highest summits receive snowfall that stays on the ground for more than a few days at a time.

The climate of the entire range is subject to both maritime and continental influences. Weather systems along the Atlantic Coast bring humid maritime conditions to the Appalachians, whereas the westerly winds that prevail at midlatitudes expose the range to the dry continental climate of the central United States. The strength of the *westerlies* varies with season (winds are stronger in winter) and with latitude (winds are stronger farther north).

In summer, the clockwise circulation around the Bermuda–Azores High carries warm humid air northward and northeastward from the Gulf of Mexico into the interior of the continent. This weak, low-level circulation brings sultry weather to most of the Appalachians (especially to the southern half) and traps pollutants and natural hydrocarbon emissions from the forests, producing a dense *haze* over the entire region. The Great Smoky Mountains were named for this bluish haze. Isolated thunderstorms often break out in the warm, humid air, with lines or clusters of thunderstorms sometimes developing during frontal passages. Tropical storms and hurricanes moving in off the Atlantic can also set off thunderstorms and can bring heavy rain and flooding to the coastal regions and the eastern slopes of the Appalachians.

In winter, the Appalachians are exposed to storms moving in from the west and to storms moving northward along the Atlantic coast to the east. Major storms form at intervals in the lee of the Rocky Mountains and then travel eastward and northeastward along the west side of the Appalachians. These storms, fed by moisture from the Gulf of Mexico, produce heavy precipitation as they are lifted over the western slopes of the Appalachians. Between storms, Arctic outbreaks that travel south off the Canadian plains are trapped between the Rockies and the Appalachians and may extend across much of the central and eastern United States. The northern and central Appalachians experience clear, bitterly cold weather during these outbreaks. To the east of the Appalachians, winter storms often track northward along the Atlantic coastline. Strong, moist, easterly flows generated by these storms can produce heavy precipitation on the east side of the range. The northern end of the Appalachians is subject not only to major storms that blow in from the southwest and to Atlantic storms that travel up the seacoast, but also to locally generated storms associated with shifts in the position of the *jet stream*.

2.2. The Coast Range, the Alaska Range, the Cascade Range, and the Sierra Nevada

Four mountain ranges, the Coast Range, the Alaska Range, the Cascade Range, and the Sierra Nevada, parallel the West Coast of North America and, from southeast Alaska southward, lie perpendicular to the prevailing westerly winds of the midlatitudes (figure 2.1). The Coast Range extends along the West Coast of North America from Kodiak, Alaska, to southern California. The Alaska Range, the largest mountain range in Alaska, arcs northward from the Coast Range into south-central Alaska and extends from the Alaska Peninsula into the Yukon Territory. The Cascade Range (figure 2.3) extends from British Columbia southward to Mount Lassen in northern California and is approximately 120 mi (200 km) inland from the coast. The mountain chain is continued by the Sierra Nevada, which run south-southeastward to Tehachapi, California, just north of Los Angeles.

The Coast Range varies considerably in elevation from north to south. The highest peaks and ridges are in mainland Alaska, southeast Alaska, and Canada. The Olympic Mountains west of Seattle, Washington, are the highest mountains in the portion of the Coast Range within the contiguous United States. Mt. Olympus (7965 ft or 2428 m), the highest peak in the Olympic Mountains, is located just 35 mi (55 km) from the coast. The Olympics support a temperate rain forest at their base and extensive glaciers and alpine tundra above 5000 ft (1500 m). There is a marked decrease in elevations in the Coast Range southward from the Olympics.

The Alaska Range includes Denali, which at 20,320 ft (6194 m) is the highest peak on the North American continent. Other major peaks include Mt. Foraker and Mt. Hunter. Because of the high latitude and the high elevations, the range is extensively glaciated.

The highest peaks in the Cascade Range are volcanoes, with Washington's Mt. Ranier the highest at 14,410 ft (4392 m). Other major peaks

Figure 2.3 Digital topographic image of the western United States showing the major mountain ranges and peaks. The color scheme is based on annual total precipitation amounts and altitude. (Provided by H. Foote, Battelle Pacific Northwest Laboratories)

include Glacier Peak, Mt. Baker, Mt. St. Helens, and Mt. Adams in Washington; Mt. Hood, Mt. Jefferson, and the Three Sisters in Oregon; and Mt. Lassen and Mt. Shasta in California. The mean crest height of the Cascades is considerably below the elevations of these isolated volcanoes. In the northern Cascades, the somewhat higher elevations and heavy winter snowfalls produce the most extensive glaciation in the contiguous United States.

Immediately to the south of the Cascades, the Sierra Nevada form a high, steep-sided barrier that is parallel to the lower elevation Coast Range and the coastline of the Pacific Ocean. The Sierra Nevada is separated from the Coast Range for most of its length by California's Central Valley. The highest point in the Sierra Nevada, Mt. Whitney at 14,495 ft (4418 m), is the highest peak in the contiguous 48 states.

These four mountain ranges form significant barriers to maritime air masses moving into the continent from the Gulf of Alaska and the northern Pacific. Moist air carried inland off the Pacific Ocean is lifted, first over the Coast Range and then over the Alaska Range, the Cascades, or the Sierra Nevada, producing heavy precipitation on the windward slopes of the mountain ranges. Amounts of precipitation on the western side of the Coast Range vary with the altitude of the range's crest, with heavier precipitation falling in the north where the mountains are higher. The southeast coast of Alaska receives between 100 and 200 inches (254 to 508 cm) of precipitation per year. Port Walter on Baranof Island has the highest average annual precipitation in the continental United States: 221 inches (561 cm). The western slope of the Olympic Mountains receives 150 inches (381 cm) of precipitation annually, the Oregon coast 120 inches (305 cm), and the northern coast of California 50 inches (127 cm).

Areas of low precipitation, or *rain shadows* (section 8.5.1), are found in the lee of mountain ranges. At Sequim, Washington, in the lee of the Olympics, precipitation totals only about 16 inches (41 cm) per year. In Oregon's Willamette Valley, only 40 inches (100 cm) of precipitation fall, about one-third as much as along the coast. At Richland, Washington, 110 miles (175 km) east of the Cascade crest, precipitation totals only 6 inches (15 cm) per year. This amount of precipitation supports a shrub/steppe ecosystem, which consists primarily of sagebrush and grasses. Desert and semiarid ecosystems are found on the valley floors in the rain shadow of the Sierra Nevada. Some points in Death Valley, only 80 miles (130 km) east of Mt. Whitney, receive less than 2 inches (5 cm) of precipitation per year.

Rainfall along the Coast Range and in the Cascades and Sierra Nevada is highly seasonal and is influenced by the presence and interaction of the Aleutian Low and the Pacific High. These two semipermanent pressure systems move in tandem, with a northward shift during summer and a southward shift during winter. During summer the high intensifies, whereas during winter the low intensifies. The coast thus feels the presence of the Aleutian Low during winter and the Pacific High during summer.

In winter, storms develop in the vicinity of the Aleutian Low and move south and east off the Aleutian Islands and Gulf of Alaska to bring nearly continuous drizzle, rain, cloudiness, and moderate to strong coastal winds to the entire coastline of southeast Alaska, western Canada, and the Pacific Northwest. Southwesterly winds to the south of the traveling low pressure storm systems bring the heaviest precipitation. The maritime influence moderates the temperatures so that these storms produce rainfall rather than snowfall throughout the winter at coastal elevations. Rain can fall at higher elevations, too, sometimes affecting ski areas in the Cascades. Forecasters closely monitor the height of the *freezing level* to determine whether precipitation will fall as snow or rain. The high Sierra Nevada are generally not affected by the continuous storms from Alaska, but they do experience occasional storms brought in by westerlies off the Pacific Ocean during winter. These storms bring copious amounts of precipitation to the western side of the range, with winter snow accumulating to form a snowpack at elevations above 5000 ft (1524 m). Snow in the Sierra Nevada and in the Cascades is usually wet.

The weather changes abruptly in the spring. The Pacific High moves northward, intensifies, and brings milder, drier weather to the Coast Range, the Cascade Range, and the Sierra Nevada. The clockwise circulation exposes the coast to winds out of the northwest, and the subsidence associated with the high pressure center suppresses cloudiness and precipitation.

2.3. The Rocky Mountains

The Rocky Mountains extend over 3000 miles (4800 km) from northwestern Alaska to northern New Mexico and are the largest mountain range in North America. The Rockies can be divided into two climatic zones, the Northern Rockies and the Southern Rockies. The Northern Rockies extend from Alaska's Brooks Range to central Wyoming and include the Cariboo, Monashee, Selkirk, and Purcell Mountains of Alberta and British Columbia, the Bitterroot Mountains of western Montana and Idaho, and the Teton, Bighorn, Absaroka, and Wind River Ranges of Wyoming, as well as Jasper, Banff, Glacier, Yellowstone, and Teton National Parks. The Southern Rockies extend from central Wyoming into New Mexico. According to Reifsnyder (1980), the Southern Rockies have nearly 900 peaks above 11,000 ft (3353 m) and contain three-quarters of the land mass above 10,000 ft (3048 m) in the contiguous United States. The highest mountains in the Rockies are in Colorado, where there are 54 peaks above 14,000 ft (4267 m). The Northern Rockies of Canada are heavily glaciated and support extensive coniferous forests. The Southern Rockies contain only remnants of glaciers and more open forests that are often limited to mountainsides because of low precipitation amounts in the valleys. The wide range of life zones (prairie, foothills, montane, subalpine, and alpine) found in the Rocky Mountains reflects the influence of altitude and latitude on climate.

The Northern and Southern Rockies share some climatic characteristics. Located in the interior of the continent, the Rocky Mountains have a typical continental climate with large daily and seasonal temperature ranges. The winters are cold; the summers are moderate and dry. Most of the moisture that reaches the Rockies is brought by the midlatitude westerlies. However, because much of the Pacific moisture is released as storms are carried over the Coast Range, the Alaska Range, the Cascade Range, and the Sierra Nevada, precipitation amounts in the Rockies are lower than in the ranges farther west. The highest mountains receive most of the precipitation, generally 40 inches or 100 cm annually, with some high-altitude sites receiving as much as 100 inches or 250 cm annually. The lower elevation intermountain valleys receive only meager amounts of precipitation, approximately 12 inches or 30 cm annually. Because winter temperatures are low at high altitudes in the Rockies, precipitation often falls as *powder snow*. Powder snow has a significantly lower water content (2–7% by volume) than the heavier snow that falls in the Cascades, where the water content of snow is often above 10%.

In winter, the Northern Rockies are subject to storms that form near the Aleutians and move across the region from northwest to southeast. In summer, the northern part of the Northern Rockies, and occasionally the southern part, is affected by low pressure storms that travel rapidly across the region from west to east, setting off isolated afternoon thunderstorms. These thunderstorms also move rapidly and produce only light precipitation, which increases the hazard of lightning-ignited forest fires.

At altitudes above the Southern Rockies, winds are generally from the west. The strength of the winds varies between winter and summer as the position of the jet stream (section 5.2.1.3) shifts. During winter, when the jet stream is over the Southern Rockies, winds aloft in the region are strong. During summer, when the jet stream is farther north, the westerlies aloft are weak, and weak southerly winds prevail at the ground on both the east and west sides of the Rockies. East of the mountains, the southerly winds are driven by the Bermuda–Azores High, which brings moist air northward from the Gulf of Mexico into the central part of the country. Enough low-level moisture reaches the eastern foothills of the Southern Rockies to produce afternoon thunderstorms over Colorado's Front Range and the foothills. As the thunderstorms grow into higher levels of the atmosphere, they encounter the prevailing westerly winds, which carry the storms eastward over the plains. West of the mountains, summertime winds are also southerly. These seasonal *monsoon* winds (section 8.5.3) are driven by strong summertime heating in the Great Basin and a summertime westward extension of the Bermuda–Azores High. The southerly winds bring moist air northward from the Gulf of California into the southern end of the Great Basin. This low-level moisture is carried over the Mogollon Rim in northern Arizona and into the Southern Rockies, producing episodes of mid- and high-level cloudiness and feeding the development of summer thunderstorms.

2.4. Between the Mountains

Between the various mountain ranges of North America there are large plateaus, basins, plains, and valleys where the climate is affected by the surrounding mountains. These intermountain areas vary in altitude, latitude, size, and topography, but their climates are generally characterized by limited precipitation and extreme temperatures because the surrounding mountains cast rain shadows and isolate these areas from moderating maritime influences.

The largest intermountain region in North America is the Great Plains between the Rocky Mountains and the Appalachians. In the far northwest of the continent, the interior of Alaska and the Rocky Mountain Trench separate the Coast Range from the Rockies. To the south, California's Central Valley lies between the Coast Range and the Sierra Nevada. The Rockies are separated from the Sierra Nevada by the Great Basin and from the Cascades by the Columbia Basin.

The Great Plains extend northward all the way to the Arctic Ocean and Hudson Bay. Because there are no major mountain features on this broad plain, arctic and polar air masses that develop in winter over the frozen Arctic Ocean and on the plains of northern Canada can easily move southward, bringing cold weather to the central plains, the eastern slope of the Rocky Mountains, and the western slope of the Appalachians. The frontal boundary between the arctic air traveling southward across this plain and the maritime tropical air from the Gulf of Mexico is often found in the central or southern United States. In the summer, the climate of the central and southern plains is dominated by the flow of warm, moist air from the south. In the winter, it is dominated by cold, dry air from the north.

The interior of Alaska, between the Brooks Range and the Coast Range, is known for temperature extremes caused by the seasonal changes in incoming solar radiation, or *isolation*, at high latitudes. The sun does not rise in the winter half-year at latitudes above the Arctic Circle, and polar night temperatures can be very low. Alaska's official low temperature record of −80°F (−62.2°C) was set at Prospect Creek in Alaska's interior on 23 January 1971. In the summer half-year, the sun remains above the horizon for much of the day. Alaska's high temperature record of 100°F (37.8°C) was recorded not far from Prospect Creek, at Fort Yukon, on 27 June 1915. The mountains surrounding the Alaskan interior not only prevent maritime moderation of the temperatures, but also reduce the amount of annual precipitation in the interior to 12–16 inches (30–41 cm), less than a tenth of the precipitation in the coastal areas of southeast Alaska. Because of the extremely low temperatures of the long polar night, a continuous layer of *permafrost* exists over the northern third of Alaska. Discontinuous or isolated patches of permafrost are found over the middle third of the state.

The Rocky Mountain Trench is a long, narrow valley between the Coast Range and the Canadian Rockies. The Yukon and Fraser Plateaus are located at the north and south ends of the trench, respectively. Because of the rain shadow cast by the Coast Range, the trench receives considerably less precipitation than the coastal areas to the west. The trench is not only drier but also colder than locations on the coast. Cold air from Arctic outbreaks collects in the trench and, when the pool of cold air is deep enough, it drains westward through gaps in the Coast Range. If winds aloft are from the east when the cold air is draining to the west, the *gap winds* can be especially strong and can be a serious hazard for ships in the Inside Passage, the narrow inland seaway in southeast Alaska and Canada.

The Central Valley of California, a 500-mi-long basin (800 km) that is oriented north–south, is actually two valleys, the Sacramento Valley to the north and the San Joaquin Valley to the south. Both valleys drain into the Pacific through the Caracena Straits near San Francisco, the Central Valley's only outlet. Precipitation in the valley is limited because it is in the rain shadow of both the Coast Range to the west and the Sierra Nevada to the east. Episodes of nighttime fog are common in low-lying sections of the Central Valley when cold, moist maritime air flows eastward off the Pacific into the valley through the Caracena Straits.

Figure 2.4 Road signs in southern Wyoming warn of the strong wind hazard. The signs are posted every 5 miles (8 km) and are a reminder of the long wind fetch in southern Wyoming.

The Great Basin, 200,000 mi^2 (518,000 km^2) of desert that extends south from eastern Oregon almost to the Gulf of California, lies between the Sierra Nevada and the Rockies. It is surrounded by high mountains, and many lower ranges rise within it. The western portion of the Great Basin is composed of a large number of short, linear, north–south mountain ranges separated by broad basins of alluvium. Nevada, in the center of the Great Basin, counts 413 of these distinctive ranges (Collier, 1990). Differences in elevation within the Great Basin are dramatic. Mt. Whitney, at 14,495 ft (4,418 m), rises on the western rim of the Great Basin. Only 80 mi (130 km) away, Death Valley's Badwater is 282 ft (86 m) below sea level, the lowest elevation in the Western Hemisphere. The Great Basin is in the rain shadows of the Sierra Nevada to the west and the Rockies to the east and thus receives relatively little precipitation. The extremely limited precipitation in Death Valley (about 2 inches or 5 cm per year) has already been mentioned. When summer precipitation does fall over the Great Basin, it often originates in high clouds and evaporates before reaching the ground. No river systems drain the Great Basin, so precipitation sinks into the alluvium on the floors of the dry lake beds between the mountain chains. The Rocky Mountains protect much of the Great Basin from the arctic and polar air masses that move south across the Great Plains in winter. Winter temperatures within the basin are, nonetheless, low for their latitude because of the basin's high altitude. In contrast, summer temperatures can be high. Death Valley holds the Western Hemisphere record for the highest recorded temperature of 134°F (56.7°C).

The Columbia Plateau extends from the Cascade Range to the Rockies and covers eastern Washington, eastern Oregon, and western Idaho. Land is fertile in the Columbia Plateau but must be irrigated because of low precipitation amounts in the rain shadows of the Cascades and the Rockies.

Plateaus, basins, plains, and valleys not only separate major mountain ranges but can also interrupt a single range. For example, the plains of southern Wyoming separate the Colorado Rockies from the Teton, Wind River, and Bighorn Mountain ranges in northern Wyoming. Westerly winds are channeled through this gap in the Rockies, bringing strong winds (figure 2.4) to the region. Extensive snow fences are needed to protect highways and railways from blowing snow. Strong winds are also experienced in southern Idaho, where strong westerly winds funnel through the broad, curving, flat-bottomed Snake River Plain, increasing the difficulty of controlling range fires ignited by lightning.

The Colorado Plateau, located in the Southern Rockies, is a broad, rough uplands region of plateaus and deep valleys that stretches over parts of Colorado, Utah, Arizona, and New Mexico. It includes the Grand Canyon, Bryce Canyon, Zion Canyon, Cedar Breaks, Monument Valley, the Painted Desert, and the Petrified Forest. Because the plateau is surrounded by high peaks and mesas, precipitation is limited and temperatures range widely.

Part II

AN INTRODUCTION TO THE ATMOSPHERE

Atmospheric Scales of Motion and Atmospheric Composition

3

3.1. Atmospheric Scales of Motion

Weather phenomena occur over a very broad range of scales of space and time, from the global circulation systems that extend around the earth's circumference to the small eddies that cause cigarette smoke to swirl and mix with clear air. Each circulation can be described in terms of its approximate horizontal diameter and lifetime (figure 3.1). Large-scale weather systems, such as hemispheric wave patterns called *Rossby waves*, monsoons, high and low pressure centers, and fronts, are called *synoptic-scale* weather systems. Temperature, humidity, pressure, and wind measurements collected simultaneously all over the world are used to analyze and forecast the evolution of these systems, which have diameters greater than 200 km (125 mi) and lifetimes of days to months. *Mesoscale* weather events include diurnal wind systems such as mountain wind systems, like breezes, sea breezes, thunderstorms, and other phenomena with horizontal scales that range from 2 to 200 km (1 to 125 mi) and lifetimes that range from hours to days. Mesoscale meteorologists use networks of surface-based instruments, balloon-borne *sounding* systems, remote sensing systems (e.g., *radar*, *lidar*, and *sodar*), and aircraft to make observations on these scales. *Microscale* meteorology focuses on local or small-scale atmospheric phenomena with diameters below 2 km (1 mi) and lifetimes from seconds to hours, including gusts and *turbulence*, dust devils, thermals, and certain cloud types. Microscale studies are usually confined to the layer of air from the earth's surface to an altitude where surface effects become negligible (approximately 1000 feet or 300 m at night and 5000 feet or 1500 m during the day). A fourth and less rigorously defined term, the *regional scale*, denotes circulations and weather events occurring on horizontal scales from 500 to 5000 km (310 to 3100 mi). The regional scale is thus smaller than synoptic scale, but larger than mesoscale. The term

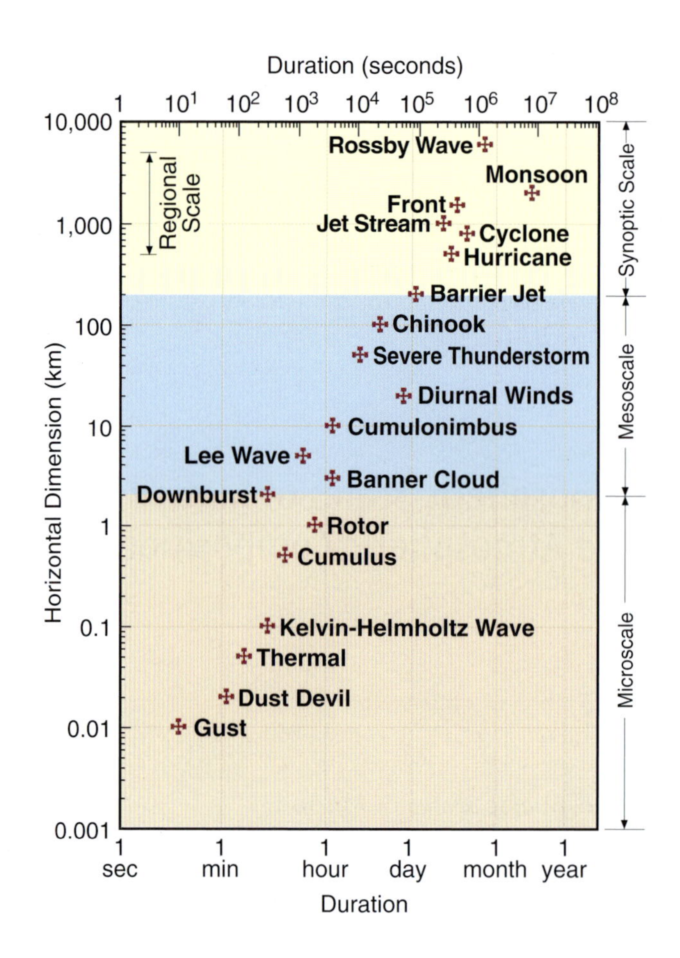

Figure 3.1 The horizontal dimensions and lifetimes of atmospheric phenomena illustrate the broad range of atmospheric space and time scales.

is often used to describe events that occur within more or less homogeneous physiographic provinces (e.g., the Pacific Northwest region).

Weather phenomena are classified as synoptic scale, mesoscale, or microscale.

Major mountain ranges impact the weather on the synoptic scale. They anchor large-scale pressure systems in the Northern Hemisphere, cause low and high pressure weather systems to form, and produce large-scale seasonal wind systems in Asia and North America. Standard weather charts and large-scale numerical models are used to identify and forecast these regional and hemispheric scale phenomena.

3.2. Atmospheric Composition

3.2.1. *Permanent and Variable Gases*

The atmosphere, when dry, is mainly a mixture of three permanent gases: nitrogen (78% by volume), oxygen (21%), and argon (1%). These gases are well mixed in the lower part of the atmosphere, so there is virtually no difference in the mixture between sea level and mountain elevations.

In addition to the permanent gases, other gases are present in the atmosphere with concentrations varying significantly from time to time and from place to place. The two main variable gases are water vapor and carbon dioxide. Water vapor, a colorless, odorless gas, is always present in air, with concentrations varying from near zero to 4% by volume. Carbon dioxide makes up, on average, only 0.034% of the atmosphere by volume, but its concentration is steadily increasing as a result of human activities. Water vapor, carbon dioxide, and a variety of other variable gases that are present in minute amounts in the earth's atmosphere absorb radiation emitted by the earth's surface and reradiate it back toward the ground, thus keeping the earth's surface temperature much higher than it would be if there were no atmosphere. There is concern that increasing amounts of the radiatively active gases may enhance this *greenhouse effect*, causing long-term warming of the earth and its atmosphere.

Water vapor content in the atmosphere varies from near zero to 4% by volume.

3.2.2. *Aerosols*

The atmosphere contains not only permanent and variable gases, but also small solid and liquid particles, called *aerosols*. Aerosols can be either natural or man-made. Natural aerosols include rain, snow, ice particles, dust, pollen, and sea spray. Man-made aerosols include the carbon particles that result from the burning of organic materials, other *air pollutants*, and dust from mechanical disturbance of soils. Regardless of the type of aerosol, the large particles settle out quickly, whereas smaller particles can remain suspended in the atmosphere for days. (A formula for computing the settling speeds of liquid droplets is provided in appendix A.)

Aerosols are solid or liquid particles, either natural or man-made, in the atmosphere.

Aerosols, including liquid water and solid water, affect the transmission of light and therefore affect *visibility*. Although particles in the atmosphere absorb some light, they reduce visibility primarily by scattering (i.e., deflecting the direction of travel) of light. Submicrometer particles (i.e., particles that are less than 1 micrometer or 0.00004 inches in diameter) are the most effective scatterers of sunlight. Larger particles are less effective scatterers of light, but they affect visibility through the processes of absorption, reflection, diffraction, and refraction. Distant objects are discerned by their contrast in color or brightness to surrounding objects. This contrast is reduced when extraneous light is scattered toward an observer by atmospheric particles in or near the line of sight, causing the observer to perceive a reduction in visibility. Visibility is affected by man-made aerosols, such as the carbon particles produced by agricultural burning or by the use of wood-burning stoves, and by natural aerosols, such as the *terpenes* emitted by evergreen forests.

Aerosols affect precipitation by serving as nuclei for the condensation of water vapor or the nucleation of ice particles. They can also affect air pollution by serving as sites where chemical reactions transform air pollutants into secondary chemical forms.

Some aerosols are known to have an adverse effect on human health. Large aerosol particles are generally removed from inhaled air in the upper respiratory tract, but smaller aerosols with diameters below about 2.5 micrometers (0.0001 in.) can be deposited directly in the lungs. Because some aerosols contain carcinogens, air quality regulations have been established to reduce the *ambient* concentrations of these small respirable aerosols. These aerosols and the regulations that apply to their control are discussed in more detail in chapter 12. Information on the measurement of aerosols can be found in the references listed in appendix C.

Relative humidity, the ratio of the actual water vapor content of the air to the maximum amount of water vapor the air could hold at the same temperature, is usually between 20% and 100%.

3.2.3. *Humidity*

The *humidity*, or water vapor content, of the air varies over time and space. The amount of water vapor that air can hold depends on air temperature. Warmer air can hold more water vapor than cooler air. Thus, high humidity is usually found in the warm equatorial regions.

Atmospheric humidity can be specified numerically in several ways, but it is most often expressed as *relative humidity*. Relative humidity is the ratio of the actual water vapor content of air to the water vapor content of saturated air at the same ambient temperature. Relative humidity is usually given as a percentage, with typical values in the range of 20–100%. Because the water vapor content at saturation varies with temperature and actual water vapor content does not, there is diurnal variation in relative humidity (figure 3.2). Relative humidity usually reaches its maximum value just before sunrise, when the temperature is lowest, and its minimum value in the mid- to late afternoon, when the temperature is highest. (Other humidity variables—*saturation vapor pressure, mixing ratio, specific humidity*, and *dew-point temperature*—can be calculated using formulas presented in appendix A.)

Relative humidity can be determined in the field using humidity measurement instruments called *hygrometers*. The most popular instrument for field measurements is the *sling psychrometer*, a device that holds a matched pair of liquid-in-glass thermometers or other temperature sen-

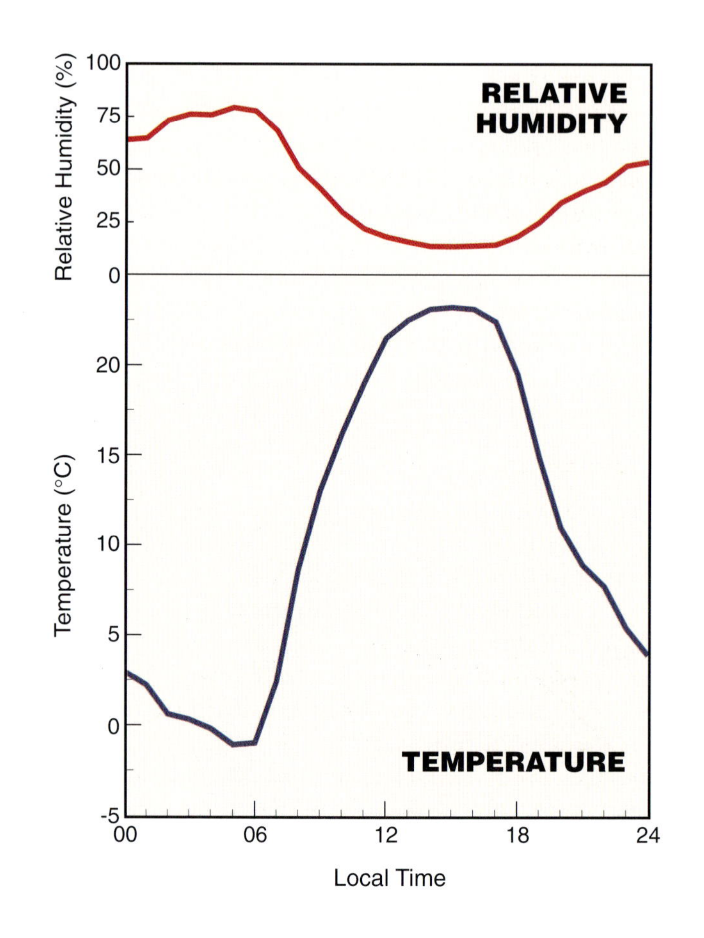

Figure 3.2 Relative humidity varies inversely with temperature, reaching a maximum when the temperature is lowest. This diurnal variation in relative humidity occurs even when the total amount of water vapor in the air remains constant.

sors. The bulb of one of the thermometers (the "wet bulb") is covered by a clean muslin wick wetted with distilled water. The bulb of the other thermometer is dry. The *psychrometer* is rotated or slung repeatedly around a pivot or handle. Evaporation causes the *wet bulb temperature* to fall below the *dry bulb temperature*. The dry bulb temperature and the dry bulb/wet bulb temperature difference, called the *wet bulb depression*, stabilize after several minutes of slinging. Manufacturers of sling psychrometers provide sets of psychrometric tables for calculating relative humidity from the wet bulb and dry bulb temperature measurements. A psychrometric table for use at altitudes near 5000 ft (1500 m) is provided in appendix B. Sling psychrometers are starting to be replaced by handheld electronic hygrometers, which are easier to use and are becoming less expensive and more widely available. An electronic hygrometer uses a fan to draw a continuous sample of air past the sensors, which are shielded from solar radiation.

Most hygrometers provide relative humidity values. Some also calculate other humidity variables from equations similar to those given in appendix A. Research-quality hygrometers are more expensive but also more accurate, especially when ambient humidity is either high (above about 92%) or low (below about 20%). Further information on hygrometers is provided in the references listed in appendix C.

3.2.4. *Water Phase Changes in the Atmosphere*

The atmosphere contains not only water vapor, but also liquid water in the form of fogs and clouds, and solid water in the form of hail, snow, and other ice particles. Water changes routinely from one phase to another within the atmosphere as clouds form and evaporate, ice particles melt or sublimate, cloud droplets freeze, or vapor is deposited as ice. These phase changes release or store large quantities of heat, called *latent heat*. Heat must be supplied to change water from a less dispersed to a more dispersed phase (from solid to liquid, solid to vapor, or liquid to vapor), and heat is liberated when the reverse phase change occurs. The amount of heat required to evaporate liquid water is equal to the amount of heat liberated when the water vapor condenses. This balance is significant for at-

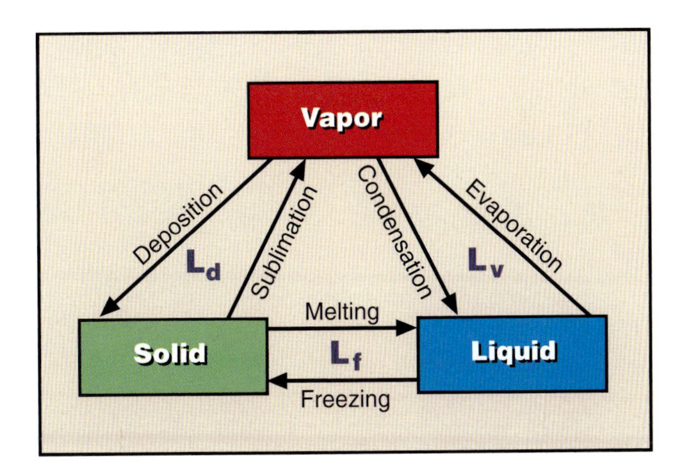

Figure 3.3 Transformations between the solid, liquid, and gaseous phases of water result in the release or storage of large quantities of heat. Heat must be supplied when the transformation is from the less dispersed to the more dispersed phase. The quantity of heat associated with individual phase changes is given in Joules per unit mass of water. (L_v = 2.5 MJ/kg is the latent heat of vaporization; L_f = 0.334 MJ/kg is the latent heat of freezing; and L_d = 2.83 MJ/kg is the latent heat of deposition.)

Latent heat is released or stored when water changes phase.

mospheric processes because a phase change in one direction can occur at one geographical location, and the opposite phase change can take place at a different location after the water has been transported a long distance. For example, the same quantity of heat needed to evaporate water in a desert is released when the water vapor condenses to form a cloud over a distant mountainside. Because heat release and storage allow the transport of heat energy from one location in the atmosphere to another, they are often as important to meteorology as the phase changes themselves.

The nomenclature used to describe transformations between the liquid, solid, and gaseous phases of water is depicted in figure 3.3. L_v, L_d, and L_f are the latent heats of condensation, deposition, and freezing, respectively.

Atmospheric Structure and the Earth's Boundary Layer

4

4.1. Vertical Structure of the Atmosphere

4.1.1. *Variations in Temperature, Pressure, and Density with Height*

The earth's atmosphere is divided into four layers: the *troposphere, stratosphere, mesosphere, and thermosphere* (figure 4.1). These layers are defined by alternating decreases and increases in air temperature with height. The boundaries between the layers are called the *tropopause, stratopause, and mesopause*. The troposphere, the lowest layer of the atmosphere, supports life on the planet and is the layer in which "weather" occurs. It extends about 7 mi (11 km) above sea level and is characterized by a mean temperature decrease with height ($-\Delta T/\Delta z$) of about 3.5°F per 1000 ft, or 6.5°C per km. This decrease explains the lower temperatures encountered at higher elevations in the mountains.

Although the mean temperature decreases with height in the troposphere, the atmospheric structure, particularly at the base of the troposphere, varies significantly over time as the earth warms during the day and cools at night, as the seasons change, and as weather systems move through the atmosphere.

The mean temperature decrease with height in the troposphere is about 3.5°F per 1000 ft, or 6.5°C per km.

The vertical structure of the atmosphere is characterized by an exponential decrease in air density and pressure with height. Air density is the mass per unit volume of the atmosphere as expressed, for example, in kilograms per cubic meter or pounds per cubic foot. Air pressure is the force exerted on a unit area by the weight of the air molecules above the measurement point as expressed, for example, in *millibars* or pounds per square inch (section 5.1.1). Air pressure at any given level is thus a measurement of the weight of a column of air above that level. Although there is no "edge" to the earth's atmosphere, approximately 99.9% of the air

The vertical structure of the atmosphere is characterized by an exponential decrease in air density and pressure with height.

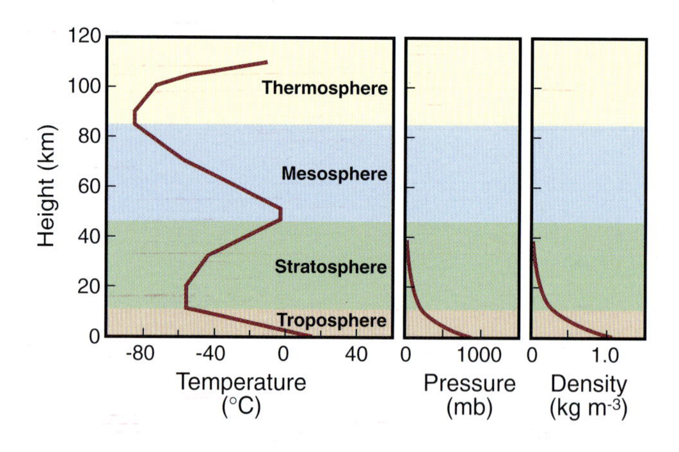

Figure 4.1 The major layers of the earth's atmosphere are distinguished by the alternating increase and decrease of temperature with height. Pressure and density decrease continuously with height. "Weather" occurs in the troposphere, the lowest layer of the atmosphere, in which temperature decreases with height.

molecules (and therefore the weight of the atmosphere) are found below 31 mi (50 km).

Temperature, density, and pressure are interrelated, so that a change in one will result in changes in the other two. The mathematical description of this relationship is called the gas law (appendix A). The gas law allows any one of these variables to be calculated if values for the other two variables are known.

4.1.2. *The Standard Atmosphere*

Figure 4.1 presented mean profiles of temperature, density, and pressure using an engineering approximation of the atmosphere called the *Standard Atmosphere*. The Standard Atmosphere is representative of average annual conditions in the midlatitudes, although it is not calculated from data at any one location. The temperature, density, and pressure data plotted in the figure are tabulated as a function of height above mean sea level (MSL) in table 4.1. Figure 4.2 shows the altitudes of several mountain peaks and their corresponding pressures in the Standard Atmosphere.

The Standard Atmosphere represents average annual conditions—temperature, density, and pressure—at midlatitudes.

Table 4.1 The Standard Atmosphere

Height		Pressure	Temperature		Density
(ft)	(m)	(mb)	(°F)	(°C)	(kg m^{-3})
0	0	1013.25	59.0	15.0	1.225
370	110	1000	57.7	14.3	1.212
3240	990	900	47.5	8.6	1.113
4780	1460	850	41.9	5.5	1.063
6390	1950	800	36.1	2.3	1.012
9880	3010	700	23.7	−4.6	.908
13790	4200	600	9.9	−12.3	.801
18280	5570	500	−6.2	−21.2	.691
23560	7180	400	−25.1	−31.7	.577
30050	9160	300	−48.1	−44.5	.457
38660	11790	200	−67.0	−55.0	.319
53170	16210	100	−67.0	−55.0	.161
102000	31100	10	−67.0	−55.0	.016

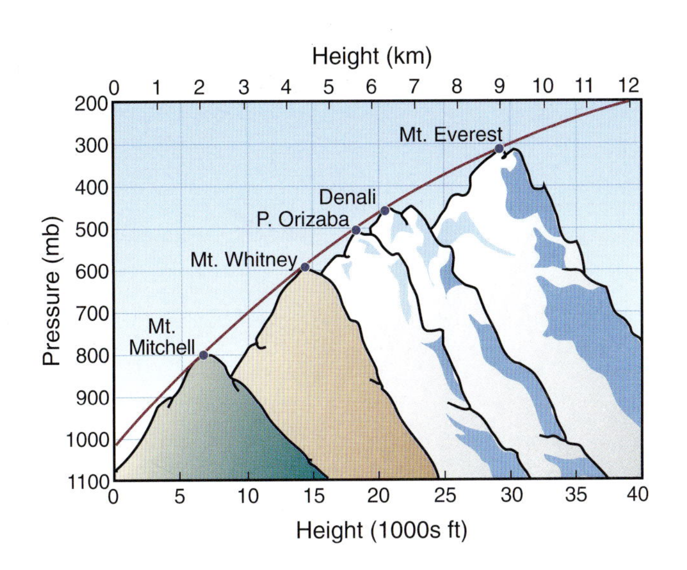

Figure 4.2 Pressure decreases with height. The correspondence between the heights above mean sea level of various well known mountain peaks and their pressures in the Standard Atmosphere is shown.

4.1.3. *Observations of the Vertical Structure of the Atmosphere*

Observations of the structure of the atmosphere are made using *radiosondes* (figure 4.3), upper air sounding devices consisting of weather instruments and a radio transmitter carried by a helium-filled balloon to about 19 miles (30 km) above the earth. Temperature, humidity, and pressure data are collected as the balloon ascends. The set of temperature data from a single ascent is called an *environmental temperature sounding*. A radiosonde that is tracked by radar or navigational positioning systems to determine wind speed and wind direction is called a *rawinsonde*. More information on radiosondes is given in section 9.3.1.

4.2. Temperature

4.2.1. *Variations in Temperature with Time, Horizontal Distance, and Height*

Air *temperature*, a familiar and easily observable weather parameter, varies with time, horizontal distance, and height above the earth's surface. Air temperature is directly related to other basic parameters, including pressure and *stability*, the degree of resistance of a layer of air to vertical motion (section 4.3).

Temporal changes in air temperature near the surface of the earth are caused by the changing seasons, the alternation of night and day, and traveling weather systems. Seasonal and diurnal temperature changes can be large or small, depending on latitude, elevation, topography, and proximity to the moderating influences of nearby oceans or lakes. Temperatures can change abruptly and significantly when traveling weather systems *advect*, or transport, colder or warmer air into a region. Temperature variations, particularly diurnal variations, are generally greater near the surface of the earth than higher in the atmosphere.

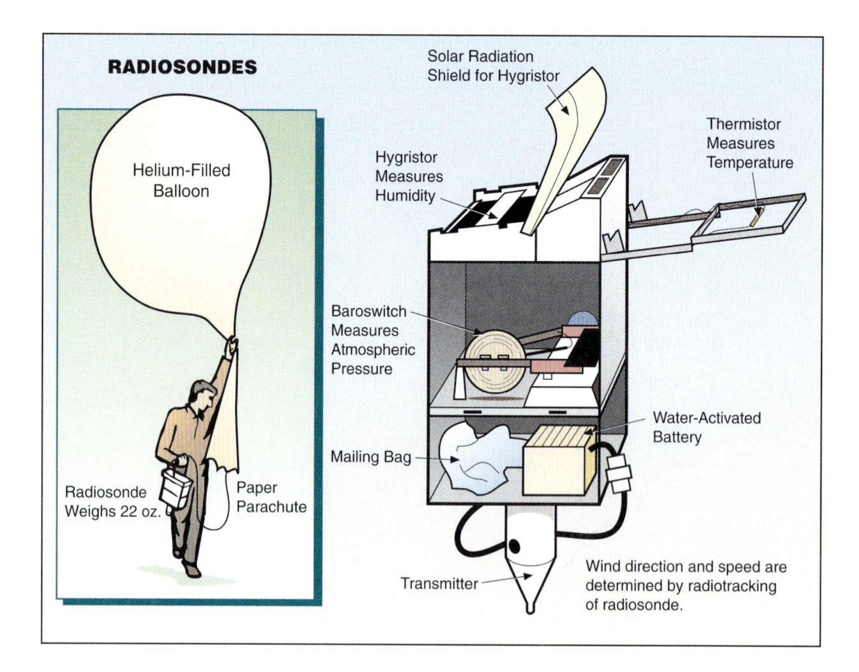

Figure 4.3 The radiosonde is an expendable upper air sounding system that measures temperature, humidity, and pressure. Radiosondes can be returned using the mailing bag provided; they are then reconditioned and reused.

Changes in air temperature from one location on the earth's surface to another are due to variations in climate (chapter 1), the movement of air masses and fronts, and local conditions, including topography and ground cover. These horizontal differences in temperature cause pressure differences (section 5.1) within the atmosphere that drive winds (chapter 11).

A typical vertical temperature structure in the atmosphere is described by the Standard Atmosphere (table 4.1). Individual environmental temperature soundings made with radiosondes often exhibit wide deviations from the Standard Atmosphere. Although temperature usually decreases with height in the troposphere, there are sometimes layers within the troposphere in which the temperature remains constant or increases with height. A layer in which the temperature is constant with height is called an *isothermal* layer. A layer in which temperature increases with height is called a temperature inversion. A *surface-based temperature inversion* is an inversion layer with its base at the ground. Such layers form when the ground cools faster than the air above (e.g., at night or in winter), when cold air is brought in near the ground by air currents (e.g., in low-lying valleys), or when the atmosphere warms more rapidly than the surface (e.g., when warm air is brought in aloft by winds). *Elevated temperature inversions* have their bases above the ground and are usually formed when air above surface high pressure centers sinks or when warmer air flows over cold air. The environmental temperature sounding in figure 4.4 illustrates short-term deviations in temperature structure in the troposphere associated with actual weather systems. This early morning sounding shows a temperature inversion in a shallow layer (about 300 m or 1000 ft deep) near the ground.

A temperature inversion is an atmospheric layer in which temperature increases with height.

Environmental temperature profiles are used to identify the freezing level, the height at which the temperature first drops below freezing, and

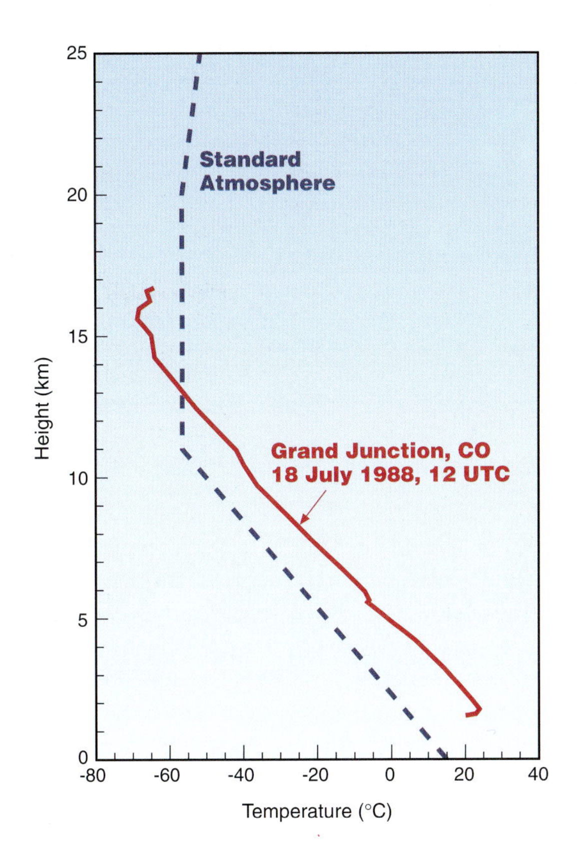

Figure 4.4 Individual temperature soundings differ from the mean sounding represented by the Standard Atmosphere (dashed curve). This sounding (solid curve) shows a shallow, surface-based temperature inversion at Grand Junction, Colorado, on 18 July 1988 at 1200 UTC (0500 LST). Temperatures above this *inversion* decreased until the tropopause was reached at 15.5 km or 9.6 miles.

are thus helpful in forecasting road conditions and skiing conditions. The range of elevations at which precipitation will fall as snow and the altitude at which the snowpack will melt is estimated from individual radiosonde soundings. Precipitation that forms and reaches the ground above the freezing level will fall as snow. Snow may also reach the ground at elevations up to a thousand feet below the freezing level because it takes some time at temperatures above freezing to melt snow completely. Forecasting sleet and ice storms, which occur when precipitation melts and then refreezes, requires a detailed study of the temperature structure (figure 4.5).

4.2.2. *Temperature and Safety Concerns*

Exposure to either extremely low or extremely high temperatures is an important safety issue. A person in a low-temperature environment, especially if improperly dressed, may be subject to *frostbite* (the freezing of body tissues, usually on the extremities) or *hypothermia* (a rapid, progressive mental and physical collapse caused by the lowering of body temperature). Frostbite and hypothermia are potential problems even at temperatures above freezing if winds are strong because wind increases the rate of heat loss from exposed skin. An indicator of the danger posed by the combination of low temperatures and strong winds is the *wind chill equivalent temperature*, a hypothetical air temperature in calm conditions that would cause the same heat loss from exposed skin that occurs for the actual wind and temperature conditions. Wind chill equivalent temperatures are use-

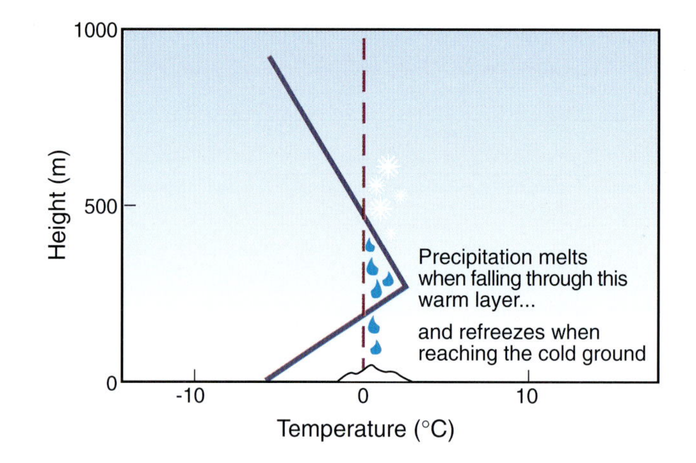

Figure 4.5 Ice storms occur when precipitation melts as it falls through an elevated warm layer of the atmosphere, where temperatures are above freezing, and then refreezes when it reaches the ground.

ful as indicators of cold weather safety hazards but are not actual temperatures. For example, water will not freeze when the wind chill equivalent temperature drops below 32°F. Water freezes only when the actual air temperature drops to 32°F or below. Wind chill equivalent temperatures are given in table 4.2; a calculation formula is provided in appendix A.

High temperatures are also a safety concern because they can lead to *heat exhaustion* or *heatstroke*. The effects of high temperature are aggravated by high humidity and low wind speeds. The rate of heat loss by perspiration is reduced when ambient humidity is high, and perspiration cools less effectively when winds are calm. Various indices, similar to the one for *wind chill*, have been developed to draw attention to the combined effect of high temperature and humidity. Table 4.3 presents one of these indices, called the *Heat Index*.

It is so hot in some places that the people there have to live in other places.
—elementary school student

4.2.3. *Measuring Temperature*

Temperature can be measured with a variety of sensors. The most familiar temperature sensor is the mercury-in-glass or alcohol-in-glass thermometer, but temperature can also be measured using *thermistors*, thermocouples, platinum resistance elements, and certain types of

Table 4.2 Wind Chill Equivalent Temperature (°F) as a Function of Air Temperature and Wind Speed

Wind Speed			Air Temperature (°F)									
mph	m/s	km/h	50	40	30	20	10	0	−10	−20	−30	−40
0	.0	.0	50	40	30	20	10	0	−10	−20	−30	−40
5	2.2	8.0	43	31	20	8	−4	−16	−27	−39	−51	−62
10	4.5	16.1	38	26	13	0	−13	−26	−38	−51	−64	−77
15	6.7	24.1	35	21	8	−6	−19	−33	−47	−60	−74	−88
20	8.9	32.2	32	18	4	−11	−25	−39	−53	−68	−82	−96
25	11.2	40.2	30	15	0	−15	−30	−44	−59	−74	−89	−104
30	13.4	48.3	28	12	−3	−18	−34	−49	−64	−80	−95	−110
35	15.6	56.3	26	10	−6	−21	−37	−53	−69	−85	−100	−116
40	17.9	64.4	24	8	−8	−24	−40	−57	−73	−89	−105	−121

Table 4.3 Heat Index (°F) or Apparent Temperature, as Determined from Actual Air Temperature (°F) and Relative Humidity (%)

Actual Air Temperature °F	Relative Humidity (%)										
	0	10	20	30	40	50	60	70	80	90	100
140	125										
135	120										
130	117	131									
125	111	123	141								
120	107	116	130	148							
115	103	111	120	135	151						
110	99	105	112	123	137	150					
105	95	100	105	113	123	135	149				
100	91	95	99	104	110	120	132	144			
95	87	90	93	96	101	107	114	124	136		
90	83	85	87	90	93	96	100	106	113	122	
85	78	80	82	84	86	88	90	93	97	102	108
80	73	75	77	78	79	81	82	85	86	88	91
75	69	70	72	73	74	75	76	77	78	79	80
70	64	65	66	67	68	69	70	70	71	71	72

THE BOILING POINT OF WATER

The temperature at which water boils is affected by altitude. As altitude increases, the boiling temperature decreases, resulting in longer cooking times.

The Boiling Point of Water and Its Effect on Cooking Times with Altitude (from Reifsnyder, 1980)

Altitude		Boiling Temperature		Cooking Time
(ft)	(m)	(°F)	(°C)	(min)
0	0	212	100	10
1000	305	210	98.9	11
2000	610	208	97.8	12
5000	1525	203	95.0	15
7500	2285	198	92.2	18
10000	3050	194	90.0	20
15000	4570	185	85.0	25

temperature-sensitive semiconductors. A self-contained spring-wound recording instrument called a *hygrothermograph* is often used to record both humidity and temperature as ink traces on a paper chart. Temperature sensors (as well as other weather sensors) are often housed in small louvered shelters that allow air to flow freely around the sensors while protecting them from exposure to direct sunlight. Other installations use fans to force air over temperature sensors that are protected from sunlight by a radiation shield. Appendix C lists sources for further information on the characteristics, selection, and effective placement of temperature sensors.

Air temperature is recorded in degrees Fahrenheit or degrees Celsius. The conversion between the two scales is straightforward. Conversion formulas are provided in appendix D, and a comparison of several points on the Fahrenheit and Celsius scales is given in table 4.4. In the United States, air temperatures at the ground are recorded and reported in degrees Fahrenheit, whereas upper air temperatures are reported in degrees Celsius. All Canadian and Mexican temperature observations, including surface observations, are in degrees Celsius. Because some *surface weather charts* plot the data in the original units, glaring discontinuities can occur at the borders of the United States and Canada, and the United States and Mexico.

4.3. Atmospheric Stability

In a stable atmosphere, vertical motions are suppressed and clouds are horizontally layered. In an unstable atmosphere, vertical motions are enhanced and clouds develop vertically.

Atmospheric stability is the resistance of the atmosphere to vertical motion. In a *stable* atmosphere, vertical motions are suppressed. Clouds are horizontally layered, and vertical exchange between air at the ground (which may contain air pollutants or moisture) and air aloft is minimized, thus reducing the vertical dispersion of air pollution. In an *unstable* atmosphere, vertical motions are enhanced. Clouds develop vertically, and the vertical dispersion of air pollution is enhanced. Observations that indicate whether the atmosphere is stable or unstable are listed in table 4.5.

4.3.1. *Determining Atmospheric Stability*

The stability of an atmospheric layer is determined by considering the behavior of a hypothetical parcel of air taken from a point in the layer, lifted a very small distance, and then released. Before lifting, the parcel is the same temperature as its surroundings. It is assumed that the displacement is an *adiabatic process,* that is, that the parcel does not exchange heat with the air around it during the displacement. The temperature of the parcel after lifting thus does not depend on the parcel's environment, but rather on thermodynamic processes caused by the pressure change associated with lifting. Atmospheric pressure decreases with height, so that the parcel, when lifted, encounters lower pressures and expands. This expansion causes the parcel to cool. (In contrast, a parcel that is lowered in the atmosphere is brought to a higher pressure, causing it to compress and warm.) If, after lifting, the parcel is warmer (and

Table 4.4 Comparison of Several Points on the Fahrenheit and Celsius Temperature Scales

Points for Comparison	Fahrenheit	Celsius
Temperature scale equivalence	−40	−40
Begin "sub-zero" temperatures	0	−17.8
Freezing point of water	32	0
Room temperature	72	22.2
Body temperature	98.6	37
Boiling point of water (standard conditions)	212	100

Table 4.5 Observations that Indicate Atmospheric Stability

Stable	Unstable
Clouds in layers with little vertical development (*stratiform* clouds, section 7.1.1), mountain and *lee wave* clouds	Clouds grow vertically (*cumuliform* clouds, section 7.1.1)
On the local scale, smoke from elevated stacks remains elevated and disperses mostly horizontally.	On the local scale, smoke *plumes* disperse well vertically and horizontally.
On the regional scale, smoke from multiple sources forms stacked layers of pollution in the atmosphere.	On the regional scale, pollution from multiple sources mixes together in a layer near the ground. The layer is shallow in the morning and deepens during the day.
Poor visibility due to smoke, haze, or fog	Good visibility
Steady winds, usually light	Gusty winds
Drizzle or light rain	Showery precipitation, thunderstorms

therefore lighter and less dense) than its new surroundings, it continues to rise, indicating that the atmospheric layer is unstable. If the parcel is colder (and therefore heavier and denser) than its new surroundings, it sinks back to its point of origin, indicating that the atmospheric layer is stable. If the parcel is the same temperature as its new surroundings, it remains at the point where it was released, indicating that the atmospheric layer is *neutral*.

Rising motions in the atmosphere produce expansion and cooling; sinking motions produce compression and warming.

The rate at which temperature decreases with height is called the *lapse rate*. This term can refer either to the cooling of a hypothetical parcel lifted adiabatically or to the actual decrease of temperature with height in an environmental temperature sounding. The lapse rate of a lifted parcel depends on whether the parcel is saturated (relative humidity of 100%) or unsaturated (relative humidity less than 100% and contains no liquid water). Radiosondes (section 4.1.3) are used to measure the moisture status (saturated or unsaturated) of the atmosphere.

The lapse rate is the rate of temperature decrease with height.

If an unsaturated parcel is lifted, it cools at the constant rate, called the *dry adiabatic lapse rate* or DALR, of 5.4°F per 1000 ft, or, equivalently, 9.8°C per km. (If the parcel is lowered, it warms at this same rate.)

If a saturated parcel is lifted, it cools at the *moist adiabatic lapse rate* or MALR. Unlike the DALR, the MALR is not constant, but is dependent on the parcel's temperature and pressure. Table 4.6 shows values of the MALR at different pressures and temperatures. The MALR is always

Table 4.6 The Moist Adiabatic Lapse Rate (MALR) at Different Pressures and Temperatures in °C per Kilometer and °F per 1000 Feet

	Temperature (°C)					Temperature (°F)				
Pressure (mb)	−40	−20	0	20	40	−40	−20	0	20	40
1000	9.5	8.6	6.4	4.3	3.0	5.2	4.7	3.5	2.4	1.6
800	9.4	8.3	6.0	3.9	2.8	5.2	4.6	3.3	2.2	1.5
600	9.3	7.9	5.4	3.5	2.6	5.1	4.4	3.0	1.9	1.4
400	9.1	7.3	4.6	3.0	2.4	5.0	4.0	2.5	1.6	1.3
200	8.6	6.0	3.4	2.5	2.0	4.7	3.3	1.9	1.4	1.1

lower than the DALR. Because the parcel is saturated, some of the cooling is offset by latent heat that is released into the parcel as water vapor condenses to form liquid cloud droplets (section 3.2.4). The MALR approaches the DALR at low temperatures when the amount of water vapor in the saturated parcel is very small.

The relationship between the DALR, the MALR, and stability are illustrated for several parcels in the temperature sounding in figure 4.6. The yellow dots represent parcels of air at five levels (A–E) of the environmental temperature sounding (ETS). If a parcel is unsaturated, it is lifted to the top of the red line; if it is saturated, it is lifted to the top of the green line. After lifting, the temperature of the parcel is compared to the actual temperature at that level, as indicated by the solid blue line. Thus, if the parcel after lifting is to the left of the blue line, it is cooler than its new environment and will sink when it is released, indicating that the atmosphere at the parcel's original level is stable. If the parcel after lifting is to the right of the blue line, it is warmer than its new environment and will continue to rise after release, indicating atmospheric instability. If the parcel has the same temperature as its new environment, the atmosphere at the parcel's original level is neutral. The outcome of the parcel method is the same for all parcels that come from a segment of the sounding in which the environmental lapse rate is constant.

A more direct approach to determining the atmospheric stability in a layer of air is to compare the slope of the environmental temperature profile (the blue line in figure 4.6) with the DALR and the MALR. The *environmental temperature lapse rate* (ETLR), the slope of the blue line, is com-

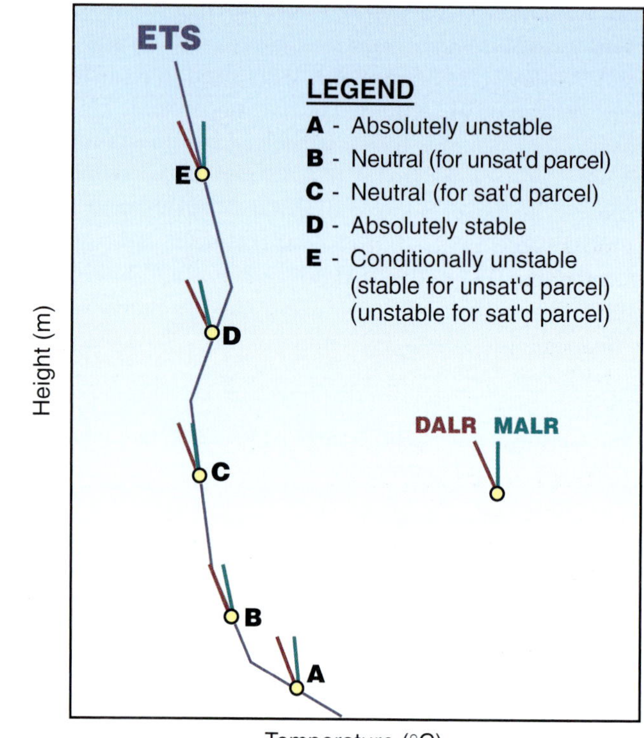

Figure 4.6 Stability can be determined from an environmental temperature sounding by lifting parcels of environmental air (A–E) from various heights. The stability of the atmosphere at levels A–E is given in the figure for both saturated and unsaturated parcels. The atmosphere at A is *absolutely unstable*, that is, it is unstable for both saturated and unsaturated parcels. The atmosphere at D is *absolutely stable*, that is, it is stable for both saturated and unsaturated parcels. The atmosphere at E is *conditionally unstable*. It is stable for unsaturated parcels but unstable for saturated parcels.

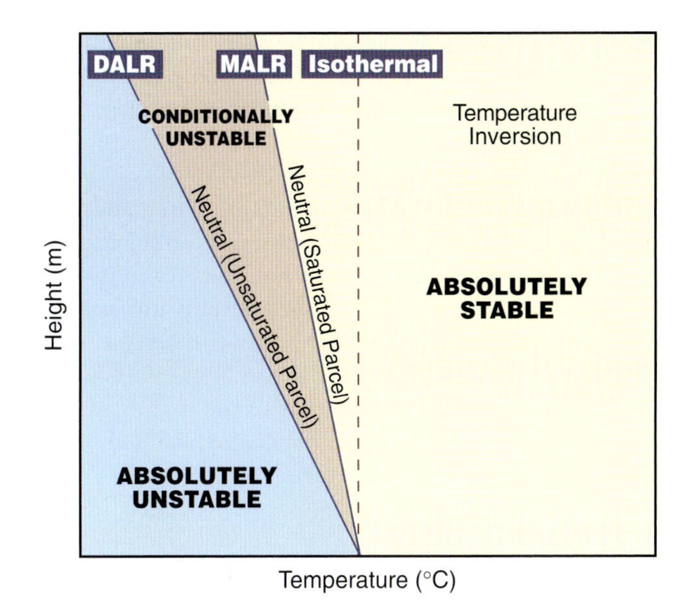

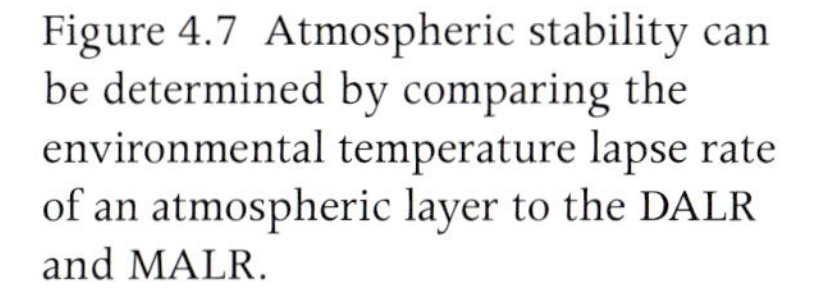
Figure 4.7 Atmospheric stability can be determined by comparing the environmental temperature lapse rate of an atmospheric layer to the DALR and MALR.

pared to the DALR if the layer is unsaturated or to the MALR if the layer is saturated. This comparison can be explained using figure 4.7. The atmosphere is absolutely stable if the ETLR is lower than the moist adiabatic lapse rate, that is, if the ETLR is in the yellow zone. The atmosphere is absolutely unstable if the ETLR is greater than the dry adiabatic lapse rate, that is, if the ETLR is in the blue zone. When the ETLR is between the dry adiabatic lapse rate and the moist adiabatic lapse rate (beige zone), the atmosphere is considered conditionally unstable because it is stable if the air is unsaturated but unstable if the air is saturated. When the ETLR for unsaturated parcels is equal to the dry adiabatic lapse rate or when the ETLR for saturated parcels is equal to the moist adiabatic lapse rate, the atmosphere is considered neutral.

4.3.2. *Variations in Atmospheric Stability*

The stability of the atmosphere is constantly changing, with variations over time and horizontal distance and from one layer of the atmosphere to the next. The troposphere is generally stable because the mean lapse rate of the troposphere, 3.5°F per 1000 ft or 6.5°C/km (Standard Atmosphere, table 4.1), is less than the dry adiabatic lapse rate.

The stability of a layer of air is affected by any change in the atmosphere that changes the environmental temperature profile of the layer. A layer at any height in the atmosphere becomes more stable (i.e., stabilizes) when the upper part of the layer warms relative to the lower part or when the lower part cools relative to the upper part. It becomes less stable (destabilizes) when the upper part cools relative to the lower part or when the lower part warms relative to the upper part. Relative changes in temperature difference across a layer can occur when horizontal winds bring colder or warmer air into part of the layer, or when air sinks and warms or rises and cools. These vertical motions are often associated with high and low pressure centers (section 5.1). The layer of air just above the earth's surface tends to stabilize during the night and to destabilize during the day. Nighttime stabilization is due to outgoing radiation that cools the ground, and daytime destabilization results from solar radiation that heats the ground.

Stability in the layer of the atmosphere close to the earth's surface is also affected by the flow of air over a mountain barrier, and, conversely, the flow of air over a mountain barrier is affected by stability. When air rises over a mountain barrier, stability is reduced. When it sinks on the lee side, stability is enhanced. Stable air, resistant to vertical motion, tends to flow around a mountain barrier rather than over it. If winds are strong, stable air can be carried over the mountain barrier, but it will try to regain its equilibrium altitude in the lee of the barrier, creating wave patterns in the atmosphere. If sufficient moisture is present, characteristic clouds (section 7.1.4) form in the waves.

4.4. The Atmospheric Boundary Layer and the Surface Energy Budget

The *atmospheric boundary layer* (ABL) is the lowest layer of the atmosphere in which temperature and humidity are affected by the transfer of heat and moisture to and from the surface of the earth. The ABL is usually warmed by an upward transfer of heat from the earth's surface during the day and cooled by a downward transfer of heat from the atmosphere to the earth's surface during the night. The transfer of moisture between the atmosphere and the ground results in drying of the atmosphere when moisture condenses onto the soil or vegetation and moistening when evaporation or transpiration occur. The ABL can also be defined as the layer of the atmosphere affected by the frictional drag of the earth's surface.

The layer of air near the earth's surface becomes more stable during the night and less stable during the day.

The amount of heat and moisture transferred between the atmosphere and the underlying surface and the direction (upward or downward) of the transfer are determined by the *surface energy budget*. The surface energy budget is a simple mathematical statement that the net solar and terrestrial radiation at the earth's surface (R) must be transformed into (1) *latent heat flux, L*, used to evaporate (or condense) water, (2) *ground heat flux, G*, used to warm (or cool) the ground, and (3) *sensible heat flux, H*, used to warm (or cool) the atmosphere, and that the total amount of energy cannot increase or decrease during the transformation. Radiation specialists are primarily interested in R, soil scientists in G, and meteorologists in the *fluxes* of heat and moisture into or out of the atmosphere at its lower boundary, H and L.

The surface energy budget is expressed by the simple equation

$$R + H + L + G = 0$$

where all fluxes toward the surface (whether from the atmosphere or ground) are positive and all fluxes away from the surface are negative. The four fluxes must add to zero at any instant in time, so that a change in the value (either positive or negative) of one of the fluxes must be balanced by an equal but opposite net change in the numerical values of the other terms.

The surface energy budget states that the net solar and terrestrial radiation at the earth's surface must be used to evaporate (or condense) water, to heat (or cool) the ground, or to heat (or cool) the atmosphere.

4.4.1. *The Net All-Wave Radiation Term* R

Shortwave radiation is radiation from the sun; longwave radiation is radiation from objects or gasses at terrestrial temperatures.

All objects emit radiation. The wavelength of the radiation depends on the temperature of the radiating body. The sun, with a temperature of 6000°C (approximately 11,000°F), emits most of its radiation in the wavelength range 0.15–3 micrometers. Human vision responds only to the portion of solar radiation in the *visible spectrum*, with wavelengths between 0.36 and 0.75 micrometers. Terrestrial objects (including the earth's atmosphere), which have much lower temperatures than the sun, radiate energy at longer wavelengths (3–100 micrometers). At the earth's distance from the sun, the intensities of solar and terrestrial radiation are comparable, but there is virtually no overlap in wavelength between the two. Thus, radiation from the sun is termed *shortwave radiation*, whereas radiation emitted by objects or gases at normal terrestrial temperatures is called *longwave radiation*.

Both shortwave and longwave radiation can be directed upward (from the ground to the atmosphere) or downward (from the atmosphere to the ground). There are thus four components in the *net all-wave radiation* term R:

- incoming shortwave radiation
- outgoing shortwave radiation, which is the fraction of incoming shortwave radiation reflected back into space
- incoming longwave radiation emitted by the gases and clouds in the atmosphere
- outgoing longwave radiation emitted by the earth's surface and objects on it

4.4.2. *Diurnal Variations in* R *and in the Surface Energy Budget*

Diurnal variations in the four components of the net all-wave radiation term R result in a strong diurnal oscillation in R (figure 4.8). Both solar terms be-

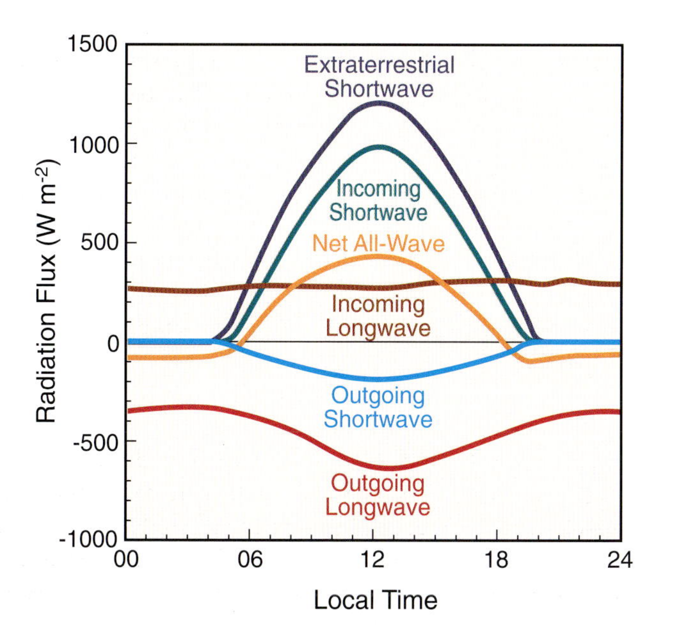

Figure 4.8 Typical diurnal evolution of the four components of R at the earth's surface under cloudless conditions on a June day at a semiarid site near the Columbia River in eastern Oregon. The *extraterrestrial shortwave radiation*, also shown, is the theoretical radiation that would be received outside the earth's atmosphere on a plane that is parallel to the earth's surface below.

Shortwave radiation begins at sunrise and ends at sunset, causing sign changes in R about an hour after sunrise and an hour before sunset.

gin at sunrise, end at sunset, and peak at midday. Incoming and outgoing longwave radiation continue both night and day, peaking in early afternoon. The net all-wave radiation is calculated numerically as the sum of the four components. At night, when the solar radiation terms are zero, R is driven by a net loss of longwave radiation because the outgoing longwave radiation is larger than the incoming longwave radiation. R changes sign about an hour after sunrise when the net loss of longwave radiation is overpowered by a net gain in shortwave radiation. R changes sign again about an hour before sunset when the amount of incoming solar radiation is reduced.

Because the terms of the surface energy budget equation must always sum to zero, diurnal variations in R result in diurnal variations in ground heat flux, latent heat flux, and sensible heat flux. Figure 4.9 illustrates the continuous changes in the budget components that occur over a 24-hour period. At night, a radiation deficit is countered by an upward flux of heat from the ground, a downward flux of heat from the air, and condensation of moisture at the surface. During the day, the radiation excess at the surface (R) heats the ground (G) and the atmosphere (H) and evaporates water (L) from the surface. Figure 4.10 presents a snapshot of typical surface energy budget components at noon and midnight under cloudless conditions over a land surface.

Measurements of surface heat and radiation fluxes are usually performed only in research programs because the instruments are expensive and the data often require extensive processing. Soil heat fluxes are often evaluated with soil heat flux plates or through soil temperature profile measurements supplemented with laboratory determinations of the physical properties of the soil. Radiative fluxes are measured with radiometers. Sensible and latent heat fluxes are often determined by measuring the correlation between vertical velocity and temperature deviations and between vertical velocity and moisture deviations on very short time scales (rising warm or moist air currents represent upward heat or moisture fluxes) or through relationships between vertical *gradients* of

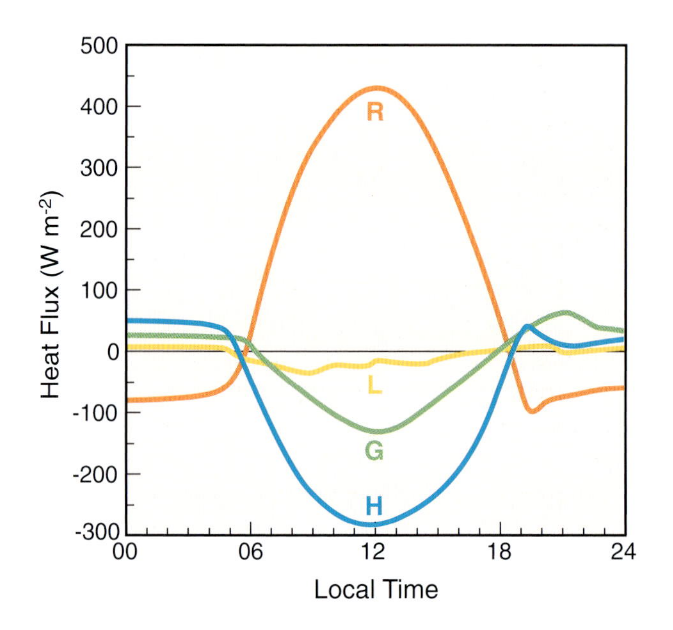

Figure 4.9 The surface energy budget terms vary diurnally. The data shown are for a June day at a semiarid site near the Columbia River in eastern Oregon. The net radiation curve in this figure is repeated from figure 4.8, but is plotted on a different scale.

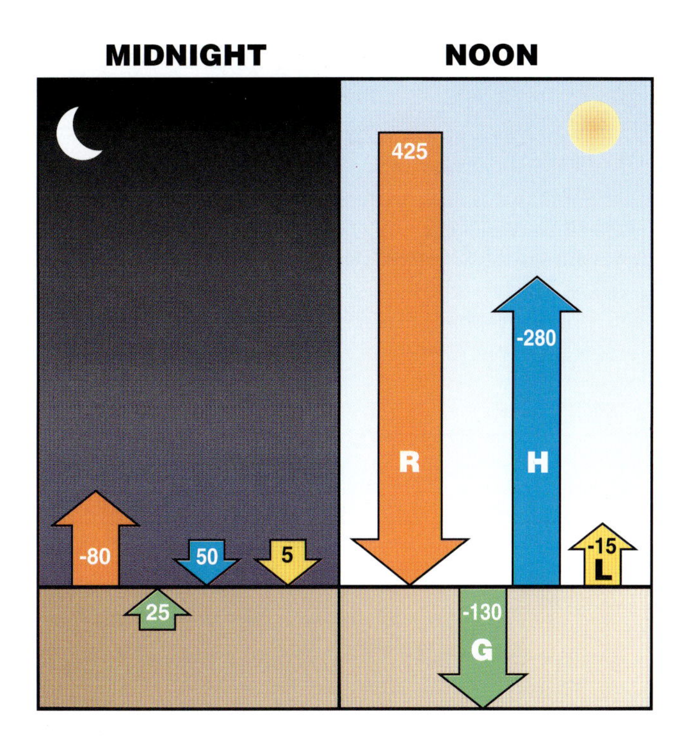

Figure 4.10 Typical values of the surface energy budget components at midnight and noon. The numerical values shown are from figure 4.9.

temperature and humidity near the ground. Appendix C provides a list of references to further information on measurement equipment.

4.4.3. *Factors That Cause Spatial and Nondiurnal Temporal Variations in* R *and the Surface Energy Budget*

The previous description of the surface energy budget assumes clear, undisturbed weather conditions typical of a high pressure region in summer. However, the surface energy budget varies significantly over time and space. It is affected by variations in shortwave and longwave radiation from site to site (section 11.2 and appendix E) and also responds to daily, seasonal, and spatial changes in atmospheric conditions and in ground cover and soil conditions.

The amount of solar radiation received at a given site is affected by latitude, altitude, day of the year, time of day, cloud cover, aerosol content of the atmosphere, slope and orientation of the surface, shading by surrounding terrain, and surface *albedo* (or reflectivity). The albedo of a surface (figure 4.11) determines the fractional amount of incoming solar radiation that is reflected (i.e., converted to outgoing shortwave radiation). A fresh snow surface or a thick cloud is a good reflector and therefore has a high albedo; forests and newly plowed fields are poor reflectors. Water is a poor reflector, too, except when the incoming radiation strikes it at a shallow angle.

The surface energy budget varies significantly over time and space, responding to changes in atmospheric conditions and to variations in terrain and ground cover.

The wavelength and the flux of longwave radiation from an object are determined by the object's temperature. The warmer the object, the shorter the wavelength and the greater the heat flux. Heat flux is also affected by the object's *emissivity*, which is a measure of the object's effec-

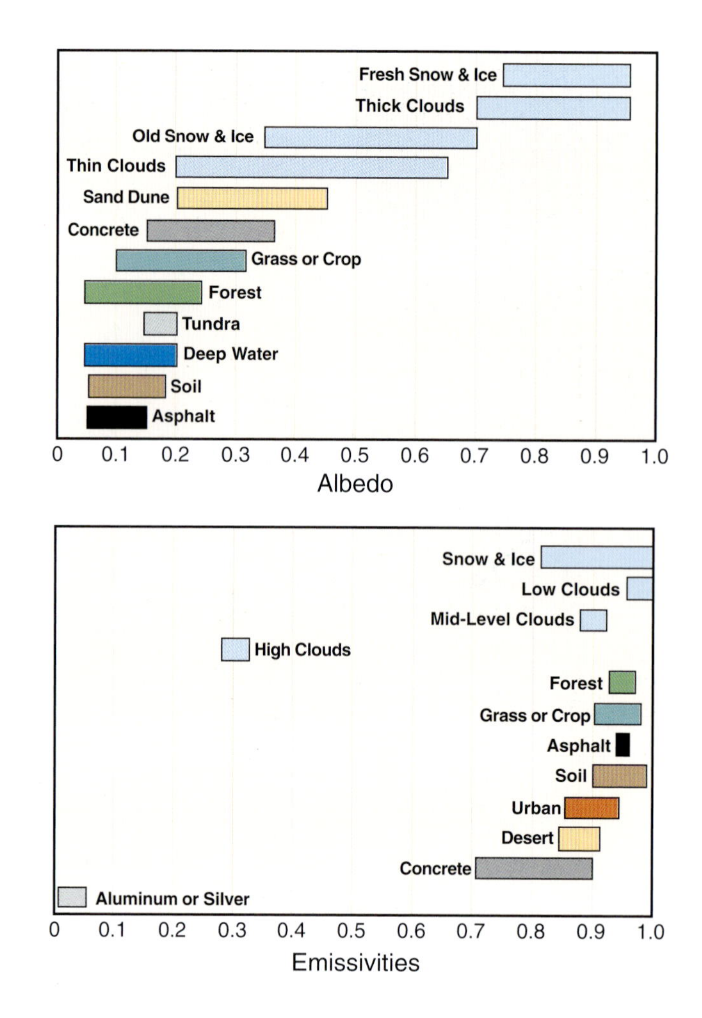

Figure 4.11 (*top*) The solar reflectivity or albedo of most natural materials at the earth's surface is between 0.05 and 0.30. Bright surfaces like fresh snow and thick clouds have relatively high albedos. Clouds can reflect sunlight back to space before it reaches the ground.

Figure 4.12 (*bottom*) Emissivities are between 0.85 and 1.00 for most natural materials. High clouds, polished aluminum, and silver have lower emissivities.

tiveness as a radiator compared to a perfect radiator. Most natural solid materials are good radiators, with emissivities (figure 4.12) between 0.85 and 1.0. Snow and clouds have high emissivities and therefore are very efficient radiators. The emissivities of atmospheric gases are lower and are dependent on water vapor and carbon dioxide content and on the depth of the layer considered.

When R changes, the other terms of the surface energy budget also change. The relative changes in the other terms, ground heat flux, latent heat flux, and sensible heat flux, are influenced by atmospheric temperature and humidity profiles, winds, the soil temperature profile, the soil moisture availability, and vegetative cover.

Spatial variations in energy fluxes within a region can be reduced temporarily if clouds cover the region, reducing the differences in solar radiation between sunny and shady sites, or if precipitation falls over the entire region, equalizing the moisture content of the soil.

4.4.4. *Diurnal Evolution of ABL Temperature Structure over the Plains*

The evolution of the vertical temperature structure in the atmospheric boundary layer (ABL) over flat terrain (and over mountainous terrain,

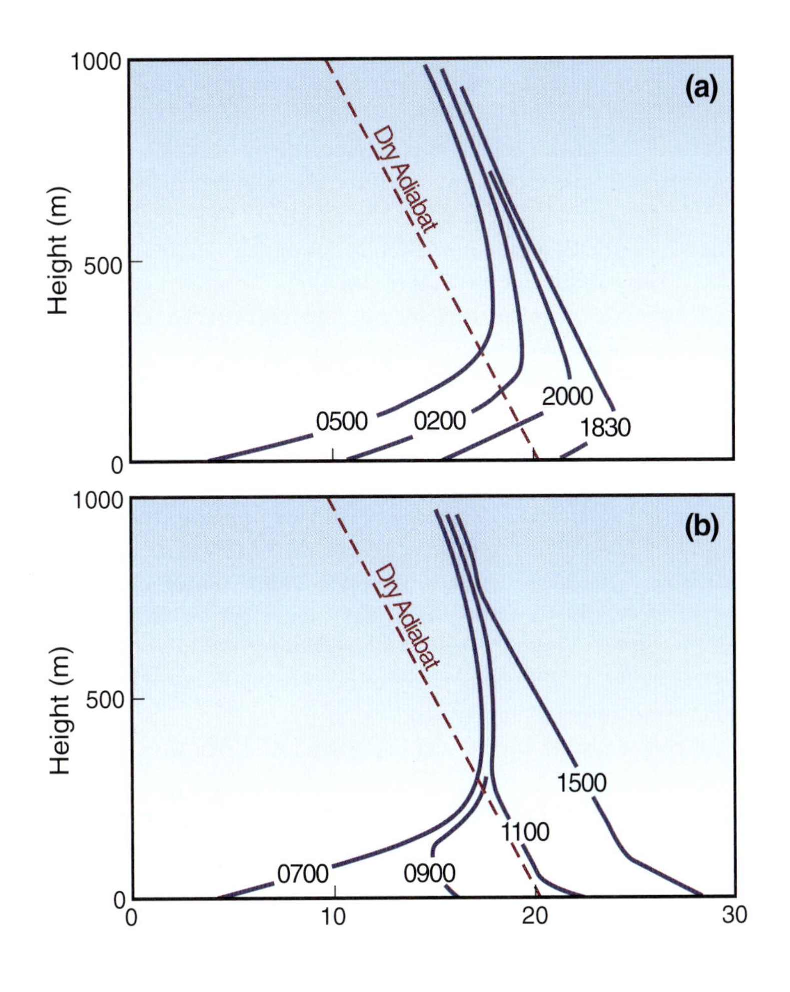

Figure 4.13 (*top*) Typical evolution of the vertical temperature structure in the *boundary layer* over the plains during (a) night and (b) day. The numbers indicate the times of the soundings. The area between successive soundings on a temperature–height plot is proportional to the heat added to the atmosphere between soundings.

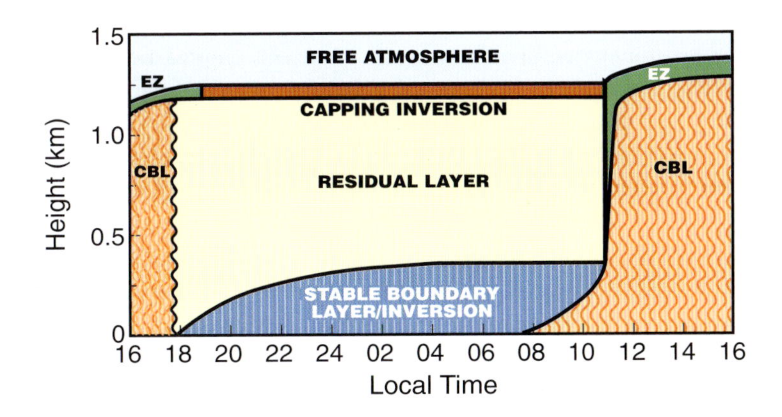

Figure 4.14 (*bottom*) Schematic diagram of the diurnal evolution of the boundary layer over a plain in fair weather during summer. (BL = convective boundary layer, EZ = entrainment zone. Adapted from Stull, 1995.)

section 11.1) is regular if skies are clear or partly cloudy in response to the cycle of upward sensible heat flux during daytime and downward sensible heat flux at night (figure 4.13).

The depth of the ABL varies, depending on atmospheric stability and sensible heat flux. Over land in winter, the ABL is typically less than 1000 ft (300 m) deep at night and less than 3000 ft (1000 m) deep during daytime. On summer days, the ABL reaches depths of 3000–6000 ft (1000–2000 m) and can exceed 10000 ft (3000 m) over arid regions of the western United States. In complex terrain, the depth of the ABL may vary over short distances. For example, the morning ABL over valley floors and sidewalls, constrained by a strong overlying valley inversion, may be much shallower than the ABL over the adjacent ridge tops.

At night, the net loss of longwave radiation cools the earth's surface, which in turn cools the air immediately above it. This layer of cold air is called the *stable boundary layer* (SBL) or the nighttime atmospheric boundary layer. Through the course of the night, more and more heat is extracted from the atmosphere by a downward sensible heat flux toward the surface, and the stable boundary layer grows deeper and deeper, reaching its maximum depth of several hundred meters around sunrise (figure 4.13a). Cooling near the ground may be strong enough to produce a temperature inversion.

A stable boundary layer develops at night; a convective boundary layer develops during the day.

During daytime (figure 4.13b), solar radiation passes largely unimpeded through the atmosphere, is received at the ground, and is converted to sensible heat flux, which heats the air adjacent to the surface. Thus, the earth's atmosphere is heated not directly by solar radiation but from below by sensible heat flux. In the warmed layer, called the *convective boundary layer* (CBL), convective thermals or plumes mix the atmosphere vertically. The temperature of the *mixed layer* increases as the upward flux of sensible heat continues, and the layer deepens.

A schematic of the layers involved in boundary layer evolution over flat terrain is shown in figure 4.14. In the late afternoon (1600 local time), a deep convective boundary layer is present, capped by an *entrainment zone* where air from the free atmosphere is incorporated into the underlying mixed layer, enabling the mixed layer to grow in depth. Near sunset (1800 local time), the sensible heat flux reverses, and the ground begins to cool. A stable boundary layer develops at the ground, ending convection. Turbulence and mixing cease in the elevated remnants of the convective boundary layer, now called the *residual layer*. The residual layer persists through the night, bounded at its top by a *capping inversion*, which is a remnant of the daytime entrainment zone. Shortly after sunrise (0730 local time), when sensible heat flux reverses, a convective boundary layer begins to grow upward from the ground at the base of the stable layer, eventually breaking through the stable layer and growing into or through the previous day's residual layer. The regular diurnal evolution of atmospheric layers with differing stabilities causes regular diurnal changes in the dispersion of air pollution. During the night, pollutants are often trapped in the stable boundary layer. During the day, pollutants are mixed vertically by convective plumes.

Pressure and Winds

5

5.1. Atmospheric Pressure

Atmospheric pressure at a given point in the atmosphere is the weight of a vertical column of air above that level. Differences in pressure from one location to another cause both horizontal motions (winds) and vertical motions (convection and subsidence) in the atmosphere. Vertical motions, whether associated with high and low pressure centers or with other meteorological processes, are the most important motions for producing weather because they determine whether clouds and precipitation form or dissipate. The location of high and low pressure centers is a key feature on weather maps, providing information about wind direction, wind speed, cloud cover, and precipitation.

Pressure-driven winds carry air from areas where pressure is high to areas where pressure is low. However, the winds do not blow directly from a high pressure center to a low pressure center. Because of the effects of the rotation of the earth and friction, winds blow clockwise out of a high pressure center and counterclockwise into a low pressure center in the Northern Hemisphere. These wind directions are reversed in the Southern Hemisphere. The strength of the wind is proportional to the pressure difference between the two regions. When the pressure difference or pressure gradient is strong, wind speeds are high; when the pressure gradient is weak, wind speeds are low.

Winds blow clockwise out of a high pressure center and counterclockwise into a low pressure center in the Northern Hemisphere.

As air flows out of a high pressure center, air from higher in the atmosphere sinks to replace it. This subsidence produces warming and the dissipation of clouds and precipitation. As air converges in a low pressure center, it rises and cools. If the air is sufficiently moist, cooling can cause the moisture to condense and form clouds. Further lifting of the air can produce precipitation. Thus, rising pressure readings at a given location indicate the approach of a high pressure center and fair weather, whereas falling pressure readings indicate the approach of a low pressure center and stormy weather. The vertical motions caused by the *divergence* of air out of a high pressure center or the *convergence* of air into a low pressure center are generally weak, with air rising or sinking at a rate of several cm per second, and they cannot be measured by routine weather observations.

Air sinks over a high pressure center and rises over a low pressure center.

The horizontal and vertical motions associated with high and low pressure centers in the Northern Hemisphere are analogous to the movement of a screw. When a screw is turned clockwise, it sinks. When it is turned counterclockwise, it rises.

A high pressure center is associated with the dissipation of clouds, whereas a low pressure center is associated with the buildup of clouds.

LEE CYCLOGENESIS AND ANTICYCLOGENESIS

Mountain barriers influence the development of low and high pressure centers. *Cyclogenesis* (the formation or intensification of low pressure storm systems or *cyclones*) often occurs in the lee of major mountain barriers. Areas of North America where long north–south-oriented mountain ranges are in the path of high-level jet streams and have high relief on their leeward sides are preferred zones for the formation and strengthening of low pressure centers. The best-known locations in North America for the formation of winter cyclones (figure Ia) are in the lee of the Colorado Rockies in southeastern Colorado; in the lee of the Northern Rockies from central Montana to northern Alberta; and over the Atlantic Ocean in the lee of the Appalachians between Cape Hatteras and Cape Cod. A weak cyclogenesis center (not shown in the figure) is also found in the lee of the Sierra Nevada. During the summer (figure Ib), the cyclogenesis center in southeastern Colorado becomes inactive, but the Appalachian and northern Rocky Mountain cyclogenesis centers continue to be active, and an additional storm formation center appears in northeastern British Columbia.

Once cyclones form in the lee of the Rocky Mountains, they are usually carried to the east and northeast as they develop. Thus, cyclones that form in southern Colorado or northern New Mexico, called *Colorado Lows*, follow trajectories into the upper Midwest, where they are the main producers of *blizzard* conditions and severe winter storms.

Anticyclogenesis, the formation or intensification of high pressure centers or *anticyclones*, is also closely tied to mountain barriers. Figures IIa and IIb show areas where winter and summer anticyclones form in relationship to the Rocky Mountains and their typical paths after formation. In winter, two major centers of anticyclogenesis are found in North America—one in mountains of the Yukon Territory and one in the Rocky Mountain Trench. Anticyclones forming over the Yukon drive southeastward into the central and eastern United States or travel eastward from the Rocky Mountains across southern Canada. Anticyclones that form in the Rocky Mountain Trench travel southeastward along the crest of the Rocky Mountains. In summer, anticyclones form near the Arctic Ocean coastline in north-central Canada and move southeastward through the Great Lakes or form east of the Canadian Rockies and travel east-southeastward over the Ohio River Valley.

(*continued*)

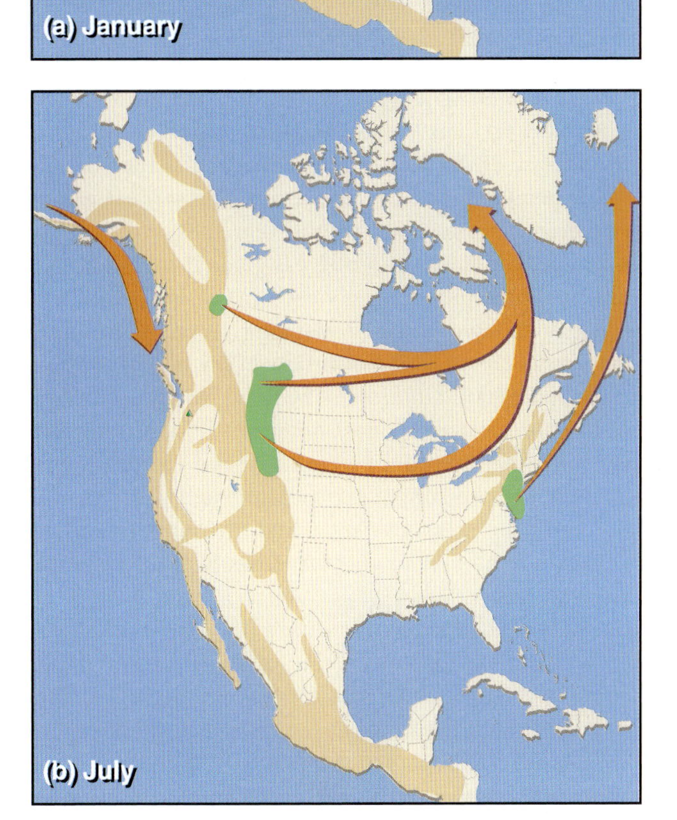

Figure I Locations (green areas) of frequent (a) January and (b) July cyclogenesis in North America and the most likely paths of the cyclones. (Adapted from Zishka and Smith, 1980)

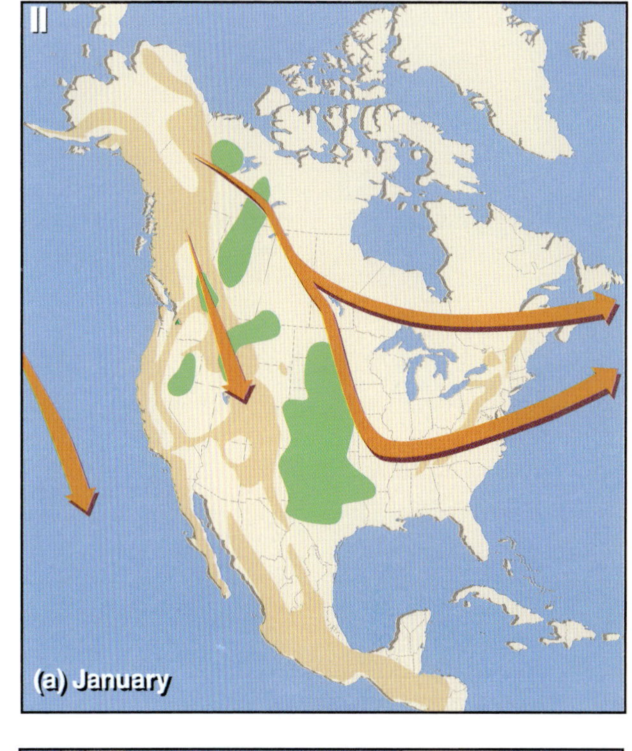

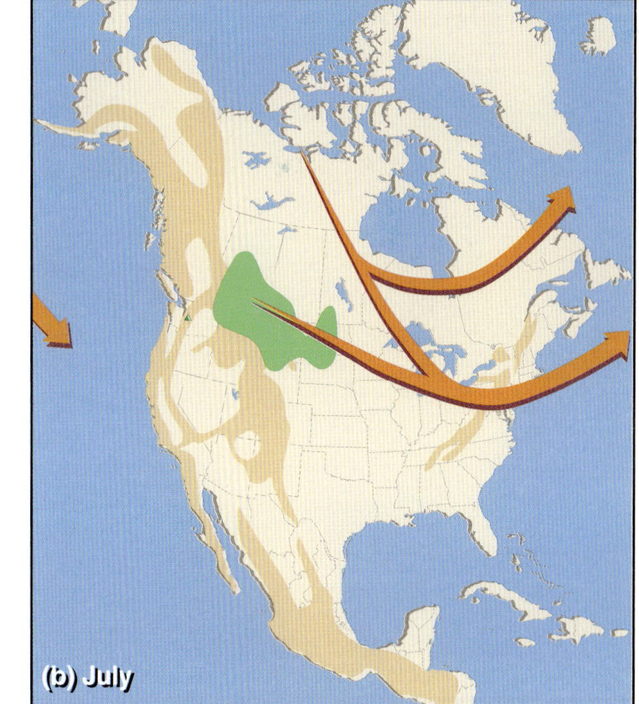

Figure II Locations (green areas) of frequent (a) January and (b) July anticyclogenesis in North America and the most likely paths of the anticyclones. (Adapted from Zishka and Smith, 1980)

5.1.1. *Measuring Atmospheric Pressure*

Atmospheric pressure is measured with a *barometer* and is therefore often referred to as *barometric pressure*. The most common types of barometers are the *mercury barometer* and the *aneroid barometer*. Information on barometers is referenced in appendix C.

The weight of a column of air can be expressed in terms of the height of the mercury column in a mercury barometer. In English units, the total weight of a 1-inch-square column of air extending from sea level to the top of the atmosphere, when averaged over the globe for long time periods, is 14.7 pounds. The mean sea level pressure of 14.7 pounds per square inch forces the mercury to rise 29.92 inches or 760 mm. The higher the atmospheric pressure, the higher the column of mercury will rise. Most pressure data on international weather maps are given not in inches but in millibars. In these metric units, mean sea level pressure is 1013.25 millibars (abbreviated mb). Conversion factors between different pressure units are given in appendix D.

5.1.2. *Pressure and Height*

Pressure decreases with height. A height change of 10 meters is equivalent to a pressure change of about 1 mb.

Pressure decreases continuously with height because, as the measurement point moves higher in the atmosphere, the weight of air above it decreases. The relationship between pressure changes and height changes is affected by air temperature and atmospheric density, but as a general rule, a height change of 10 m equals a pressure change of about 1 mb. Pressure changes much more rapidly with height than with horizontal distance. The pressure difference between the base of the Seattle Space Needle and the observation deck at 605 feet (184 m) is approximately 22 mb. This is often greater than the pressure difference between Seattle and Denver. The pressure at different elevations in the Standard Atmosphere is given in table 4.1.

A *barometer* indicates pressure changes over time at the height of a measurement site. An alternative way of visualizing pressure changes is to think of them in terms of the variation over time of the height of a constant pressure surface that initially intersects the fixed height of the measurement site. This pressure surface is quasi-horizontal, but it is not flat, resembling a topographic map in rolling terrain. As the surface of constant pressure descends, the pressure at the measurement site drops; as the pressure surface rises, the pressure rises. Thus, an alternative to describing pressure changes at a fixed height is to describe height changes of a constant pressure surface.

Meteorologists usually reference fixed standard pressure levels rather than fixed heights. By common agreement, the standard pressure levels for the troposphere are at 1000, 850, 700, 500, and 300 mb. In the Standard Atmosphere (table 4.1), these pressure levels are found, respectively, at 370, 4780, 9880, 18,280 and 30,050 ft (110, 1460, 3010, 5570 and 9160 m) above mean sea level. As the atmosphere deviates from the conditions of the Standard Atmosphere, the height of the pressure levels shift above or below these values. Because pressure is a measure of the weight of the air column above the measurement point, and because sea level pressure is near 1000 mb, approximately 100, 85, 70, 50, and 30% of the atmospheric mass remains in a column above these standard levels, respectively.

ALTIMETERS ARE BAROMETERS

Altimeters, which are used to determine distance above sea level or some other reference point, are barometers that are calibrated to indicate altitude. The height scale is based on the relationship between pressure and height in the Standard Atmosphere. An altimeter, when sensing a pressure of 900 mb, for example, will report a height of 3240 ft (table 4.1). Because the pressure at a fixed height level (or, equivalently, the height of a given pressure level) varies with time, and because atmospheric structure is never represented perfectly by the Standard Atmosphere, the absolute height indicated by the altimeter will nearly always be somewhat in error. Errors in altimeter readings can be corrected by adjusting the altimeter using surface barometric observations. An altimeter that has been set at a location where the elevation is known will report heights accurately as it is carried to higher and lower elevations, as long as traveling weather disturbances do not bring generally higher or lower atmospheric pressures into the region during the period of the measurements. When such disturbances occur, adjustments to the altimeter readings are necessary to correct the resulting height errors.

Pilots flying below 18,000 feet are required to obtain pressure data adjusted to mean sea level and to calibrate their altimeters every 100 miles to ensure that aircraft fly at known heights relative to other aircraft and the underlying terrain. The phase "from high to low, look out below" reminds pilots that, if they fly from a high pressure area into a low pressure area, they will be flying at an altitude lower than that indicated by their altimeters.

Altimeters can also be used to monitor changes in atmospheric pressure. An altimeter kept at a fixed height will, over time, indicate variations in height, which in fact reflect changes in pressure. An increase in height indicates falling pressure; a decrease in height indicates rising pressure.

The standard pressure levels can be visualized as a set of concentric spheres encircling the earth. The 1000-mb sphere intersects low mountains that rise from sea level in coastal areas. The 850-mb sphere intersects the foothills of the Rockies at Denver and passes over the Great Plains to cut through a few of the highest summits in the Appalachians.

PRESSURE ON MOUNT EVEREST

Mt. Everest (29,028 ft or 8848 m), the highest point on earth, has a mean pressure near 314 mb at its summit (extrapolated from table 4.1). Only about 31% of the mass of the atmosphere is still present above this level, making it difficult for human lungs to extract the necessary oxygen for work. Mt. Everest has been climbed without supplemental oxygen, but most climbers use bottled oxygen above the 7300-m (24,000-ft) level. For comparison, private pilots in the United States are required to use supplemental oxygen on flights that remain at elevations above the 3810-m (12,500-ft) level for more than 30 minutes.

The 700-mb level passes over the Appalachians, intersects many of the ridges in the Rockies, and touches the highest peaks in the Sierra Nevada and Cascades. The 500-mb sphere is above all of the mountains in the contiguous United States, although Pico de Orizaba in Mexico nearly reaches this sphere, and Mt. Logan, Denali, and a few other Alaskan peaks protrude through it. Mt. Everest is the only peak that approaches the height of the 300-mb sphere, which is above all of the earth's mountains, although still within the troposphere (figure 4.2).

5.1.3. *Pressure Analyses*

Pressure analyses are performed using data collected at the ground and at each of the standard pressure levels. The data are plotted on maps, called surface weather charts and *upper air weather charts*. Surface weather charts are constant height charts on which the pressures at sea level are plotted. Upper air weather charts are constant pressure charts. They are used to plot the heights above sea level of the standard pressure surfaces. A single pressure chart, either a surface chart or an upper air chart, identifies traveling high and low pressure weather disturbances and, because pressure differences drive winds, provides information about spatial variations in wind speed and direction. Comparing charts from two or more atmospheric levels provides information about the vertical structure of the atmosphere. Because the atmosphere is constantly changing both horizontally and vertically, and because characteristics of the upper atmosphere have a significant impact on weather at the ground, this three-dimensional picture is the basis of weather forecasts.

Upper air weather charts are prepared for each of the standard pressure levels (850, 700, 500, and 300 mb). For a 500-mb chart, for example, the heights at which that pressure level is encountered on individual radiosonde ascents are plotted on an upper air weather chart. Contours are then drawn through points of equal height, creating the pattern of highs and lows familiar from television weather reports. Low heights on a constant pressure surface are equivalent to low pressures on a constant height surface, so the high and low height centers on these charts are often referred to as high and low pressure centers. High and low pressure weather systems, also called anticyclones and cyclones, respectively, are indicated by the letters H and L. Not all contours on a weather chart are closed. Elongated regions of low pressure, which may or may not include the closed circulation of a low pressure center, are called *troughs*. Elongated regions of high pressure are called *ridges* and may or may not include a high pressure center with a closed circulation. An example of a 500-mb analysis is shown in figure 5.1. The contours on this chart are the heights above sea level of the 500-mb pressure surface.

Upper air weather charts are analyzed for the 850, 700, 500, and 300 mb pressure levels.

Whereas an upper air weather chart is an analysis of the heights above sea level of a constant pressure surface, a surface weather chart is an analysis of pressure at sea level height. Pressure measurements from surface stations are plotted on the chart, and points of equal pressure are connected by lines called *isobars*, which are labeled in millibars. For example, suppose there are many ships sailing on a large ocean and an observer on each ship measures the pressure at sea level height at the same time. All of the pressure observations could then be plotted on a map of the

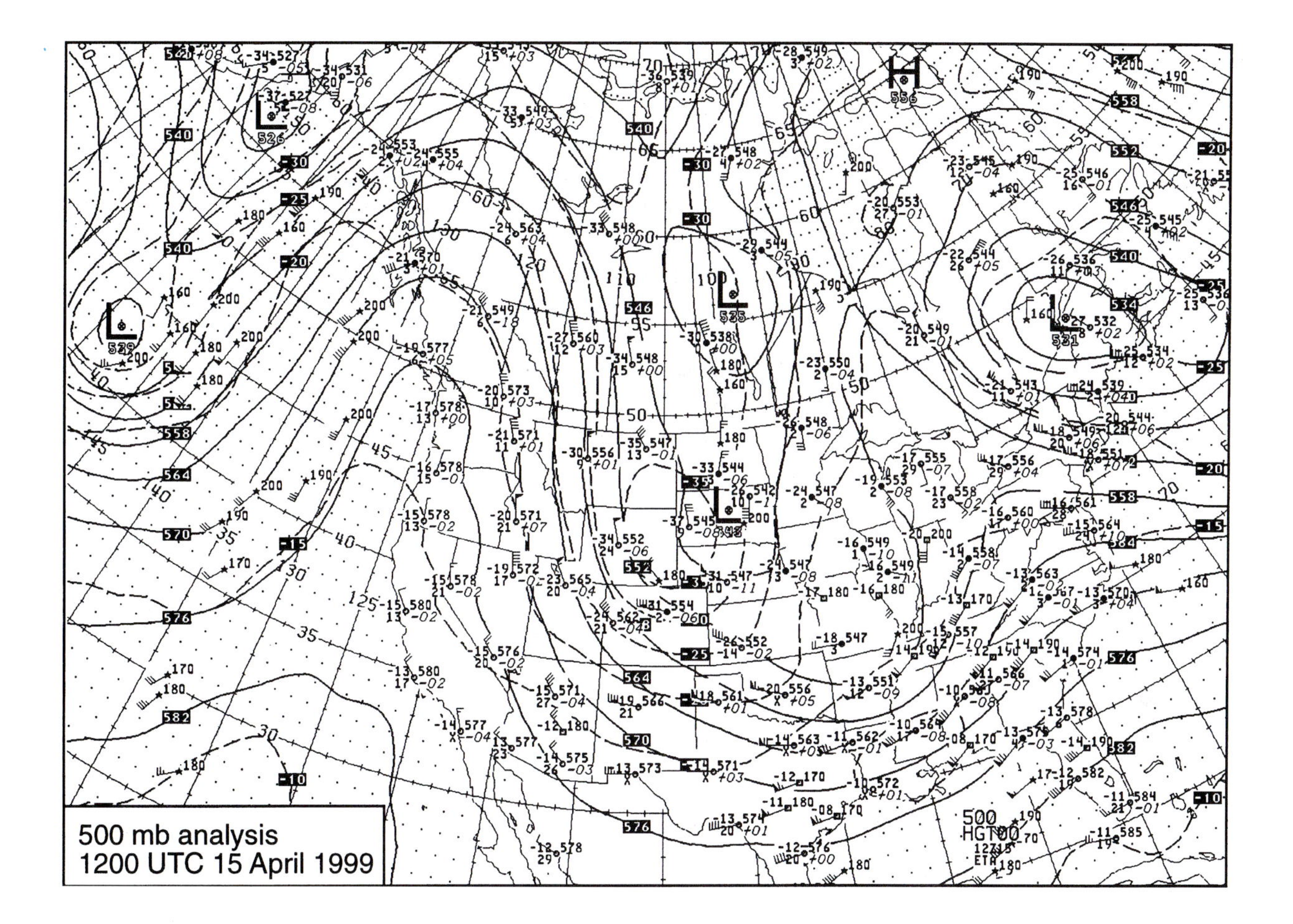

ocean area, and isobars could be drawn through points of equal pressure. This map of isobars would look like a topographic map in which the contours are labeled with pressure values rather than height values. There would be some closed contours (like mountain peaks on the topographic map) indicating high pressure and some closed contours (like a basin area on a topographic map) indicating low pressure. Because of the limited area of coverage of the map, many contours would run through the map without closing back on themselves.

The analysis of pressure at the earth's surface is more complicated than this example implies because pressure measurements are taken not only at sea level, but at many different elevations. Because pressure decreases rapidly with altitude (figure 4.2), the relatively small horizontal changes in pressure associated with traveling high and low pressure systems are masked by the larger altitudinal changes in pressure. The measured pressures from all stations must therefore be extrapolated to a common altitude, namely, sea level. The analysis is then performed on a map representing this height surface. Sea level, of course, is below ground in inland areas of the country, so downward extrapolations are necessary. Because the pressure at one point depends on the weight of the air in the column above, estimation of sea level pressure from the measured station pressures requires an estimate of the mean temperature in the (nonexistent) atmospheric column between the station altitude and sea level. Estimates

Figure 5.1 This upper air chart for 15 April 1999 at 1200 UTC shows height contours (solid lines labeled in tens of meters above sea level) and temperature contours or *isotherms* (dashed lines labeled in °C) at the 500-mb pressure level. The station model or key to values plotted at individual rawinsonde stations includes temperature in °C plotted to the upper left, dew-point depression in °C to the lower left, 500 mb heights in tens of meters to the upper right, and 12-hour height change in tens of meters to the lower right. Winds are plotted in knots following the scheme in figure 9.4. Note how the winds generally blow along the height contours. The surface chart for the same date and time is shown in figure 9.1.

are based on the air temperature at the measurement station. Potential errors are larger for stations at very high elevations, especially when a strong temperature inversion produces unusually low temperatures at the measurement station.

Isotherms and isobars are even more important than their names sound.
—elementary school student

5.1.4. *Hemispheric Pressure Waves*

A view of a 500-mb analysis from a point above the north pole and covering the entire Northern Hemisphere typically shows a series of troughs and ridges that make a wavelike pattern around the globe (figure 5.2). Unlike ocean waves, which are primarily vertical, these Rossby waves are primarily horizontal, representing the north–south meandering of wind currents. Rossby waves encircle the hemisphere and usually travel from west to east around the globe at speeds of about 10 to 15 mph (4 to 7 m/s). On the rare occasions when the pattern shifts westward, the pattern is said to be *retrograding*. There are usually four to seven waves around the hemisphere, but the number of waves, their position, and their amplitude can change over periods of several days. Although the Rossby wave patterns are relatively slow moving, winds flow through them more or less parallel to the height contours at speeds that are typically 20 to 50 mph (9 to 22 m/s). Wind speeds can exceed 100 mph (45 m/s) in jet streams found in the middle and upper troposphere.

Rossby waves guide the movement of storms, which usually follow the flow direction through the wave but at about half the flow speed, and determine whether surface temperatures will be cold or warm. Cool, cloudy weather is usually found under the cold, low pressure air in *long-wave troughs*, whereas warm, clear weather is found under the warm, high pressure air in *long-wave ridges*. General weather conditions over a region (cool or warm, cloudy or clear) for periods of several days or longer can be predicted reasonably well from Rossby wave trough and ridge positions, which can be forecast with computer models.

Small-scale waves or ripples, known as *short-wave troughs* and *short-wave ridges*, are superimposed on the Rossby waves and can be advected or transported through them at variable rates of speed. A dense network of upper air weather observations is needed to observe short-wave troughs and ridges. Sparse data, particularly over oceans, can result in forecasting errors. When advected into Rossby wave troughs, short-wave troughs cause the troughs to *deepen* (strengthen), and short-wave ridges cause the troughs to *fill* (decay). Rossby wave ridges *build* (strengthen) with the *advection* of short-wave ridges and decay with the advection of short-wave troughs. The deepening of a short-wave or long-wave trough or the advection of a trough into a region (*troughing*) produces deteriorating weather; the building of a short-wave or long-wave ridge or the advection of a ridge into a region (*ridging*) produces improving weather.

Cool, cloudy weather is usually found in Rossby wave troughs. Warm, clear weather is associated with Rossby wave ridges.

Climatological summaries of the mean positions of Rossby waves, such as the mean January position in figure 5.3, show that the waves tend to be anchored to the underlying terrain by major mountain barriers, with wave troughs in the lee of the Tibetan Plateau and the Rocky Mountains. Areas of low surface pressure, associated with rising motions, cloudiness, and precipitation, usually occur in advance of the wave troughs, so that areas east of the Rocky Mountains and the Tibetan Plateau frequently ex-

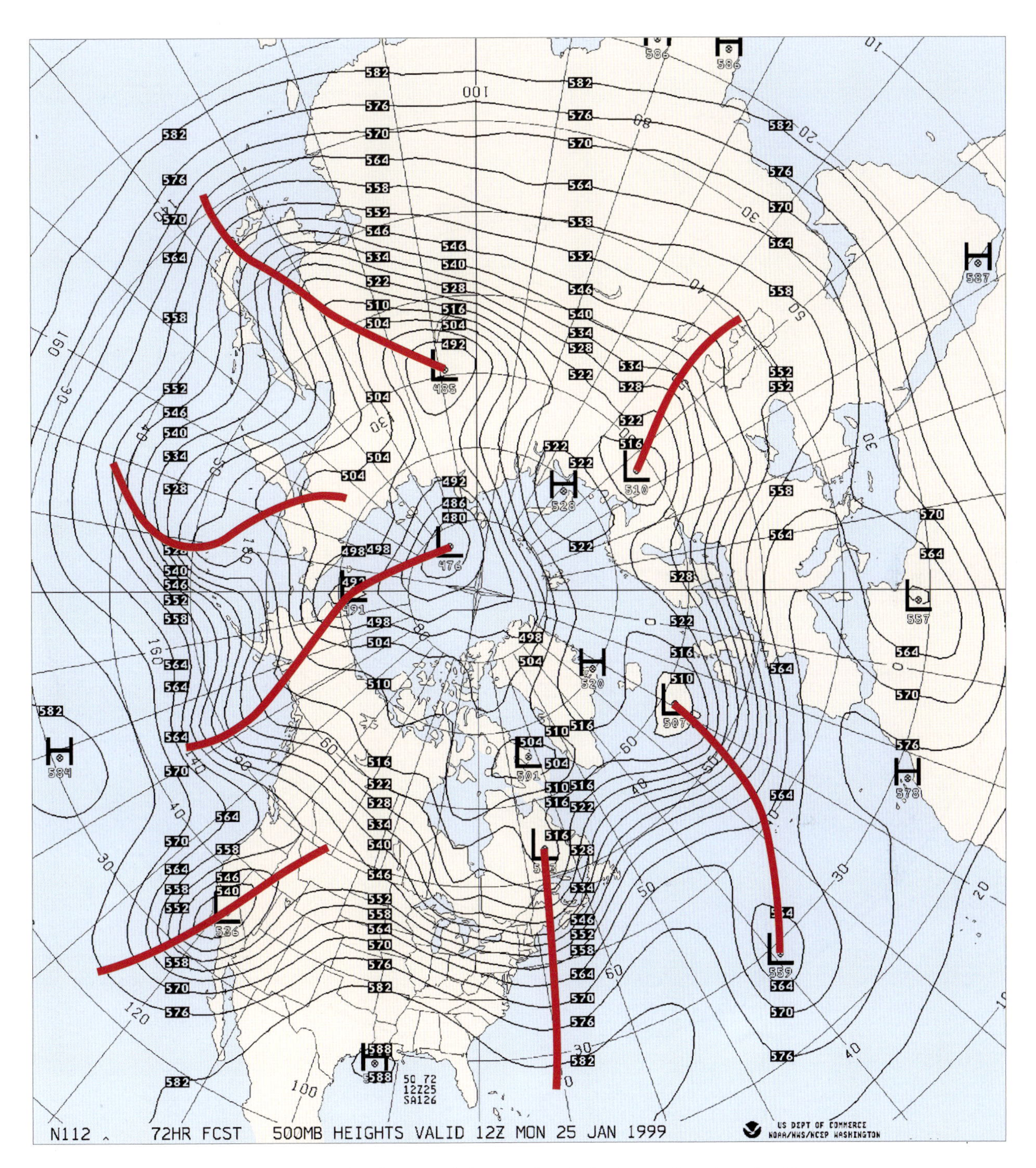

Figure 5.2 Example of a 500-mb forecast hemispheric weather chart for 25 January 1999, 1200 UTC. Major wave trough positions are indicated by red lines; solid lines are 500-mb height contours labeled in decameters (i.e., tens of meters). The movement and continuous development and decay of these hemispheric pressure patterns cause a succession of troughs and ridges to pass over a given location on the globe. The prediction of these pressure patterns is the key to making successful long-term forecasts. (Weather chart provided by European Center for Medium-Range Weather Forecasting)

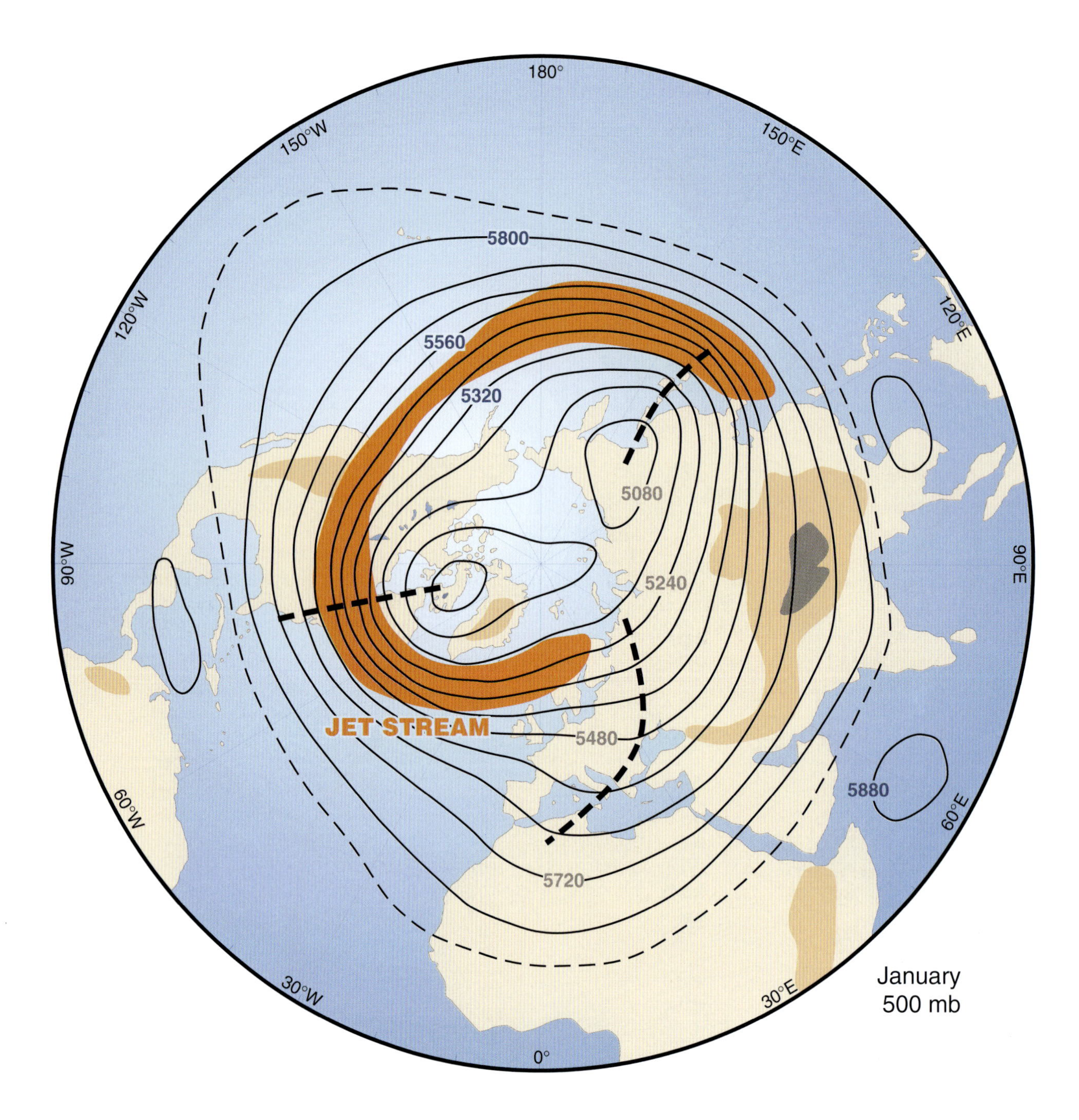

Figure 5.3 Mean height in meters of the 500-mb pressure surface for the month of January. Major mountain areas are tan or brown and the mean position of the jet stream is orange. Trough axes are shown in the lee of the Rocky Mountains, the Tibetan Plateau, and the Alps. (Adapted from Wallace and Hobbs, 1977)

perience surface low pressure centers, frontal passages, and unsettled weather.

Troughs and ridges, as well as the jet stream (section 5.2.1.3) and wind patterns that accompany them, produce upper tropospheric areas of horizontal convergence or divergence that are key to determining where high and low pressure systems will form and grow at the earth's surface. The stratosphere is strongly stable, so that convergence or divergence in the upper troposphere is compensated primarily by corresponding divergence or convergence patterns in the lower troposphere. Thus, when

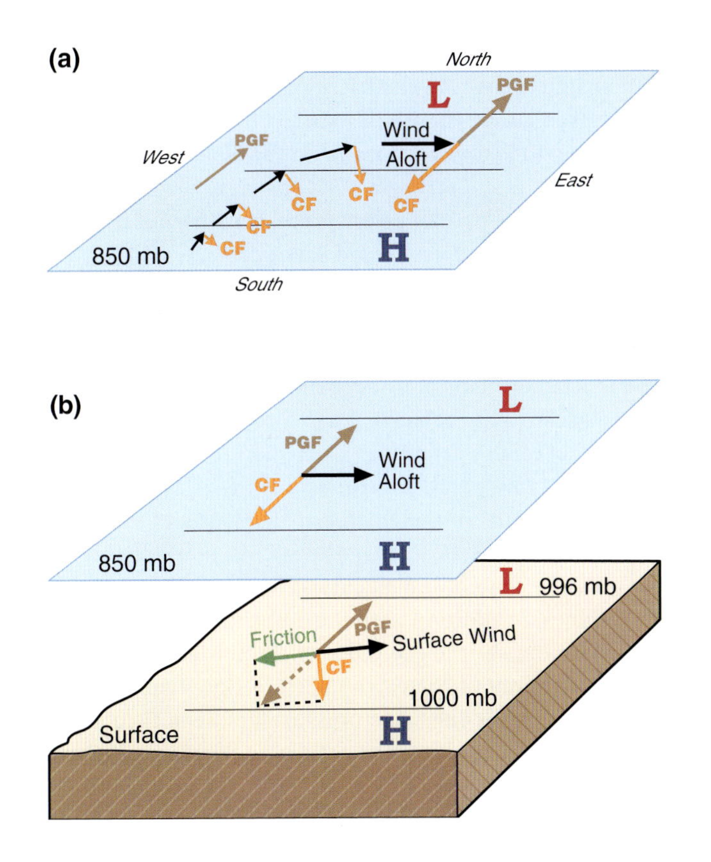

Figure 5.4 Differences in the balance of forces (a) aloft and (b) at the ground result in different wind directions and speeds relative to the height contours (aloft) and pressure contours (at the ground). The forces are shown as colored arrows, the winds are shown as black arrows, and the solid lines are height or pressure contours.

changes in wind speed or direction in the upper atmosphere cause air to converge, the air must sink toward the surface where it must then diverge. The sinking motions produce warming, drying, and the build-up of a surface high pressure center from which air diverges in a clockwise direction. When divergence occurs aloft, the diverging air is replaced by air that converges counterclockwise into a low pressure center at the surface and rises above the low pressure center to produce cooling, condensation, and often precipitation. Although there are many factors that can produce divergence aloft, the strongest divergence usually occurs on the east side of a trough where the flow changes from counterclockwise to clockwise. Thus, low pressure storms at the surface frequently develop and grow when a Rossby wave trough or short-wave trough approaches from the west.

5.1.5. *High and Low Pressure Centers and Wind Direction*

The direction the wind blows, both near the ground and aloft, is determined by a balance of forces (figure 5.4). The primary force driving the wind is the *pressure gradient force*, which is produced by horizontal differences in pressure and is proportional to the pressure gradient. The pressure gradient force is directed from high pressure to low pressure, crossing isobars at right angles. A second force, the *Coriolis force*, an apparent force that results from the earth's rotation and is proportional to wind speed and latitude, gradually deflects the wind 90° to the right of its direction of travel in the Northern Hemisphere (and to the left in the Southern Hemisphere). Above the frictional influence of the earth (about 3300 ft. or 1000 m above the earth's surface), the balance of these two forces causes the winds to blow parallel to the isobars or pressure height contours.

Figure 5.4 shows the path of an air parcel starting out from rest at upper levels. Air begins to flow toward low pressure, but is turned progressively to the right by the Coriolis force (CF). When the pressure gradient and Coriolis forces exactly balance, winds blow parallel to the isobars, a shift of 90° to the right of the pressure gradient force, and a steady wind is attained. This balance of the pressure gradient and Coriolis forces, called a geostrophic balance, is a theoretical approximation of conditions above the earth's frictional layer. The resulting winds, called *geostrophic winds*, blow parallel to the isobars with speeds proportional to the pressure gradient. In the Northern Hemisphere, the low pressure center is always on the left so that the winds blow counterclockwise around low pressure centers and clockwise around high pressure centers.

Figure 5.4 shows, in addition to the equilibrium balance of forces aloft, the balance of forces near the ground, which includes a third force, friction. The frictional drag of the earth's surface slows the wind, resulting in lower wind speeds for the same pressure gradient than would occur higher in the atmosphere above the influence of friction. (Compare the 850 mb and ground surfaces in figure 5.4b.) The lower wind speeds weaken the Coriolis force, causing the surface winds to turn to the left (in the Northern Hemisphere) across the isobars at angles of 10° to 35°, with greater turning over rougher terrain. Thus, winds near the ground blow more directly into low pressure centers. The balance of these forces can be disrupted in complex terrain, where terrain channeling (section 10.4) can force the wind to turn and blow along the axis of a valley, through passes, or along other terrain channels.

The relationship between high and low pressure fields and the direction of winds aloft in the Northern Hemisphere was first stated by Buys-Ballot, a Dutch scientist, in 1857. The Buys-Ballot Rule states that if the wind blows into your back, low pressure will be on your left (and high pressure will be on your right). The direction of winds aloft can be determined by observing the movement of clouds or jet aircraft condensation trails (*contrails*) above the earth's frictional layer. By applying the Buys-Ballot Rule at intervals of several hours, an observer in the field can track the movement of high and low pressure systems and make a rough weather forecast (figure 5.5).

If the wind aloft blows into your back, low pressure will be on your left (and high pressure will be on your right).

5.2. Winds

Winds exist over a range of scales, from the trade winds and midlatitude westerlies that are characteristic of the earth's *general circulation* to small eddies around obstacles. Wind speeds and directions vary with time, horizontal distance, and height above the earth's surface. Winds affect human comfort, aviation safety, aerial spraying, the behavior of forest fires, and the transport and dispersion of smoke and other pollutants.

The wind is like the air, only pushier.
—elementary school student

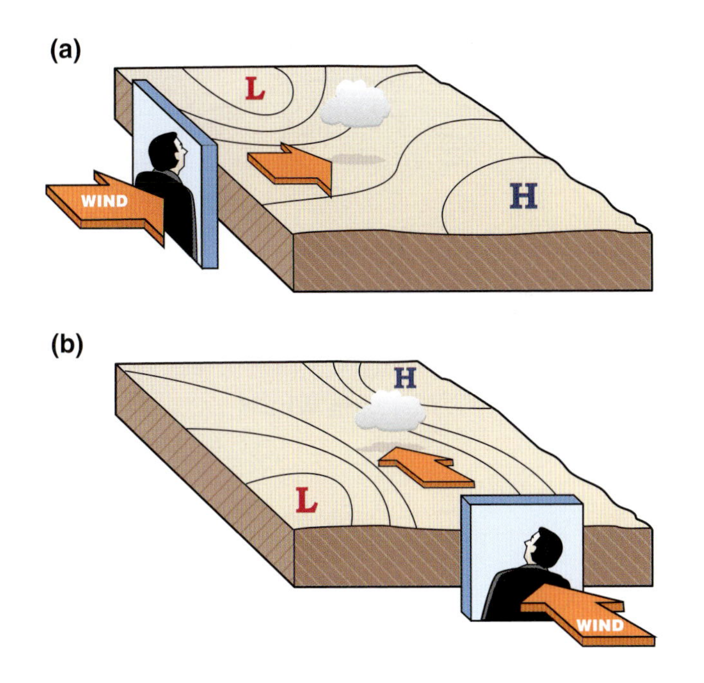

Figure 5.5 By monitoring changes in wind direction above a mountainous area, an observer can visualize changes in the pressure pattern and the movements of high and low pressure centers. In part (a), the winds aloft are from the west and the low will be to the observer's north. In part (b), some hours later, the winds aloft are from the south and the low has driven southward so it is now west of the observer.

Winds over rough terrain (mountains, coastal areas, forests, urban areas) are particularly complex. Winds in mountainous areas are generally of two types. *Terrain-forced flows* enter the mountain region from its surroundings and are modified or channeled by the mountains (chapter 10). *Diurnal mountain winds* are generated within the mountain area through local temperature contrasts that form within the mountain region or between the mountains and the nearby plains (chapter 11).

5.2.1. *General Circulation*

5.2.1.1. *Pole–Equator Temperature Contrasts.* On an annual basis, the earth as a whole loses as much longwave radiation to space as it gains via shortwave radiation. There are, however, significant latitudinal differences in the amounts of outgoing longwave radiation and incoming shortwave radiation. In equatorial regions, the shortwave gain is greater than the longwave loss, whereas in the polar regions the shortwave gain is smaller than the longwave loss. The tropics thus experience a net gain in energy, and the polar regions experience a net loss. The latitudinal temperature differences that result from this uneven distribution of net energy on the globe drive a continuous circulation of wind and ocean currents, called the general circulation. The general circulation moderates the temperature difference between the poles and the equator by transporting warm air (and water) toward the poles and cold air (and water) toward the equator.

The general circulation is driven by latitudinal temperature differences produced by the uneven distribution of solar radiation on the globe.

5.2.1.2. *Vertical Motions, Pressure Belts, and Wind Belts.* On a nonrotating globe, the unequal heating between equator and pole would cause a single large circulation cell in which air would rise over the equator, travel toward the poles aloft, subside over the poles, and return to the equator

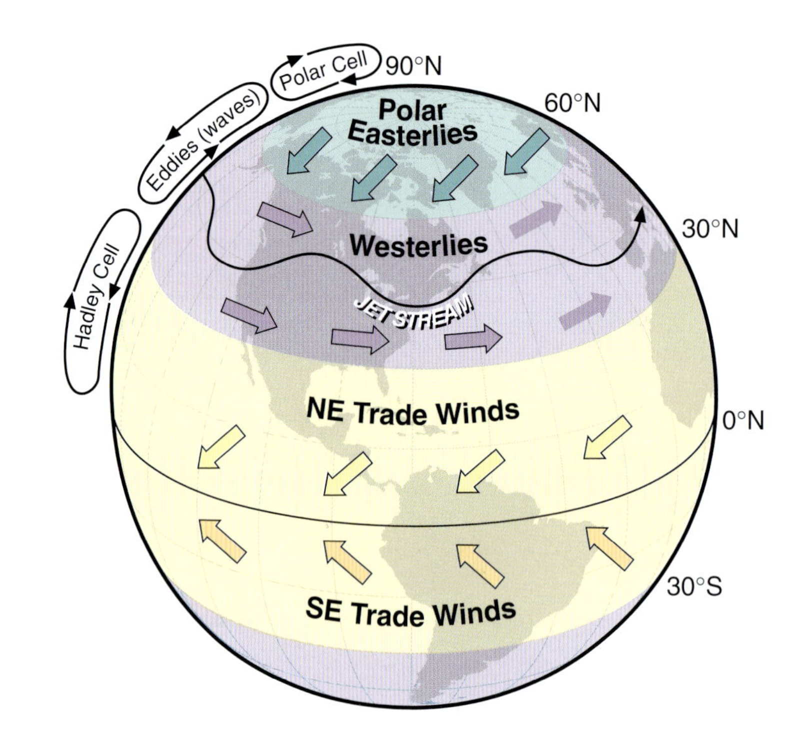

Figure 5.6 Vertical cellular structure and mean wind directions associated with the general circulation in the Northern Hemisphere

at low altitudes. This circulation transfers air from the high pressure regions at the poles to the low pressure belt at the equator. Because of the earth's rotation, this idealized circulation splits into three cells in each hemisphere. Figure 5.6 focuses on the three cells in the Northern Hemisphere. The locations of the cells correspond to the alternating belts of high and low pressure in figure 1.3. The rising air over the equator does not reach the poles, but rather sinks at about 30° latitude, where a belt of high pressure is located. The circulation in this Hadley cell is closed by air flowing southward near the earth's surface. Similarly, the air moving southward from the pole rises when it reaches the subpolar low pressure belt at about 60° latitude, forming a second polar cell with warm air traveling northward aloft and cold air returning southward near the ground. Between these two cells, a third cell with a reverse circulation carries air northward near the surface and southward aloft. Similar cells occur in the Southern Hemisphere.

The three cells of the general circulation correspond not only to the belts of high and low pressure that encircle the globe, but also to bands of wind. Easterly winds predominate from the equator to 30° latitude and from 60° latitude to the poles. Westerly winds predominate between 30° and 60° latitude.

Easterly winds prevail in the equatorial and polar regions; westerly winds prevail in the midlatitudes.

The easterly winds between the equator and 30° latitude are called *trade winds* because in the past they allowed commercial sailing ships to travel westward across the Atlantic. Winds are northeasterly in the Northern Hemisphere and southeasterly in the Southern Hemisphere so that the winds converge near the equator, forming an *intertropical convergence zone* or ITCZ where intense convection and heavy precipitation are often found. The easterly trade winds are responsible for the general

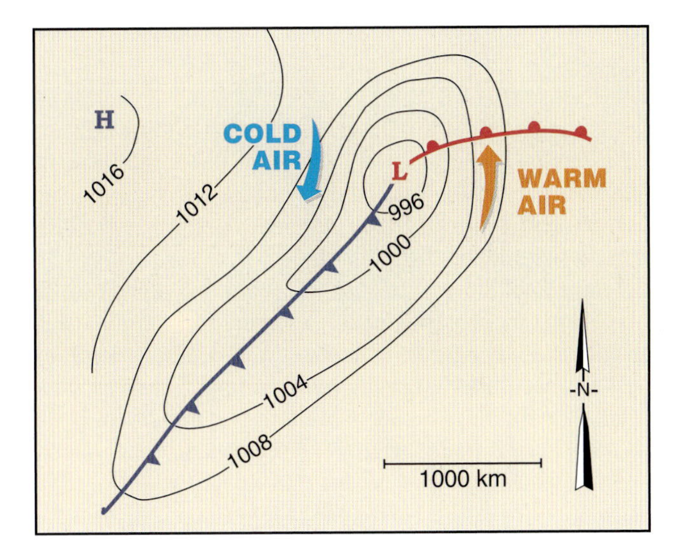

Figure 5.7 Air flows counterclockwise around traveling low pressure centers in the Northern Hemisphere, carrying warm air northward and cold air southward; this decreases the north–south temperature gradient between the pole and the equator. Lows are indicated on weather maps by the letter L, highs by the letter H.

westward movement of hurricanes that develop in equatorial air masses. The hurricanes that hit the Caribbean Islands, the south Atlantic coast of the United States, and the Gulf Coast arrive from the east. It is only when they drift far enough north that they come under the influence of the midlatitude westerlies, and their trajectories recurve to be carried back eastward.

The westerly winds between 30° and 60° latitude follow the oscillating Rossby waves, but wind direction changes frequently in the lower levels of the troposphere as high and low pressure centers or troughs and ridges migrate from west to east, mixing the cold air from the north with warm air from the south (figure 5.7).

5.2.1.3. *Jet Streams.* Jet streams are strong currents of air in the Rossby wave pattern that are produced by the pressure gradient between the poles and the equator and are strengthened by temperature differences across frontal boundaries. Jet streams may be thousands of miles long, hundreds of miles wide, and thousands of feet deep. Jet stream winds are strongest in the upper troposphere at altitudes between 30,000 and 40,000 ft (9000 to 12,000 m), where they may exceed 180 mph (75m/s). A *jet* is characterized by high wind speeds concentrated in a narrow vertical band, with lower speeds both above and below.

Jet streams are strong currents of air in the Rossby wave pattern.

The location, strength, and orientation of jet streams vary with the season and from day to day within a season. Multiple jet streams are observed on many weather charts. In summer, a *polar front jet* is typically found over central Canada where north–south temperature contrasts are maximized. In winter, the temperature and pressure patterns move southward, the pressure gradient becomes stronger, and a jet stream is typically positioned over the northern and central United States (figure 5.8). It is not uncommon for two jet streams to form in winter: a *subtropical jet* at latitudes near 30°N and a polar front jet at latitudes near the United States and Canadian border. When the strong westerly winds in the lower troposphere associated with jet streams cross the north–south

EL NIÑO

Changes occur periodically in the general circulation. The best known of these changes is *El Niño*, a rearrangement of the circulation pattern in the tropical Pacific that impacts regions far from the tropical Pacific. El Niño is familiar to residents of North America because of the changes it brings in precipitation in the midlatitudes.

El Niño (Spanish for "Christ child") is a term originally applied by fishermen along the coast of Equador and Peru to a warm coastal ocean current that appears near Christmastime and lasts for several months. Fishing is adversely affected by the warm current, which can sometimes persist into late spring, affecting the fishermen's livelihood. The term El Niño is now applied more generally to changes in global atmospheric and oceanic circulations that produce the coastal warm water current and have far-reaching effects on global climate.

El Niño occurs at irregular intervals and varies in strength. In the last 40 years, there have been nine El Niño events that have temporarily raised coastal water temperatures from 1 or 2°F to more than 10°F in 1982–1983.

During normal years, easterlies carry surface waters westward from the equatorial coast of Central and South America. As the surface waters move away, they are replaced by cold, nutrient-rich waters from below in a process called *upwelling*. The upwelling keeps the surface waters of the central Pacific cool. Heavy rainfall is confined to the warm waters surrounding Indonesia at the western end of the Pacific. During El Niño, the easterlies weaken, allowing the central Pacific to warm and the areas of rainfall to migrate eastward. This shift in the location of tropical rainfall affects wind patterns over much of the globe.

El Niño affects the United States and Canada primarily in winter. A ridge of high pressure typically forms over North America's west coast during El Niño winters, keeping temperatures above normal in western Canada and parts of the northern United States and steering storms that would normally pass through Washington and Oregon northward into southeastern Alaska. El Niño also creates a favorable climate for storms to develop in the Gulf of Mexico, causing wet weather over much of the southern United States. Because the weather in the United States is affected by many factors, the impact of a weak El Niño can be subtle.

You don't need to be a weatherman to know which way the wind blows.
—Bob Dylan

Rocky Mountains, precipitation is often produced on the windward side of the mountains as the air is lifted over the mountains. *Lenticular clouds* may form over and downwind of the mountains, and *downslope windstorms* may develop on the leeward side of the mountains.

5.2.2. *Designating Wind Direction and Wind Speed*

Wind directions are named for the cardinal direction from which the wind blows.

Wind directions are named for the cardinal direction from which the wind blows. Thus a north wind blows from north to south and a west wind blows from west to east. Similarly, a sea breeze blows from sea to land and

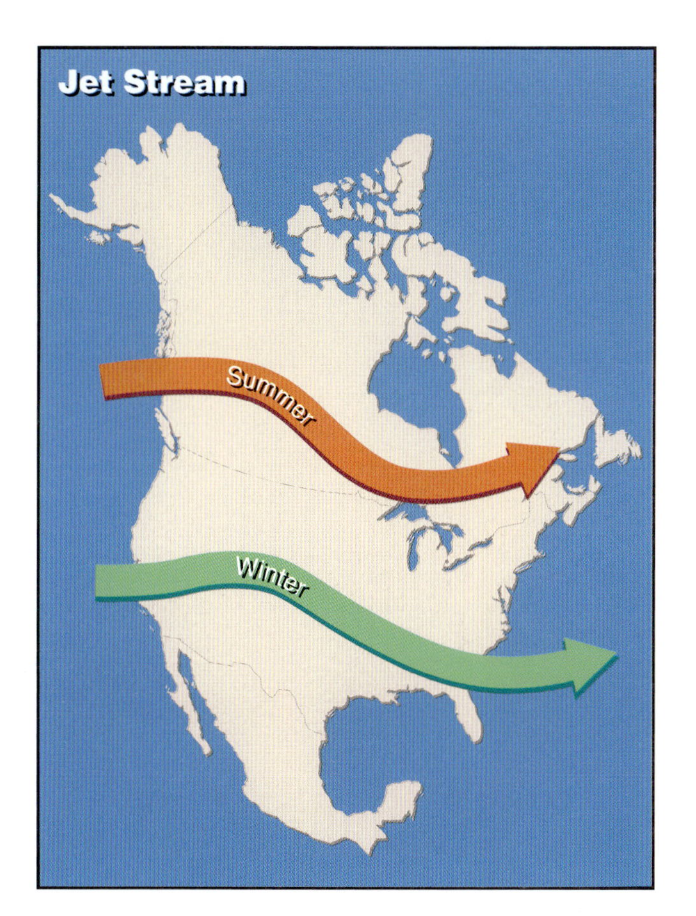

Figure 5.8 The mean position of the jet stream in winter is over the central United States, but it shifts north into Canada in the summer.

a mountain breeze blows from mountain to valley. When the cardinal directions are too imprecise a measure, wind direction can be specified by a bearing in degrees. For example, the wind may be blowing from 90° (east) or from 225° (southwest). Data reported in degrees reference true north rather than magnetic north.

Wind speed data are reported in miles per hour, meters per second, kilometers per hour, or nautical miles per hour (knots). Because the United States does not use the international metric system but shares meteorological data with other countries, conversions from one scale to another must often be made. Conversion formulas are provided in appendix D, and approximate conversions are given in table 5.1.

5.2.3. *Wind Observations*

Observations of wind speed and direction can be made with a number of instruments, including cup- and propeller-type *anemometers*, sonic and laser anemometers, wind vanes, clear air radars, sodars, and balloons tracked optically or by radio or radar. Hand-held cup, propeller, pith ball, and fan-blade anemometers are commonly used in the field to measure wind speed. Selecting a suitable wind sensor and siting it properly are important. When data are to be collected in a mountainous area where

Table 5.1 Wind Conversions

mph	m/s	km/h	knots
1	0.4	1.6	0.9
2	0.9	3.2	1.7
3	1.3	4.8	2.6
4	1.8	6.4	3.5
5	2.2	8.0	4.3
6	2.7	9.7	5.2
7	3.1	11.3	6.1
8	3.6	12.9	6.9
9	4.0	14.5	7.8
10	4.5	16.1	8.7
12	5.4	19.3	10.4
14	6.3	22.5	12.2
16	7.2	25.7	13.9
18	8.0	29.0	15.6
20	8.9	32.2	17.4
30	13.4	48.3	26.1
40	17.9	64.4	34.7
50	22.4	80.5	43.4
60	26.8	96.6	52.1
100	44.7	160.9	86.8
150	67.1	241.4	130.3
200	89.4	321.9	173.7

winds are complex and anemometers can be subject to strong winds or *icing*, an experienced meteorologist should be consulted. Appendix C lists sources of information on wind and other weather instruments. Many of these sources discuss instrument placement.

When wind instruments are unavailable, wind speeds can be estimated by observing the visible effects of wind on land objects or on the surface of the sea. An updated and modified version of the wind scale developed in the early 19th century by Admiral Beaufort of the British Navy is given in table 5.2. Beaufort's original wind speed scale was based on the amount of canvas that a full-rigged frigate could carry.

Both wind speed and wind direction can often be estimated by observing the movement of clouds, the formation of distinctive types of clouds (e.g., the *banner cloud*, section 7.1.4.6) and the movement of blowing dust or blowing snow. Snowdrifts, cornices, *sastrugi* (long wavelike ridges of hard snow formed on a snowpack), and flagged trees provide information on the direction (and, to some extent, the speed) of recurring, strong, wintertime winds. Sand dunes provide similar indications of winds during all seasons. Snowdrifts and sand dunes indicate where strong winds that have entrained dust, sand, or snow have been slowed, resulting in the deposition of entrained material.

A blizzard is when it snows sideways.
—elementary school student

Snowdrifts form downwind from shrubs, lines of trees, ridges, or other obstacles when strong winds carry large quantities of snow past the obstacle. The snow is deposited primarily in the lee of the obstacle in elongated pillowlike drifts. The snowdrift grows in depth and extent from the obstacle as more snow is deposited. Falling snow is not necessary for the buildup of snowdrifts because strong winds can remove snow from the existing snowpack to create drifts. In fact, strong wind

Table 5.2 The Beaufort Wind Scale: A Method to Estimate Wind Speeds on Land and Sea

Beaufort Number	Wind Speed (mph)	Seaman's Term	Effects at Sea	Effects on Land
0	<1	Calm	Sea like mirror	Calm; smoke rises vertically
1	1–3	Light air	Ripples with appearance of "fish scales"; no foam crests	Smoke drift indicates wind direction; vanes do not move
2	4–7	Light breeze	Small wavelets; crests of glass appearance not breaking	Wind felt on face; leaves rustle; vanes begin to move
3	8–12	Gentle breeze	Large wavelets; crests begin to break; scattered whitecaps	Leaves, small twigs in constant motion; light flags extended
4	13–18	Moderate breeze	Small waves; becoming longer; numerous whitecaps	Dust, leaves and loose paper raised up; small branches move
5	19–24	Fresh breeze	Moderate waves; becoming longer; many whitecaps; some spray	Small trees in leaf begin to sway
6	25–31	Strong breeze	Larger waves forming; white-caps everywhere; more spray	Large branches of trees in motion; whistling heard in wires
7	32–38	Moderate gale	Sea heaps up; white foam from breaking waves begins to be blown in streaks	Whole trees in motion; resistance felt in walking against wind
8	39–46	Fresh gale	Moderately high waves of greater length; foam is blown in well-marked streaks	Twigs and small branches broken off trees
9	47–54	Strong gale	High waves; sea begins to roll; dense streaks of foam; spray may reduce visibility	Slight structural damage occurs; slate blown from roofs
10	55–63	Whole gale	Very high waves with overhanging crests; sea takes white appearance; visibility reduced	Seldom experienced on land; trees broken; structural damage occurs
11	64–72	Storm	Exceptionally high waves; sea covered with white foam patches	Very rarely experienced on land; usually with widespread damage
12	>73	Hurricane force	Air filled with foam; sea completely white with driving spray; visibility greatly reduced	Violence and destruction

events following snowstorms (often with winds from a different direction than during the snowfall) are a major factor in snowdrift buildup. Snow fences (figure 5.9) are used to form snowdrifts upwind of roadways, railroad tracks, and buildings, thus preventing blowing snow from reaching these structures. They can also be used to accumulate snowdrifts as a water resource.

Snow cornices, masses of snow or ice projecting over the lee side of a ridge, are formed by windblown snow. A typical progression of cornice buildup is shown in figure 5.10. Depending on wind and snow conditions, cornices are sometimes soft and easily broken and sometimes extremely hard and firmly attached to the underlying rock. They are espe-

Figure 5.9 This snow fence protects Interstate 80 in southern Wyoming from drifting and blowing snow. A well-formed *snow pillow* is seen in the lee of the fence in this December photograph. (U.S. Forest Service photograph provided by R. L. Jairell)

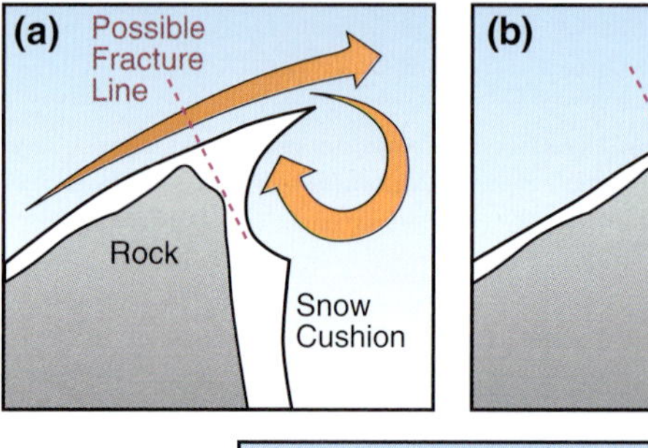

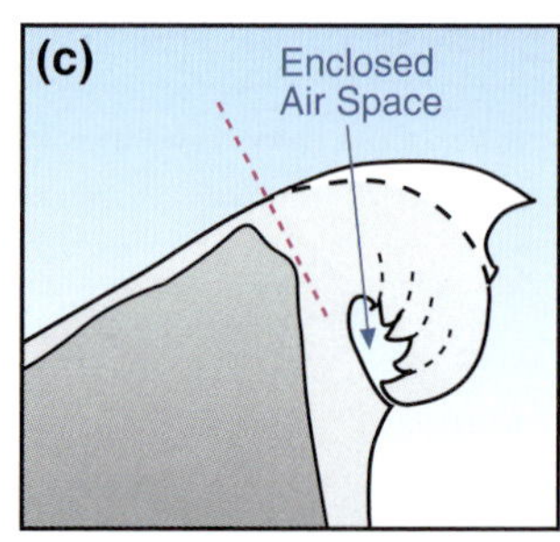

Figure 5.10 Three stages of cornice buildup: (a) young, (b) middle, and (c) mature. (Adapted from Graydon and Hanson, 1997)

Figure 5.11 Long, wavelike ridges of hard snow, called sastrugi, form perpendicular to the direction of the wind. The wind here is blowing from right to left. (Photo © B. Martner)

cially subject to fracture and collapse when additional weight is added to the cornice, for example, during windstorms that deposit new snow on the cornice or when a climber walks on the cornice's crest. Climbers must be careful not to walk on the crest when snow conditions are unstable or when a fracture crack has developed between the mature cornice and the underlying ridge. The collapse of a cornice results in the entire snow mass falling over the lee side of the ridge, which may trigger avalanches on lee slopes.

Sastrugi, a distinctive pattern of undulating step-like ridges of hard snow on the snowpack surface (figure 5.11), is produced by ice particles that bounce along the snow surface as they are carried along by strong winds. The ridges are generally oriented perpendicular to the mean direction of the blowing snow (although this depends somewhat on the uniformity of snowpack characteristics), with the vertical steps on the upwind side of the ridges. Sastrugi form only when winds are of sufficient strength (usually above 25 mph) to carry ice particles along the surface.

Flagged trees are trees that are deformed by winds. Figure 5.12 shows an example of a flagged tree with branches growing on only one side of the trunk. Wind-driven snow and desiccating wintertime winds prevent branches from growing on the upwind side of the tree. Flagged trees indicate the direction of strong winter winds and are not necessarily representative of the wind direction during other seasons or even during wintertime when winds are weak. Figure 5.13 shows the effects of in-

Figure 5.12 A flagged tree near timberline in Rocky Mountain National Park, Colorado, indicates strong winds blowing from left to right. (Photo © C. Whiteman)

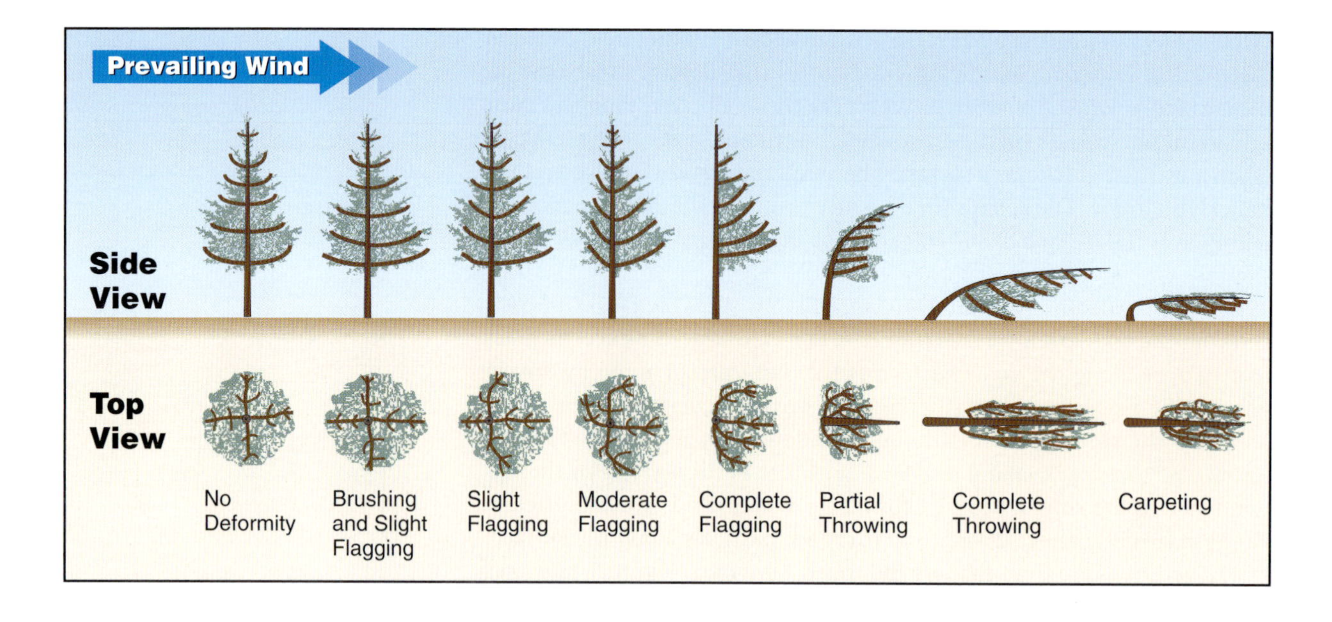

Figure 5.13 Increasing wind speeds produce increasing tree deformities. These deformities can be classified into eight categories based on shape. (Adapted from Justus, 1985)

creasingly strong winds on trees. Branch deformities are affected by a number of factors other than wind speed, including the tree species, and the amount of protection that the tree receives from snow cover during the strong wind episodes.

Sand dunes and dune fields provide an indication of past wind directions and speeds. Sand is entrained from the ground during strong winds and is carried with the wind. The entrained sand falls to the ground when winds are slowed. Deposition zones for sand dunes are often found in low-lying areas upwind of low passes (e.g., at the Great Sand Dunes National Monument in Colorado's San Luis Valley), where a *flow separation* (section 10.1.1) occurs on the steep mountainside leading up to the pass. Depositions may also be found behind terrain obstructions (e.g., at the Stovepipe Wells dune field in Death Valley, Nevada) that generate slow-moving eddies. During sandstorms, sand is deposited on the lee side of the dune, which is usually the steepest side. The slope angles there are often at or near the angle of repose of sand (33° to 37°). When additional sand is deposited on the upper parts of these slopes, small sand avalanches occur and the dune advances. It is sometimes difficult to identify the lee side of the dune (and thus the direction of movement) because wind directions vary from storm to storm and dunes may not develop an idealized shape.

5.2.4. *Vertical Wind Structure and Its Evolution*

The layers of horizontal winds in the atmosphere can be observed by tracking ascending weather balloons by theodolite, by radar, or by using aircraft or satellite navigational aids such as Long Range Navigation (*Loran*) or the Global Positioning System (GPS). The horizontal movement of the balloons during their ascent is caused by the horizontal winds, which force the balloon to change direction as it climbs from one layer to

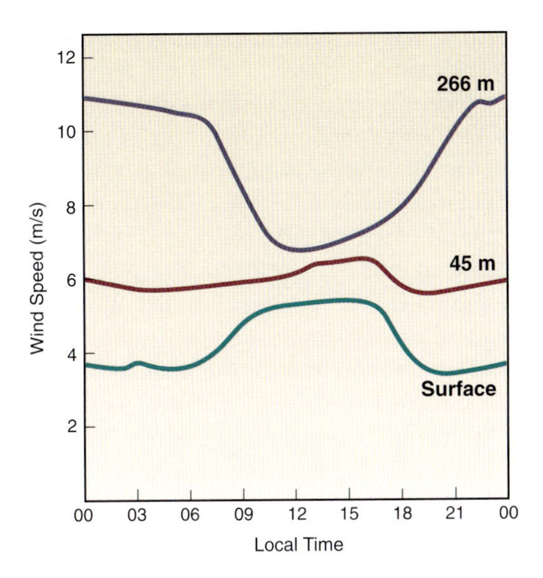

Figure 5.14 The diurnal wind speed variation at the ground is out of phase with the variation higher in the boundary layer. (Adapted from Crawford and Hudson, 1973)

the next. Standard photogrammetric techniques are used to determine the mean horizontal wind between successive balloon positions. Vertically pointing clear air radars, sodars, tethered balloons, and airplanes also provide vertical wind profiles. Tower-mounted anemometers

TERRAIN ROUGHNESS AND FRICTIONAL DRAG ON THE WIND

The effect of frictional drag on wind speed varies depending on the roughness of the terrain. The rougher the underlying surface, the more frictional drag on the winds. Frictional drag is greatest near the ground and decreases with height above the surface. The wind speed at one height can be estimated from a formula provided in appendix A when the wind has been measured at a different height and a parameter accounting for the roughness of the underlying surface, z_0, is assigned a value from the following table.

Davenport–Wieringa Roughness Length Classification

z_0 (m)	Classification	Description
0.0002	very smooth	sea, paved areas, snow-covered flat plain, tide flat, smooth desert
0.005	smooth	beaches, pack ice, snow-covered fields
0.03	open	grass prairie or farm fields, tundra, airports, heather
0.1	roughly open	cultivated area with low crops and occasional obstacles (single bushes)
0.25	rough	high crops, crops of varied height, scattered obstacles such as trees or hedgerows, vineyards
0.5	very rough	mixed farm fields and forest clumps, orchards, scattered buildings
1.0	closed	regular coverage with large obstacles with open spaces roughly equal to obstacle heights, suburban houses, villages, mature forests
≥ 2	chaotic	centers of large towns and cities, irregular forests with scattered clearings

[a]From Stull, 1995.

and wind vanes indicate the vertical variation of winds in the shallow layer of air near the earth's surface.

Wind speeds generally increase with height through the midtroposphere. Thus, winds on exposed mountain peaks and ridges are higher than winds at lower elevations. Wind speeds in the lowest 10% of the boundary layer (approximately 300 feet or 100 m) generally increase quite rapidly (logarithmically) with height because of the rapid decrease with height of the frictional drag of the earth's surface.

Winds generally increase with height.

Wind speed and its rate of increase with height undergo regular diurnal changes caused by day–night changes in atmospheric stability. At night, the atmosphere is more stable. The increased resistance to vertical motions isolates surface winds from the stronger winds aloft and protects winds aloft from the frictional drag of the earth's surface. Thus, surface winds are light, and higher wind speeds are usually found only above the surface-based temperature inversion (figure 5.14). During the day, the atmosphere is less stable. The generally stronger winds aloft are mixed downward to the surface, increasing surface wind speeds. Convection also communicates the frictional drag at the earth's surface through a deep boundary layer, slowing the winds higher in the boundary layer. This diurnal cycle of wind speeds is disrupted in winter because diurnal changes in stability are suppressed. The surface can remain decoupled from the higher speeds aloft for long periods when stability is strong or can be coupled to the strong winds aloft for long periods when stability is weak.

Air Masses and Fronts

6

6.1. Air Mass Source Regions and Trajectories

An air mass is a regional-scale volume of air with horizontal layers of uniform temperature and humidity. Air masses form during episodes of high pressure when weak winds allow air to remain for several days over a flat area with uniform surface characteristics. The characteristics of the underlying surface determine the characteristics of the air mass, which is given a two-letter identifier. Air masses are identified by their locations of origin (maritime "m" or continental "c") and by their characteristics (tropical "T" or polar "P"). Tropical air masses form in high pressure areas in warm, tropical regions. When a tropical air mass is formed over oceans (mT), it is warm, moist, and usually unstable. When formed over land (cT), it is hot and dry, with unstable air near the surface and stable air aloft. Polar air masses form in high pressure areas in the polar and subpolar regions. A polar air mass that forms over water (mP) is cool, moist, and unstable. A polar air mass that forms over land (cP) is cold, dry, and stable. An extremely cold polar air mass that forms in winter over arctic ice and snow surfaces is called an arctic air mass (cA). The distinction between arctic and polar air masses is not always clear because an arctic air mass that travels over a warm surface may be warmer near the surface than a polar air mass, although it is still colder aloft.

Air masses form over flat regions with uniform surface conditions where air stagnates for long periods.

Source regions for air masses and typical trajectories affecting North America are shown in figure 6.1. Polar air masses that originate over the flat, ice- and snow-covered regions east of the Rocky Mountains in northern and central Canada and Alaska, and arctic air masses that originate over the ice-covered Arctic Ocean influence winter weather.

The midlatitudes are not a good air mass source region. The exposure to traveling weather systems is too great, the range of temperature and

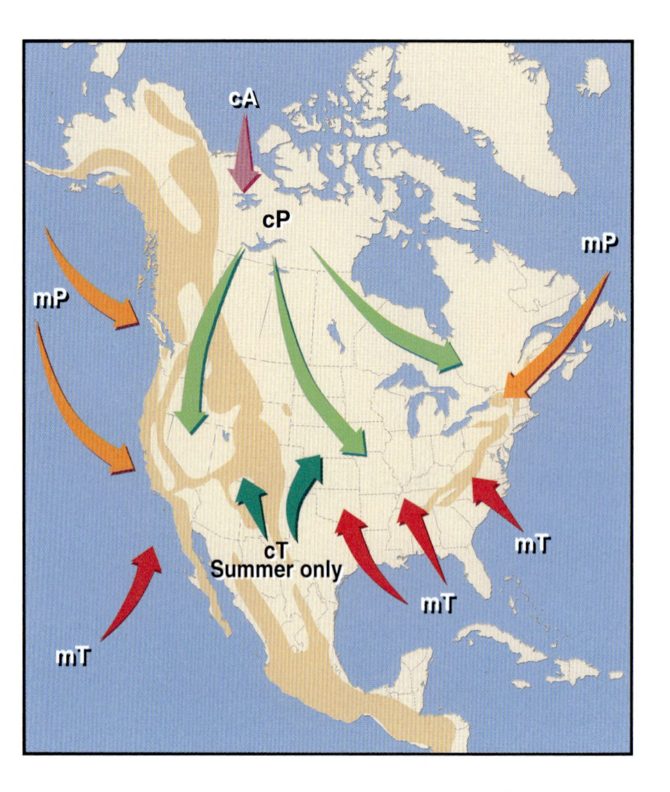

Figure 6.1 Source regions of North American air masses. (Adapted from Ahrens, 1994.) (See text for key to abbreviations.)

humidity too wide, and, in the United States, the topography is too varied. Instead, the midlatitudes are a region where clashing air masses meet. Cold air masses are usually driven southward from the subpolar regions, whereas warm air is forced northward from tropical regions.

6.2. Fronts

Fronts are boundaries between air masses.

Boundaries between air masses are called fronts. The term *frontal zone* is more descriptive, however, as the boundaries are often indistinct. Weather can change very rapidly near frontal zones because, as the front moves, one air mass at the surface can be replaced by another with quite different characteristics. The three types of fronts (figure 6.2) are *cold fronts*, *warm fronts*, and *occluded fronts*. Frontal zones (whether cold, warm, or occluded) that remain at approximately the same location for a period of time before moving on or dissipating are called *stationary fronts*. Fronts are represented on weather maps by the lines and symbols in figure 6.2. A cold front is indicated by a blue line with blue triangles pointing in the direction of travel of the front; a warm front is indicated by a red line with red semicircles pointing in the direction of travel; an occluded front is indicated by a purple line with alternating purple triangles and semicircles pointing in the direction of travel; and a stationary front is indicated by alternating blue and red line segments with triangles and semicircles pointing in opposite directions.

All three types of fronts are associated with surface low pressure centers. Most low pressure centers are associated with two to four fronts. The fronts move counterclockwise around the low as the low moves and

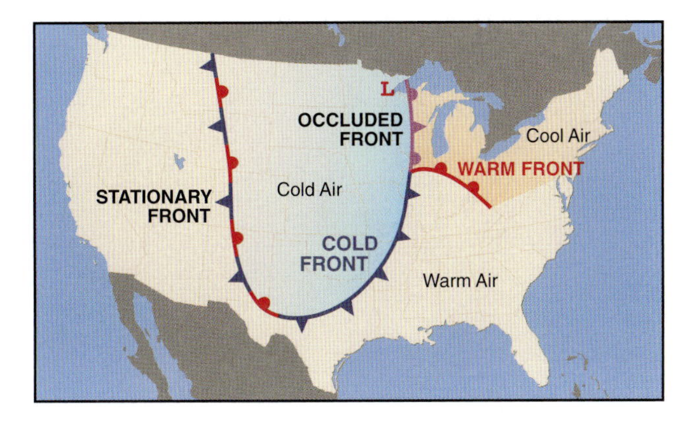

Figure 6.2 Fronts are boundaries between warm and cold air masses. The cold front, warm front, and occluded front positions are typical of a wintertime weather map. The western boundary of the cold air in the north and central Plains area has become stationary where its westward movement is stalled at the steep eastern slope of the Rockies.

evolves. The development of fronts and their relationship to low pressure systems can be visualized by imagining a continuous front, called the *polar front*, separating the cold air mass that builds up over the pole from the warmer air to the south. The front forms a wavy boundary at about 60°N latitude, with lobes of cold air occasionally intruding into more southerly regions and lobes of warm air intruding northward into the polar region (figure 6.3). In reality, the polar front is not continuous because there are always areas where cold air from the north and warmer air from the south are so well mixed that no sharp frontal boundary can be identified.

Polar front development and cyclogenesis (section 5.1) are illustrated in figure 6.4. Initially, the cold polar and warm equatorial air masses are separated by a near-vertical boundary (figure 6.4a), with easterly winds in the cold air and westerly winds in the warm air. A wavelike perturbation is initiated at the boundary when rising motions occur in the synoptic-scale environment (often due to divergence aloft associated with passing troughs or jet streams), and a weak surface low pressure center develops on the perturbed boundary (figure 6.4b). The counterclockwise circulation around the low causes the cold air west of the low to accelerate southward, the warm air east of the low to move northward, and the vertical cold–warm air boundary to tilt toward the cold air. As the wave on the boundary continues to develop, the low deepens, winds around the low become stronger, and the sloping cold air boundary west of the low (the cold front) moves actively into the *warm sector*, lifting the warm air ahead of it as it progresses (figure 6.4c). The northward flow in the warm sector lifts air up the sloping surface of the cold air mass east of the low. This sloping surface defines the warm front. The cold air ahead of the warm front is warmed somewhat by latent heat release as clouds form and by warm precipitation that falls through the front. The combination of southerly winds in the now cool air mass north of the warm front and the drag of the air being lifted over the warm frontal surface causes the remnants of the cool air north of the warm front to retreat northward. The low continues to deepen, the winds around the low strengthen, and the cold front, traveling faster than the warm front, eventually overtakes the warm front to the south of the low (figure 6.4d). The cold air runs under the warm front, lifting it off the ground and forming an occluded front.

Warm, cold, and occluded fronts meet in low pressure centers and move with the traveling lows.

The approach of a front can often be recognized by observing the pro-

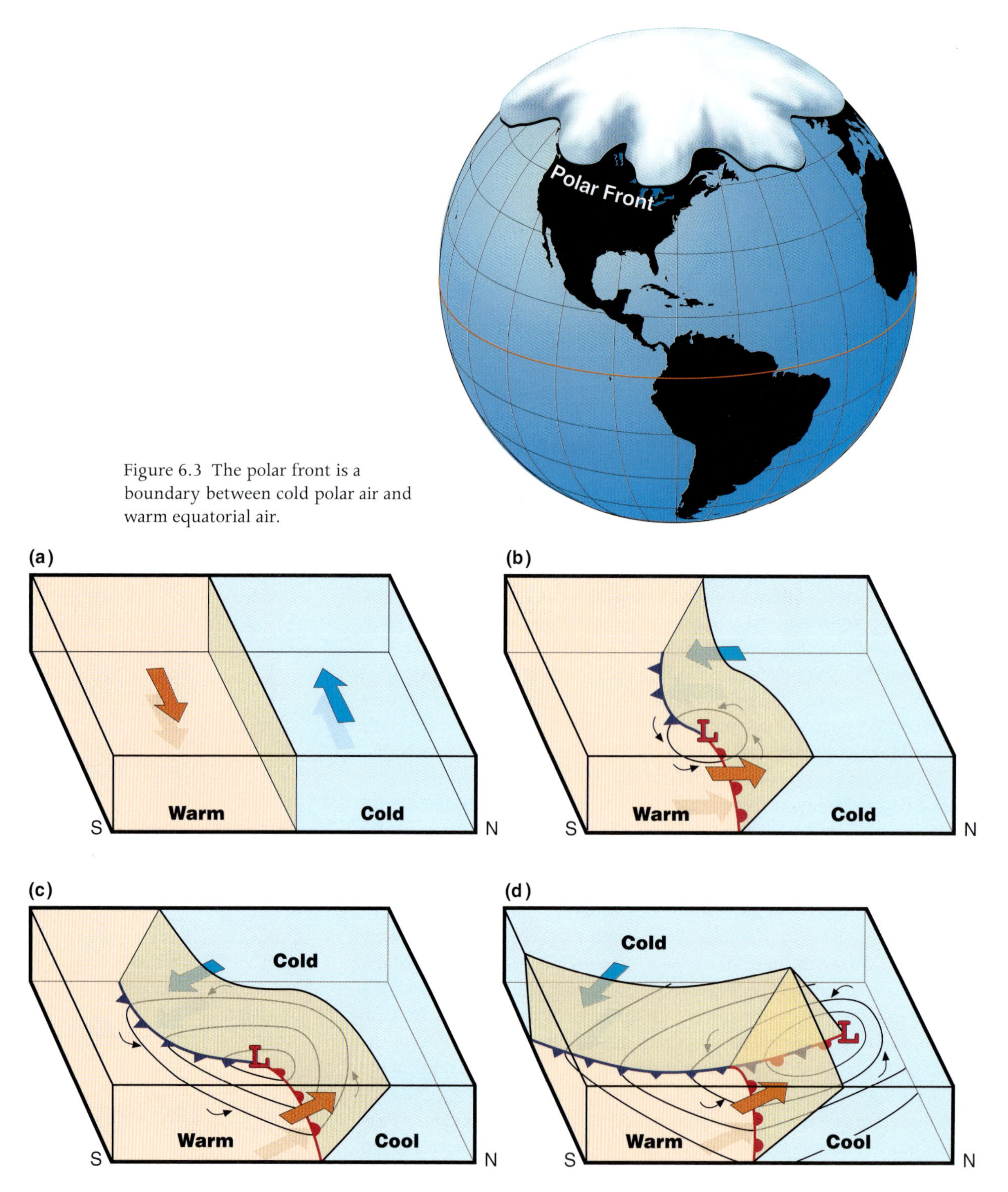

Figure 6.3 The polar front is a boundary between cold polar air and warm equatorial air.

Figure 6.4 Development of frontal systems and their relationship to cyclogenesis at the polar front.

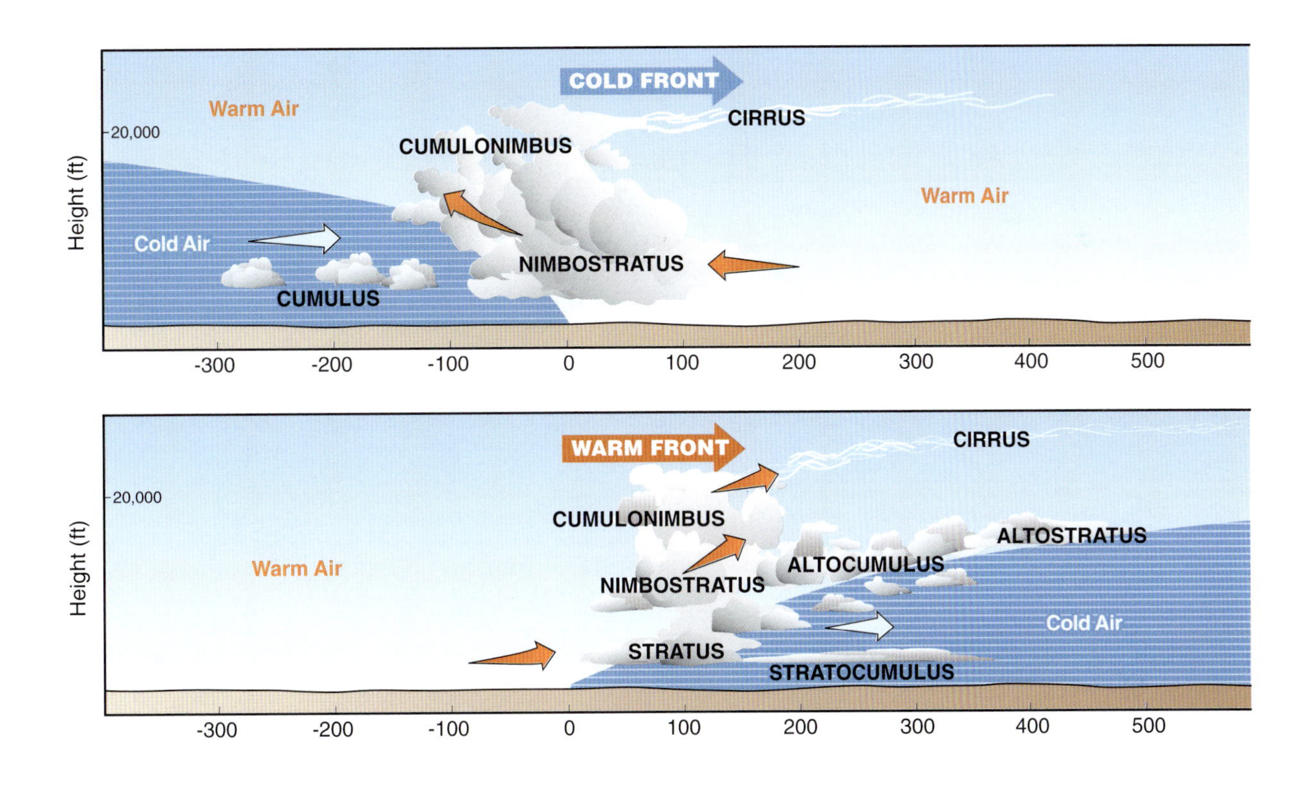

Figure 6.5 (*top*) Cold fronts and (*bottom*) warm fronts are associated with different types of clouds. The progression of cloud types and the rate of cloud development and thickening are often good visual indicators of an approaching front. The cross sections shown are perpendicular to cold and warm frontal boundaries at the surface. (Adapted from Slade, 1968)

gression of cloud types. Figure 6.5 shows the progression of cold and warm frontal clouds. Stationary fronts are characterized by variable cloudiness (or no clouds at all); occluded fronts are often accompanied by extensive stratiform cloudiness.

A cold frontal passage occurs when cold air moves forward, usually toward the south, southeast, or east, into warmer air, forcing the warm air to rise up the steep, sloping surface of the cold air mass. The slope on the leading edge of the cold air mass is usually between 1:50 and 1:150. The warm air is lifted most forcefully when the frontal surface is steep and is moving rapidly (15–30 mph or 7–13 m/s). The rapid lifting of the warm (and usually moist) air over the sloping frontal surface causes *convective clouds* (clouds with strong vertical development) to form along the frontal boundary. These clouds, which are discussed further in the next section, bring intense, showery precipitation, gusty winds, and often thunder. A cold frontal passage also brings strong winds caused by the strong frontal pressure gradient, a shift in wind direction, and, of course, dropping temperatures. The frontal passage is usually marked by the passage of a pressure trough, so that barometer readings fall as the front approaches and then rise as the front passes.

Cold fronts replace warm air with colder air. Warm fronts replace cold air with warmer air. An occluded front has characteristics of both warm and cold fronts.

A warm frontal passage occurs when a warm air mass is driven northward and is forced to rise over cool air that is retreating to the north. The rate of movement of a warm front is about half that of a cold front (i.e., about 7–15 mph or 3–7 m/s). The trailing edge of the cool air is not as steep as at the sloping surface of a cold front (usually between 1:150 and 1:300), so that the lifting of air associated with a warm front is slower than the lifting associated with a cold front. Thus, the clouds along a warm

Table 6.1 Weather Conditions Characteristic of the Approach and Passage of Fronts and Storm Systems

	Warm Front		Cold Front		Occluded Front		
Phenomenon	Approach	Passage	Approach	Passage	Approach	Passage	Observer North of Frontal System[a]
Pressure	falls steadily	levels off or falls unsteadily	falls slowly, or rapidly if storm is intensifying	rises sharply	falls steadily	rises, often not as sharply as for a cold front	falls slowly, then rises slowly as system passes
Wind	SE quadrant; speed increases	veers to S quadrant	S quadrant; may be squally at times	sharp veer to SW quadrant; speed increases; gusty	E quadrant; may veer slowly to SE quadrant; speed increases	veers to SW quadrant; speed decreases	NE quadrant; backs through N to NW quadrant
Clouds[b]	Ci; Cs; As; Ns; thickening	Sc; sometimes Cb; clearing trend	Cu or Ac; Cb in squall line	Cb; sometimes few clouds; clearing trend	Ci; Cs; As; Ns	slow clearing; Sc; Ac	Ci; Cs; As; Ns; Sc; Cu
Precipitation	steady rain or snow starts as clouds thicken; intensifies as front approaches	precipitation tapers off; may be showery	none or showery; intense showers or hail in pre-front squall line	showery; perhaps thunderstorms; rapid clearing	steady rain or snow starts as clouds thicken; intensifies as front approaches precipitation	precipitation tapers off slowly	rain or snow starting as clouds thicken and lower; slow clearing
Temperature	increases slowly	rise slightly	little change or rises slowly	sharp drop	slow rise	slow fall	steady or slow decrease
Humidity	increases	increases; may level off	steady or slight increase	sharp drop	slow increase	slow decrease	increase; slow decrease as storm passes
Visibility	becomes poorer	becomes better	fair; may become poor in squalls	sharp rise; becomes excellent	becomes poorer	becomes better	becomes poor; slow betterment

From Reifsnyder, 1980.

[a]This indicates changes in the weather at a site north of the frontal systems. Changes there are due not to the passage of a front, but to the passage of the low pressure storm center associated with the frontal systems.

[b]See table 7.1 for cloud abbreviations.

front do not have the vertical development of clouds along a cold front. Instead, they are stratiform, or layered, and extend over a broad frontal band. The precipitation in warm frontal zones is generally light and widespread rather than showery and intermittent. Stratiform clouds, which thicken and decrease in altitude with time, widespread light precipitation, passage of a pressure trough, a shift in wind direction, and rising temperatures all indicate a warm frontal passage.

An occluded front develops when a fast-moving cold front overtakes a slower warm front. The vertical structure in an occluded front is complicated and cannot be measured using surface observations alone. It can therefore be difficult to locate. If the cold front contains air that is colder than the retreating cool air ahead of the warm front, it moves under the warm front, forcing it to rise off the ground. This process is called a *cold occlusion*. A *warm occlusion* occurs if the cold front contains air that is warmer than the cool retreating air ahead of the warm front. The cold front leaves the ground and rises up and over the denser air ahead of the warm front. The approach of an occluded front is characterized by the stratiform clouds and light precipitation associated with a warm front. The frontal passage is marked by heavier, showery precipitation, characteristic of a cold front, and by a shifting of wind direction. The pressure trough and wind shift are often less distinct than for a cold or warm front.

A mountain barrier can affect the movement of an approaching front (section 10.3.3). A front may be slowed or stopped as it approaches a mountain barrier and may accelerate after it passes the crest of the mountains. Mountain barriers can channel or turn approaching fronts, causing departures from the weather patterns that are seen with fronts over flat or homogeneous terrain. Table 6.1 describes weather conditions associated with the passage of fronts and storm systems.

Clouds and Fogs

7

7.1. Clouds

Clouds are visual indicators of physical processes taking place in the atmosphere. They provide information about winds, stability, moisture content, and traveling weather systems that can be used in short-term (several hour) and long-term (24–36-hour) weather forecasts and in field assessments of weather conditions.

Most clouds form when moist air is lifted and cooled. Lifting occurs when air rises over low pressure areas, is carried up inclined frontal surfaces, is carried upward by convective currents that originate at the heated ground, or flows over mountains.

I am not sure how clouds get formed. But the clouds know how to do it, and that is the important thing.
—elementary school student

7.1.1. *Classification of Clouds*

Clouds form when moist air is lifted and cooled.

Clouds are classified according to their appearance (table 7.1) and the approximate altitude of their bases (table 7.2) following a simple scheme that was developed in 1803 by an English druggist, Luke Howard. This scheme can be expanded to describe certain clouds more precisely (World Meteorological Organization, 1987). The cloud types listed in table 7.1 are pictured in figure 7.1.

The two basic cloud classifications are stratiform and cumuliform. Stratiform clouds form horizontal layers, with the horizontal dimension much greater than the vertical dimension. Cumuliform clouds, on the other hand, have roughly equal horizontal and vertical dimensions, giving them the appearance of a cotton ball. Stratiform clouds are given the stem name *stratus*, and cumuliform clouds are given the stem name *cumulus*.

Stratiform and cumuliform clouds are assigned to one of three height classifications: high, middle, or low. High clouds are composed of ice

Table 7.1 Cloud Classification Chart

Cloud Form	Cloud Height: Low	Middle	High
Stratiform	stratus (St)	altostratus (As)	cirrostratus (Cs)
Cumuliform	cumulus (Cu)	altocumulus (Ac)	cirrocumulus (Cc)
Mixed	stratocumulus (Sc)	—	—
Fibrous			cirrus (Ci)
Cumuliform clouds of great vertical development	towering cumulus (TCU)	towering cumulus (TCU)	—
	cumulonimbus (Cb)	cumulonimbus (Cb)	cumulonimbus (Cb)
Precipitating, without lightning	nimbostratus (Ns)	nimbostratus (Ns)	

Clouds are classified by height and appearance. Stratiform clouds develop more horizontally than vertically; cumuliform clouds have roughly equal horizontal and vertical dimensions.

crystals rather than water droplets and are named using the prefix *cirro-* (*cirrocumulus* or *cirrostratus*). Clouds at midlevels, which may be composed of either water or ice particles, are given the prefix *alto-* (*altocumulus* or *altostratus*). Low clouds are given no prefix (stratus or cumulus). When a stratus deck is composed of an array of individual cumulus elements or lumps, the term *stratocumulus* is used. *Cirrus* are stringy or fibrous high clouds that are neither stratiform nor cumuliform. *Towering cumulus* extend through low and middle levels of the atmosphere. A thunderstorm cloud, called a *cumulonimbus*, extends through low, middle, and high levels of the atmosphere and is accompanied by lightning and showery precipitation. Clouds that bring widespread, light precipitation but not lightning, thunder, or hail are called *nimbostratus*.

Several different types of clouds can be present at any given time. The identification of clouds is a challenge even for an experienced professional weather observer. An amateur weather observer can gain experience in recognizing cloud types and understanding their significance by comparing observations to one or more of the illustrated cloud classification schemes referenced in appendix F and by noting or recording cloud observations and subsequent changes in the weather.

Because cloud classification systems are based on cloud appearance, they are of little use on dark nights when the only indication of the presence of clouds is the absence of stars. However, some types of clouds and the heights of their bases can be determined from their rates of movement, opacity, precipitation, or optical effects, such as rings around the moon. Aircraft, *ceilometers*, or radars can provide precise observations of

Table 7.2 Approximate Cloud Base Heights for the Midlatitudes

Cloud Height Category	Cloud Base Height (above ground level)	
Low	below 6500 ft	below 2000 m
Middle[a]	6500–23,000 ft	2000–7000 m
High	above 16,000 ft	above 5000 m

[a]Note the overlap between middle and high cloud base heights. The boundary between middle and high clouds is higher near the equator and lower in the polar regions.

TCU

Cb

Ns

Cs & Contrails

Cc

Ci

As

Ac

St

Cu

Sc

Figure 7.1 Photographs illustrating the principal cloud types shown in the cloud classification chart of table 7.1. (Ns © Ronald L. Holle, 1987; Sc © Ronald L. Holle, 1977; remaining photos © C. Whiteman)

cloud height even at night. In the absence of other information, the nighttime cloud cover is sometimes assumed to be composed of the same types of clouds that were observed late in the day.

7.1.2. *Clouds Associated with Fronts*

Warm and cold frontal passages are each associated with particular cloud types (section 6.2). The approach of a front can be simulated in figure 6.5 by placing an observer at the ground 600 miles ahead of the front and considering the progression of clouds that the observer would see if the front moved from left to right toward the observer. The rate of progression of the clouds depends on the speed of movement of the front (say, 25 mph or 11 m/s for a cold front and 10 mph or 4.5 m/s for a warm front).

The observer has plenty of advance warning of the approach of a warm front. After the first high-level cirrus clouds appear, the sky cover increases as middle-level clouds (altostratus and altocumulus) cover more of the sky, thicken, and lower. Low-level clouds (stratocumulus, stratus, and often fog) soon follow. Precipitation from the middle and low clouds is often widespread and persistent. It can be supplemented by showery precipitation from cumulonimbus and nimbostratus clouds that occur just in advance of the frontal passage.

The approach of a warm or cold front is often indicated by the progression of clouds that precede the front.

A cold front, like a warm front, is often preceded by cirrus clouds, which may be blown off the "anvil" of a thunderstorm (cumulonimbus) at the frontal zone. However, the extensive stratiform middle-level cloudiness associated with a warm front is absent, and the next cloud indicator of a cold frontal passage are the deep cumuliform (nimbostratus or cumulonimbus) clouds at the frontal zone itself. These clouds produce showery, sometimes heavy, precipitation. Cumulus clouds may also be present in the cold air that follows frontal passage.

The progression of cloud types as a front approaches a mountainous area is more complicated than over the plains. Additional clouds form as air is lifted over the mountains. Low clouds produced as a front approaches the mountains are often not visible from lee slopes, so that the increasing sky coverage, thickening, and lowering of high- and middle-level clouds are often the best indicators of increasing moisture associated with an approaching front.

7.1.3. *The Influence of Mountains on Circulations around Pressure Centers and Cloud Formation*

The circulation around high and low pressure centers can force air to flow up or down mountainsides when the pressure centers are in the vicinity of mountains. Forced rising motions cool the ascending air and enhance the formation of clouds and precipitation, whereas sinking motions warm the air and cause clouds and precipitation to dissipate. These effects are illustrated in figure 7.2 for low pressure centers on opposite sides of a mountain barrier. Because of the counterclockwise flow around a low pressure center, a low on the west side of a north–south barrier forces air to the south and east of the low up the barrier, whereas air to the north and east of the low is forced down the barrier. Clouds and precipitation are concentrated over the mountains somewhat to the south of the low

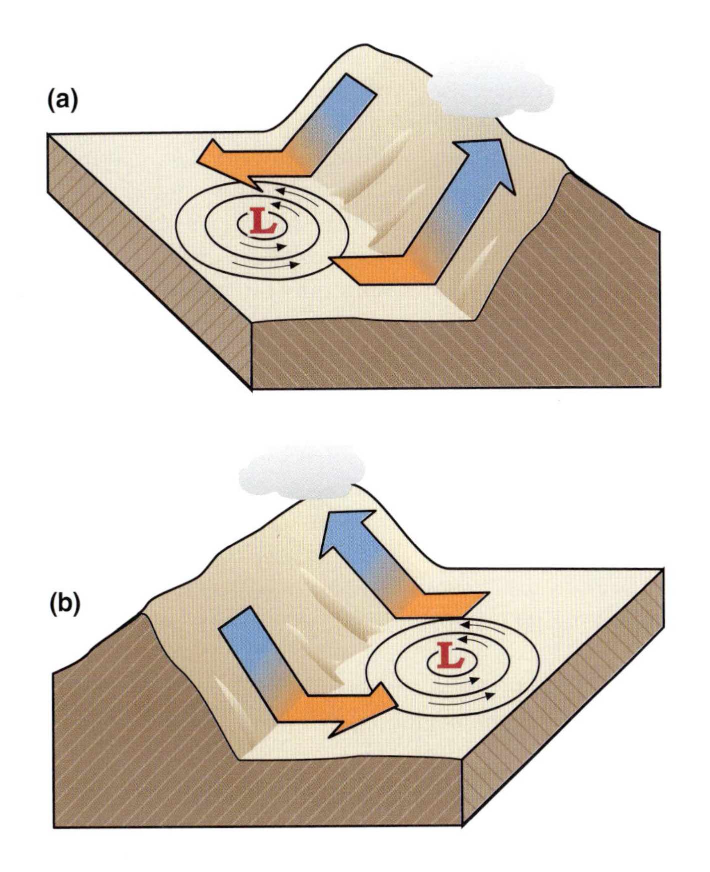

Figure 7.2 When a low pressure center approaches and crosses a mountain barrier, areas of enhanced precipitation occur where the counterclockwise circulation around the low is forced up steep mountain slopes. For a mountain ridgeline that is oriented north to south, a low pressure center crossing a mountain barrier from west to east produces enhanced cloudiness and precipitation (a) to its south on the west side of the mountains, but (b) to its north on the east side of the mountains. Thus the location of enhanced precipitation shifts abruptly when the low crosses the barrier. Corresponding drying, warming, and dissipation of clouds occur in locations where the circulation causes air to descend steep mountain slopes.

pressure center. Once the low crosses the barrier, however, the air to the north and west of the low is forced up the barrier, and the air to the south and west is forced to descend. The concentration of clouds and precipitation remains over the mountains, but shifts northward. Because the rising and sinking motions caused by the barrier affect the locations of clouds and precipitation associated with the low, exact predictions of the movement of a low pressure center are important for making precipitation forecasts in mountain areas. Of course, real terrain is more complicated than that shown in figure 7.2. Topographic details are important because the steeper the underlying terrain, the higher the precipitation rate when air is forced directly up the slope.

7.1.4. *Clouds that Form over Mountains*

Several types of clouds form primarily or only in mountainous terrain and are good visual indicators of mountain winds. Strong, warm *downslope winds* on the lee side of a mountain range, called *chinooks* in the western United States and *foehn* in the European Alps, are often associated with the *foehn wall* and the *chinook arch*, as well as *rotor*, lenticular, and *cap clouds*. During any given wind episode, none, some, or all of these clouds may be observed. Other special clouds often form in mountain areas when chinooks may not be present. These include banner clouds, *billow clouds*, fractocumulus, *fractostratus*, and *jet stream cirrus*.

Mountains can enhance rising and sinking motions associated with high and low pressure centers, thus enhancing or suppressing cloud formation.

7.1.4.1. *Foehn (Chinook) Wall Clouds.* A foehn wall is the leeward edge of an extensive cloud sheet that forms on ridge tops as moist air is lifted up the windward side of a long mountain barrier (figures 7.3 and 7.4). The cloud base is below the ridge on the windward side, and its upper and lower surfaces are more or less parallel to the underlying large-scale topography. Although the *foehn wall cloud* appears stationary, air rushes through it continuously, with moisture condensing to form the cloud as it ascends the windward side of the mountain barrier and evaporating as it descends the leeward side. The foehn wall can persist for several hours or days and is a common wintertime feature along the Front Range of the Colorado Rockies and in many other mountain regions.

A foehn wall forms when winds and moisture increase on the windward side of a mountain range, thus it often indicates that a storm is approaching. The wall can produce a nearly continuous light snowfall on the upper lee slopes. It is associated with strong, gusty winds on both the windward and leeward sides of the barrier, especially near mountain passes, indicating strong cross-barrier winds. The turbulence within the cloud is a serious hazard to aircraft.

7.1.4.2. *Chinook Arch Clouds.* A chinook arch forms in the lee of long mountain barriers at the beginning of a chinook (or foehn) windstorm

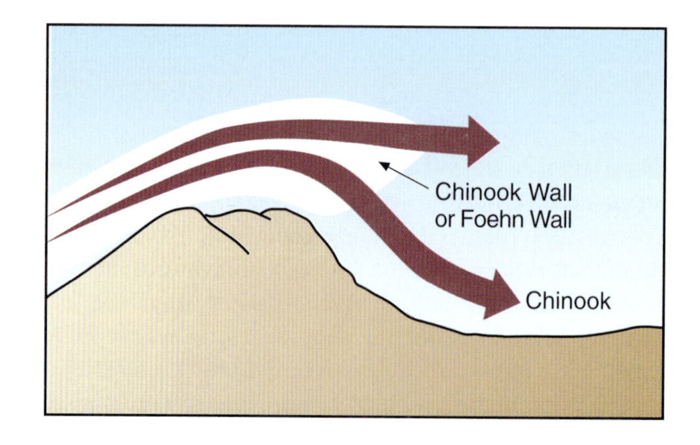

Figure 7.3 (*top*) Diagram of a chinook wall or foehn wall cloud

Figure 7.4 (*bottom*) Photograph of a wintertime chinook wall cloud in the Indian Peaks west of Ward, Colorado, on 2 October 1982. The cloud has formed along the crest of the north–south *Continental Divide* and is viewed here from the east. (Photo © Ronald L. Holle, 1982)

(section 10.2.2), but may disappear once the windstorm begins on the lee slope. An extensive altostratus layer forms downwind of the mountains over the plains. When viewed from the plains, the edge of the cloud deck nearest the mountains appears to form an arch over the mountains with a layer of clear air separating the cloud from the mountains under the center of the arch, but with the ends of the arch seeming to converge with the mountains on the distant horizons. Figure 7.5 is a photograph of a chinook arch cloud, and figure 7.6 is a satellite photograph of an extensive chinook arch cloud.

7.1.4.3. *Rotor Clouds.* A rotor cloud is a stratocumulus or altocumulus cloud that forms downwind of a mountain barrier in the crest of a large-amplitude atmospheric wave (figure 7.7). (Waves form in the atmosphere downwind of an obstacle, just as waves form in a river downstream from a protruding rock.) A rotor cloud indicates strong winds and turbulence. Rotor clouds typically form on a line parallel to the mountain barrier with cloud bases near mountaintop level, and they appear to rotate about a horizontal axis parallel to the mountains and somewhat below the cloud base. Significantly higher winds at the top of a rotor cloud than at the base cause the top of the cloud to appear to roll over ahead of the cloud base. This turbulence can be a severe hazard to aircraft. A glider being used to investigate *mountain waves* during the Sierra Wave Project in the early 1950s was demolished while flying through heavy turbulence in a rotor zone near Bishop, California. The pilot was able to parachute to safety.

7.1.4.4. *Lenticular Clouds.* Lenticular clouds (figure 7.8a–g) are shaped like horizontally oriented lenses. They have a smooth, sharp outline, are quasi-stationary, and form in the crests of waves in the atmosphere. Lenticular clouds can form at any height in the atmosphere. (Depending on their height, they are called cirrocumulus lenticularis, altocumulus lenticularis, or stratocumulus lenticularis.) They can form one above the other and are often present above rotors.

Figure 7.5 This chinook arch cloud seen at Calgary, Alberta, was formed by westerly flow over the Canadian Rockies. Its leading edge (the lower edge in the photo) parallels the north–south-oriented Rocky Mountains to the west. (Photo © P. Lester)

Figure 7.6 (*at right*) This satellite photograph shows a long chinook arch cloud that formed immediately to the east of the Canadian Rockies. The sharp leading edge of the cloud parallels the mountains on the southwestern border of Alberta and extends south-southeastward into Montana. (Photo © P. Lester)

Figure 7.7 (*bottom*) A rotor cloud photographed by P-38 fighter pilot Robert Symons looking south along the eastern lee slope of the Sierra Nevada. Flow is from right to left. Dust is being swept up off the floor of the Owens Valley and lifted vertically into the rotor cloud on the left of the picture. Higher lee wave clouds are seen above and to the right of the rotor cloud, and a chinook wall cloud is visible over the crest of the Sierra Nevada at the right of the photo. (U.S. Air Force photo provided by R. A. Houze)

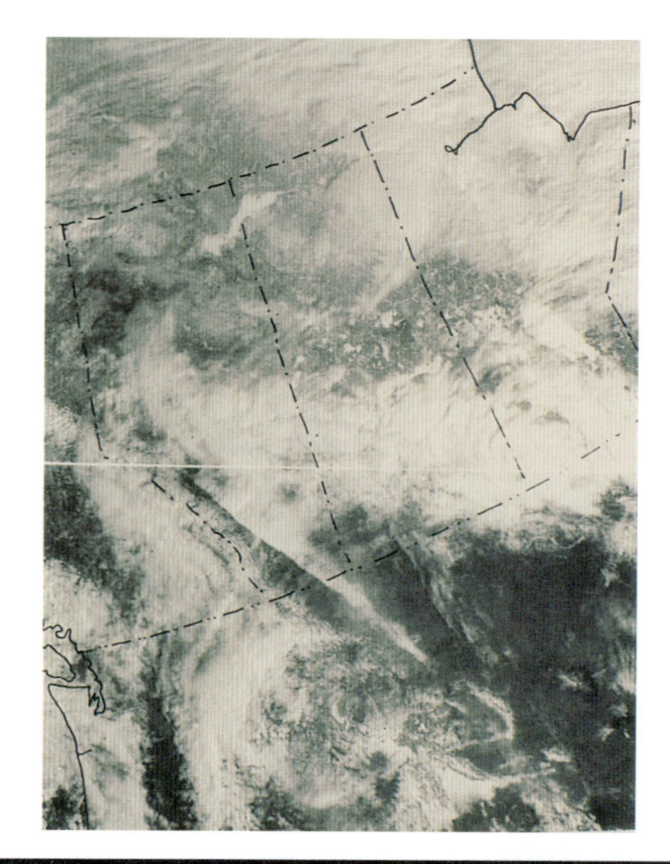

Air flows continuously through a lenticular cloud, with condensation forming the cloud as the air rises into the wave crest and evaporation dissipating the cloud as the air descends from the wave crest.

Lenticular clouds indicate moisture in moderately stable to strongly stable layers and are associated with strong winds in the mountains, especially near the mountain crests. Changes in the fraction of lenticular cloudiness or the thickness of individual clouds indicate changes in atmospheric moisture as traveling weather systems approach or recede.

FORMATION OF LENTICULAR CLOUDS

The formation of lenticular clouds can be illustrated by focusing on a single, initially unsaturated air parcel. The parcel is carried through the wave attaining successive positions A through G as shown in figure I. At position A, below the cloud, the parcel is unsaturated. It is lifted in a trajectory toward the crest of the wave, causing it to cool. If the parcel has sufficient water vapor, this cooling causes condensation on tiny aerosol particles called *cloud condensation nuclei* within the parcel, and the upwind or leading edge of the wave cloud is formed (position B). The water droplets in the cloud are initially small, but as the parcel is lifted farther to positions C and D, additional moisture becomes available, and the water droplets increase in size. At D, the parcel reaches the crest of the wave and begins to descend. The cloud of water droplets then begins to evaporate as the parcel descends and warms (E and F). At F, all water droplets evaporate and the parcel becomes unsaturated, forming the downwind or trailing edge of the wave cloud. As it descends toward G, the unsaturated parcel continues to warm. The net result of a succession of parcels being carried through the wave is the formation of the cloud on the upwind side of the wave and the dissipation of the cloud on the downwind side of the wave. The wave cloud appears to be stationary (i.e., "standing"), but it is actually produced by the continuous stream of air parcels passing through the wave. The narrow distribution of droplet sizes in the cloud, the (usually) shallow depth of the cloud, and the limited period of time available for the droplets to freeze or fall as they are carried rapidly through the cloud, results in a cloud that is thin and symmetrical.

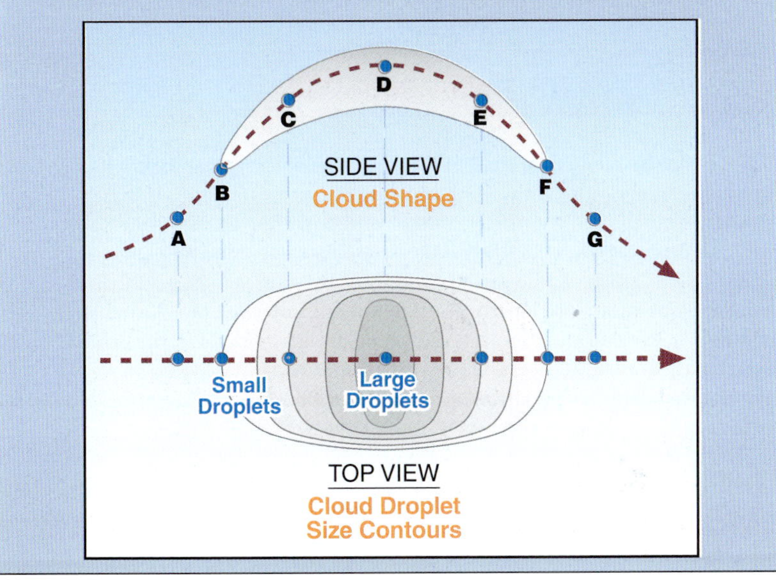

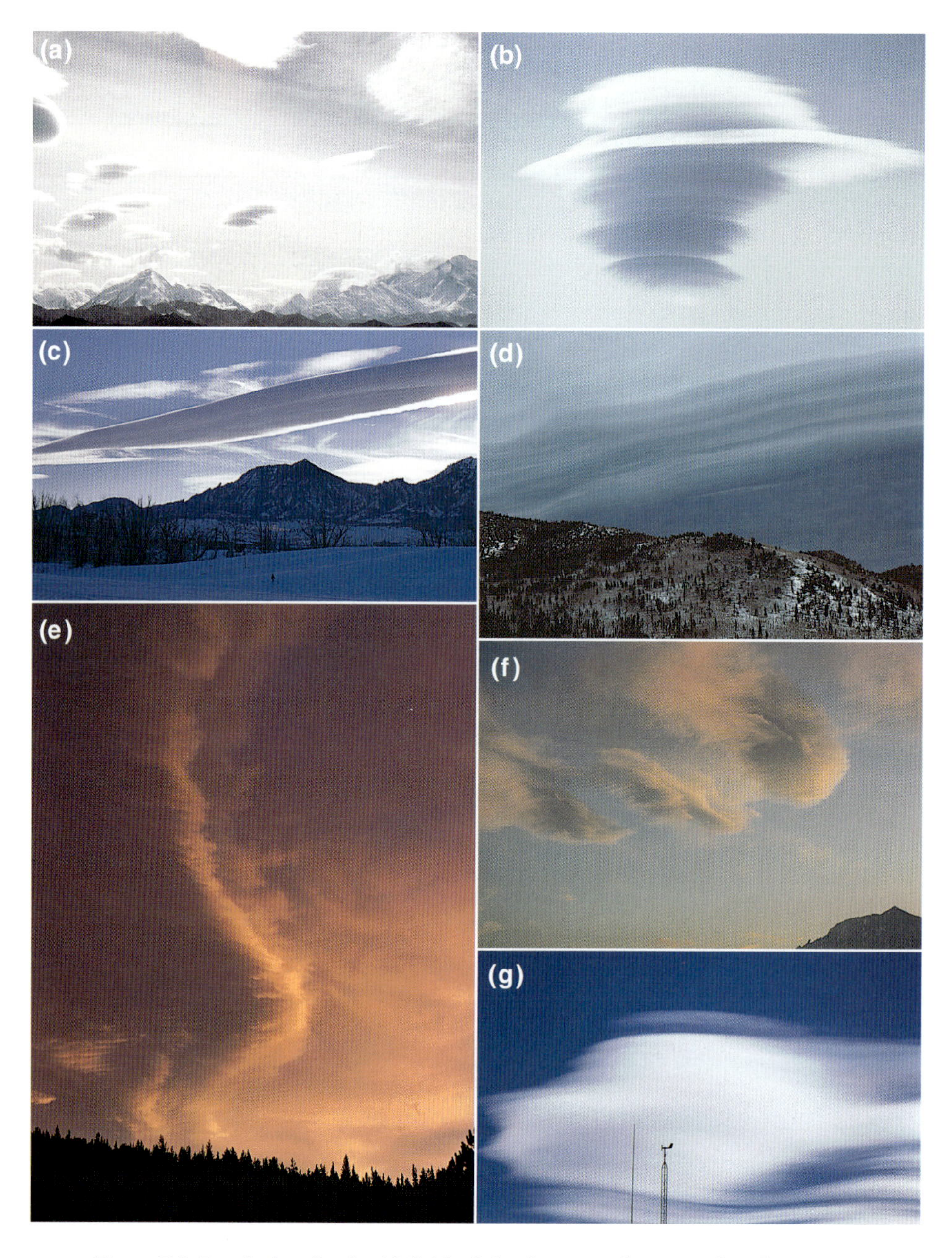

Figure 7.8 Lenticular clouds: (a) field of clouds at Denali National Park, Alaska; (b) lenticular clouds in the form of a pile of plates, Denali National Park, Alaska; (c) elongated lenticular cloud oriented parallel to the Front Range over Boulder, Colorado; (d) lenticular clouds with distinctive leading and trailing edges. The clouds can be easily recognized by these edges, even when the lenticular shape cannot be clearly seen; (e) the trailing edge of a lenticular cloud at sunset over Colorado's Front Range; (f) lenticular clouds in the crests of a lee wave near Boulder, Colorado. The three lenticulars are equally spaced, indicating the wavelength of the flow; (g) lenticular cloud in a broad area of rising motion over Laramie, Wyoming. (Photos a, b, and g © B. Martner; photos c–f © C. Whiteman)

Figure 7.9 Cap cloud in Grand Tetons National Park, Wyoming. The cap cloud is best formed over the peak on the right side of the photograph. (Photo © B. Martner)

7.1.4.5. *Cap Clouds.* A cap cloud (figure 7.9) is a lenticular cloud that forms over a mountain peak with the cloud base below the peak's summit. A cap cloud is sometimes surmounted by a stack of lenticular clouds. Like lenticular clouds, cap clouds are stationary, forming as humid air is lifted over the peak. Also like lenticular clouds, they form when there is stable stratification, a smooth, nonturbulent flow, moderate to strong winds, and high humidity. Cap clouds are quite common in maritime mountain ranges where there is a good supply of low-level moisture, and they are often the precursors of moist air masses that approach the mountains. Changes in cloud coverage or thickness indicate the rate at which moisture is increasing and can aid in timing the approach of a storm.

7.1.4.6. *Banner Clouds.* Banner clouds are plumes that form downwind from the upper lee slopes of isolated, steep-sided mountain peaks with sharp ridges. Banner clouds serve as a natural wind vane, indicating the presence of strong winds and humidity at mountaintop level. A banner cloud is sometimes the only cloud present on a fine weather day.

A banner cloud forms as the air flows over and around a peak, producing a low pressure area on the lee side of the peak somewhat below its summit. This low pressure center causes upward vertical motion on the lee slope, and an eddy forms downwind of the peak, rotating about a horizontal axis perpendicular to the approaching flow. If the air on the lee side of the peak is sufficiently moist, condensation will occur in the rising branch of the eddy, producing the cloud. Observations of the flight patterns of birds and of snow plumes blown off summit ridges suggest that horizontal-axis eddies are often present in the lee of mountain peaks even when there is insufficient moisture for banner clouds to form.

The most widely recognized banner cloud is the one formed in the lee of Switzerland's Matterhorn (figure 7.10), but banner clouds form on any isolated, sharp-ridged peak when there is sufficient moisture in the air.

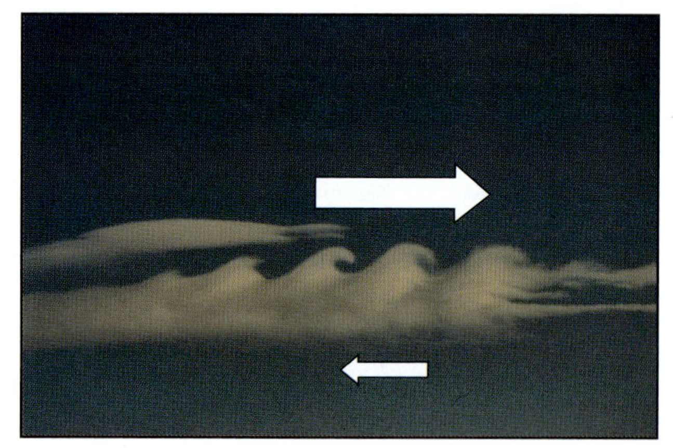

Figure 7.10 (*top*) The Matterhorn banner cloud forms on the lee side of the peak. The winds are blowing from right to left. (Photo © C. Whiteman)

Figure 7.11 (*right*) Billow clouds are produced when strong shear occurs across a sharp change in temperature or density in a cloudy atmosphere. Here, the breaking waves in the atmosphere are like breaking waves in an ocean. (Photo © B. Martner)

In the United States, banner clouds are observed more frequently in the coastal ranges than in the dry Rocky Mountains and are well known, for example, on Mount Rainier and Denali.

7.1.4.7. *Billow Clouds.* Billow clouds form when vertical *wind shear* (change of wind speed, wind direction, or both, with height) occurs across a sharp change in temperature in a cloudy atmosphere (figures 7.11 and 7.12). The wind shear may cause the waves to curl over and break, much like breaking ocean waves. Billow clouds are most frequently superimposed on lenticular clouds, sometimes propagating across the top of the cloud, as in figure 7.12. They are less frequently seen as rows of waves that form on the upper surfaces of cloud decks at the top of surface-based temperature inversions. The atmospheric waves in which billow clouds form are called *Kelvin–Helmholtz waves.* These waves, which also occur when clouds are not present to make them apparent, cause clear air turbulence and pose a significant hazard to aircraft.

Figure 7.12 Billows on the upper surface of an iridescent lenticular cloud over Boulder, Colorado. The lines of billows in different parts of the cloud have different orientations, and the amplitude of the billows increases where the lines intersect. The billows crest and break, and shadows are cast on the top of the lenticular cloud by the cresting billows. (National Oceanic and Atmospheric Administration photo)

Figure 7.13 Fractocumulus clouds are formed in the vicinity of peaks by turbulent motions. (Photo © C. Whiteman)

Figure 7.14 Jet stream cirrus as seen from a peak in central Colorado. (Photo © C. Whiteman)

OPTICAL EFFECTS IN CLOUDS

Optical effects are often seen in clouds, such as lenticulars, that are thin and have uniform droplet sizes and concentrations. The effects, which can best be seen with sunglasses, include *coronas* (concentric rings of alternating green and pink around the sun or moon) and *iridescence* (bands of alternating green and pink that contour the shape of the cloud). If cloud droplets freeze, *halos* may be observed. Halos are bright circles or arcs exhibiting prismatic coloration ranging from red inside to blue outside. They are centered on the sun or moon and are caused by the refraction or reflection of light by suspended ice crystals.

The *Brocken specter* (or *glory*) is an optical phenomenon that today is seen most frequently by airline passengers, but that was originally described by mountain climbers. The Brocken specter is named for the highest mountain in Germany's Harz Range where, according to legend, witches meet to dance on Walpurgis Night, the last night in April. A Brocken specter can be seen by a person standing on a cloudless summit with the sun low in the sky at his or her back and a fog or stratus deck present in the valley below. The observer's oversized shadow is cast onto the cloud layer below with rings of color surrounding the head of the shadow. This phenomenon is now frequently observed by airline passengers who see rings of color surrounding the shadow of the airplane cast onto a cloud below.

7.1.4.8. *Fractocumulus and Fractostratus Clouds.* Fractocumulus and fractostratus clouds (figure 7.13) are ragged cumulus and stratus clouds, small and irregular in appearance. They can be produced by condensation and mixing in near-saturated air after rainstorms. The mixing process is promoted by instability and is suppressed in a stable atmosphere. In mountains, fractocumulus and fractostratus are often the remnants of clouds that are torn apart by turbulence in the vicinity of terrain obstacles. They are thus good indicators of strong turbulent winds and instability that may be hazardous to aircraft.

7.1.4.9. *Jet Stream Cirrus.* Jet stream clouds (figure 7.14) indicate significant wind shear and often the presence of a jet stream. The wind shear can be seen when ice crystal streamers fall from the base of cirrus clouds into slower or faster moving air below. The wind shear distorts the *fallstreaks* into curved wisps.

7.2. Fogs

Fogs are clouds that are based at the ground. In the United States, heavy fogs occur most frequently on coastlines or in coastal mountainous areas where good moisture sources are present (figure 7.15), such as on the west side of the Cascade Mountains, in Oregon's Willamette Valley, in the Sierra Nevada, and on the east side of the Appalachian Mountains. Fog occurs

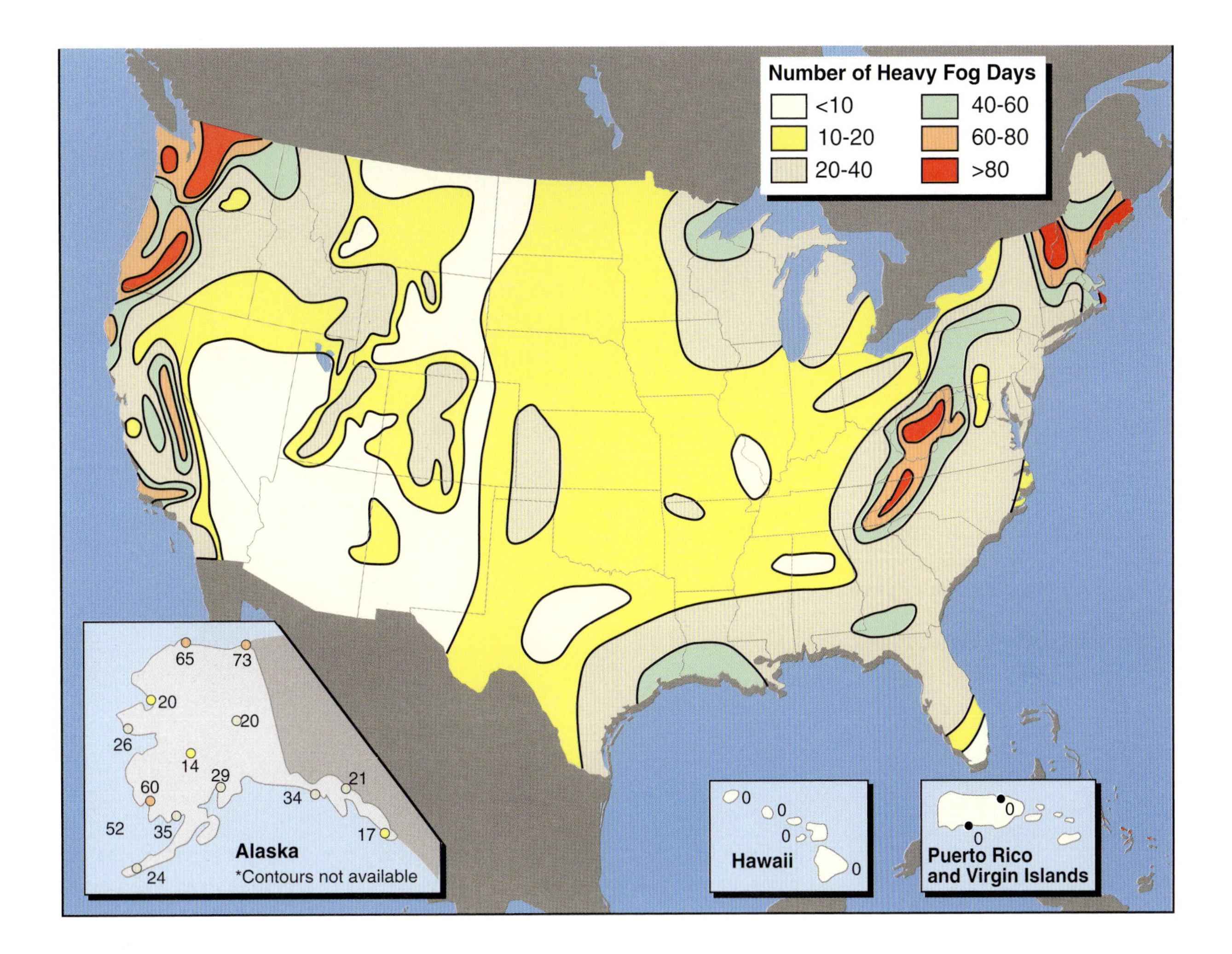

Figure 7.15 Average number of days per year with heavy fog.

less frequently in the Rocky Mountains and in the continental intermountain areas.

One cubic meter of fog typically contains only about 0.1 g of liquid water dispersed over 1 to 10 million droplets with diameters between 1 and 20 micrometers. These cloud droplets can significantly restrict visibility, in some cases reducing it to zero, and can therefore be hazardous to all forms of transportation. Fogs composed of water droplets are designated by names that describe the processes that lead to their formation (*upslope fog, radiation fog, advection fog, evaporation-mixing fog*). Fogs composed of ice crystals are called ice fogs, regardless of the processes leading to their formation.

An upslope fog develops when moist air is cooled by being lifted up terrain slopes. Upslope fogs are commonly seen over the upper sections of valley sidewalls following rainstorms, when moist air is lifted and cooled by winds blowing up the slopes. These fogs are often accompanied by strong isolation, which evaporates water from the moist ground and produces a combined evaporation–upslope fog. When the

Figure 7.16 Valley fog as seen in the distance from Mount Rainier, Washington, in the early morning of a late summer day. The unusual snow cup formations in the foreground are called *nieve penitente*. (Photo © C. Whiteman)

sidewall moisture sources are nonuniform or the winds are perturbed by individual trees or forest stands, fog tendrils rather than extensive fog banks form. Widespread upslope fog occurs when moist air is lifted up gently rising topography. This occurs occasionally over the gentle slope of the Great Plains as air is lifted northward and westward toward the Rocky Mountains, producing a relative maximum in fog frequency in northern Texas, western Oklahoma, and western Kansas (figure 7.15).

Radiation fog develops when nighttime outgoing longwave radiation cools the near-surface air below its dew-point temperature. This occurs frequently in the valleys and basins of continental mountainous areas as a result of nighttime or winter radiative cooling, especially if the air has been moistened by evaporation. These fogs are common, for example, in the Columbia Basin, the Salt Lake Valley and in many shallow urban basins throughout the country. Satellite photos show that the fog is often confined to the branching dendritic patterns of the drainage basin. The top of the fog deck, when seen from higher ground, usually marks the top of a nocturnal temperature inversion (figure 7.16). Radiation fogs are most frequent around sunrise, when nighttime cooling is at its maximum and relative humidity reaches its maximum value. Dissipation of valley fogs often involves an intermediate stage in which the fog lifts to form low stratus. This process is furthered by rising wind speeds and by solar heating of the slopes (either through direct heating of the slopes above the fog or through solar radiation that penetrates the fog to heat the underlying slopes).

An advection fog is produced when warm air flows over a cold surface and cools from below until saturation is reached. Advection fogs are characteristic of the Pacific Coast during the summer. As mentioned in section 1.4, the Japanese Current draws surface water away from the Pacific Coast, allowing much colder water from below to rise to the surface. Warm air moving in from the west thus passes over cold water along the coast, resulting in frequent episodes of fog.

An evaporation-mixing fog results when water evaporates and mixes with the adjacent air, raising the mixture's dew point. If sufficient moisture is present, the air becomes saturated and fog forms. *Steam fog*, which develops when a cold air mass flows over a warm body of water, is an example of an evaporation-mixing fog. The vertical tendrils that are seen with this fog are typical of the mixing process, which occurs in the form of vertical plumes and is promoted by atmospheric instability. Another example of an evaporation-mixing fog is the fog that condenses from breath on cold winter days. The air leaving the lungs is at body temperature (98.6°F or 37°C) and is nearly saturated. When the breath mixes with cold winter air, the resulting mixture becomes saturated, producing a fog.

Ice fog, which consists of small ice crystals rather than water droplets, can form in extremely cold air, usually at temperatures below −20°F (−29°C) and can be quite dense and persistent. Ice fogs are usually localized and form frequently in valleys that have open sources of water vapor (e.g., nonfrozen streams or herds of animals). Ice fogs form frequently in winter in the Fraser Plateau, the Yukon Plateau, the Northwest Territories, and in the interior of Alaska.

Precipitation

8

Precipitation is often the primary weather factor affecting outdoor activities. Precipitation that is of an unexpected type or intensity or that comes at an unexpected time or recurs more frequently than expected can disrupt both recreational and natural resource management plans. Heavy rain or snowfall can interfere with travel and threaten safety.

8.1. Types of Precipitation

Precipitation is water, whether in liquid or solid form, that falls from the atmosphere and reaches the ground. Table 8.1, adapted from Federal Meteorological Handbook No. 1 (National Weather Service, 1995), describes the different types of precipitation particles, collectively called *hydrometeors*.

International guidelines for the reporting of precipitation do not include a category for *sleet*. Meteorologists in the United States use the term to describe tiny ice pellets that form when rain or partially melted snowflakes refreeze before reaching the ground. These particles bounce when they strike the ground and produce tapping sounds when they hit windows. Colloquial usage of the term, often used by the news media, coincides with British usage, which defines sleet as a mixture of rain and snow.

Snow pellets, or graupel, are common in high mountain areas in summer. Graupel are low density particles (i.e., not solid ice) formed when a small ice particle (an ice crystal, snowflake, ice pellet, or small hailstone) falls through a cloud of *supercooled* (section 8.4) water droplets. The tiny droplets freeze as they impact the larger ice particle, building it into a rounded mass containing air inclusions (figure 8.1). This coating of granular ice particles is called *rime*, and the particle is said to be rimed.

Table 8.1 Precipitation Types

Type	Description
Drizzle	Fine drops with diameters of less than 0.02 inch (0.5 mm) that fall uniformly and are very close together. Drizzle appears to float while following air currents, but, unlike fog droplets, it falls to the ground. If the drizzle freezes upon impact and forms a *glaze* on the ground or other exposed objects, it is called *freezing drizzle.*
Rain	Drops usually larger than 0.02 inch (0.5 mm) that, in contrast to drizzle, are widely separated. If rain freezes upon impact and forms a glaze on the ground or other exposed objects, it is called *freezing rain.*
Snow	Branched ice crystals in the form of six-pointed stars
Ice crystals	Unbranched ice crystals in the form of six-sided needles, columns, or plates; sometimes called diamond dust
Hail	Small balls of ice falling separately or frozen together in irregular lumps
Snow grains	Very small, white, and opaque grains of ice
Small hail and/or *snow pellets*	White, opaque grains of ice from about 0.08 to 0.2 inch (2 to 5 mm) in diameter. The grains are round or sometimes conical. Snow pellets are also called *graupel.*
Ice pellets	Transparent or translucent pellets of ice, which are round or irregular, rarely conical, and which have a diameter of 0.2 inch (5 mm) or less. There are two main types of ice pellets: (1) hard grains of ice consisting of frozen raindrops or largely melted and refrozen *snowflakes* and (2) pellets of snow encased in a thin layer of ice. The layer of ice forms when water droplets intercepted by the snow pellets freeze or when the pellets partially melt and refreeze.

Graupel is usually produced in deep convective clouds that extend above the freezing level. Whereas graupel reaches the ground at high elevations, it usually melts to form rain before reaching the ground at lower elevations.

As falling snow accumulates, a snowpack develops that can be described in terms of water content and density. The water content of snow is usually expressed as *specific gravity,* a number obtained in this application by dividing the water-depth equivalent of snow by the actual snow depth. For example, a 10-inch (25-cm) snowfall, if melted, might produce 1 inch (2.5 cm) of water. The specific gravity of the snow would be 0.10 (1 inch divided by 10 inches). Specific gravity is numerically equivalent to the actual density (mass per unit volume) of the snow if expressed in units of g cm^{-3}. The densest (wettest) snow has a specific gravity near 0.40 and the lightest snow ("wild snow") has a specific gravity near 0.01. Most new snow densities in the United States are in the range of 0.04 to 0.10 g cm^{-3}, with peak frequencies between 0.06 and 0.09 g cm^{-3} (Doesken and Judson, 1996).

The density of newly fallen snow usually varies directly with temperature over a range of temperatures between the freezing point and about 10°F. Over this temperature range, warmer snow has higher density than colder snow. When temperatures are near the freezing point, rime can form on falling snowflakes, increasing snow densities. Below this temperature range, snow densities increase because of changes in crystalline structure. Snow density is affected not only by the shape of the snow crystals (e.g., needles, plates, dendrites, columns, and scrolls or cups)

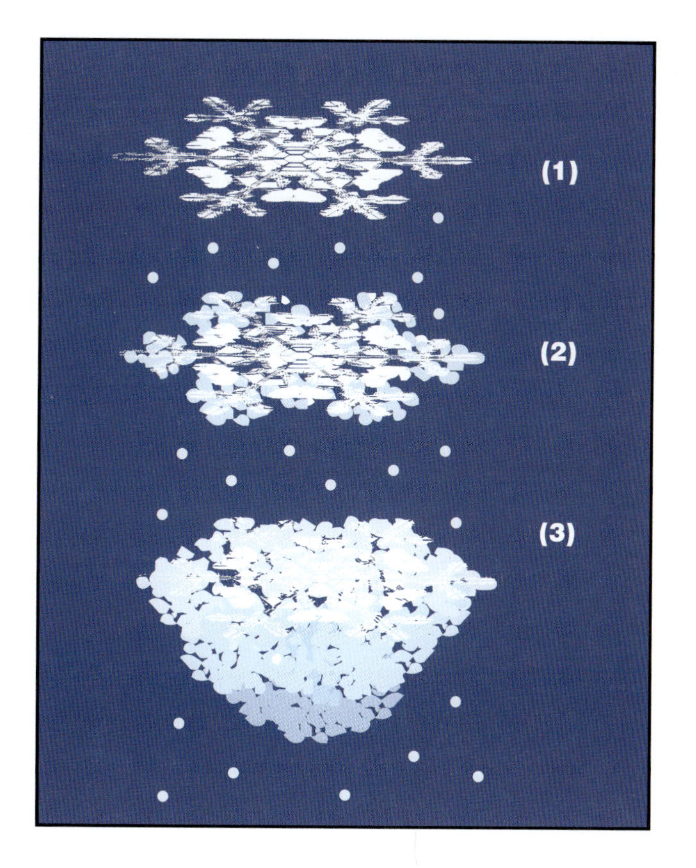

Figure 8.1 Snow pellets (also called graupel) form when supercooled water droplets freeze on an ice particle as it falls through the atmosphere, as shown successively in steps (1) through (3). Graupel is a common form of precipitation in mountainous areas.

when they are formed at the ambient temperature and humidity in the cloud, but by surface temperature and by wind (which breaks up the delicate branches of some crystals or snowflakes, increasing their density). Thus, high snow densities are associated with warm surface temperatures, small crystals, and strong winds that break the branched crystals into small pieces. Low snow densities are associated with low temperatures, highly branched crystals, and weak winds.

8.2. Intensity of Precipitation

There are three categories of precipitation intensity: light, moderate, and heavy. The classification criteria depend on precipitation type, as shown in table 8.2.

8.3. Measuring Precipitation

Rain gauges are used to measure precipitation amounts. A standard, manually operated rain gauge funnels precipitation into a small diameter tube, thereby increasing the depth of the collected precipitation per unit of collected rain. A calibrated measuring stick is used to determine the rainfall amount, with a measurement resolution of 0.01 inch. Recording

Table 8.2 How Heavy Is the Precipitation? Precipitation intensity can be determined from the observed rate of fall of the precipitation or by other methods.

Precipitation Intensity	Based on	Rain[a]	Ice Pellets	Snow or Drizzle, Occurring Alone
Light	rate of fall	Up to 0.10 inch per hour; maximum 0.01 inch in 6 minutes	same as for rain	
	other	Scattered drops that, regardless of duration, do not completely wet an exposed surface; individual drops are easily seen	Scattered pellets that do not completely cover an exposed surface regardless of duration; visibility is not affected	Visibility greater than ½ mile
Moderate	rate of fall	0.11 inch to 0.30 inch per hour; 0.01 inch to 0.03 inch in 6 minutes	same as for rain	
	other	Individual drops are not clearly identifiable; spray is observable just above pavements and other hard surfaces	Slow accumulation on ground. Visibility reduced by ice pellets to less than 7 miles	Visibility greater than ¼ mile but less than or equal to ½ mile
Heavy	rate of fall	More than 0.30 inch per hour; more than 0.03 inch in 6 minutes	same as for rain	
	other	Rain appears to fall in sheets; individual drops are not identifiable; heavy spray to height of several inches is observed over hard surfaces	Rapid accumulation on ground; visibility reduced by ice pellets to less than 3 miles	Visibility less than or equal to ¼ mile

[a]English units are given here because the definitions are those used by the U.S. National Weather Service.

Adapted from National Weather Service, 1995.

rain gauges record the weight of collected water as a function of time or count the number of volume units of rainfall as a function of time, thus allowing rainfall rates or intensity to be determined.

Snow accumulations can be measured with a rain gauge that is protected from winds and turbulence by a slatted wind screen, or measurements of snow accumulations can be taken at several sites and then averaged to compensate for drifting. Annual precipitation reports include the water equivalent of snow, which is determined by melting collected snow.

Rain gauges must be shielded from the effects of wind and turbulence around obstacles to provide accurate measurements. More information on the various types of precipitation measurement equipment and discussions of equipment exposure are provided in references listed in appendix C.

8.4. Formation of Precipitation

When unsaturated air is lifted (figure 8.2), it cools at the thermodynamically determined rate of 5.4°F per 1000 ft (9.8°C/km), the dry adiabatic lapse rate (section 4.3.1). When the air becomes so cold that it can no longer maintain water as vapor, the vapor condenses onto cloud condensation nuclei to form cloud droplets. These aerosol particles are always present in air, often at concentrations of approximately 1000 per cubic centimeter. Water condenses preferentially on the largest and most *hygroscopic* of these aerosols. Condensation releases heat, which adds to the

Cloud droplets can exist as liquid water droplets at temperatures below freezing.

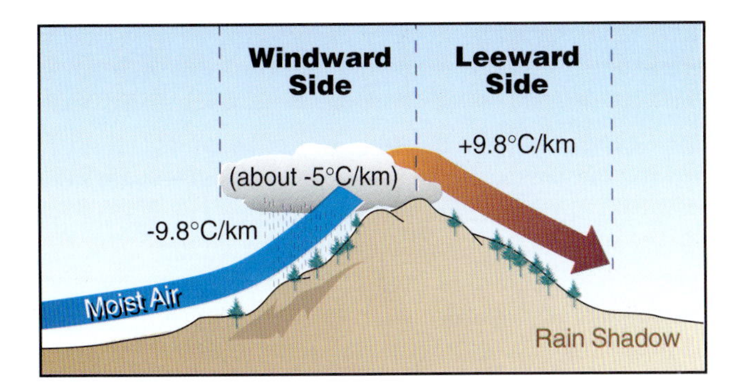

Figure 8.2 Air cools as it is lifted over mountains, resulting in clouds and precipitation. The cooling of unsaturated air is initially at the DALR, but, once the air attains saturation, the cooling rate decreases to the MALR. If condensed water falls as precipitation on the windward side of the mountain, descent of the unsaturated air on the lee side will cause the air to warm at the DALR, producing higher temperatures in the lee than on the windward side of the mountain.

buoyancy of the air, causing it to rise more rapidly. However, because the air is now saturated, it cools at a slower rate, the moist adiabatic lapse rate (section 4.3.1). Eventually, the lifting process cools the cloudy air to temperatures below freezing. Cloud droplets rarely freeze at 0°C but instead become supercooled. That is, they exist as liquid water droplets at temperatures below freezing.

Clouds are composed of a very large number of cloud droplets, with hundreds of droplets per cubic centimeter. The cloud droplets are so small, only about 20 micrometers in diameter, that they follow air motions within the cloud rather than falling out of the cloud. Thus, clouds composed solely of cloud droplets do not produce precipitation. Before a cloud can produce precipitation, rain drops or ice particles must form that are many times larger than cloud droplets. A typical rain drop has a diameter of about 2000 micrometers (2 mm) and is equivalent in volume to 1,000,000 cloud droplets.

Two processes allow precipitation-size particles to develop within clouds. The first of these occurs in clouds with temperatures above −15°C and requires cloud droplets of different sizes to collide and coalesce into larger droplets. This process is enhanced when cloud droplets are of differing size and fall at different speeds, with the larger droplets collecting smaller droplets until they become large enough to fall from the cloud. The initial formation of droplets of different sizes is facilitated by the presence of hygroscopic aerosols of differing sizes. Because there is a broader spectrum of aerosol sizes over the oceans, this *warm rain process* is common over coastal areas.

Precipitation can be produced by the warm rain process or by the ice crystal process. Over interior mountainous regions, the ice crystal process is more important.

A second process, called the *ice crystal process*, requires the coexistence of water droplets and ice crystals in a cloud at temperatures below freezing. This process produces most of the precipitation that forms in temperate continental regions. The introduction of ice crystals into a cloud of water droplets can occur when ice crystals fall from higher clouds, when supercooled cloud droplets within the cloud freeze, or when subfreezing temperatures initiate the formation of ice particles on small aerosols called *ice nuclei*. Ice nuclei are much less numerous than cloud condensation nuclei. For example, only one ice nucleus may be present for every 1,000,000 cloud droplets at −10°C (14°F). Larger numbers of ice nuclei become active at lower temperatures. Nonetheless, the relative scarcity of ice nuclei explains why cloud droplets do not freeze at temperatures below freezing.

OROGRAPHIC LIFTING PRODUCES RECORD SNOWFALL AND RAINFALL

Paradise Ranger Station on the south side of Mt. Rainier in Washington received the most snow ever recorded in the United States during the winter of 1971–1972: 1122 inches (28.5 m). Snow falls at Paradise when the prevailing wintertime winds lift moist air from the Pacific Ocean up the steep sides of the volcanic cone of Mt. Rainier. Orographic lifting also produces the world's highest annual precipitation record (500 inches) on the island of Kauai, where the northeast trade winds lift Pacific moisture up the northeast side of the island.

When cloud droplets and ice particles coexist in a cloud, water vapor diffuses from the cloud droplets to the ice particles because the saturation vapor pressure over water is greater than that over ice (figure A.1). The ice particles grow very rapidly at the expense of the cloud droplets and eventually fall from the cloud. As the particles fall, they may gain additional mass by colliding with supercooled cloud droplets, which freeze on the surface of the ice particles. The particles may reach the ground as solid particles or, if they fall through the freezing level, they may melt and reach the ground as raindrops. At higher elevations, precipitation often falls as snow, ice crystals, graupel, or sleet. It is only on warm days, when the freezing level is well above the ground, that these solid forms of precipitation melt and reach the mountainsides as raindrops.

A cloud composed of cloud droplets, if lifted farther, is brought to lower temperatures where more and more ice nuclei become activated and more ice particles are formed. If the ice particles fall through the cloud they will initiate the sudden freezing of the supercooled cloud droplets, causing the cloud to glaciate or freeze. *Glaciation* causes a pronounced change in the appearance of the cloud. Unglaciated or water clouds have clearly defined edges and form towers and turrets (figure 8.3). Glaciated clouds, composed of ice crystals, have a diffuse or filmy appearance.

Figure 8.3 The altocumulus clouds in the foreground and background of this photo have sharp outlines and high-contrast edges, indicating that they are composed of water droplets. A left-to-right strip of clouds in the middle of the altocumulus deck exhibits the filmy, diffuse appearance of ice clouds. Here, the altocumulus have become glaciated, probably because ice crystals have fallen into the altocumulus from a higher cloud. Light from the sun, which is above the picture, is reflected upward off the ice crystals, producing an optical effect called an undersun. (Photo © C. Whiteman)

Glaciation is also associated with optical effects. A cloud exhibiting a rainbow is composed of liquid water droplets. Optical effects such as halos, *undersuns*, and sundogs are invariably found in ice clouds. Some optical effects—coronas, Brocken specters, and iridescence—can be observed in both ice and water clouds.

Clouds consisting of ice crystals have a diffuse, filmy appearance.

Just as clouds composed of cloud droplets look different from those composed of ice particles, rain falling from a cloud often looks different from precipitation falling in a frozen form. Rain appears dark, whereas frozen precipitation is whitish under most lighting conditions. Figure 8.4 shows the whitish, filmy appearance of a graupel shower. When cloud bases are high and the subcloud air is dry, a precipitation shaft may descend from the base of a cloud but evaporate before reaching the ground below. The streaks of precipitation, either water or ice, that dissipate before reaching the ground (figure 8.5) are called *virga*.

8.5. Spatial and Temporal Distribution of Precipitation

8.5.1. *Effects of Terrain Height on Spatial Variation of Precipitation*

Terrain height is a key factor affecting annual precipitation amounts. Figure 8.6 shows the spatial distribution of mean annual precipitation across the United States based on observations for the years 1931–1960. The increase of precipitation with height is apparent for the Olympic Mountains, Coast Range, Cascade Range, Sierra Nevada, Rockies, Black Hills, and Appalachians. A more detailed look at the variation of mean annual precipitation with elevation is shown for Washington in figure 8.7.

In mountainous areas, precipitation amounts depend primarily on elevation.

Precipitation distribution is affected not only by terrain height but also by proximity to moisture sources, terrain relief, and terrain *aspect* (i.e., the direction a slope is facing) relative to the direction of the approaching wind. A west–east cross section (figure 8.8) through Seattle, Washington,

Figure 8.4 Graupel shower on Trail Ridge Road in Rocky Mountain National Park, Colorado. The whitish appearance of the precipitation is an indication that the precipitation is in frozen form. (Photo © C. Whiteman)

Figure 8.5 Precipitation and virga over the plain west of Mesa Verde National Park, Colorado. Much of the precipitation evaporates before reaching the plain. (Photo © C. Whiteman)

illustrates the relationship between precipitation and underlying terrain. The prevailing westerly winds lift moist Pacific air masses first up the Olympic Mountains and then over the Cascades. The highest precipitation amounts are found slightly upwind of the crests of the mountain ranges, where the relief and, consequently, upward motions are strongest. As air descends the lee sides of the mountain barriers, it warms and dries, creating a rain shadow on the east sides of both mountain ranges. Annual precipitation reaches nearly 150 inches (381 cm) in the Olympics and 110 inches (279 cm) in the Cascades but drops to only 14 inches (36 cm) at Sequim, in the rain shadow of the Olympics, and just 6 inches (15 cm) in eastern Washington, in the rain shadow of the Cascades.

Maps of precipitation distribution in mountainous areas are of limited value when detailed information about a particular site is needed. The accuracy of available measurements is affected by wind, frequent freezing and thawing, and heavy precipitation. In addition, maps are usually based on interpolations between widely dispersed measurement stations. The interpolations rely heavily on terrain height, but may not take terrain steepness, proximity to moisture sources, or the measurement site's exposure to prevailing winds into account.

Switzerland's Rhone Valley, a major interior valley that parallels the crestline of the Alps, has one of the driest climates in Switzerland, despite its high altitude. The valley is surrounded by ridges and high peaks that remove the moisture from air being carried into the valley from any direction. As the air descends into the valley, it warms, and clouds and precipitation dissipate. Similar situations exist in U.S. mountain ranges, affecting precipitation amounts in locations such as California's Central Valley, the Great Basin, North and South Parks, Colorado, and Jackson Hole, Wyoming.

8.5.2. *Day to Day and Diurnal Variations in Precipitation*

Precipitation amounts vary from day to day, depending on a number of factors, including the passage of synoptic-scale weather systems, avail-

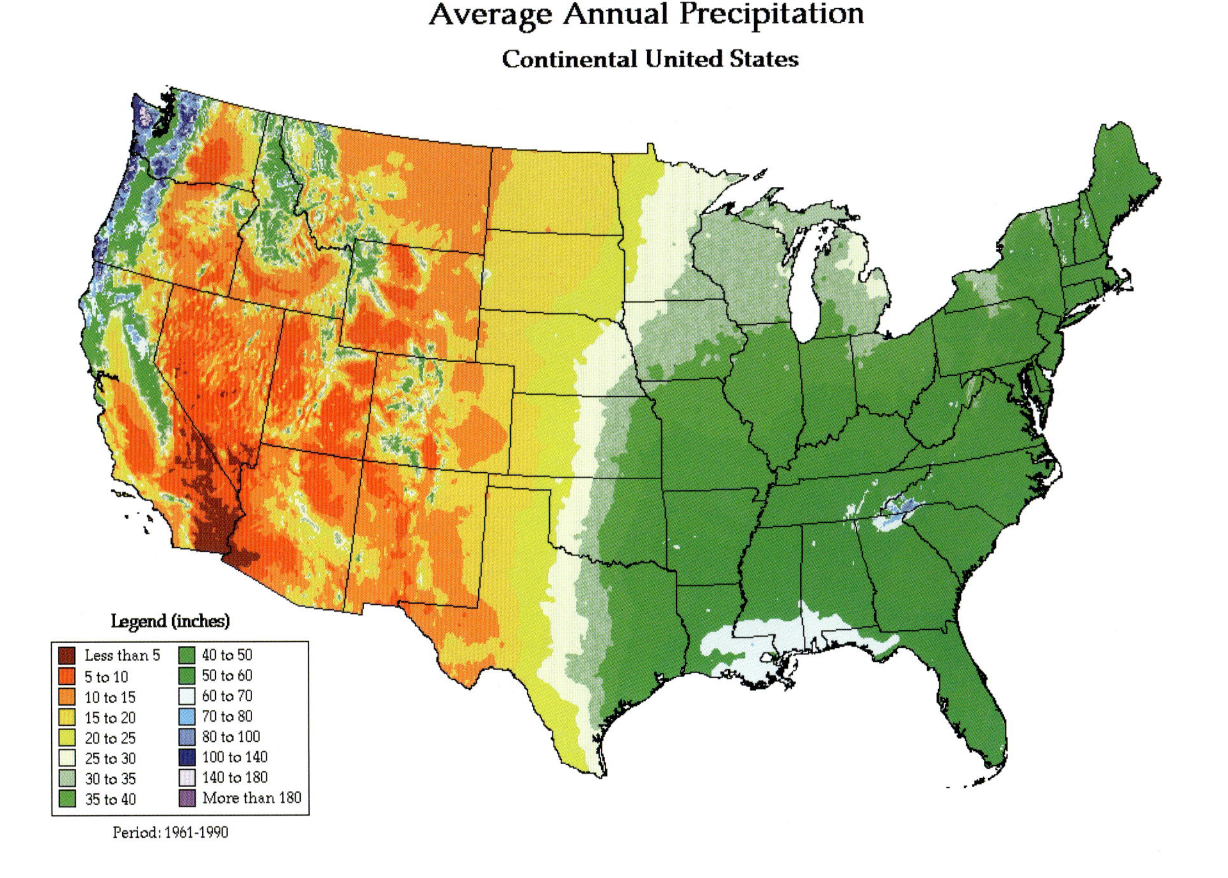

ability of low-level moisture, cloudiness, wind speed and direction, and the exposure and degree of insolation of individual sites. Because so many factors affect precipitation, good quantitative precipitation forecasts are one of the most difficult forecasts in meteorology. Forecasting precipitation is especially difficult in summer when isolated thunderstorms can produce heavy rains in isolated locations while the surroundings receive no precipitation at all.

Figure 8.6 Average annual precipitation in inches, as determined from the PRISM model by Chris Daly, based on 1961–1990 normals from NOAA Cooperative stations and NRCS SNOTEL sites. Modeling sponsored by USDA-NRCS Water and Climate Center, Portland, Oregon. Available from George Taylor, Oregon State Climatologist, Oregon Climate Service.

Diurnal variations in precipitation often reflect diurnal variations in thunderstorm activity. When averaged over a long period of time, thunderstorm frequency over most of the Rockies is highest in the early to late afternoon (figure 8.9), except for a nighttime maximum in southern Arizona, which is apparently associated with moist air brought northward from the Gulf of California by the monsoon. Over the central part of the country, maximum thunderstorm frequency occurs at night in connection with a strong, southerly current of air, the *low-level jet* (section 10.6). The low-level jet develops just above the surface of the sloping Great Plains and brings moist air northward from the Gulf of Mexico. The relatively high frequency of severe summer thunderstorms east of the Rockies is promoted by the presence of this low-level moisture.

Diurnal variations in precipitation often reflect diurnal variations in thunderstorm activity.

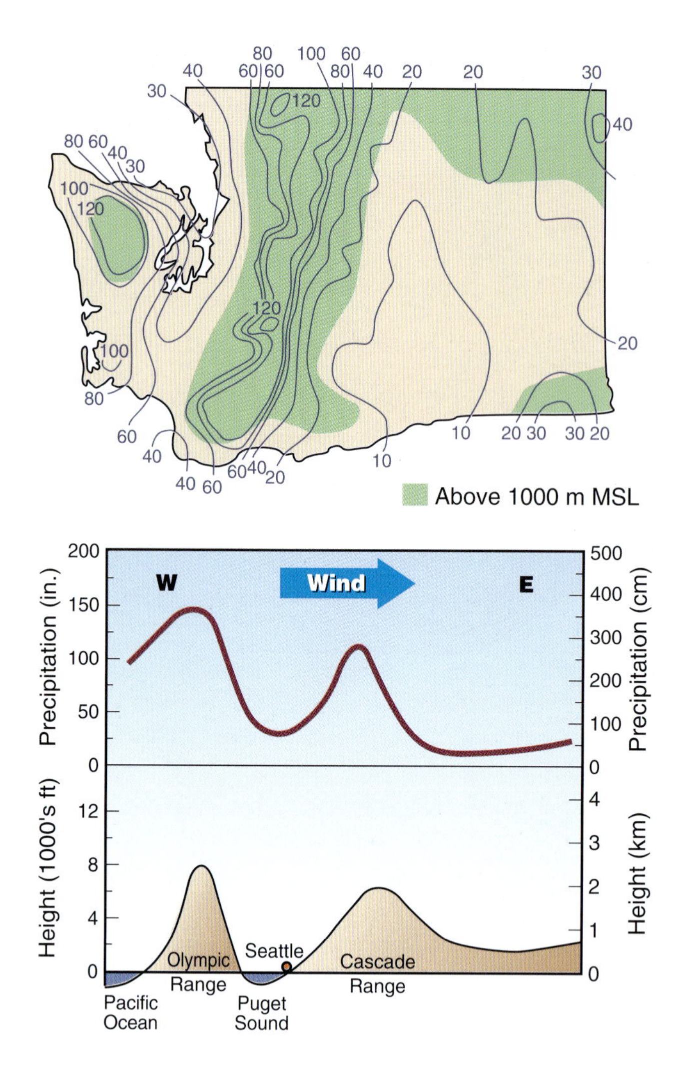

Figure 8.7 (*top*) The mean annual precipitation distribution in Washington illustrates the close relationship between precipitation amount and elevation. The green areas represent the Olympic Mountains (west), the north–south-oriented Cascades (central), and the Blue Mountains (southeast). The contours are labeled with the annual precipitation in inches.

Figure 8.8 (*bottom*) This west–east cross section through Seattle, Washington, shows the increase of precipitation with elevation and the precipitation maxima on the upper windward slopes of the mountains. (Adapted from Riehl, 1965)

8.5.3. *Seasonal Variation of Precipitation and the Monsoon*

Precipitation at a given site often varies seasonally (figure 8.10). The Pacific Northwest, for example, receives much of its precipitation in winter when strong westerly winds carry moist air inland over the mountains. Most inland areas of the mountainous west, however, receive their highest precipitation amounts from late spring or summer convection.

> ON THE GREAT PLAINS: THE DRY LINE
>
> The *dry line* is the roughly north–south boundary between moist air within the Mississippi Valley, flowing northward at low levels from the Gulf of Mexico, and dry air on the western side of the Great Plains, descending from the Mexican Plateau and Southern Rockies. Thunderstorm formation and propagation in west Texas and Oklahoma are often closely associated with diurnal east–west movements of the dry line. The dry line tends to move eastward during the morning and return westward in the evening.

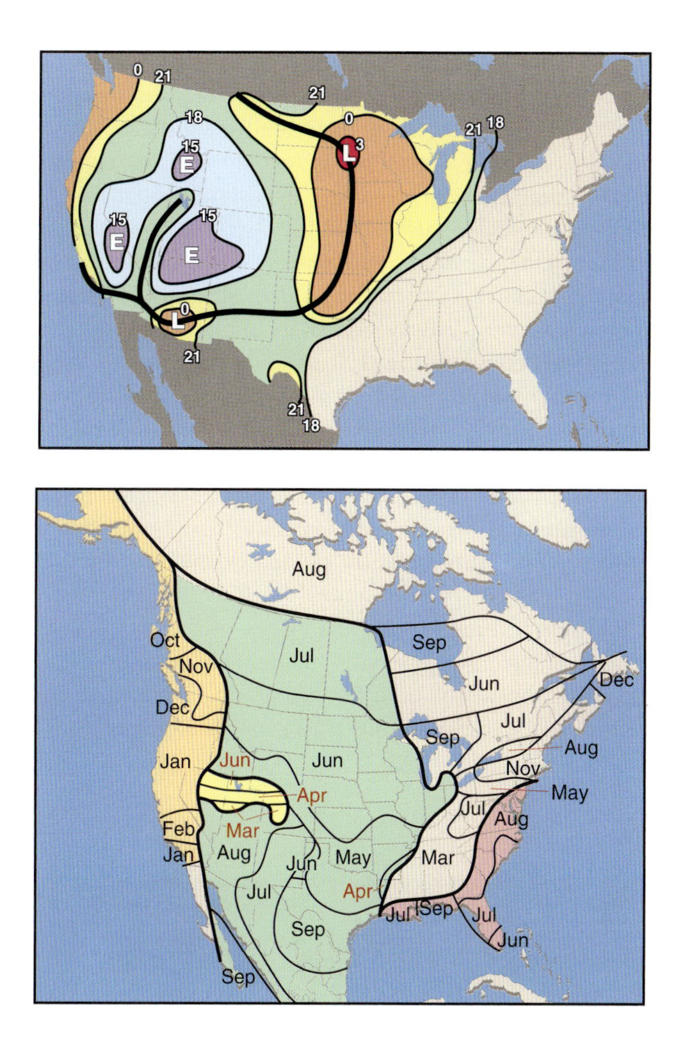

Figure 8.9 (*top*) Local time of day of maximum thunderstorm occurrence. The thin lines are labeled with the hour of the day and the heavy line shows the axis of late thunderstorm occurrence. E = early, L = late. Data were not analyzed for much of the eastern and southern United States. (Data from Wallace, 1975; adapted from Reiter and Tang, 1984)

Figure 8.10 (*bottom*) Geographic distribution of the month of peak precipitation. Colors are used to separate the country into several more or less homogeneous seasonal divisions. (Adapted from Reiter and Tang, 1984)

The month of maximum precipitation in the Appalachians varies with latitude.

Seasonal or monsoonal wind direction reversals between continental and adjacent ocean areas can lead to large seasonal differences in precipitation. Because the annual temperature variation over large land areas is much greater than the temperature variation over ocean surfaces, temperatures in the winter are lower over land than over the ocean. The cold air over land is heavy and dense and thus sinks, causing high pressure areas to develop over the continent and causing winds to blow from land to sea (section 11.4.4). The situation is reversed in the summer when temperatures over land are higher than over the ocean and winds blow from the sea to the land. The summer inflows into the continental areas can lead to heavy precipitation when the moist marine air flows onshore and is lifted up steep mountainsides.

The strongest monsoons develop on the edge of the Tropics where large plateau or upland areas are located near oceans. The Himalayan Monsoon of Southeast Asia is the most familiar example. During winter and spring, the high pressure that forms over the Himalayas causes air to flow southward and descend toward the Indian Ocean, resulting in a winter–spring

Peak precipitation occurs at different times of year in different parts of the country.

dry season. When the Himalayas are heated in the spring and summer, however, a low pressure center forms over the mountains and the pressure gradient reverses, bringing moisture-laden southeast winds into the Himalayas. The resulting monsoon rains replenish water supplies for the Indian subcontinent but can also cause devastating floods.

The Mexican Monsoon (also called the Arizona, Southwest, or North American Monsoon) affects the southwestern United States. Spring and early summer are usually dry in the Southwest, but a change occurs in midsummer when dry westerly winds are replaced by moist southerly winds. The winds come up the Gulf of California, where they pick up moisture from the warm (~29°C or 84°F) water before moving across Arizona and New Mexico (figure 8.11). The moist, low-level wind feeds afternoon thunderstorms that develop over mountain barriers on the east side of the Great Basin, particularly on Arizona's Mogollon Rim (figure 8.12). The moisture is distributed vertically in the atmosphere by the thunderstorms and is subsequently carried throughout the intermountain area and the Southern and Central Rockies by the prevailing mid-altitude and high-altitude winds. The inflow of this moisture is responsible for the summertime precipitation maxima in Arizona, New Mexico, and the southern halves of Utah and Colorado. The almost daily thunderstorms that occur in the southern and central Rockies from late June

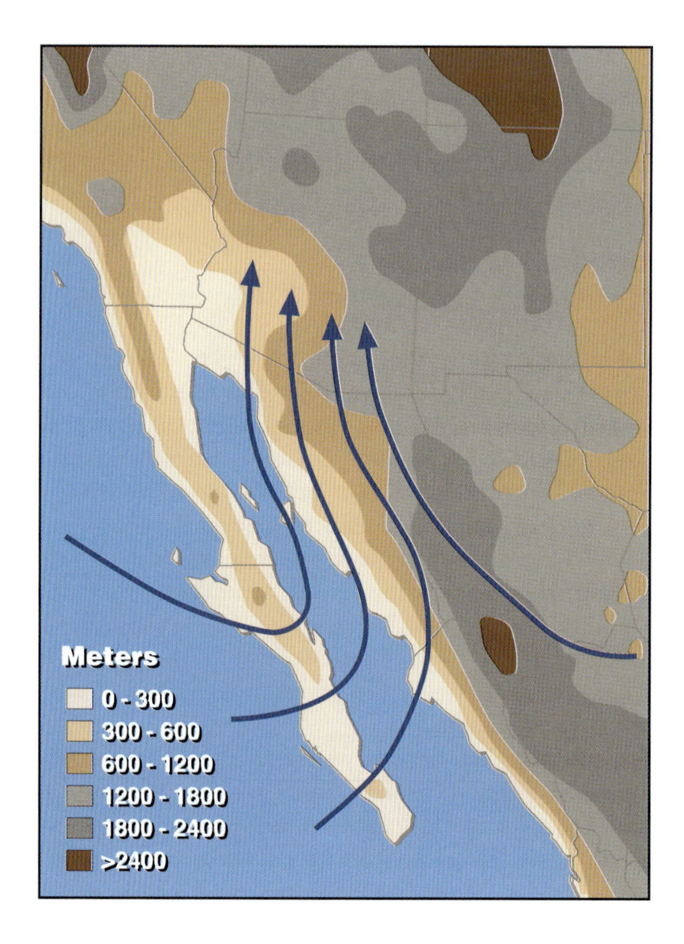

Figure 8.11 Low-level moisture is carried up the Gulf of California into the southwestern United States in the Mexican Monsoon. Shown are typical wind *streamlines* at 450 m (1476 ft) above ground level during the monsoon. Maximum wind speeds exceed 7 m/s or 15.7 mph. Terrain altitudes are shown in the legend. (Adapted from Stensrud et al., 1995)

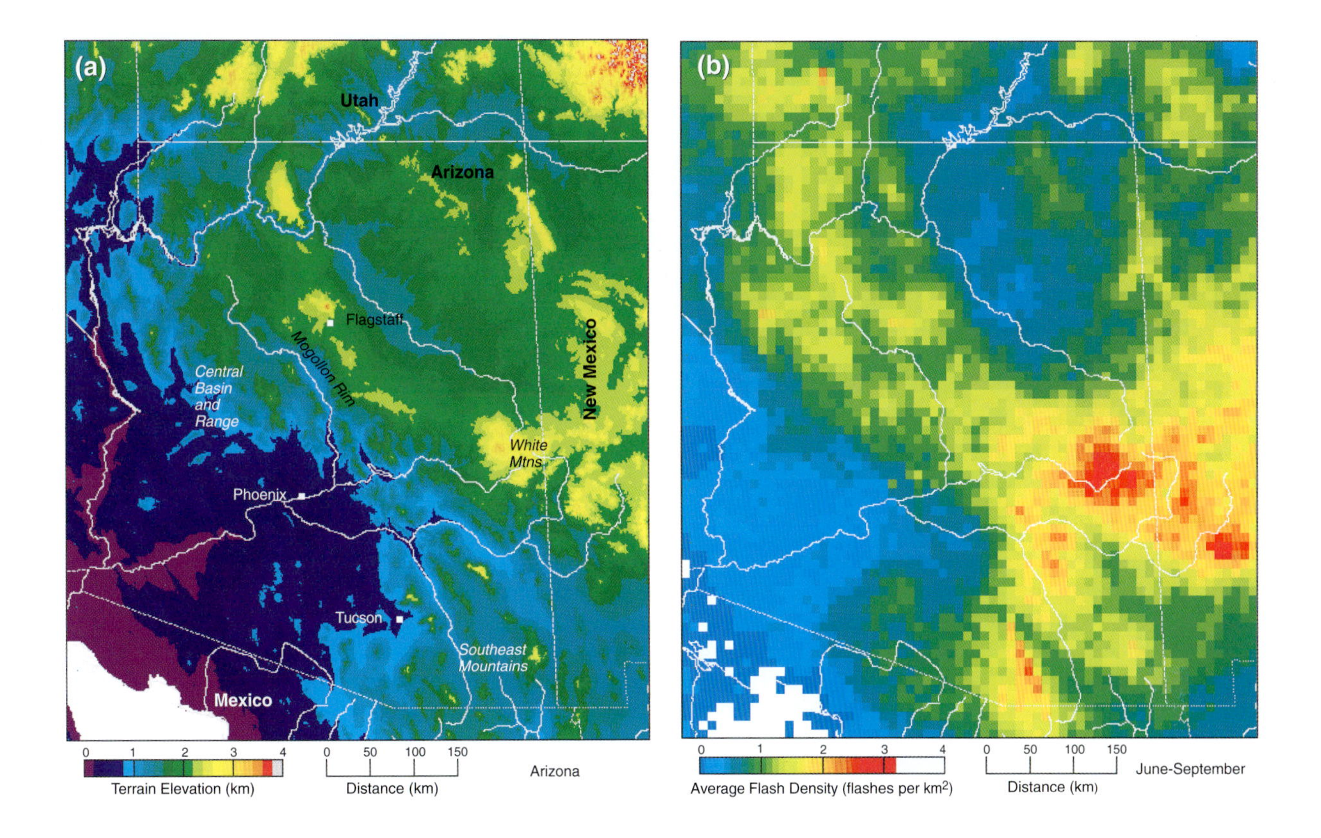

Figure 8.12 The Mexican Monsoon carries low-level moisture into Arizona from the south in summer, producing thunderstorms over the higher terrain. This is illustrated by (a) a topographic map of Arizona, showing the major terrain features, and (b) a composite of mean lightning *flash density* (*flashes* per square kilometer) for June through September as averaged for the years 1984–1993 (1985 and 1986 omitted). Parts (a) and (b) of this figure cover the same geographical area. The maximum flash density is on the Mogollon Rim and in the White Mountains. (From López et al., 1997)

through the end of the summer and the sheets of thick cirrus that can be observed in the afternoon or evening, remnants of thunderstorms that formed upwind earlier in the day, indicate the presence of low-level monsoonal moisture.

8.5.4. *Year to Year Variations in Precipitation*

The year to year variability in precipitation can be very large in continental interiors, especially in dry areas where rain comes predominantly from isolated thunderstorms and where the number of thunderstorms affecting a given location and the amounts of precipitation per thunderstorm can be so variable. Parts of the country affected by frequent traveling storms or by stratiform precipitation usually have a much lower year-to-year variability. Nonetheless, the extreme natural variability of precipitation leads to occasional droughts and floods in most parts of the country.

8.6. Icing

Summertime precipitation in the Southwest and in the Southern Rockies is enhanced by the Mexican or Southwest Monsoon.

When a cloud of supercooled water droplets contacts a solid surface, such as an airplane wing, a tree, or a power line, the water droplets freeze suddenly in a process called icing. If the supercooled droplets are large enough, they spread before freezing to form a layer of *clear ice*. If the water droplets are small, they usually freeze before they spread across the

Figure 8.13 Trees are covered with ice after a freezing rain episode in Fort Collins, Colorado. (Photo © C. Whiteman)

surface, forming a rough and brittle coating of white rime. When strong winds carry many supercooled droplets, rime can deposit as long feathery structures on the upwind sides of flow obstacles. Supercooled drizzle and rain droplets that freeze when they come into contact with cold surfaces, such as streets, trees, and power lines, can form either clear ice or rime. Freezing drizzle and freezing rain can create extremely dangerous traveling conditions. The aftermath of a freezing rainstorm is shown in figure 8.13.

Clear ice forms when supercooled droplets spread before freezing. Rime forms when supercooled droplets freeze on contact with cold surfaces.

Rime is common in areas that have frequent episodes of supercooled clouds with bases below mountain summits, strong winds, and large collecting surfaces, such as evergreen trees or other vertically oriented surfaces. Rime accumulations on middle latitude mountains and in polar regions can be significant and can break tree branches or even power lines. In the eastern Cascades of Washington, rime accumulations augment the annual precipitation in forested areas above 5000 ft (1500 m) by 2–5 inches (50–125 mm) (Berndt and Fowler, 1969).

When rime is deposited on weather instruments (figure 8.14), the accuracy of the temperature, humidity, wind, and other data are affected and sensors may be damaged. Robust instruments must be chosen for locations that are subject to frequently and heavy rime deposition. Wind instruments must be heated to reduce rime accumulation. Anemometer cups that are coated with ice cause the instrument to report wind speeds that are too low. Wind vanes that are coated with ice are slower to respond to changing wind directions and fail completely when frozen in position.

8.7. Mountain Thunderstorms

A *thunderstorm* is a local storm that produces lightning and thunder. It can consist of a single cumulonimbus cloud (section 7.1.1), a cluster of clouds, or a line of clouds. Thunderstorms form when moist, unstable air near the surface is lifted. This lifting can be caused by convection from the heated ground, by the forcing of air up frontal or terrain surfaces, or by the upward motion produced by the horizontal convergence of air

Figure 8.14 Rime deposited on the summit of Mount Washington, New Hampshire, is sometimes so heavy that it obscures almost all the man-made structures on the peak, including the Mount Washington Observatory. (Mt. Washington Observatory photo)

streams. Because thunderstorms are well developed vertically, have a relatively limited horizontal extent, and are carried along by winds, they are transient phenomena that last anywhere from 10 minutes to several hours. Thunderstorms are often accompanied by showery rain and gusty winds and may also bring hail or snow. Thunderstorms occur most frequently during summer, but they are not unknown in the winter in midlatitudes when thunder can sometimes be heard during snowstorms.

Thunderstorms are a common phenomenon in mountainous regions. They are particularly common in the Southern Rocky Mountains during the summer monsoon (section 8.5.3). At Raton, New Mexico, the average number of July thunderstorm days is 29. Colorado Springs, Colorado, averages 26 thunderstorm days in July. In the Northern Rockies, the number of July thunderstorm days is much lower. Missoula, Montana, averages 10, whereas Butte has 15.

The development of a mountain thunderstorm can be triggered by the same mechanisms that trigger thunderstorms over flat terrain. In addition, however, there are two types of terrain-driven mechanisms that can lead to the development of thunderstorms (figure 8.15). Terrain-forced

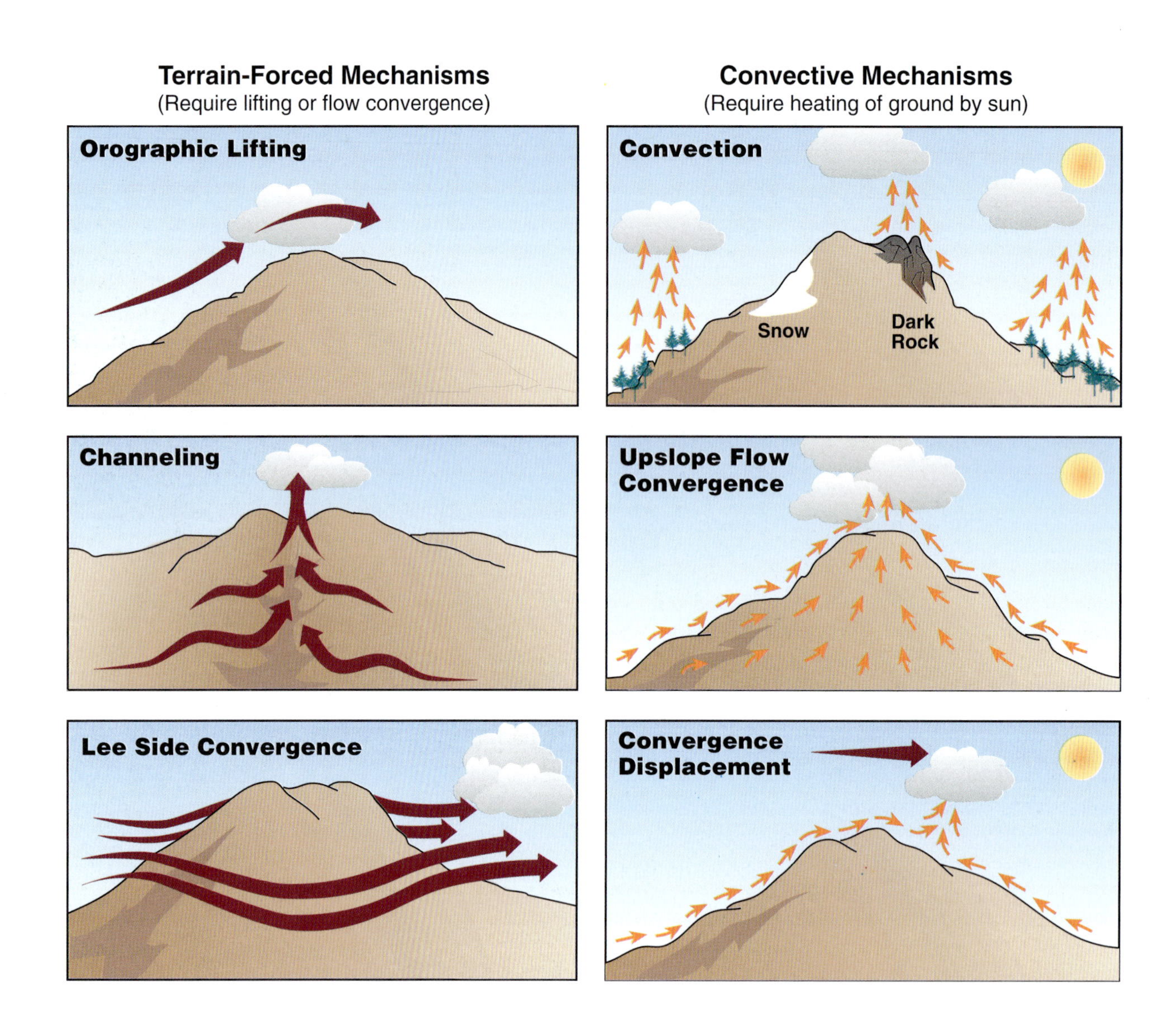

Figure 8.15 Terrain-related mechanisms can produce lifting motions that may trigger thunderstorm formation if sufficient moisture is present and the air is convectively unstable or may contribute to a general enhancement of cloudiness and precipitation.

mechanisms include orographic lifting, channeling, and lee side convergence. Convective mechanisms include convection over surfaces with low albedos and upslope flow convergence over ridges or mountaintops. The convergence zone can be displaced by strong synoptic winds. Terrain-related convergence zones trigger storms rather regularly day after day over the same terrain and are a well-known feature of the meteorology of some regions (figure 8.16). Storms that form routinely over ridges and mountaintops are carried by the prevailing winds at heights midway through the storm's depth (typically 20,000–30,000 ft or 6000–9000 m above sea level). Thunderstorms may also occur randomly over complex terrain, depending on local stability or moisture content and on the presence of a local trigger. They may also be associated with traveling storm systems that affect some stations and not others.

Thunderstorms form routinely over ridges and mountaintops and are then carried by the prevailing winds at midstorm height.

Thunderstorms are categorized as ordinary and severe. *Severe thunderstorms* last longer than ordinary thunderstorms and can produce strong winds, flooding, hail or tornadoes. Ordinary thunderstorms are short lived and do not produce strong winds or large hail. Ordinary thun-

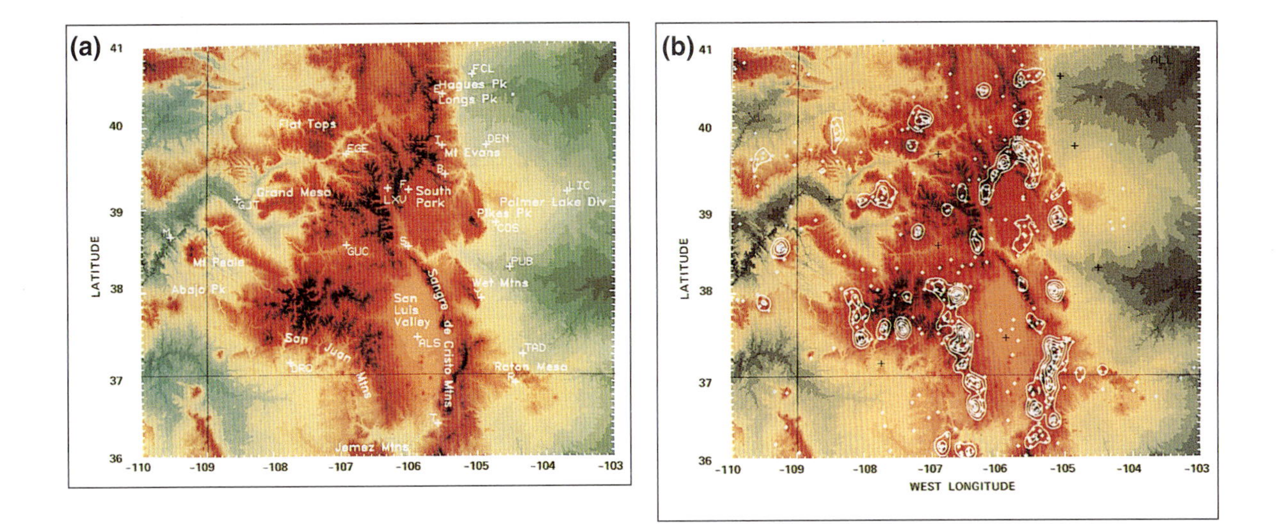

derstorms that form in summer can develop in lines along fronts, or they can be widely scattered throughout an air mass (*air mass thunderstorms*). Frontal thunderstorm development depends primarily on synoptic-scale processes (e.g., the rate of movement of the front, its steepness, the moisture content and stability of the air it displaces). The successful prediction of these thunderstorms depends on the accurate forecasting of frontal positions. The structure and life cycle of these storms is variable because there are often interactions between adjacent cells in a cluster or line of cells. Air mass thunderstorms, in contrast, can occur anywhere within an air mass and are often isolated from other similar thunderstorms. The structure and evolution of air mass thunderstorms is thus more regular.

Figure 8.16 (a) Digital elevation model of the Southern Rocky Mountains with longitude on the x-axis and latitude on the y-axis and (b) density contours for thunderstorm initiation sites as determined from geostationary satellite images for the summer months of 1983 through 1985. Thunderstorms form preferentially over the mountaintops and ridges. The color scheme represents elevations; the lowest elevations (black) are on the east and west sides of the figures. Elevation increases in this order: green, yellow, orange, red. The highest elevations are black plotted against a red background. (From Schaaf et al., 1986)

8.7.1. *Life Cycle of Mountain Air Mass Thunderstorms*

Air mass thunderstorms undergo a three-stage life cycle (figure 8.17). In the *cumulus stage*, warm, moist air rises in a buoyant plume or in a series of convective updrafts. When the air becomes saturated, a convective cloud begins to grow. As the warm air plume continues to rise, more water vapor condenses, releasing the latent heat of vaporization. This heat enhances convection, and cloud turrets form. The cloud edges during this stage are sharp and distinct, indicating that the cloud is composed primarily of water droplets. The convective cloud continues to grow upward, eventually growing above the freezing level where supercooled water droplets and ice crystals coexist.

Air mass thunderstorms go through a three-stage life cycle defined primarily by vertical motion fields in the cloud.

The *mature stage* of a mountain thunderstorm is characterized by the presence of both updrafts and downdrafts within the cloud. The downdrafts are initiated by the downward drag of falling precipitation. Air is entrained into the rain shaft (not shown), and the downdraft is strengthened by evaporative cooling as the precipitation falls into the subsaturated air below the cloud base. The cold descending air in the downdraft often reaches the ground before the precipitation. If the freezing level is

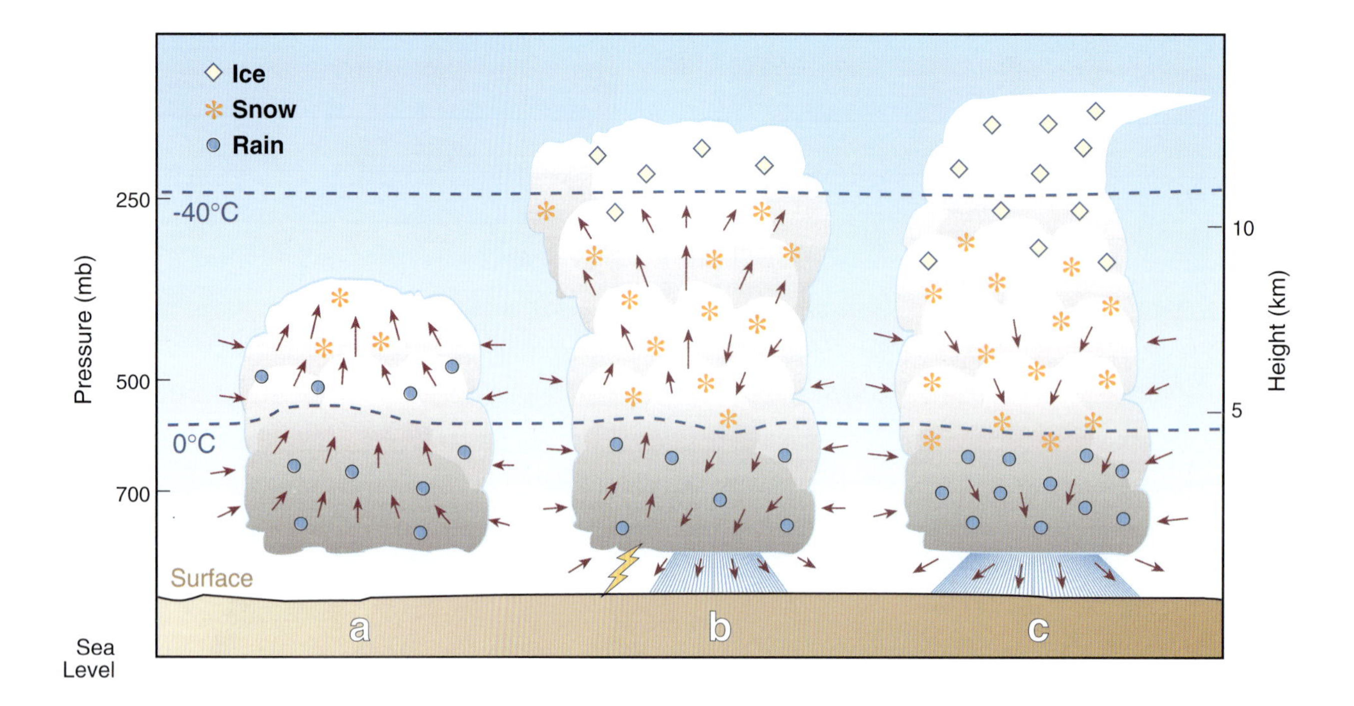

Figure 8.17 The three stages of the life cycle of air mass thunderstorms: (a) cumulus stage, (b) mature stage, and (c) decaying stage. Arrows indicate wind directions. (Adapted from Byers and Braham, 1949)

sufficiently high, the precipitation melts and reaches the ground as rain; otherwise, graupel showers occur. Cloud to ground lightning usually begins when precipitation first falls from the cloud base. As the top of the cloud approaches the tropopause, it starts to flatten out, forming an anvil shape.

The *decaying stage* of the mountain thunderstorm is characterized by downdrafts throughout the cloud. Decay often begins when the supercooled cloud droplets freeze and the cloud becomes glaciated. Glaciation typically appears first in the anvil, which becomes more pronounced in this stage. The glaciated cloud appears filmy, or diffuse, with indistinct cloud edges. The cloud begins to collapse because no additional latent heat is released after the cloud droplets freeze and because the shadow of the cloud reduces insolation at the ground. Sinking motions occur with the fall of precipitation throughout the cloud. An example of a mountain thunderstorm in the decaying stage is shown in figure 8.18.

The decay of a thunderstorm can also be initiated when the precipitation within the storm becomes too heavy for the updrafts to support, when the source of moisture is cut off, or when lifting ceases. The convection necessary for the growth of the thunderstorm can be interrupted by *convective overdevelopment* of the thunderstorm, that is, the storm itself becomes so large that it cuts off insolation. However, it is more common for mountain thunderstorms to decay in the late afternoon or early evening when solar input diminishes, convection wanes, and sinking motions begin to occur over the entire mountain massif. An example of the sudden clearing that occurs with the sinking motions in the late afternoon following a day with many mountain thunderstorms is shown in figure 8.19. Not all air mass thunderstorms decay in the late afternoon or early evening. Intense thunderstorms can form their own circulations,

Figure 8.18 (*top*) This mountain thunderstorm over the San Juan Mountains of Colorado is in the decaying stage. The anvil and most of the clouds in the foreground are glaciated. Heavy precipitation is starting to fall as the storm begins to collapse. (Photo © C. Whiteman)

Figure 8.19 (*bottom*) Thunderstorms and convective clouds can dissipate suddenly in the late afternoon or early evening when convection ends and gentle sinking motions begin over the mountains, as shown here in the Colorado Rockies. (Photo © C. Whiteman)

which feed water vapor into the cloud and convert it into water and ice, thus releasing latent heat that allows the storms to persist into the night.

8.7.2. *Severe Thunderstorms*

Severe thunderstorms occasionally form in mountainous areas. The structure and life cycle of these storms are fundamentally different from air mass thunderstorms. Air mass thunderstorms are more or less erect and can self-destruct when precipitation becomes too heavy for the updrafts to support. The updrafts in severe storms, in contrast, are tilted by strong vertical wind shears. The updrafts tilt over the downdrafts, allowing heavy precipitation to fall out of the updrafts and into the downdrafts. The updrafts can thus continue without having to support heavy suspended precipitation.

Although precipitation (including hail), lightning, and winds can be associated with any thunderstorm, severe thunderstorms tend to produce

Severe thunderstorms, characterized by tilted updrafts, persist for a longer period of time than air mass thunderstorms and can produce more precipitation, more lightning, and/or stronger winds.

more precipitation, more lightning, and stronger winds. Extremely large amounts of precipitation from severe thunderstorms can cause local flooding. In mountains, where terrain channels the flow of water, rocky soil or bedrock keeps precipitation from percolating into the ground, and thunderstorm precipitation rates can be high, *flash floods* can occur. This is especially likely when a large storm is provided with a continuous, low-level moisture inflow and is anchored in place for several hours by the topography or weak upper level winds. Severe thunderstorms can occasionally spawn tornadoes and funnel clouds. A rare sighting of a high-elevation tornado on Long's Peak, Colorado, was described by Nuss (1986).

THE BIG THOMPSON FLOOD

On 31 July 1976, the eve of Colorado's centennial celebration, a flash flood tore through the Big Thompson Canyon between Loveland and Estes Park. The flood claimed 139 lives (an additional six people were never found) and caused damage in excess of $35 million.

An unusually large thunderstorm developed near Estes Park as moist air from the south and colder polar air from the north met along Colorado's Front Range. Because westerly winds aloft were weak and surface winds were from the east, the storm was held in place and did not move eastward onto the plains. As much as 14 inches of rain fell at some locations during the 6 hours that the storm remained in place.

The Big Thompson Flood and the flood in the Black Hills of South Dakota that killed 237 people on 10 June 1972 both illustrate the three key factors required for heavy flooding in complex terrain:

- moist, low-level winds lifted up the mountains to form thunderstorms
- weak, upper level winds that allow the thunderstorms to remain anchored to the mountains so that heavy precipitation amounts are confined to a single drainage
- a rocky, steep-sided canyon that channels the accumulated rain into the canyon bottom

Strong, localized downdrafts, called *downbursts*, often occur along the leading edge of a severe thunderstorm and presage the arrival of thunderstorm precipitation (figure 8.20). These cool, gusty winds result when downdrafts within the storm are strengthened by evaporative cooling as precipitation falls into the subsaturated air below the cloud base and evaporates. When the downburst hits the ground, it fans out radially in all directions (figure 8.21). A downburst extending less than 4 km is called a *microburst*. The straight-line winds from microbursts have been known to exceed 75 m/s (168 mph) and thus can be as strong as rotating tornadic winds. The strongest microbursts can also generate a roaring sound that is usually associated with tornadic winds.

Downbursts are hazardous to aircraft, especially during takeoff and landing. A plane encounters strong headwinds as it enters the downdraft and strong tailwinds as it exits. Downbursts can also cause fires to flare up or change direction suddenly and may cause forest *windthrows* or *blowdowns*, such as the one shown in figure 8.22. The damage from down-

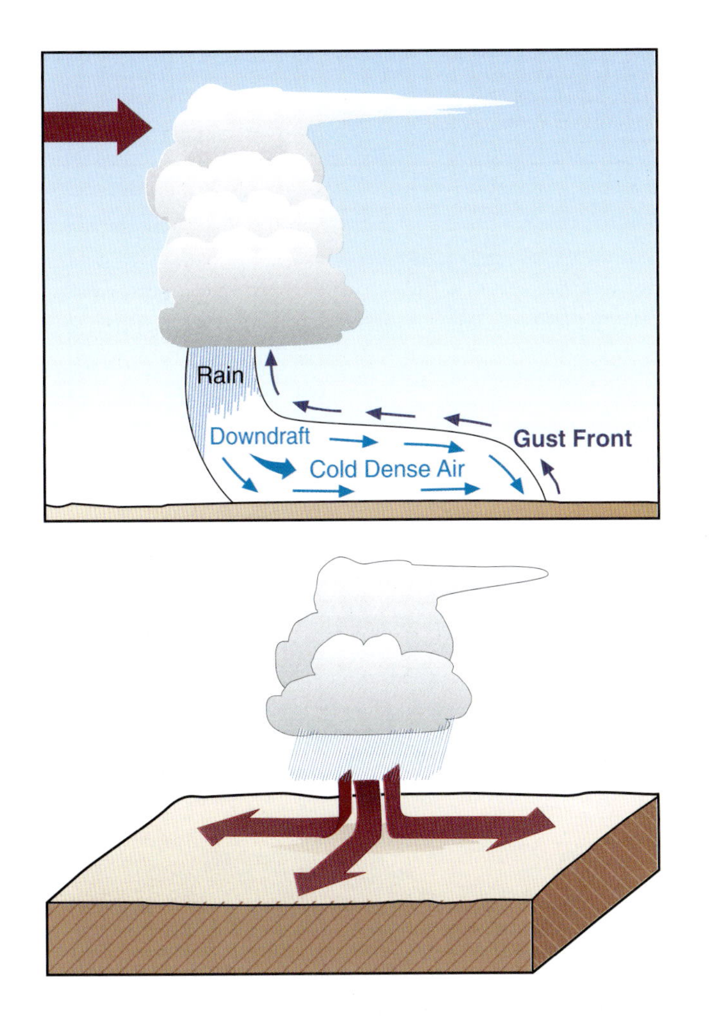

Figure 8.20 (*top*) Downdrafts often travel ahead of thunderstorms, presaging the arrival of the thunderstorm precipitation. The advancing cold air is similar to a cold front, causing the warmer air ahead of it to rise.

Figure 8.21 (*bottom*) Thunderstorm downbursts produce very strong winds whose direction can shift suddenly with distance.

burst winds occurs in a straight line pattern that is quite different from the swirling patterns of damage produced by tornadoes. (Many blowdowns are caused not by sudden downbursts but by high sustained or gusty winds. Foehn winds (section 10.2.2) are responsible for many blowdowns on the lee sides of mountain ridges. The occurrence of blowdowns depends not only on wind strength, but also on tree rooting depths, soil moisture, accretion of snow and ice on branches, and the maturity and health of the forest.)

8.7.3. *Lightning*

Lightning is an electrical spark that occurs in the atmosphere. Lightning may be confined to one cloud, or it may travel from cloud to cloud, from cloud to air, or from cloud to ground. The electrical discharge heats the air through which it travels, causing the air to expand explosively, thus producing thunder. It is estimated that 100 lightning flashes occur each second somewhere on earth, adding to nearly 8 million lightning flashes per day. Lightning flashes between cloud and ground represent only 20% of all lightning flashes, but they cause more than 100 deaths per year, set tens of thousands of fires that destroy more than $50 million of timber

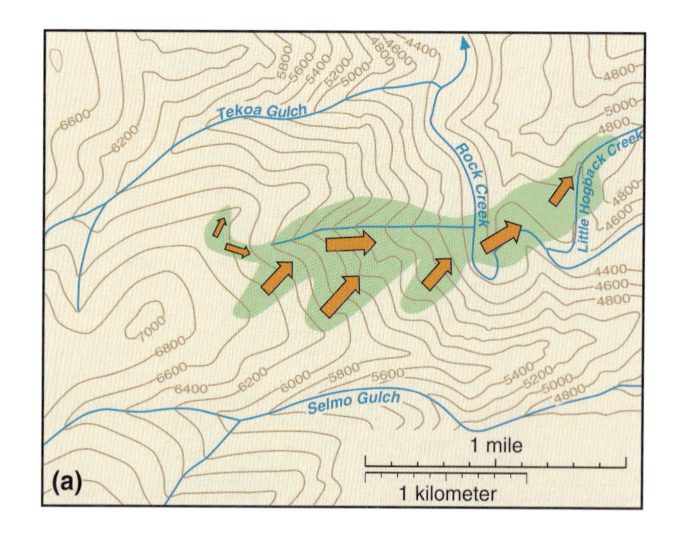

Figure 8.22 A forest windthrow in Montana's Rock Creek Valley was caused by a microburst from a severe thunderstorm on the afternoon of 16 June 1989. (a) The microburst began high on the west sidewall of the Rock Creek Valley, and the descending wind was channeled eastward 2000 ft down an unnamed drainage to the floor of Rock Creek and up the opposite sidewall. The windthrow damage ended when the momentum was dissipated as the cold air climbed Little Hogback Creek. (b) This photo shows a portion of the windthrow damage path on the west sidewall. (c) This photo shows the damage at the floor of the Rock Creek Valley, where the microburst crossed a road. (U.S. Forest Service Lolo Ranger District photo provided by D. Stack)

each year, and cause an additional $40 million in property damage per year in the United States alone. Lightning and thunder are primarily associated with thunderstorms, but they may also be produced in nimbostratus, in snowstorms, in dust storms, and even in volcanic eruptions (figure 8.23).

The distance to a lightning flash in miles can be determined by dividing the seconds that elapse between the lightning stroke and the thunder by five.

WHY THUNDER RUMBLES

Thunder is the sound produced by the rapidly expanding gases along a lightning discharge channel where air is instantaneously heated to temperatures near 10,000°C. In the immediate vicinity of a lightning stroke, thunder is heard as a sharp and loud cracking sound. Thunder is seldom heard at distances of more than 15 miles from the lightning discharge. An upper limit to the distance thunder will carry is 25 miles, and a more typical value for the range of audibility is 10 miles. At these greater distances, however, thunder has a characteristic low-pitched sound caused by the strong attenuation of the high-frequency components of the original sound. The rumbling is caused by echoing and by the varying arrival times of the sound waves that originate from different parts of the lightning channel and from the multiple strokes that occur within or near the original channel.

You can listen to thunder after lightning and tell how close you came to getting hit. If you don't hear it you got hit, so never mind. —elementary school student

As a thunderstorm grows, intense electrical fields develop within it: a large positive charge in the frozen upper part of the cloud and two charge regions—a large negatively charged region and a smaller positively charged region—in the lower part of the cloud. A thunderstorm also affects the electrical charge of the ground below it. The ground normally maintains a small negative charge with respect to the atmosphere, but when a thunderstorm drifts overhead, the negative charge at the cloud base induces a positive charge on the ground below the storm. The positive ground current follows the movement of the cloud like a shadow and

Figure 8.23 A time-exposure photograph captures several lightning flashes from a nighttime thunderstorm that moved eastward off the Colorado Front Range at Fort Collins, Colorado. (Photo © C. Whiteman)

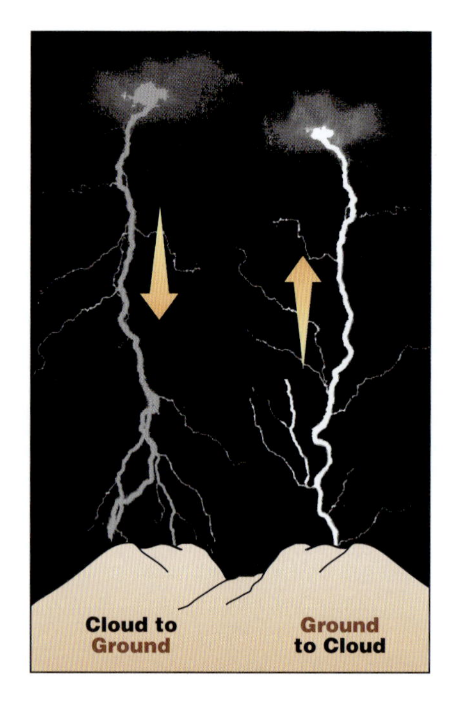

Figure 8.24 The dendritic branching patterns of lightning differ for cloud-to-ground and ground-to-cloud strokes. The rarer ground-to-cloud strokes, which are often more luminous and more damaging, can be recognized by their upward branching pattern.

concentrates on elevated objects, such as trees, buildings, and terrain projections, in an attempt to establish a current to equalize the charges between cloud base and ground. Air, however, is a good insulator, and the electrical potential between cloud and ground must build up to levels of tens to hundreds of millions of volts before the insulating properties of the air break down and an ionized conductive channel is established for current to flow between the two charges.

Lightning is usually initiated within the thunderstorm cloud when a faint, negatively charged channel called the *stepped leader* emerges from the base of the cloud and propagates toward the ground in a series of steps of about 1 microsecond in duration and 150–300 ft (50–100 m) in length. The individual steps are usually separated by pauses of about 50 microseconds. The stepped leader reaches from cloud base to ground in about a hundredth of a second. As the stepped leader approaches the ground, streamers of positive charge rush upward from objects on the ground. When one of the streamers contacts the leading edge of the stepped leader some 150–300 ft (50–100 m) above the ground, the lightning channel is opened, negative charge starts flowing to the ground, and a *return stroke*, lasting about a tenth of a second, propagates up the channel as a bright luminous pulse. Sometimes, following the initial return stroke, one or more additional leaders may propagate down the decaying lightning channel at intervals of about a tenth of a second. These leaders, called *dart leaders*, are not stepped or branched like the original leader, but are more or less direct and continuous. Like the stepped leader, however, they initiate return strokes from the ground. All of the strokes that follow essentially the same path are called a single cloud-to-ground flash.

The cloud-to-ground flashes usually transport negative charge from the cloud to the ground. However, a small fraction of cloud-to-ground lightning flashes (typically 5–15%, although variable by region) have

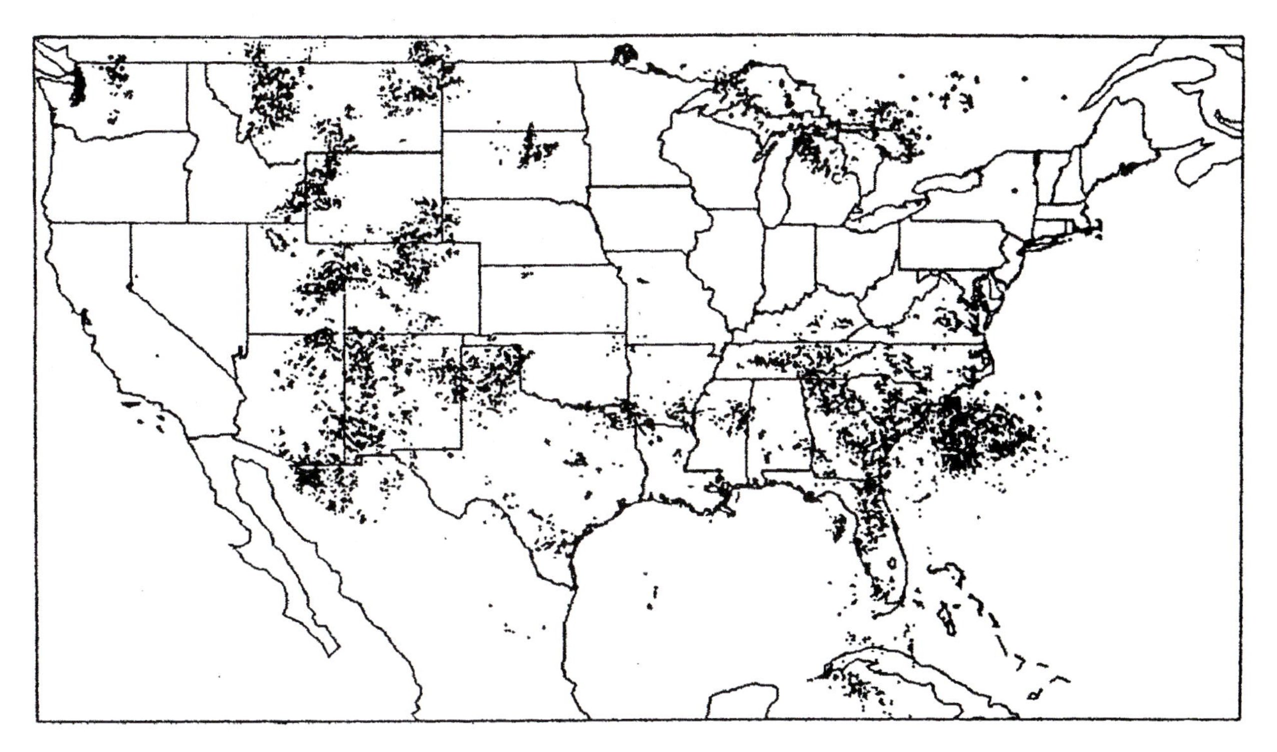

Figure 8.25 Cloud-to-ground lightning flashes over the continental United States and adjacent regions during 16 hours of a summer day (43,776 flashes). (From Holle and López, 1993)

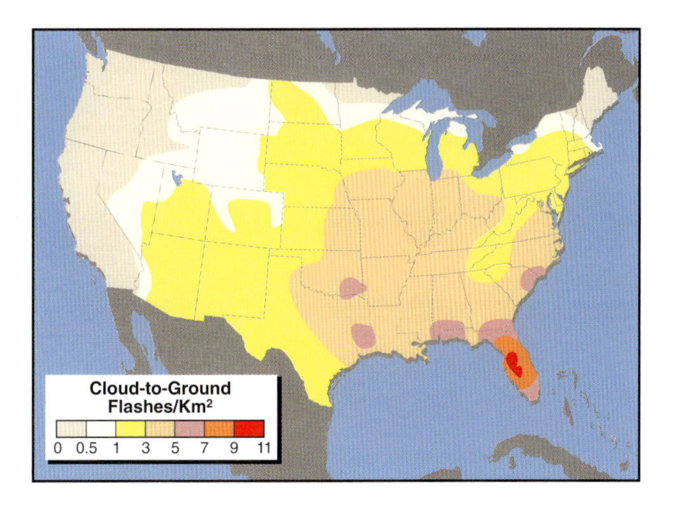

Figure 8.26 Mean annual cloud-to-ground lightning flash density during the years 1989 through 1991. Mountain areas have relatively low annual flash densities compared to the Southeast because few thunderstorms occur in the colder half of the year. Nevertheless, summer flash densities are high in the mountains, particularly at the highest elevations or areas with greatest relief where thunderstorms are often initiated. (Adapted from Orville, 1994)

stepped leaders that are positively charged. These positive strokes are solitary strokes with no subsequent leaders and can be recognized by the upward orientation of the branching pattern of their stepped leaders (figure 8.24). Positive strokes generate higher and more continuous currents than negative strokes and thus cause greater damage to power lines and may have a greater potential for igniting forest fires.

An automated system for counting cloud-to-ground lightning strokes and for determining their locations by detecting their electromagnetic emissions has been developed in the United States, providing valuable data on U.S. electrical storms. Multiple network instruments distributed

LIGHTNING SAFETY

Lightning can be expected from a growing summertime convective cloud after virga is first seen falling from the cloud or when lightning is seen or thunder heard from other clouds in roughly the same stage of development.

The distance to visible lightning can be estimated by counting the elapsed time between the sight of the lightning and the sound of the thunder. Although the light from a lightning flash reaches an observer almost instantaneously, the sound of the flash travels much more slowly. It takes approximately 5 seconds for the sound to travel a mile.

The speed and direction of movement of a thunderstorm can be estimated by tracking the positions of successive lightning flashes. For example, suppose a storm is approaching from the west and the first flash is 20 seconds (4 miles) away. If a second flash occurred 10 minutes later on the same bearing and is 15 seconds (3 miles) away, the storm is moving eastward at 6 mph and could be expected overhead in 30 minutes. Because large thunderstorms frequently have multiple cells, the successive flashes used to estimate speed must come from the same cell.

Direct strokes are most likely to strike high peaks, ridges, isolated tall trees, or other terrain projections (figure I) because a sharp, projecting point of topography concentrates electrical charge and promotes the ionization of air around it. Direct strokes can be fatal. Protection is afforded by getting off ridges and peaks and staying out from under isolated trees when lightning is imminent. Because projections serve as lightning rods which attract the lightning strokes, they offer protection from direct strokes to surrounding lower lying areas.

Most lightning injuries are caused not by direct strokes, but by *ground strokes*. A ground stroke is the current that propagates along the ground from the point where the direct stroke hits the ground. Victims may suffer burns or unconsciousness. Victims who lose vital signs may sometimes be revived by cardiopulmonary resuscitation, if action is taken quickly.

To avoid ground strokes:

- **DO** get inside a home, a large building, or an all-metal vehicle.
- **DO NOT** touch or use equipment attached to electrical or phone lines.

(continued)

LIGHTNING SAFETY (*continued*)

- If caught outdoors, **DO** squat on an insulating material (e.g., rope, rubber-soled climbing boots, pack) on lower elevation terrain that is protected from direct strokes by ridges, distant trees, or terrain projections and is out of likely low-resistance paths that could act as channels for the ground stroke (figure II, left illustration).
- **DO NOT** lean back against rock walls or other surface features that are closer to the direct stroke than your feet because a ground stroke could travel through your torso as a preferred path of lower resistance (figure II, right illustration).
- **DO NOT** take shelter under tall, isolated trees. Approximately one-quarter of all lightning fatalities are persons seeking shelter under trees.
- **DO NOT** take shelter in caves (figure III), shallow depressions (figure IIIb), under large boulders (figure IIIc), or under overhangs or in wet crevices or watercourses that lead upward (figure IIId). The ground stroke will seek the path of lowest resistance, which in these situations could be your body.

Figure I (*top*) Lighting frequency depends on local topographic relief. The highest peaks and the ridges receive the most frequent lightning strokes. (From Peterson, 1962)

Figure II (*bottom*) Ground strokes travel on the surface from the point where the direct stroke reaches the ground. Protection from ground strokes is afforded by creating a high-resistance path (left), rather than creating an alternate low-resistance path through the body (right). (From Peterson, 1962)

Figure III (*right*) Lightning hazards can occur (a) in caves, (b) in shallow depressions, (c) near isolated blocks, and (d) under sheltering overhangs or near crevices that lead upward. (From Peterson, 1962)

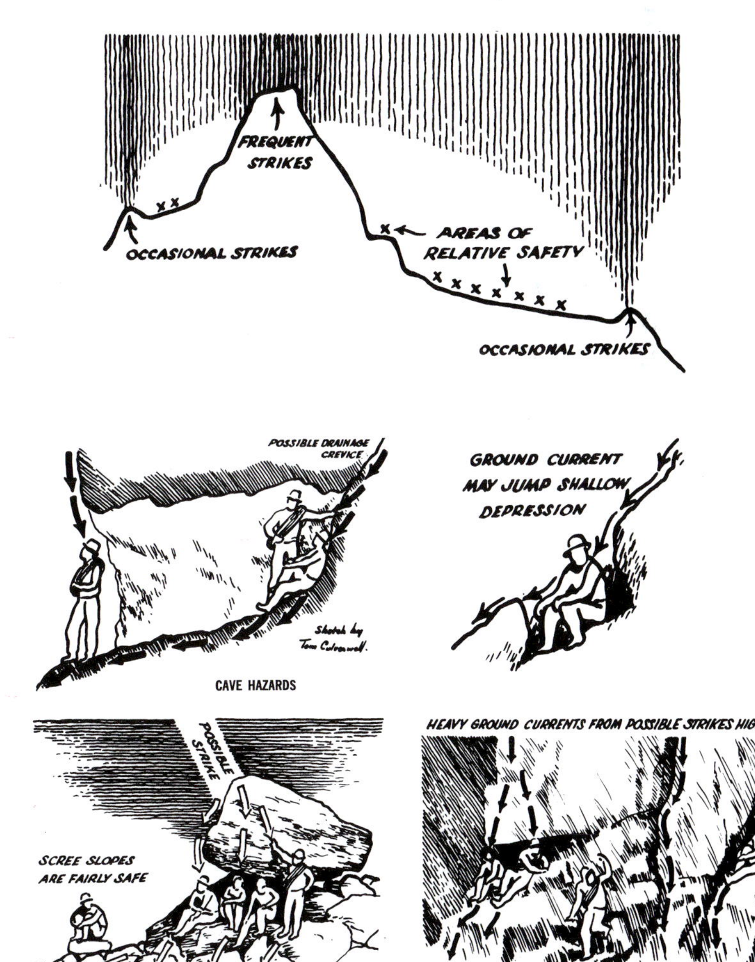

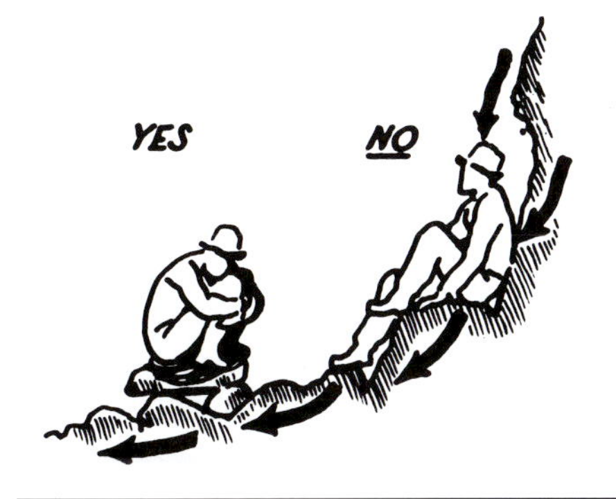

over the country at a spacing of about 150–180 miles (250–300 km) detect the characteristic waveforms that emanate from the return stroke current at a point approximately 300 ft (100 m) above the ground. Radio direction finding is used to identify the positions of the return strokes, with accuracies of 1.2–2.4 miles (2–4 km) over homogeneous country. (The accuracy of this technique in mountainous terrain, where reception is restricted by surrounding mountains, has not yet been quantified.) Data from all network instruments are collected at a central processing site, where algorithms are used to determine the times and locations of the detected strokes. Multiple strokes are combined into a count of lightning flashes. Additional information, including the number of return strokes in the flash, the time and location of the flash, and the peak signal amplitude of the first return stroke, is appended to the lightning stroke record. These data have been used to plot the number and position of lightning flashes in individual storm outbreaks (figure 8.25) and to determine the climatological characteristics of lightning flashes in the United States (figure 8.26). Because the data are processed immediately as they are collected, they are useful to weather forecasters and fire fighters, as well as to electric utility crews responsible for keeping electric power grids in operation.

9 Weather Maps, Forecasts, and Data

9.1. Weather Maps

Weather maps prepared by the National Weather Service summarize and synthesize weather data to provide a comprehensive picture of weather conditions at a given time. They are the basis of weather maps used on television to show precipitation, high and low pressure centers, and fronts. Weather maps are produced using both surface data and data from specified pressure levels. Data are plotted and contoured by computer, and analysts use satellite photos, satellite video loops, weather forecast models, and extrapolations from previous frontal and pressure system analyses to locate fronts and pressure centers. An example of a surface weather chart is presented in figure 9.1. A 500-mb chart for the same date and time was presented in figure 5.1.

Symbols are used on weather maps to indicate synoptic-scale features. High and low pressure weather systems (*highs* and *lows*) are indicated by the letters H and L, with isobars labeled in millibars. Lines indicating frontal positions (section 6.2) represent the position on the ground of boundaries between air masses. Additional meteorological variables, such as temperature, are often analyzed on the same map using dashed or colored lines.

Pressure, temperature, and other data from the reporting stations are plotted in coded form at the station locations. A *station model* specifies the positions in which different types of data are plotted relative to the station location. Figure 9.1 used an abbreviated station model. A complete station model is shown in figure 9.2. Figures 9.3–9.7 show additional symbols used in station models to indicate total sky cover, winds, pressure tendency, cloud types, and present weather types, respectively.

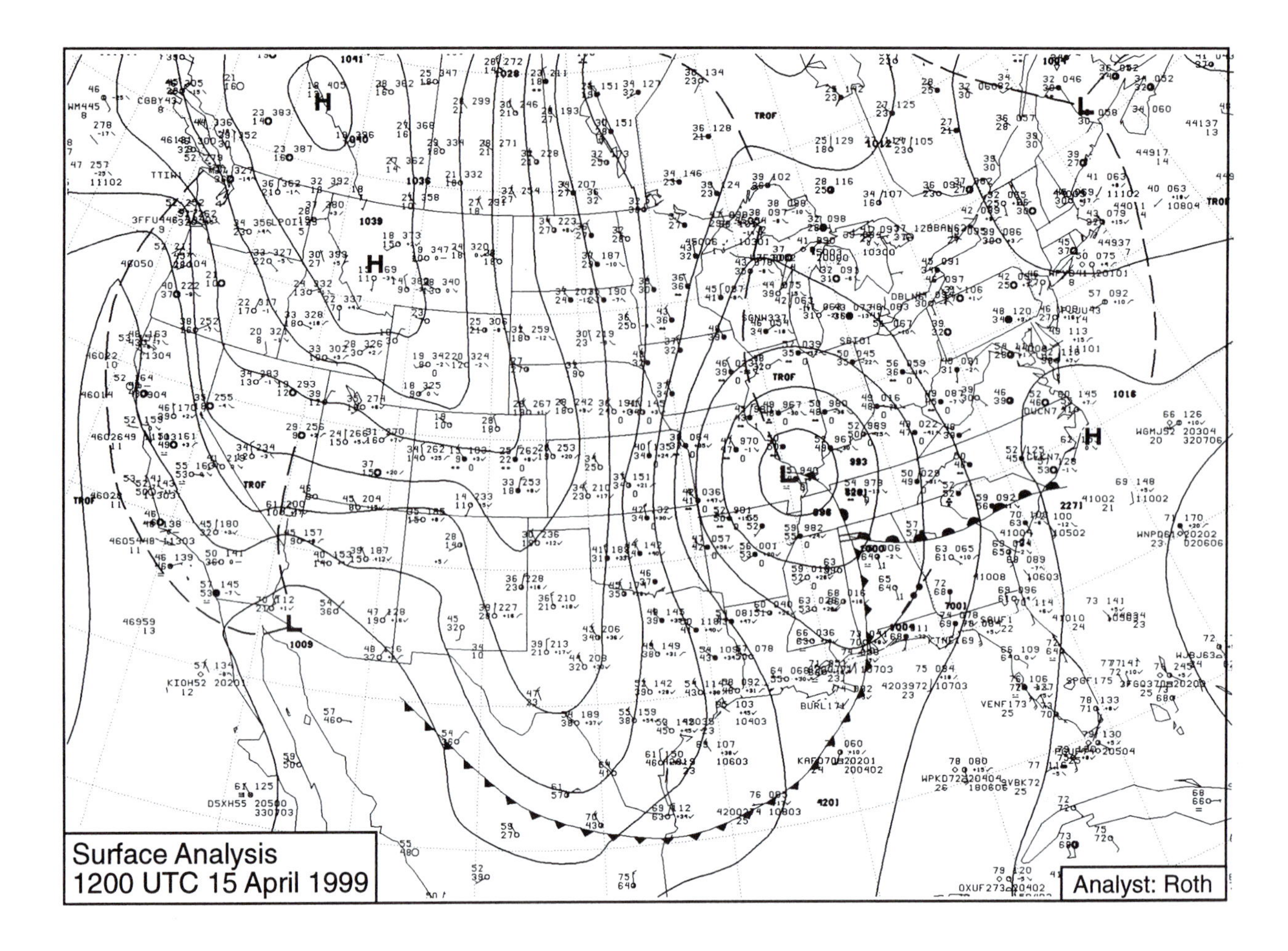

Figure 9.1 This surface weather chart for 15 April 1999 at 1200 UTC (7 A.M. EST) shows a major low pressure center over southeastern Missouri and a major high pressure center over southwestern Canada. Isobars are labeled in millibars, frontal positions are shown by the heavy lines (lines without frontal symbols are developing or decaying low pressure troughs), and weather data are plotted at selected weather stations using the station model described later in this chapter. (From National Oceanic and Atmospheric Administration)

The rates of movement of fronts and pressure centers can be determined by tracking them on successive surface charts.

9.2. Forecasting Guidelines

A surface weather chart, with the symbols indicating fronts and high and low pressure centers and the information included in the station model, provides a snapshot of synoptic-scale conditions at ground level. By overlaying charts for several pressure levels (section 5.1.3), changes with altitude can be identified and the three-dimensional structure of the atmosphere at a given point in time can be visualized. By comparing consecutive charts, the rate of movement of fronts and the rates of development of high and low pressure centers can be determined. Frontal positions and high and low pressure centers can be recorded on a transparent acetate sheet using different color grease pencils. The key to forecasting is to relate the synoptic-scale information available on weather maps to local conditions. Table 9.1 provides forecasting guidelines.

The forecasting hints in table 9.1 depend primarily on knowledge of the weather associated with highs, lows, fronts, and air masses and on information provided by weather maps on the direction and rate of movement of these synoptic-scale weather features. There are cases, however, when no synoptic-scale weather information is available and forecasts must be made in the field based on only a few weather observations. The

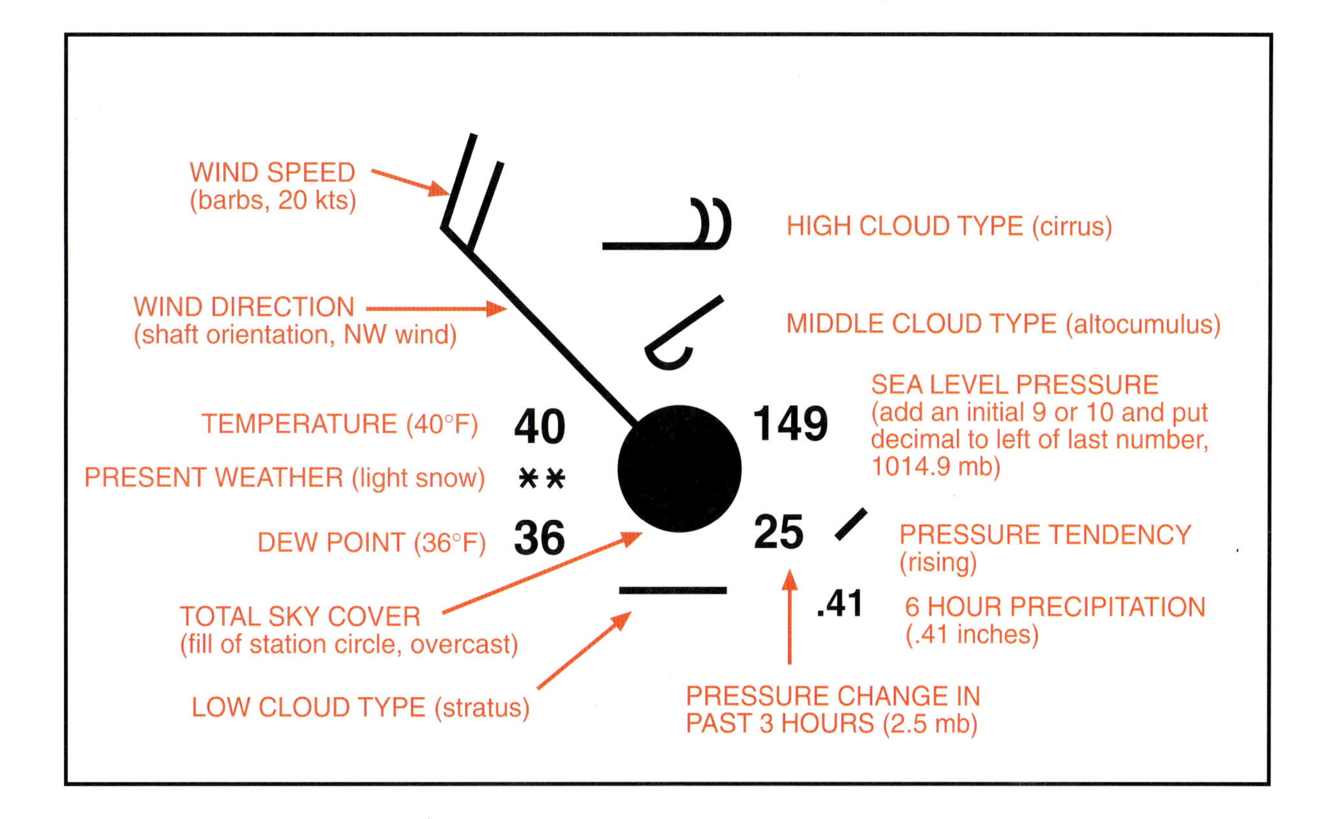

Figure 9.2 Station model used on surface weather charts. In this station model, temperature and dew point are in degrees Fahrenheit, the 6-hour precipitation is in inches, the 3-hour pressure change is in tenths of millibars, and the sea level pressure is written in tenths of millibars, but omitting the most significant digit. Keys for the symbols used for total sky cover, wind speed and direction, pressure tendency, cloud types, and present weather are presented in figures 9.3–9.7.

following Point of Interest provides guidelines for forecasts based on limited observations.

9.3. Weather Information: Data Collection and Dissemination

9.3.1. *Weather Data Collection*

You can observe an awful lot by just watchin'.

—Lawrence Peter (Yogi) Berra

Weather forecasts are based on sounding data collected by a global network of balloon-borne radiosondes. Radiosondes are released simultaneously twice per day at 0000 and 1200 Coordinated Universal Time at sites around the world, providing information on synoptic-scale weather systems. (Coordinated universal time, abbreviated UTC by international agreement, is the local time observed at Greenwich, England. The local times at sites around the world corresponding to 0000 and 1200 UTC vary depending primarily on longitude but are also affected by political boundaries. Conversions between UTC and local times can be made with the time zone map and algorithms found in appendix D.) The data include vertical profiles of temperature and humidity as a function of height or air pressure and vertical profiles of horizontal wind direction and speed obtained by tracking the balloons with radar or radio-direction-finding systems. The data thus provide a full three-dimensional picture of the atmosphere.

In the United States, radiosonde stations are typically separated by

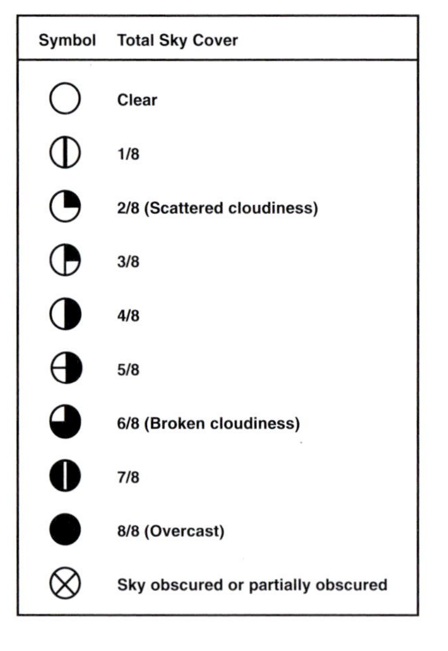

Figure 9.3 Weather map symbols for total sky cover.

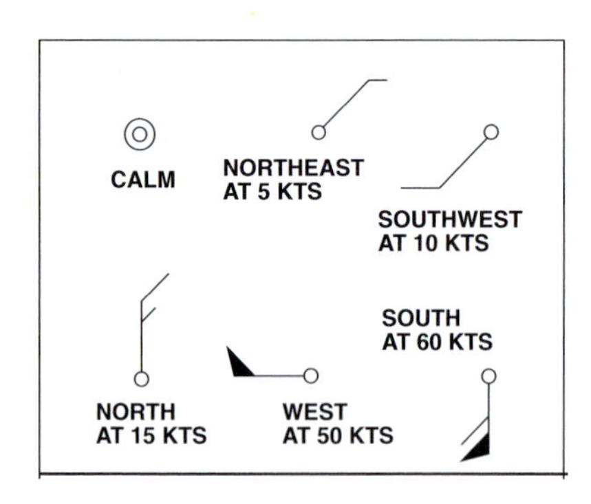

Figure 9.4 Weather map symbols used to indicate wind directions and speeds.

GUIDELINES FOR MAKING FORECASTS BASED ON LIMITED OBSERVATIONS AT A SINGLE STATION INCLUDE:

- Daily minimum temperatures are higher on cloudy nights than on clear nights.
- Low minimum temperatures are promoted by clear skies, light winds, and fresh snow cover.
- Daily maximum temperatures are higher on clear days than on cloudy days. The daily maxima usually rise slightly from day to day during a string of clear days.
- Falling pressure usually indicates the approach of a low pressure trough or low and deteriorating or stormy weather. Rising pressure usually indicates the approach of a ridge or high pressure center and fair weather.
- The appearance of cirrus, cirrostratus, and altostratus clouds, in that order, indicates an approaching warm front and the possibility of general, widespread precipitation. A line of deep convective clouds (usually preceded by cirrus or thunderstorm anvil cirrus) often indicates an approaching cold front and the possibility of isolated showers.
- A counterclockwise shifting of the surface winds from northeast to north to northwest is usually accompanied by clearing skies and cold air advection.
- A counterclockwise shifting of the surface winds from west to southwest is usually accompanied by warm air advection.
- A fair afternoon is likely after radiation fog clears (because radiation fog forms at night under clear skies).
- An unusually rapid buildup of towering cumulus clouds by middle to late morning suggests that afternoon thunderstorms or showers are likely.

distances of about 210–270 miles or 350–450 km (figure 9.8). The data from this network are therefore suitable for observing large-scale traveling weather systems of length scale greater than about 240 miles (400 km) that persist for 12 hours or more. Radiosonde station density is higher in western Europe than in the United States but is much lower in most of the rest of the world.

Radiosonde and other weather data are transmitted by the Global Telecommunications System to meteorological centers where they are analyzed and input to numerical weather forecast models. Personnel at the national meteorological centers and regional forecast offices use the numerical model outputs, any new or supplementary data that were not used by the models, and their regional expertise to produce specialized weather forecasts for public safety, aviation, or other purposes. Supplementary data may include satellite images or upper air data from radar and sodar systems or aircraft.

Surface weather networks have a denser coverage than the radiosonde network and are more suitable for observing weather systems or phenomena occurring on smaller length or time scales. Surface observations, like the upper air observations, are made at exactly the same time every-

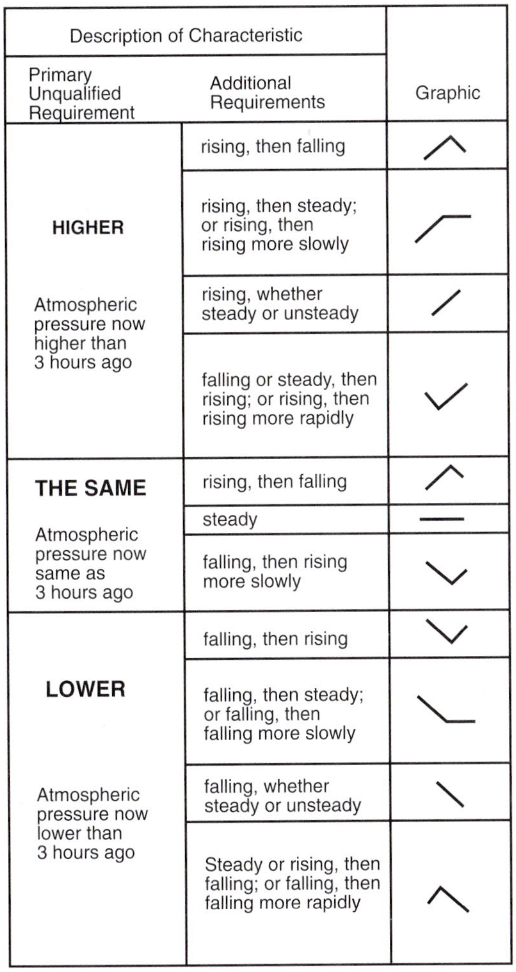

Description of Characteristic		Graphic
Primary Unqualified Requirement	Additional Requirements	
HIGHER Atmospheric pressure now higher than 3 hours ago	rising, then falling	
	rising, then steady; or rising, then rising more slowly	
	rising, whether steady or unsteady	
	falling or steady, then rising; or rising, then rising more rapidly	
THE SAME Atmospheric pressure now same as 3 hours ago	rising, then falling	
	steady	
	falling, then rising more slowly	
LOWER Atmospheric pressure now lower than 3 hours ago	falling, then rising	
	falling, then steady; or falling, then falling more slowly	
	falling, whether steady or unsteady	
	Steady or rising, then falling; or falling, then falling more rapidly	

Figure 9.5 Weather map graphics and definitions for pressure tendency, the change of pressure with time. The graphics illustrate how a pressure trace would look on a barograph.

where on the globe. By convention, they are made on the hour, with times labeled in UTC. Surface weather observations come primarily from NOAA and Federal Aviation Administration surface airways stations. Data are collected and transmitted hourly on national telecommunications networks to support the needs of aviation. Surface weather charts on which these data are plotted are analyzed eight times per day at 3-hour intervals beginning at 0000 UTC.

Other less formal networks of surface observations are located throughout the country to meet the needs of government agencies, industry, research, agriculture, or other special interest groups. These data are typical unavailable in near-real time from the Global Telecommunications System but can be obtained through Web sites, dial-up modems, or from historical data archives. Major networks that provide supplementary meteorological data in the United States and Canada have been described by Meyer and Hubbard (1992). Tucker (1997) has recently summarized information on surface weather observation networks of the western United States, including network history, station locations, types of data collected, observation heights, and data accessibility.

Radiosondes are released twice per day from a global network of stations.

Two networks of automatic meteorological stations operated by the

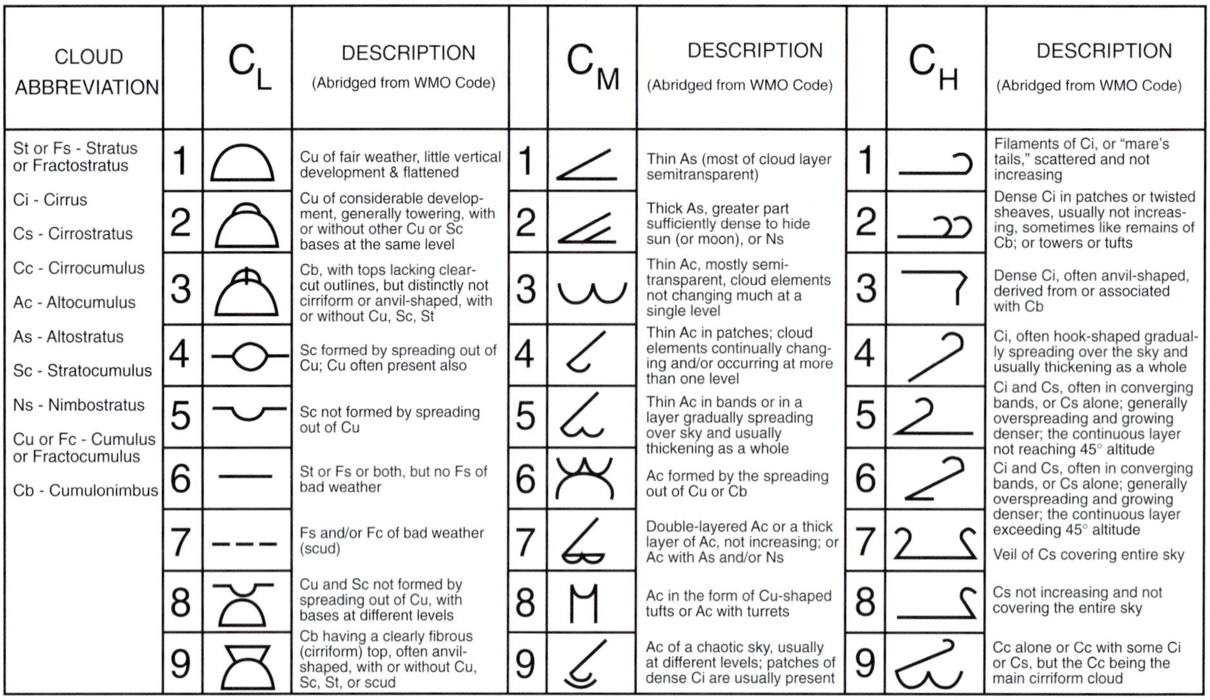

CLOUD ABBREVIATION		C_L	DESCRIPTION (Abridged from WMO Code)		C_M	DESCRIPTION (Abridged from WMO Code)		C_H	DESCRIPTION (Abridged from WMO Code)
St or Fs - Stratus or Fractostratus	1		Cu of fair weather, little vertical development & flattened	1		Thin As (most of cloud layer semitransparent)	1		Filaments of Ci, or "mare's tails," scattered and not increasing
Ci - Cirrus Cs - Cirrostratus	2		Cu of considerable development, generally towering, with or without other Cu or Sc bases at the same level	2		Thick As, greater part sufficiently dense to hide sun (or moon), or Ns	2		Dense Ci in patches or twisted sheaves, usually not increasing, sometimes like remains of Cb; or towers or tufts
Cc - Cirrocumulus Ac - Altocumulus	3		Cb, with tops lacking clear-cut outlines, but distinctly not cirriform or anvil-shaped, with or without Cu, Sc, St	3		Thin Ac, mostly semi-transparent, cloud elements not changing much at a single level	3		Dense Ci, often anvil-shaped, derived from or associated with Cb
As - Altostratus Sc - Stratocumulus	4		Sc formed by spreading out of Cu; Cu often present also	4		Thin Ac in patches; cloud elements continually changing and/or occurring at more than one level	4		Ci, often hook-shaped gradually spreading over the sky and usually thickening as a whole
Ns - Nimbostratus	5		Sc not formed by spreading out of Cu	5		Thin Ac in bands or in a layer gradually spreading over sky and usually thickening as a whole	5		Ci and Cs, often in converging bands, or Cs alone; generally overspreading and growing denser; the continuous layer not reaching 45° altitude
Cu or Fc - Cumulus or Fractocumulus Cb - Cumulonimbus	6		St or Fs or both, but no Fs of bad weather	6		Ac formed by the spreading out of Cu or Cb	6		Ci and Cs, often in converging bands, or Cs alone; generally overspreading and growing denser; the continuous layer exceeding 45° altitude
	7		Fs and/or Fc of bad weather (scud)	7		Double-layered Ac or a thick layer of Ac, not increasing; or Ac with As and/or Ns	7		Veil of Cs covering entire sky
	8		Cu and Sc not formed by spreading out of Cu, with bases at different levels	8		Ac in the form of Cu-shaped tufts or Ac with turrets	8		Cs not increasing and not covering the entire sky
	9		Cb having a clearly fibrous (cirriform) top, often anvil-shaped, with or without Cu, Sc, St, or scud	9		Ac of a chaotic sky, usually at different levels; patches of dense Ci are usually present	9		Cc alone or Cc with some Ci or Cs, but the Cc being the main cirriform cloud

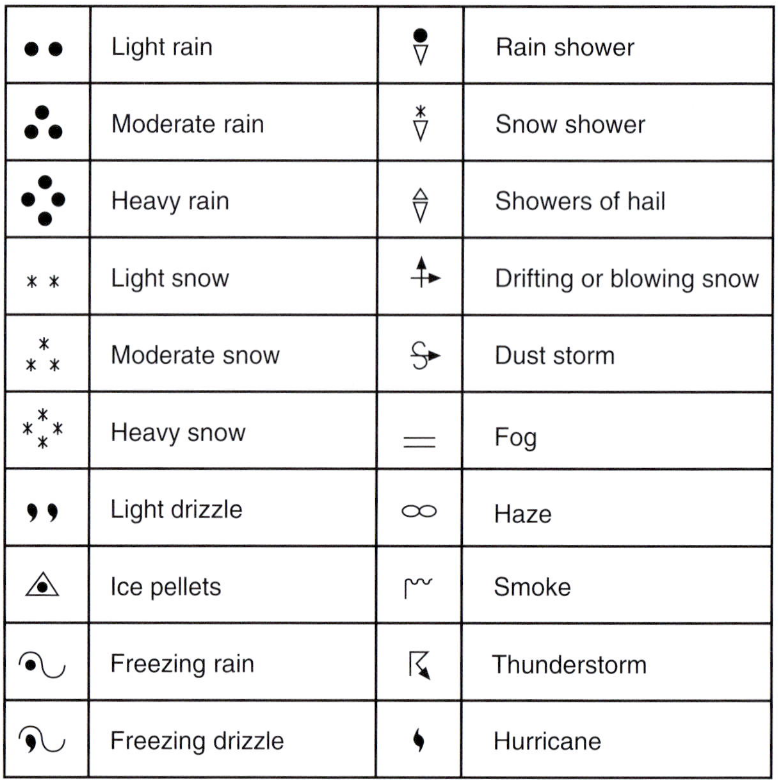

	Light rain		Rain shower
	Moderate rain		Snow shower
	Heavy rain		Showers of hail
	Light snow		Drifting or blowing snow
	Moderate snow		Dust storm
	Heavy snow		Fog
	Light drizzle		Haze
	Ice pellets		Smoke
	Freezing rain		Thunderstorm
	Freezing drizzle		Hurricane

Figure 9.6 (*top*) Weather map symbols for cloud types.

Figure 9.7 (*at right*) Weather map symbols for present weather types.

Table 9.1 Interpreting Surface Weather Maps

General weather	Assume that the types of clouds and weather normally associated with synoptic-scale systems (highs, lows, and fronts) will move along with these systems. The variation of weather parameters with the approach and passage of fronts was summarized in table 6.1. The direction and rate of movement can often be easily tracked by noting successive frontal or pressure center positions from a series of weather maps.
General weather	Assume that the current weather associated with an approaching weather system will not change as the system moves.
Clouds and precipitation	Assume that bad weather will accompany low pressure centers. The air spiraling into the low at the surface must rise as it converges over the low pressure center. Rising motions produce clouds and precipitation. Pressure gradients and, therefore, winds are stronger near lows.
Clouds and precipitation	Assume that good weather will accompany high pressure centers. The air that spirals outward from a high pressure center is replaced by air that descends above the high pressure center. Sinking motions warm the air and cause clouds and precipitation to dissipate.
Clouds and precipitation	Be aware that clouds and precipitation form when moist air is lifted up an inclined surface (e.g., a frontal zone or the side of a mountain or mountain range). Air will dry and warm when it descends, causing clouds and precipitation to dissipate.
Wind direction	Estimate wind direction by looking at the position of high and low pressure centers and the orientation of contours. Air aloft will move parallel to contour lines, with the low on the left. Air near the surface moves basically along the contours with the low on the left, but it turns slightly across the contours toward low pressure at angles of 10–35°. Thus, air spirals out of a high pressure center in a clockwise direction and into a low pressure center in a counterclockwise direction.
Wind speed	Estimate wind speed by looking at the spacing of pressure height contours. Wind speeds are strongest where the pressure height contours are most tightly packed and are weakest where the height contours are farthest apart.
Temperature	Expect a temperature change when a front passes. Passage of a cold front brings colder air; passage of a warm front brings warmer air; passage of an occluded front can bring colder or warmer air.
Stability	Evaluate stability by observing the movement of clouds at different levels in the atmosphere. Air masses tend to become less stable when the wind direction turns counterclockwise with increasing height (e.g., winds change from east to northeast to north) and more stable when winds turn clockwise with height. Instability increases the likelihood of cumulus clouds and showers, whereas stability suppresses cloud development and favors the layering of haze and smoke.
Upper air pattern	Assume that weather will deteriorate as Rossby wave troughs or short-wave troughs approach from the west. Weather will improve when Rossby wave ridges or short-wave ridges approach.

U.S. Forest Service, the Bureau of Reclamation, and the Department of the Interior are especially useful for mountain meteorology applications in the United States. The Remote Automated Weather System, RAWS, covers the entire mountainous western United States (figure 13.25). The Automated Weather Data Network, AWDN, covers the adjacent High Plains. Together the networks have over 700 stations, all of which measure wind and temperature. There is great variability in the other parameters measured because the stations were designed for different applications. Data from RAWS and AWDN are available from archives at the

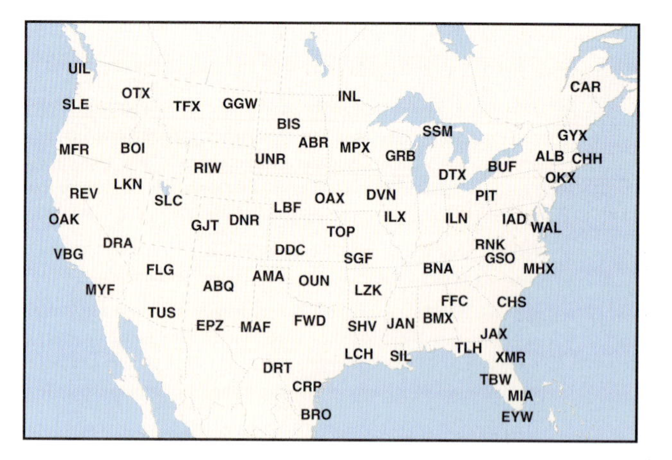

Figure 9.8 Radiosonde launch stations in the United States.

Western Regional Climate Center in Reno, Nevada, and from the High Plains Climate Center in Lincoln, Nebraska.

A large, national network of NOAA Cooperative Climate Network substations has been operated by volunteers for decades. There are presently approximately 8000 active stations in this network, but as many as 23,000 stations have been included at one time or another. Records at some stations go back to 1948. The network focuses largely on agricultural and urban interests and therefore has a meager sampling of mountain stations. Daily high and low temperature, daily rainfall, daily snowfall and snow depth, and evaporation data are available on CD-ROMs from NOAA's National Climatic Data Center in Asheville, North Carolina, and are also available from NOAA's regional climate centers.

The Natural Resources Conservation Service (NRCS) distributes weather data taken from an automatic network of 550 to 600 high-elevation snow depth monitoring sites in the mountains of the western United States and Alaska. The Snowpack Telemetry Network (SNOTEL) stations report daily snow water equivalent, daily precipitation, and maximum, minimum, and average daily temperatures. Data are available from the NRCS National Water and Climate Center in Portland, Oregon.

9.3.2. *Weather Data Dissemination*

The National Weather Service (NWS) is the source for nearly all general weather forecasts in the United States. Forecasts are available directly from regional offices of the NWS through automatic telephone answering systems, facsimile machines, teletypes, radio broadcasts on special weather frequencies outside the normal AM and FM bands, Web pages, and telephone briefings. The NWS offers a wide variety of weather analysis and forecast products. The simplest forecasts are in the form of plain-language bulletins that are read or paraphrased by television announcers, printed in newspapers, or published by on-line services. The media customize forecasts, adding graphics and commentary that appeal to a broad audience, and may have professional meteorologists on staff. Users who require specialized weather forecasts not available through media sources or directly from the NWS (e.g., forecasts for agriculture, aviation, air quality, or fighting of fires on nonfederal lands) usually obtain forecasts from corporate or private meteorologists who have access to NWS data.

The basic tools used by professional weather forecasters are now available to the general public via on-line services, generally in the form of surface and upper air weather charts, radar charts, and satellite images. The outputs of NWS and research weather simulation models are also becoming more widely available through the Internet.

9.3.3. *METARs and TAFs*

Coded weather observations and forecasts based on NWS data are provided for aviation purposes over national and international data dissemination networks. The codes, used to reduce the volume of telecommunication traffic, can be deciphered using information given here and in appendix G.

Weather observations, coded in METAR or Meteorological Aviation Routine Weather Reports (figure 9.9), and forecasts for individual loca-

Figure 9.9 How to read a METAR. (Adapted from Goyer, 1996)

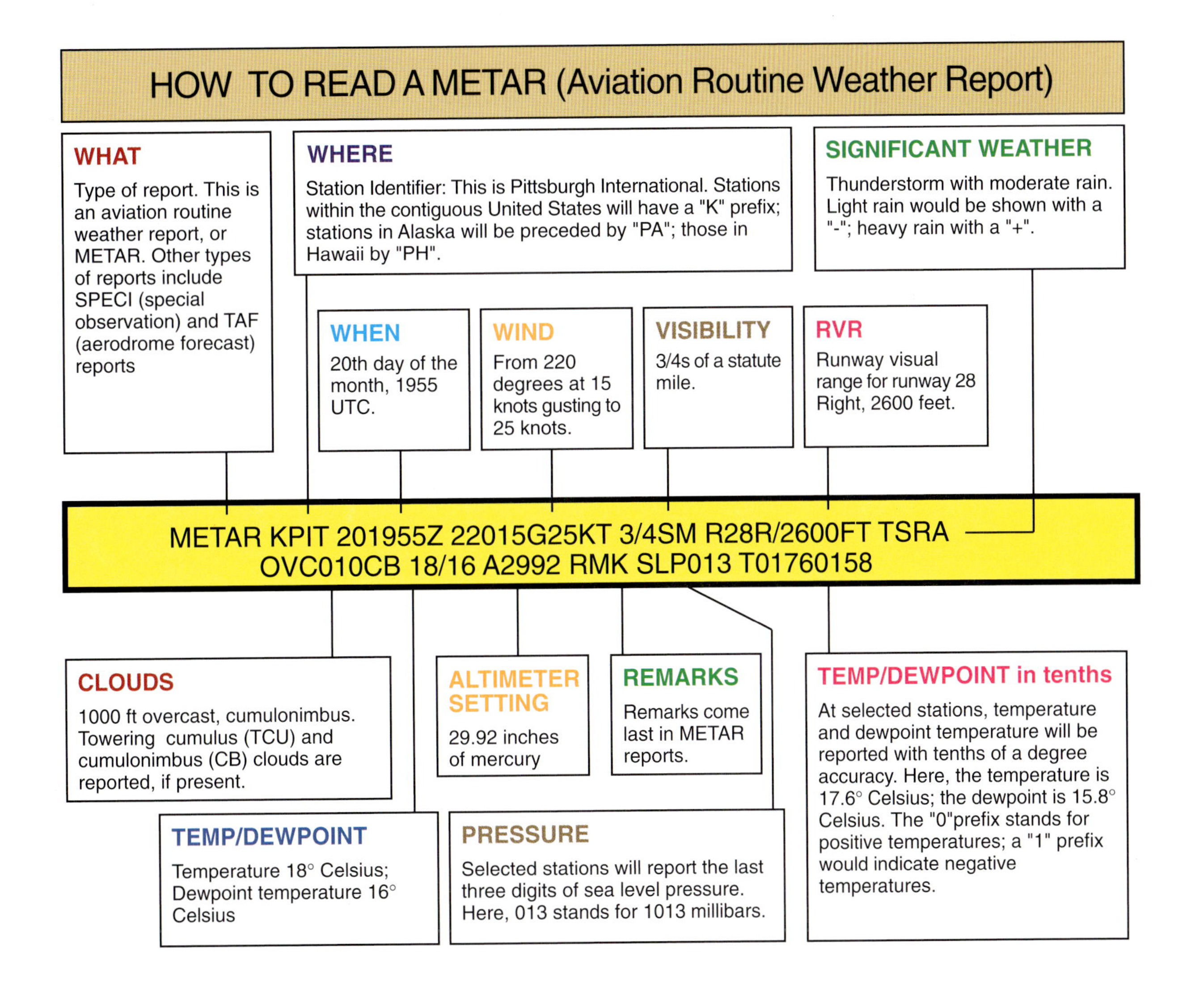

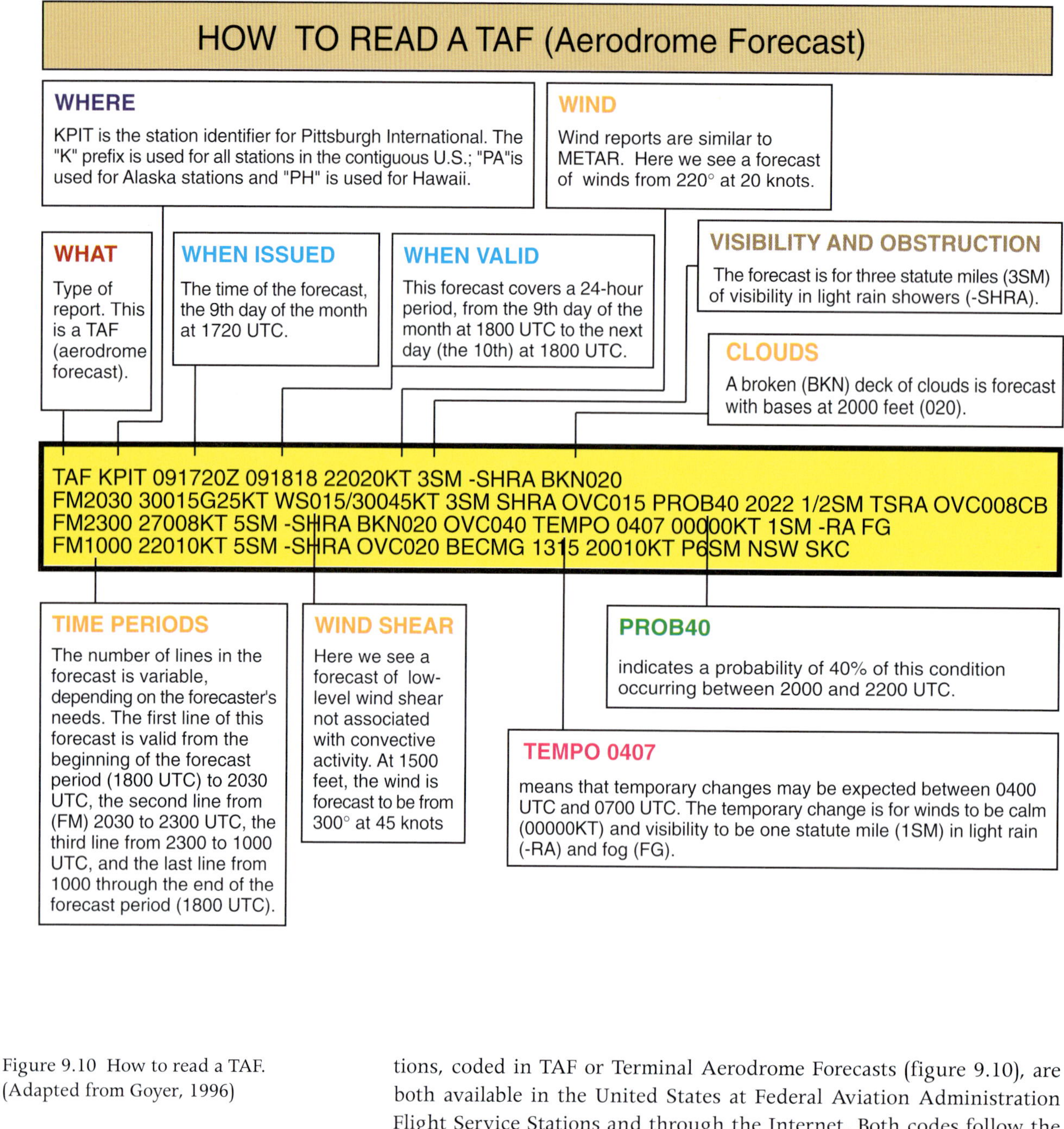

Figure 9.10 How to read a TAF. (Adapted from Goyer, 1996)

Weather observations are coded in METAR format; forecasts for aviation are coded in TAF format.

tions, coded in TAF or Terminal Aerodrome Forecasts (figure 9.10), are both available in the United States at Federal Aviation Administration Flight Service Stations and through the Internet. Both codes follow the international format for meteorological data specified by the World Meteorological Organization, except that the United States has received special permission to use English units for some weather variables. Abbreviations used in the codes are not necessarily English abbreviations.

The METAR and TAF codes designate cloud cover as follows:

clear (CLR, no clouds)
few (FEW, 1/8–2/8 of the sky covered)
scattered (SCT, 3/8–4/8)

broken (BKN, 5/8–7/8)
overcast (OVC, 8/8)

The sky cover amount (in eighths) for any given layer is the sum of the sky cover for that layer plus the sky cover for all lower layers. Cloud heights (in hundreds of feet) are also reported and forecast, and thunderstorms (cumulonimbus, abbreviated CB) or towering cumulus (TCU) clouds, which can develop into thunderstorms, are noted in the observations. Several examples of cloud reports are given in table 9.2. A detailed classification of low, middle, and high clouds is transmitted from major weather stations in METAR form every three hours—at 0000, 0300, 0600, etc. UTC.

METAR reports on cloud conditions in mountain areas have some limitations. First, there are few mountain airports, many with limited hours of operation. Second, airport cloud reports provide local cloud cover information useful to pilots landing at that airport but may be unrepresentative of conditions nearby because mountain cloud conditions are variable. Finally, the reported cloud base heights are given relative to runway height rather than sea level. Because airports in a mountainous region are at different elevations and station elevations are not included in the reports, it can be difficult to visualize cloud cover over the region unless the elevations of the stations are known.

Comments that relate to nearby mountain areas are sometimes attached to METAR reports. For example, the comment

MTNS OBSCD NW-N, LTG N

states that the mountains to the northwest through north of the airport are obscured by clouds and that lightning was seen to the north.

Appendix G provides further information for decoding the METAR and TAF codes, including standard abbreviations (table G.1), weather phenomena abbreviations (table G.2), weather descriptor abbreviations (table G.3), and cloud type abbreviations (table G.4).

Table 9.2 Interpretation of METAR Cloud Codes

Code	Interpretation
CLR	No clouds are present.
SCT100	1 cloud layer is present, with its base at 10,000 ft; 3⁄8 to 4⁄8 of the sky is covered by this layer.
FEW040 SCT070	2 cloud layers are present, with bases at 4000 and 7000 ft. The first layer covers 1⁄8 to 2⁄8 of the sky. The second layer brings the total sky coverage to less than 4⁄8.
BKN050 BKN100	2 cloud layers are present with bases at 5000 and 10,000 ft. The first layer covers 5⁄8 to 7⁄8 of the sky. Although the second layer increases sky coverage, it remains in the 5⁄8 to 7⁄8 range.
SCT020 BKN045 OVC100	3 cloud layers are present, with bases at 2000, 4500 and 10,000 ft. The first layer covers 1⁄8 to 2⁄8 of the sky, the second brings the sky coverage to 5⁄8 to 7⁄8, and the third layer brings total sky coverage to 8⁄8.

9.4. Obtaining Professional Forecasts for Major Federal Projects

Major federal field operations (e.g., aerial photography, aerial spraying, and prescribed or natural fires) may have access to customized forecasts from the National Weather Service Forecast Office (WSFO) serving the area of interest if existing forecasts are insufficiently detailed for the planned operation. Forecasts are usually based on analyses and model runs that are initialized with data collected on the synoptic scale. The forecaster, however, may be able to access real-time mesoscale data that are not carried on the Global Telecommunication System, but are pertinent for the intended use. When forecasting for a mountain area, the forecaster will rely on training and experience to produce a localized forecast based on the larger scale weather analyses and model outputs.

Arrangements for special forecast support should be made through the Meteorologist-in-Charge at the nearest WSFO. Customized forecasts may be available only from certain individual forecasters at certain times of the day. Forecasters are often under time constraints (e.g., at 3- or 6-hour intervals when they have to code and issue their regular forecasts, during the 10-min period before each hour when they have to read weather instruments and code their hourly observations, and during the times between 2300 and 2400 UTC and between 1100 and 1200 UTC when radiosondes are being prepared for launch).

Before a field operation, the client should meet with the forecaster to explain the operation and to discuss its susceptibility to different weather conditions. The forecaster must know what types of forecasts are needed and how frequently and over what time interval they are to be provided. This information should be repeated in a follow-up letter to the forecast office.

During the operation, the client should provide the forecaster with a brief description of the weather conditions at the field site to help the meteorologist calibrate the forecast. Feedback on previous forecasts should be given, and specific concerns, such as fog, strong winds, or cloud conditions, should also be mentioned. The client can request a forecast update if weather conditions deteriorate unexpectedly and may choose to monitor other sources of NWS information on days when weather is critical to the operation.

Part III

MOUNTAIN WINDS

10 Terrain-Forced Flows

Winds associated with mountainous terrain are generally of two types. Terrain-forced flows are produced when large-scale winds are modified or channeled by the underlying complex terrain. Diurnal mountain winds are produced by temperature contrasts that form within the mountains or between the mountains and the surrounding plains and are therefore also called *thermally driven circulations*. Terrain-forced flows and diurnal mountain winds are nearly always combined to some extent. Both can occur in conjunction with small-scale winds, such as thunderstorm inflows and outflows, or with large-scale winds that are not influenced by the underlying mountainous terrain.

10.1. Three Factors that Affect Terrain-Forced Flows

Terrain forcing can cause an air flow approaching a mountain barrier to be carried over or around the barrier, to be forced through gaps in the barrier, or to be *blocked* by the barrier. Three factors determine the behavior of an approaching flow in response to a mountain barrier:

- the stability of the air approaching the mountains,
- the speed of the air flow approaching the mountains, and
- the topographic characteristics of the underlying terrain.

Unstable or neutrally stable air (section 4.3) is easily carried over a mountain barrier. The behavior of stable air approaching a mountain barrier depends on the degree of stability, the speed of the approaching flow, and the terrain characteristics. The more stable the air, the more resistant it is to lifting and the greater the likelihood that it will flow around, be forced through gaps in the barrier, or be blocked by the barrier. A layer of stable air can split, with air above the *dividing streamline height* flowing over the mountain barrier and air below the dividing streamline height splitting upwind of the mountains, flowing around the barrier (figure 10.1), and reconverging on the leeward side (section 10.3.2). A very stable approaching flow may be blocked on the windward side of the barrier (section 10.5.1).

Moderate to strong cross-barrier winds are necessary to produce terrain-forced flows, which therefore occur most frequently in areas of cyclogenesis (section 5.1) or where low pressure systems (figure 1.3) or jet streams (section 5.2.1.3) are commonly found. Whereas unstable and neutral flows are easily lifted over a mountain barrier, even by moderate winds, strong cross-barrier winds are needed to carry stable air over a mountain barrier. As the speed of the cross-barrier flow increases, the amount of stable air carried over the barrier also increases, with less air channeled around or through the barrier and less air blocked upwind of the barrier.

Strong cross-barrier winds are needed to carry stable air over a mountain barrier.

10.1.1. *Topographic Characteristics of the Underlying Terrain*

A number of terrain characteristics influence both the speed and the direction of terrain-forced flows. The height and length of the mountain barrier can determine whether air goes over or around the barrier. The amount of energy required for air to flow over a high mountain ridge is much greater than that required to flow over a small hill. More energy is required to flow around an extended ridge than around an isolated peak. Thus, high wind speeds are required to carry air over a high mountain range or around an extended ridge. When stable air flows around an isolated peak or the edges of an elongated mountain range, the highest wind speeds are on the hillsides where the flow is parallel to the peak's contour lines (figure 10.2).

The shape of the vertical cross section through the mountain barrier affects wind speed on both the windward and leeward sides. Wind speed increases at the crest of a mountain, with gently inclined triangular-shaped hills producing the greatest increase in speed, flat-topped mesas producing the smallest increases, and rounded mountaintops producing

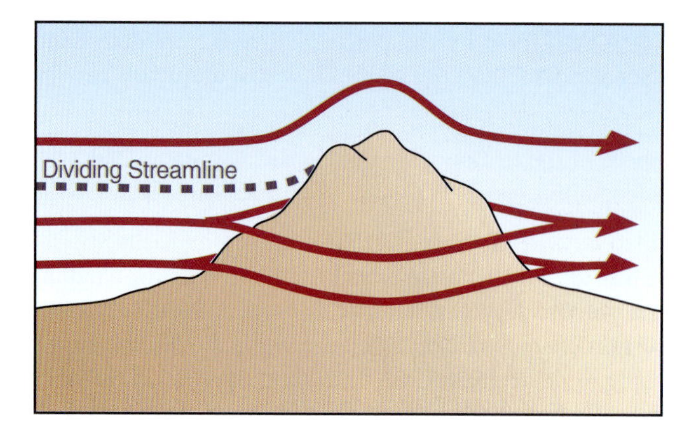

Figure 10.1 The dividing streamline height is the height of the boundary between low-level air, which splits to flow around the barrier, and upper level air, which is carried over the barrier. (Adapted from Etling, 1989)

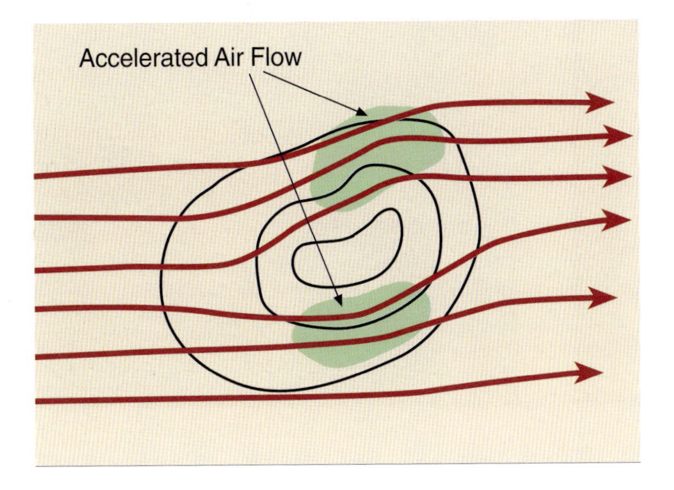

Figure 10.2 Under stable conditions, winds split around an isolated mountain, and strong wind zones are produced on the edges of the mountain that are tangent to the flow. (Adapted from Justus, 1985)

intermediate speedups. Flow separations or *separation eddies* can form over steep slopes or cliffs on either the windward or leeward sides of a barrier. These elongated, horizontal-axis eddies, which can extend along the entire length of the barrier, reduce near-ground wind speeds over the slopes (figure 10.3). The flow above these separation eddies, however, speeds up as it crosses the barrier.

The orientation of a mountain ridge relative to the approaching flow and the curve of the ridgeline (as viewed from above) affect wind direction and speed. Ridgelines that are concave to the windward side and mountain barriers oriented perpendicular to the flow cause flow across the barrier to increase and frequently generate lee waves downwind of the obstacle. Flows approaching barriers with ridgelines that are parallel to, oblique to, or convex to the approaching flow (figure 10.4) change direction to follow the underlying terrain and generate lee waves less frequently. Smaller scale features on a ridgeline can also affect the approaching flow, channeling winds through passes and gaps.

The presence of valleys and basins affects wind speed. Sites low in a valley or basin are often protected from prevailing winds by the confining topography. If winds aloft are strong, however, eddies can form within a valley or basin downwind of a ridgeline, bringing strong, gusty winds to lower elevations.

The roughness of the underlying surface affects wind speed: the rougher the surface, the greater the reduction in wind speed (section 5.2.4). Wind speeds increase when winds move from a rough surface to a smooth surface (for example, from a rough, mountainous area onto a large lake) and decrease when winds move from a smooth surface to a rough surface. The layer in which wind speeds are affected by a roughness boundary deepens with distance downwind from the roughness boundary. An abrupt increase in roughness causes winds to converge, air to rise, and clouds to form. For example, *lake-effect storms* that form in the fall and winter on the eastern shorelines of the Great Lakes result from the abrupt increase in roughness encountered by the prevailing westerly winds. The westerly flow picks up moisture as it moves across the open lakes and then rises when it reaches the shoreline and the rolling hills to the east, producing clouds and locally heavy snowfall.

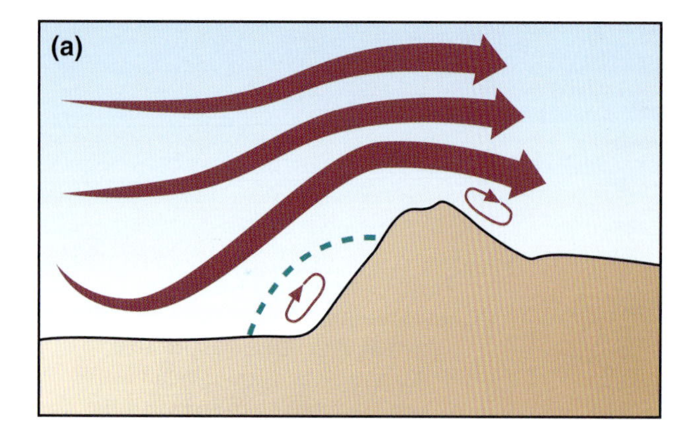

Figure 10.3 Flow separations occur on the windward or leeward faces of steep-sided hills or mountains: (a) A flow separation on the windward side occurs below the dashed green line. A smaller flow separation is also present on the lee side. (b) Sand dunes form where wind speeds drop in separation eddies on the windward face of a hill. This dune formed near Vantage, Washington, on the Columbia River. (Photo © C. Whiteman)

Terrain-forced flows are influenced by the height and length of the mountain barrier, the shape of the vertical cross section through the barrier, the orientation of the barrier relative to the approaching flow, the curve of the ridgeline, the presence of valleys and basins, the roughness of the underlying surface, and terrain obstacles.

Finally, terrain obstacles (mountains, forest patches, shelter belts, individual trees, terrain projections, buildings) generate turbulent *eddies* and/or *wakes* (section 10.1.2) when the approaching flow has sufficient speed. Eddies are swirling currents of air at variance with the main current. A wake forms when eddies are shed off an obstacle and cascade into smaller and smaller scales. Wakes may also form when air splits to flow around rising bubbles or columns of warm air or convective towers in thunderstorms. Wakes are characterized by low wind speeds, but their high turbulence can produce locally gusty winds. They can be seen when air motion tracers (clouds, smoke, or blowing snow or dust) are present and are easily identified when captured by a time-lapse video camera.

Approaching flows that are carried over mountains respond to overall or large-scale features of the topography rather than to small-scale topographic details. In fact, when a low-lying cold air mass is present on the windward and/or leeward sides of the mountain barrier, the approaching flow responds to the combination of the actual terrain and the adjacent air mass. Thus, the *effective topography* that influences the approaching flow can be higher and wider than the actual topography. Terrain-forced flows respond to both landforms (such as valleys, passes, plateaus, ridges, and basins) and roughness elements (such as peaks, terrain projections, trees, and boulders) in complex terrain. Wind speeds can vary significantly between sites exposed to prevailing winds or terrain-forced flows and sites protected from winds by the terrain. Table 10.1 includes information from chapter 6 and provides general rules on where to expect high and low wind speeds.

10.1.2. *Wakes, Eddies, and Vortices*

Eddies can form anywhere in the atmosphere, both at the ground and aloft, in relationship to obstacles and over unobstructed terrain as the result of wind shear or convection. The relative horizontal and vertical dimensions of an eddy are determined by the stability. *Isotropic* eddies (i.e., eddies that have similar vertical and horizontal dimensions) develop in neutral stability, whereas vertically suppressed eddies develop in more stable air, and vertically enhanced eddies form in unstable air.

A wake is an area extending downwind of an obstacle and is characterized by relatively slower wind speeds but increased gustiness. Winds

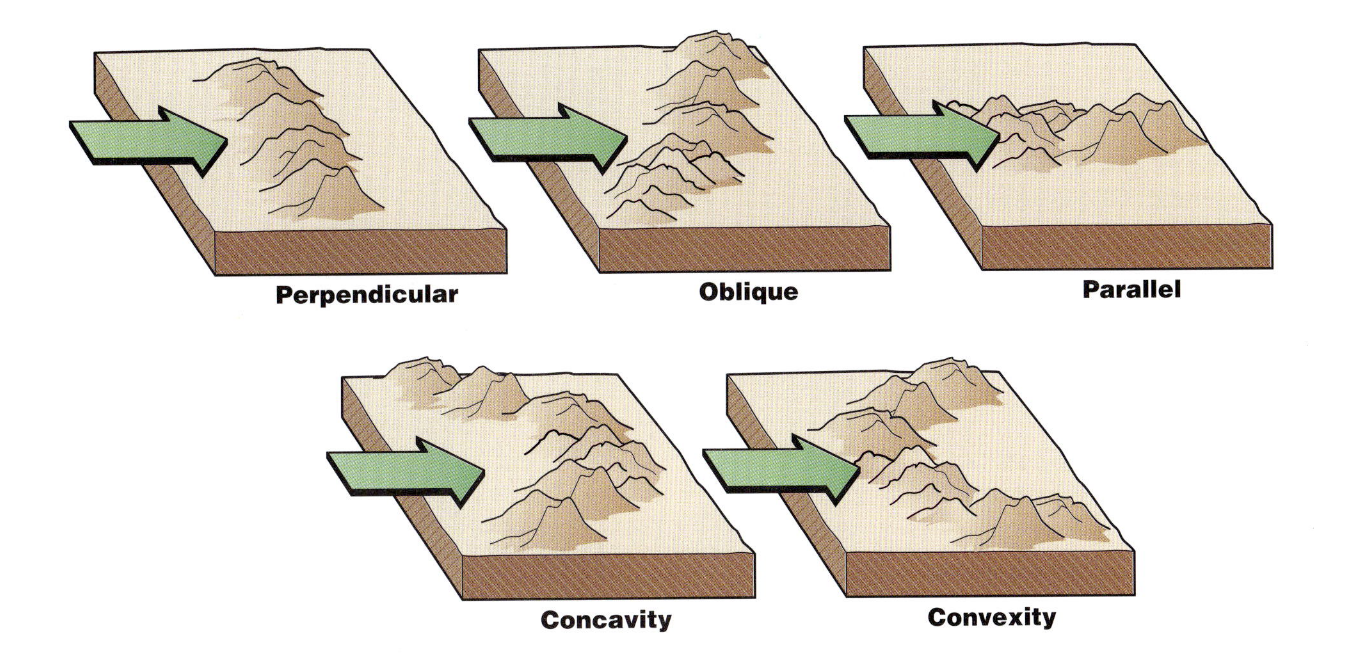

Figure 10.4 The orientation and shape of a ridgeline affect the speed and direction of a flow crossing a mountain barrier. The highest speedups occur over ridgelines that are perpendicular to the flow or have a concavity oriented into the flow. (Adapted from Justus, 1985)

are generally slowed downwind of obstacles to distances roughly 15 times the obstacle height (Angle and Sakiyama, 1991), although velocity reductions in wakes have been detected as much as 60 obstacle-heights downwind.

Vortices, whirling masses of air in the form of a column or spiral, usually rotate around either vertical or horizontal axes. Vertical-axis vortices in the atmosphere range in size from dust devils to tornadoes, waterspouts, hurricanes, and synoptic-scale high and low pressure centers. Horizontal-axis vortices are caused by vertical wind shear and turbulence and are common in the lee of extended terrain obstacles, such as mountain ridges, shelter belts, and snow fences. The reduced wind speed in the lee of shelter belts and snow fences protects crops from damage and highways from drifting snow (figure 5.9). Horizontal-axis vortices can be stretched into vertical-axis vortices by convection when the ground is heated and the atmosphere is unstable.

Eddies are common in mountainous terrain. Their presence is often indicated by the formation of particular clouds or snow cornices. Banner

Table 10.1 Landforms Associated with Strong and Weak Winds at the Ground

Expect high wind speeds at sites:

- located in gaps, passes, and gorges in areas with strong cross-gap pressure gradients
- exposed directly to strong prevailing winds, especially mountain summits, upper windward or leeward mountain slopes, high plains, or elevated plateaus
- located downwind of smooth fetches, as on downwind shores of large lakes or oceans

Expect low wind speeds at sites:

- protected from prevailing winds as at lower elevations in basins or in deep valleys oriented perpendicular to prevailing winds
- located upwind of mountain barriers or in intermountain basins where low-level air masses are blocked by the barrier
- located in areas of high surface roughness, such as forested, hilly terrain

Figure 10.5 The snow on the left of the summit ridge of Handies Peak in Colorado is the remnant of a snow cornice. The general flow, from right to left, causes a horizontal-axis eddy to form on the lee side of the ridge. In this summertime photo, moist air from an afternoon rainstorm is lifted up the slope in the rising part of the eddy where it condenses into a cloud on the left side of the ridge. A hiker on the ridge has cloudy air on one side and clear air on the other. (Photo © C. Whiteman)

clouds form in eddies in the lee of sharp isolated peaks (section 7.1.4.6). Rotor clouds (section 7.1.4.3) form in horizontal-axis eddies associated with trapped lee waves (section 7.1.4.3) some tens of miles downwind of the mountain barrier. Snow cornices can build up in winter when horizontal-axis eddies (figure 10.5) form over the lee slope. Cornices sometimes extend for some distance along the mountain crest (figure 5.10).

Large, generally isotropic vertical-axis eddies can be produced by the flow around mountains or through gaps in the topography as eddies are shed from the vertical edges of the terrain obstruction (figure 10.6). The Schultz eddy on the north side of the Caracena Straits in California's Sacramento Valley and the San Fernando and Elsinore convergence zones on the northeast sides of the Santa Monica and Santa Ana Mountains are examples of large, vertical-axis eddies formed in this way.

10.2. Flow over Mountains

An approaching flow tends to go over a mountain barrier rather than around it if the barrier is long, if the cross-barrier wind component is strong, and if the flow is unstable, near-neutral, or only weakly stable. These conditions are frequently met in the United States because the long, north–south-oriented mountain ranges lie perpendicular to the prevailing westerly winds and the jet stream. Flow over a mountain barrier can be surmised from the presence of certain cloud types, including lenticular clouds, cap clouds, banner clouds, rotors, a foehn wall, a chinook arch, and billow clouds (section 7.1.4). Blowing snow and cornice build-up on the ridge crests in winter and blowing dust at the ridge crests or on the upper lee slope in summer also indicate flow over a mountain barrier. Flow over mountains generates mountain waves and lee waves in the atmosphere and can produce downslope windstorms.

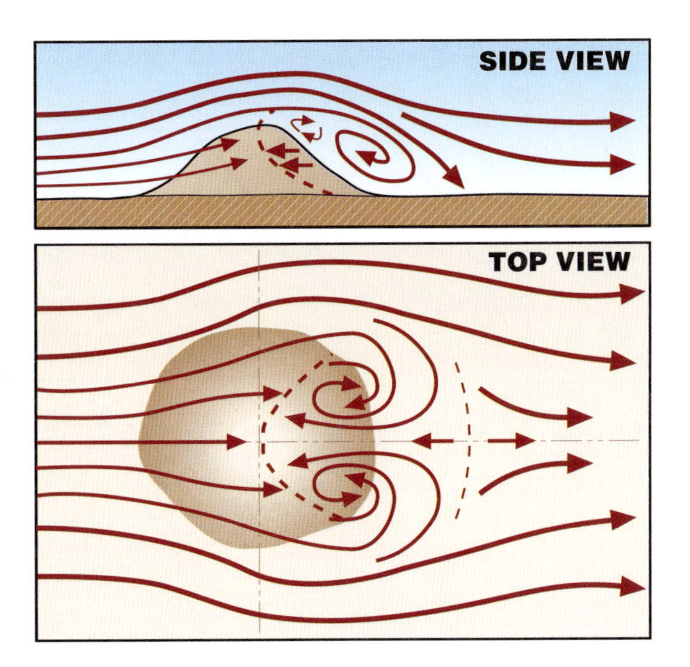

Figure 10.6 Vortex pairs and a wake are generated by the flow around a mountain. (Adapted from Orgill, 1981)

10.2.1. *Mountain Waves, Lee Waves, and Hydraulic Flows*

As stable air flows over a mountain range, *gravity waves* (i.e., vertical undulations or waves in the atmosphere created as gravity acts on local variations in air density) can be generated either over the mountains or in the lee of the mountains. Stable air that is lifted over a mountain barrier cools, becomes denser than the air around it, and, under the influence of gravity, sinks again on the lee side of the barrier to its equilibrium level. The air overshoots and oscillates about its equilibrium height (figure 10.7).

Gravity waves that form over the mountains are called mountain waves. Mountain waves have a tendency to propagate vertically and can thus be found not only at low levels over hills and mountains but throughout the troposphere and even in the stratosphere (figure 10.8). Waves that form in the lee of mountains are called lee waves. Lee waves are often confined or trapped (figure 10.9) in the lee of the barrier by a smooth, horizontal flow above. The two types of waves are collectively called *orographic waves* or simply mountain waves. Classification of waves on a particular day can be difficult because the two types can be present simultaneously and because there is a continuum between the two. In general, mountain waves are found higher in the atmosphere and tend to have longer wavelengths and smaller amplitudes than lee waves.

If there is sufficient moisture in the atmosphere, clouds form in the crests of the waves as air is lifted through the wave in its passage across and downwind of the mountain barrier. Thus, the presence of lenticular clouds is a clear indication of the presence of orographic waves. Larger clouds, such as the chinook arch, are associated with the longer wavelength mountain waves. When no clouds are present, the existence of waves can nonetheless be assumed whenever a stable air mass approaches a significant barrier with a strong cross-barrier wind component.

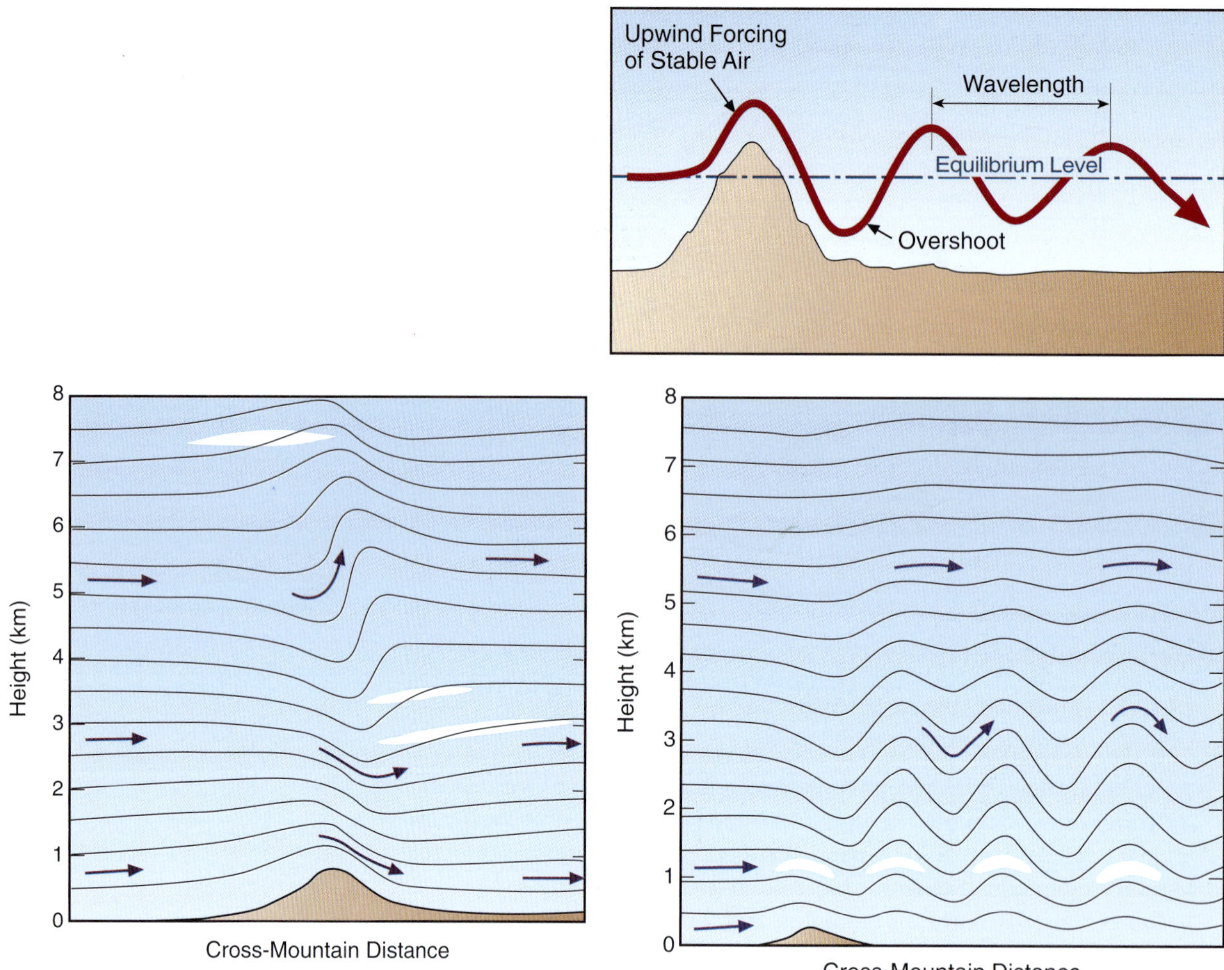

Figure 10.7 (*top*) Stably stratified air that is lifted over mountains oscillates about its equilibrium level on the lee side of the mountain, producing waves. (Adapted from Stull, 1995)

Figure 10.8 (*bottom left*) Vertically propagating mountain waves have their highest amplitudes well above the mountains. High lenticular clouds are often indicators of these waves. (Adapted from Carney et al., 1996)

Figure 10.9 (*bottom right*) Trapped lee waves reach their highest amplitudes in a confined layer on the lee side of the mountains. Regularly spaced low-altitude lenticular clouds are the best indicators of trapped lee waves. (Adapted from Carney et al., 1996)

Thickening of wave clouds and increasing sky coverage are indicative of the increasing moisture that often precedes a frontal passage. The rate of cloud development can be determined from direct observations or from hourly airport observations, to which special remarks on wave clouds (altocumulus standing lenticulars or ACSL) are often appended. The formation of lenticular clouds was discussed in section 7.1.4.4.

Orographic waves form most readily in the lee of steep, high barriers that are perpendicular to the approaching flow. The amplitude of the waves, which decays with distance from the mountain barrier, depends on the initial displacement of the flow above its equilibrium position on the windward side of the mountain. In general, the higher the barrier, the greater the amplitude of the waves. If the flow crosses more than one ridge crest, the waves generated by the first ridge can be amplified (a process called *resonance*) or canceled by the second barrier, depending on its height and distance downwind of the first barrier (figure 10.10).

The basic form of a wave (trapped or vertically propagating) and its wavelength depend on variations in the speed and stability of the approaching flow. When stable air is carried over a mountain barrier, one of three flow patterns will result, depending on the characteristics of the

THE OCTOBER 1997 FOREST BLOWDOWN IN THE CENTRAL ROCKY MOUNTAINS

The largest forest blowdown ever recorded in the Rocky Mountains occurred during a short period on the morning of 25 October 1997. Twenty thousand acres of old-growth trees were blown down on the west side of the Park Range in the Routt National Forest and Mt. Zirkel Wilderness areas northeast of Steamboat Springs, Colorado. Fallen trees were stacked as high as 30 feet (9 m), and hunters in the forest were stranded for as long as two days.

Blowdowns occur most often on the lee sides of mountain barriers, which in Colorado are usually the east side. In this case, however, a strong downslope windstorm developed in easterly flow on the west side of the Continental Divide. The storm was similar to those that occur frequently on the east side of Colorado's Front Range when the amplitude and wind speeds of a deep wave are enhanced by hydraulic flow or lee wave breaking.

The October 25 blowdown in the Park Range was preceded by the development of an intense low pressure storm system in the Four Corners area (where Colorado, New Mexico, Arizona, and Utah meet) on October 24. The counterclockwise winds flowing around the low exposed eastern and northern Colorado to easterly winds that carried moist air up the foothills of the Front Range, producing snowfall up to 4 feet deep with drifts to 15 feet. Wind gusts exceeded 100 mph (45 m/s) for 5 hours at Arapahoe Basin Ski Area (above timberline at 12,500 ft or 3810 m) in the central Colorado Rockies. Wind gusts in the Park Range north of Arapahoe Basin were probably as strong, but according to hunters, were of much shorter duration, lasting only about 30 minutes.

wind profile. If the winds are weak and nearly constant with height, shallow waves form downwind of the barrier. When winds become stronger and show a moderate increase with height, the air overturns on the lee side of the barrier, forming a standing (i.e., nonpropagating) lee eddy with its axis parallel to the ridgeline. When winds become stronger still and show a greater increase with height, deeper waves form and propagate farther downwind of the barrier. Wavelength increases when wind velocities increase or stability decreases.

Orographic waves form most readily in the lee of steep, high barriers that are perpendicular to the approaching flow.

Under certain stability, flow, and topographic conditions, a large-scale instability can cause the entire mountain wave to undergo a sudden transformation to a *hydraulic flow* (figure 10.11). As the name implies, the air flows like river water over a large boulder, with high speeds on the leeward slope, a cavity at the bottom of the slope, and severe turbulence immediately beyond the cavity. A hydraulic flow exposes the lee side of the mountain to sweeping, high-speed, turbulent winds that can cause forest blowdowns and structural damage. Damage is most likely to occur near the base of the lee slope, where the winds are strongest, or slightly downwind of the base of the slope, where turbulence levels are high.

10.2.2. *Downslope Windstorms: The Bora, the Foehn, and the Chinook*

Downslope windstorms occur on the lee side of high-relief mountain barriers when a stable air mass is carried across the mountains by strong

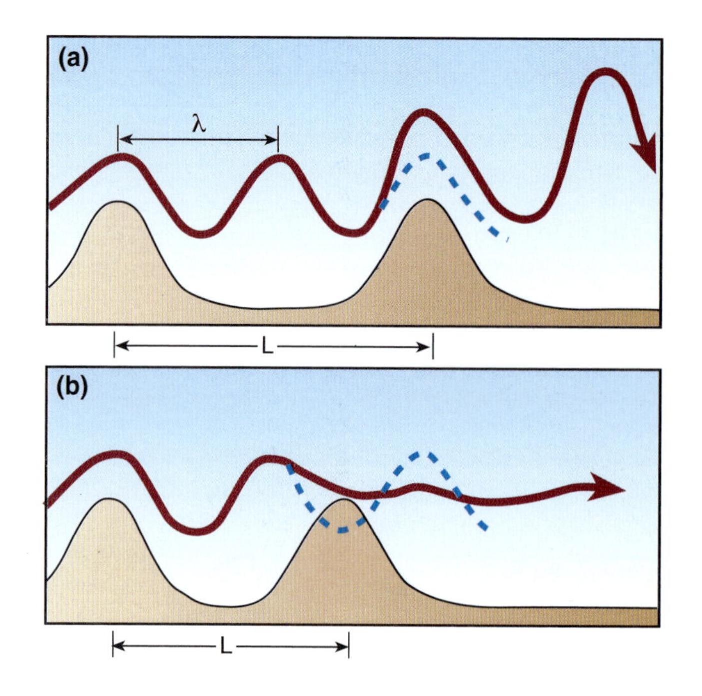

Figure 10.10 Waves can be (a) amplified or (b) canceled by successive ridges, depending on the relationship between the ridge separation distance *L* and the wavelength λ of the flow. (Adapted from Bérenger and Gerbier, 1956)

cross-barrier winds that increase in strength with height. The strong winds (greater than 50 mph or 22 m/s and sometimes exceeding 100 mph or 45 m/s) are caused by intense surface pressure gradients, with a high pressure center on the upwind side of the barrier and a low pressure trough paralleling the lee foothills. The cross-barrier surface pressure gradient is intensified as the descending air on the lee side of the barrier produces local warming and thus a decrease in pressure at the surface. The pressure gradient may be intensified further if the windstorm coincides with the arrival of a short-wave trough (section 5.1.4), which causes the surface pressure to drop on the lee side of the barrier and to rise on the windward side. The short-wave trough can also cause the winds at mountaintop level to shift direction and become more perpendicular to the barrier. Elevated inversions have been noted in many windstorms where observations were available near mountaintop level. Because elevated inversions are difficult to observe and to forecast, their presence is usually assumed when all other meteorological conditions for windstorms are met. This assumption may result in the overforecasting of windstorms, which is considered preferable to underforecasting.

Downslope windstorms occur primarily in winter on the lee side of high-relief mountain barriers and are associated with large-amplitude lee waves.

Downslope windstorms occur primarily in winter and appear to be associated with large-amplitude lee waves. The descending branch of the first wave reaches the ground at the foot of the slope because the amplitude of the first wave has been increased by resonance (section 10.2.1), by wave trapping (trapping of the vertical energy below a smooth horizontal flow at a given height), or by the development of a hydraulic flow.

Local topography influences the strength of windstorms at a given location. Winds are strong downwind of ridgelines that are high, continuous, and oriented perpendicular to the flow. A ridgeline that is concave on the upwind side also enhances strong winds (section 10.1). Steep leeward slopes, where flow separations form under normal conditions, can cause an acceleration of hydraulic flows.

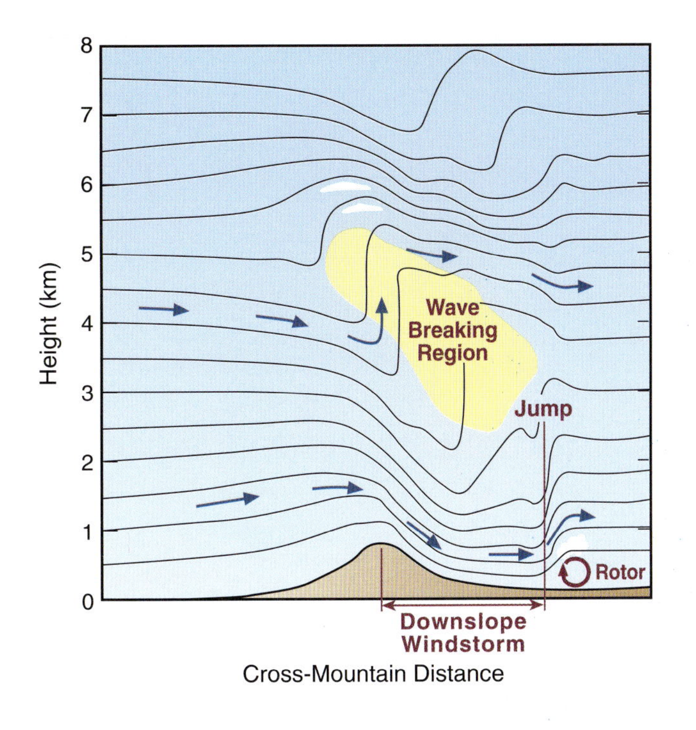

Figure 10.11 Hydraulic flow produces a distinctive flow pattern in the lee of a mountain barrier that is characterized by a region of wave-breaking aloft and a sudden jump in the streamline pattern (*hydraulic jump*) downwind of the barrier. A turbulent rotor cloud may form behind the hydraulic jump. Downslope windstorms may occur during hydraulic flow. (Adapted from Carney et al., 1996)

Downslope winds can bring either cold or warm air into the leeward foothills. A cold downslope wind is called a *bora*, after a wind that brings very cold air to the eastern coast of the Adriatic Sea in Slovenia, Croatia, and Bosnia. The bora originates in an area in central Asia where temperatures are so low that, despite adiabatic warming, the wind is still cold when it reaches the Adriatic coast. The German word *föhn* (or foehn) is used internationally to designate a warm, dry downslope wind. The warming and drying are caused by *adiabatic compression* as air descends the slopes on the leeward side of a mountain range. In the western United States, the foehn is called a chinook, after a Northwest Indian tribe. The term was first applied to a warm southwest wind that was observed at the Hudson Bay trading post at Astoria, Oregon, because it blew from "over Chinook camp" (Burrows, 1901). "Chinook" became more widely used as the adjacent country was settled and is now applied to foehn winds over the entire western part of the United States and Canada (figure 10.12). (Interestingly, the wind at Astoria could not actually have been a foehn wind because the topography southwest of Astoria is not high enough to produce significant adiabatic warming).

In the United States, chinooks are primarily a western phenomena because the topographic relief of the Appalachians is generally insufficient for the production of strong downslope winds. In the Rocky Mountains, chinooks blow most frequently from November through March, bringing relief from long periods of arctic cold east of the mountains (figure 10.13). The gusty, warm winds rapidly melt wintertime snow cover, earning them the name "snoweaters." Four factors contribute to the warmth and dryness of the chinook winds (figure 10.14):

- The air that descends the lee slope is warmed and dried by *compression heating* at the dry adiabatic rate of 5.4°F per 1000 ft or 9.8°C/km

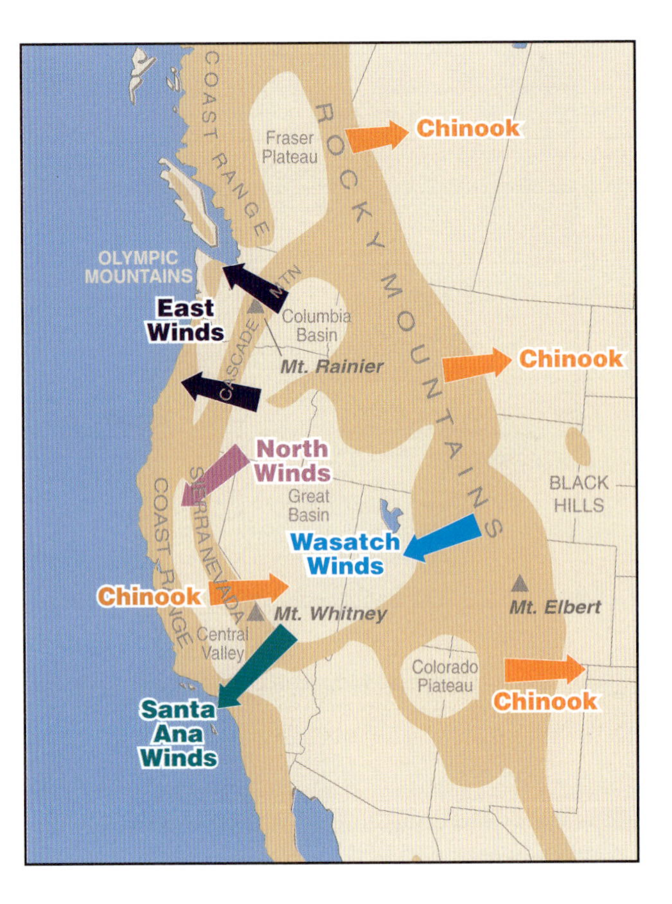

Figure 10.12 (*top*) Foehn winds in the western United States.

Figure 10.13 (*at right*) Waiting for a chinook. (Painting by C. M. Russell, used with permission from the Montana Stockgrowers Association, Helena, Montana)

as the air is brought to the lower altitudes and, thus, higher pressures at the base of the lee slope.

- When a deep flow causes air at low levels upwind of the mountain barrier to be lifted up the barrier, latent heating occurs as clouds form and precipitation falls on the windward side, thus warming the air before it descends on the lee side.
- Warm air descending the lee slopes can displace a cold, moist air mass, thus enhancing the temperature increase and the humidity decrease associated with the winds.
- The turbulent foehn flow can prevent nocturnal temperature inversions from forming on the lee side, thereby allowing nighttime temperatures to remain elevated.

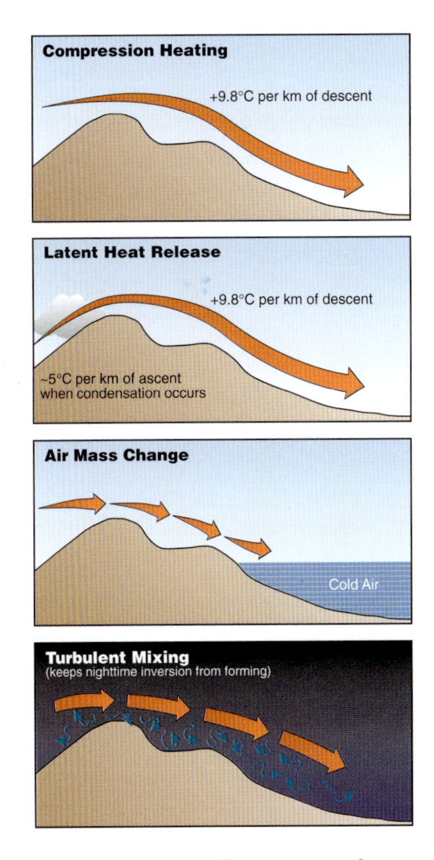

Figure 10.14 Four factors cause the warming and drying associated with chinooks. (Adapted from Beran, 1967)

Downslope windstorms can start and stop suddenly at a given location when changes in the cross-barrier flow component or the stability of the approaching flow cause the wavelength of the orographic waves to change. However, the sudden onset or cessation of winds is usually due to changes in the position of a shallow, cold air layer on the lee side of the mountains that protects the surface from the strong winds. (If the cold air layer is very shallow, strong winds aloft may be heard at the ground). Strong downslope winds can suddenly reach the ground if the upwind edge of the air mass is scoured away by the downslope winds or if waves on the upper surface of the cold air mass depress the surface near its edge. Downslope winds can also reach the ground if prevailing winds aloft drag the cold air away from the mountains or if winds within the cold air mass weaken, allowing it to slide down the topography away from the mountains. The windstorm ends abruptly if the cold air mass sloshes back into place. An abrupt cessation of downslope winds at a given site is called a *foehn pause* or *chinook pause*. Alternating strong wind break-ins and foehn pauses can cause temperatures to oscillate wildly, as illustrated by temperature records for December 1933 in Havre, Montana (see Point of Interest).

Flow speeds in downslope windstorms are highest in a narrow zone along the base of a mountain barrier. For example, cities like Boulder, Colorado, and Livingston, Montana, that are located near the mountain–plain interface of the Rocky Mountains frequently report windstorms. Downslope windstorms also affect less-populated regions along the Rocky Mountain foothills from Canada to southern Colorado but often go unreported because they do not impact large populations. The highest wind speeds occur on elevated mesas or other terrain projections on the edge of the mountains. For example, strong winds are well documented on Table Mesa in southwest Boulder and on Rocky Flats south of Boulder.

Strong, gusty downslope windstorms can cause considerable damage along the mountain–plain interface. Damaging winds rarely extend more than 15 miles (25 km) out onto the adjacent plain, although winds can still be strong at these distances. The strong winds pose a number of hazards. Open fires can be spread rapidly by the winds, and local flooding can result from the rapid melting of snow cover. Smoke, windblown dust, and strong gusts can cause poor driving conditions. In built-up areas, damage to roofs and fences, and windows broken by windblown objects are common. Buildings under construction and mobile homes are particularly susceptible to wind damage. Aviation hazards include turbulent

Flow speeds in downslope windstorms are highest in a narrow zone along the base of a mountain barrier, where the strong, gusty winds can cause considerable damage.

rotors that develop below the crest of lee waves and in the lee of the hydraulic flow cavity. Damage caused by downslope windstorms may be exacerbated by vertical-axis eddies, sometimes called *mountainadoes*. Mountainadoes develop when vertical wind shear near the ground produces horizontal eddies that are then stretched vertically to rotate about a vertical axis. If these vertical eddies are advected with the mean wind, they produce strong horizontal shears and wind gusts that are much more damaging than the general prevailing winds.

Because most of North America is within the zone of prevailing westerlies, chinook winds usually occur on the eastern sides of U.S. mountain ranges. They can, however, occur on the western sides of mountain barriers when upper level winds are from the east. The *Santa Ana winds* in the Los Angeles Basin and the *Wasatch winds* in Utah are familiar examples of northeasterly or easterly chinook winds.

The Santa Ana winds develop in the late fall and winter when a high pressure center in the Great Basin reverses the normal pressure gradient that causes winds to blow off the ocean in Southern California. The high

BATTLE OF THE CHINOOK WIND AT HAVRE, MONTANA

The thermograph record in figure I shows a series of extreme temperature variations that occurred at Havre, Montana, during the week of 15–22 December 1933 when chinook winds repeatedly broke through a shallow cold air mass. A southwest chinook struck Havre on 16 December, causing temperatures to rise 27°F (15°C) in only 5 minutes as wind velocities increased from 5 to 30 mph (2 to 13 m/s). The chinook continued until the morning of 18 December, when temperatures fell 40°F (22°C) as cold arctic air spread over Havre from the north and east. Chinook break-ins occurred again for several hours on 19 December and during the nights of 20–21 December and 21–22 December, with sudden temperature variations of 20–30°F (11 to 17°C).

The weather maps in figure II illustrate the changes in frontal position that produced alternating periods of cold and warm air temperatures at Havre. In figure IIa, a developing low pressure storm system in Alberta was accompanied by warm and cold fronts. Warm chinook winds descended the lee slopes of the Rockies in Saskatchewan, Montana, and Wyoming to push cold air eastward away from the Rockies and immerse Havre in the warm air behind (west of) the warm front. Meanwhile, a surge of cold air was moving southward along the eastern slopes of the Rockies to the west of the low. In figure IIb, the cold air surged as far south as Colorado and immersed Havre in cold air. The cold air dome was dammed from westward movement by the Rockies, as indicated by the stationary front.

The conceptualized weather maps in figure III illustrate a typical break-in of warm chinook winds on the east slope of the Rockies. In figure IIIa, a stationary front indicates the western edge of a cold air layer that is dammed against the eastern slope of the Rockies. In figure IIIc, a decrease in the easterly component of winds within the cold air mass allows it to slide southeastward down the slope of the plains, exposing the western edge of the plains to westerly chinook winds.

(continued)

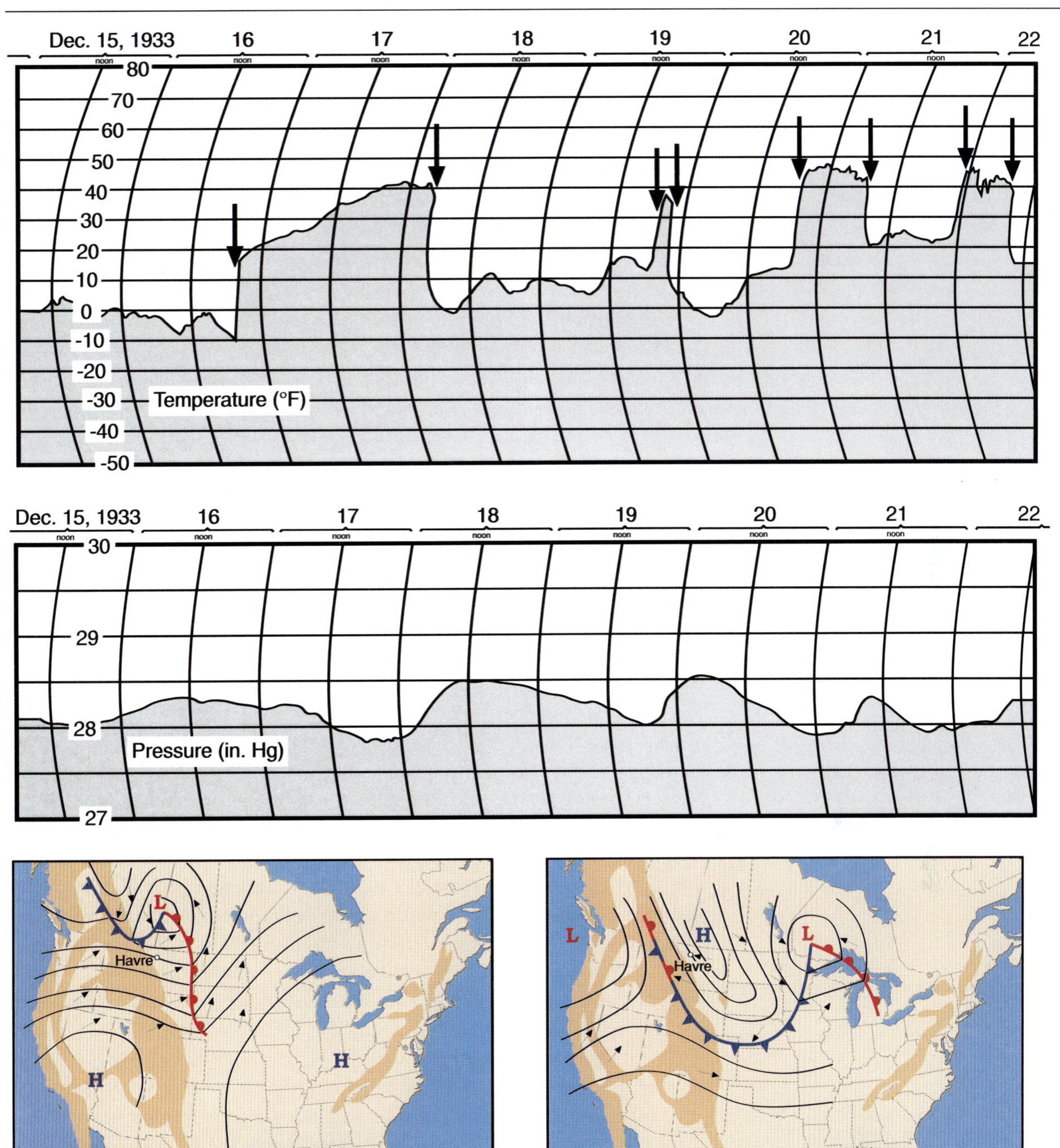

Figure I (*top*) A series of eight abrupt temperature changes occurred as warm chinook winds and shallow cold air masses alternated at Havre, Montana, during the week of 15–22 December 1933. Surface pressures increase when the cold air is in place. (Adapted from Math, 1934)

Figure II (*bottom*) Surface weather charts for (a) 18 December 1933 at 0100 UTC and (b) 19 December 1933 at 0100 UTC show the battle of the chinook wind at Havre, Montana. (Adapted from Glenn, 1961)

(*continued*)

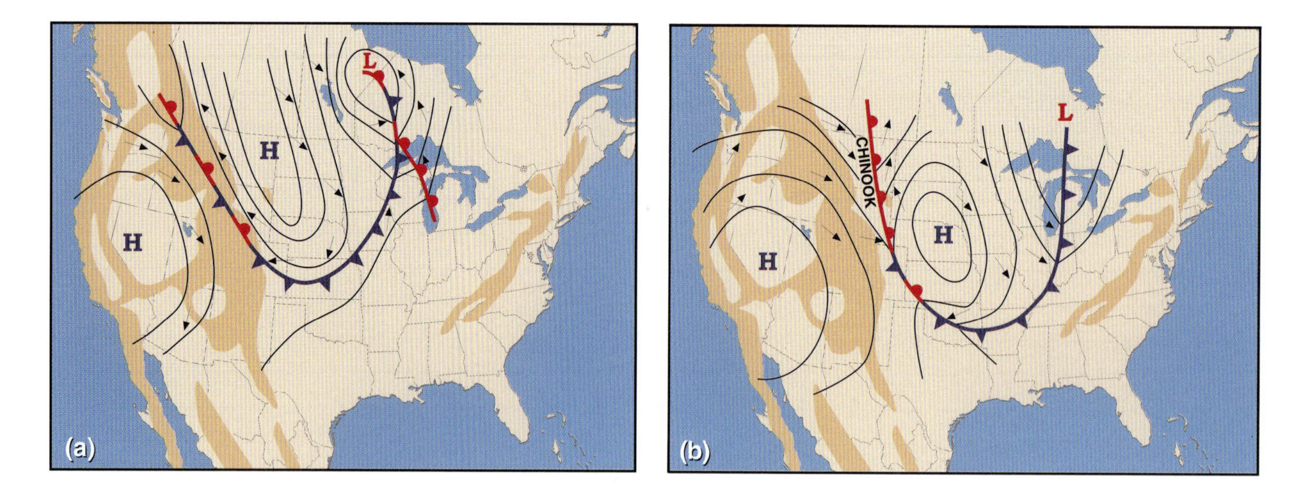

Figure III Conceptualized surface weather charts illustrating the break-in of chinook winds as an arctic air mass is carried eastward away from the mountains. (Adapted from Glenn, 1961)

pressure center channels air south and west out of the Great Basin around the southern end of the Sierra Nevada, over the Mojave Desert, and through low passes (e.g., Cajon Pass or Bannon Pass) into the Los Angeles Basin (figure 10.15). The synoptic pressure gradients leading to the Santa Ana winds (figure 10.16) are distinctive and easily recognized by forecasters. These foehn winds can be strong and gusty and can last for several days. Because the winds are dry, they desiccate the vegetation and can whip *wildfires* into firestorms that threaten property and lives.

Wasatch or *canyon winds* occur on the west side of Utah's Wasatch Mountains between Ogden and Provo when a strong east–west pressure gradient develops across the north–south Wasatch Range east of Salt Lake City. The easterly winds blow into a low pressure center in the Great Basin. Canyon winds appear to be of two types. The more common are winds that come through gaps in the ridgeline of the Wasatch Range and are channeled down the major canyons, producing localized strong and gusty winds at the canyon mouths. Less common are windstorms that affect a more or less contiguous zone along the foothills, with the strongest winds usually occurring in the lee of the major unbroken north–south sections of the Wasatch ridgeline. This second type of wind is produced by strong mountain waves or hydraulic jumps or combinations of the two. An example of a synoptic weather chart for a Wasatch wind episode is shown in figure 10.17. The wind strengths have been correlated with 850-mb height differences between Fort Bridger, Wyoming, and Salt Lake City. The greater the height difference (Salt Lake City has the lower height), the stronger the winds. (Similar "East winds" are found on the west side of the Cascades in Washington (figure 10.12).).

Gap winds that develop along the Pacific coast in Washington, Oregon, British Columbia, and southeast Alaska are sometimes enhanced by foehnlike flows when a strong east–west pressure gradient exists at and above mountaintop level between the coast and the inland areas, with low pressure along the coast. Winds are channeled westward through gaps in the coastal ranges and blow out over the Pacific Ocean or over the waterways of the Inside Passage. They are especially strong downwind of low passes or where major river valleys issue onto the seaways, such as below

WINDSTORMS AT BOULDER, COLORADO

- Boulder has between 2 and 16 windstorms per year.
- Windstorms are most frequent in Boulder from November through March (see figure I).
- Windstorms are somewhat more likely to occur at night than during the day.
- The typical windstorm lasts about 8 hours.
- The 11 January 1972 windstorm caused over 2.5 million dollars of damage in Boulder, with much of the damage in mobile home parks, housing construction areas, and residential areas where strong winds and/or blowing materials caused glass breakage and roofing damage.
- Historical records of windstorm damage in the Boulder area extend back to the 1860s.
- Parked railroad cars were reportedly blown off the tracks between

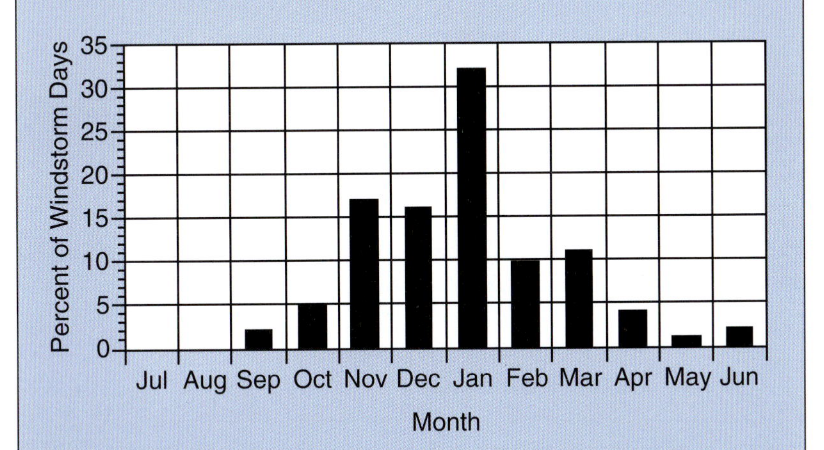

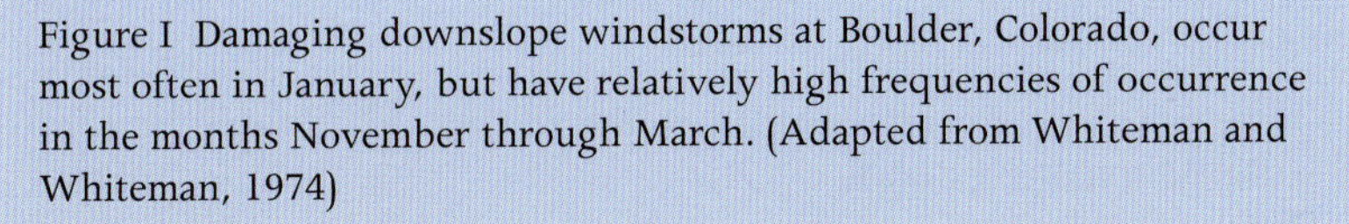
Figure I Damaging downslope windstorms at Boulder, Colorado, occur most often in January, but have relatively high frequencies of occurrence in the months November through March. (Adapted from Whiteman and Whiteman, 1974)

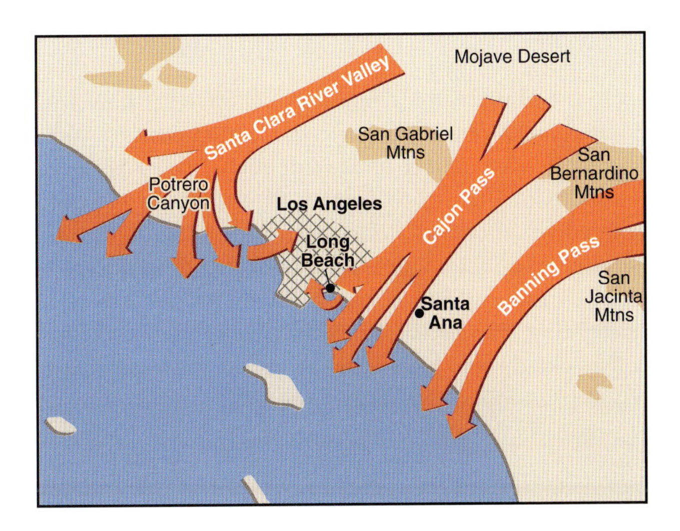

Figure 10.15 Santa Ana winds blow out of the Mojave Desert through the Santa Clara River Valley and through Cajon and Banning Passes toward the Pacific Ocean. Wildfires are very difficult to control when these strong, gusty, dry winds are at their strongest. (Adapted from Rosenthal, 1972)

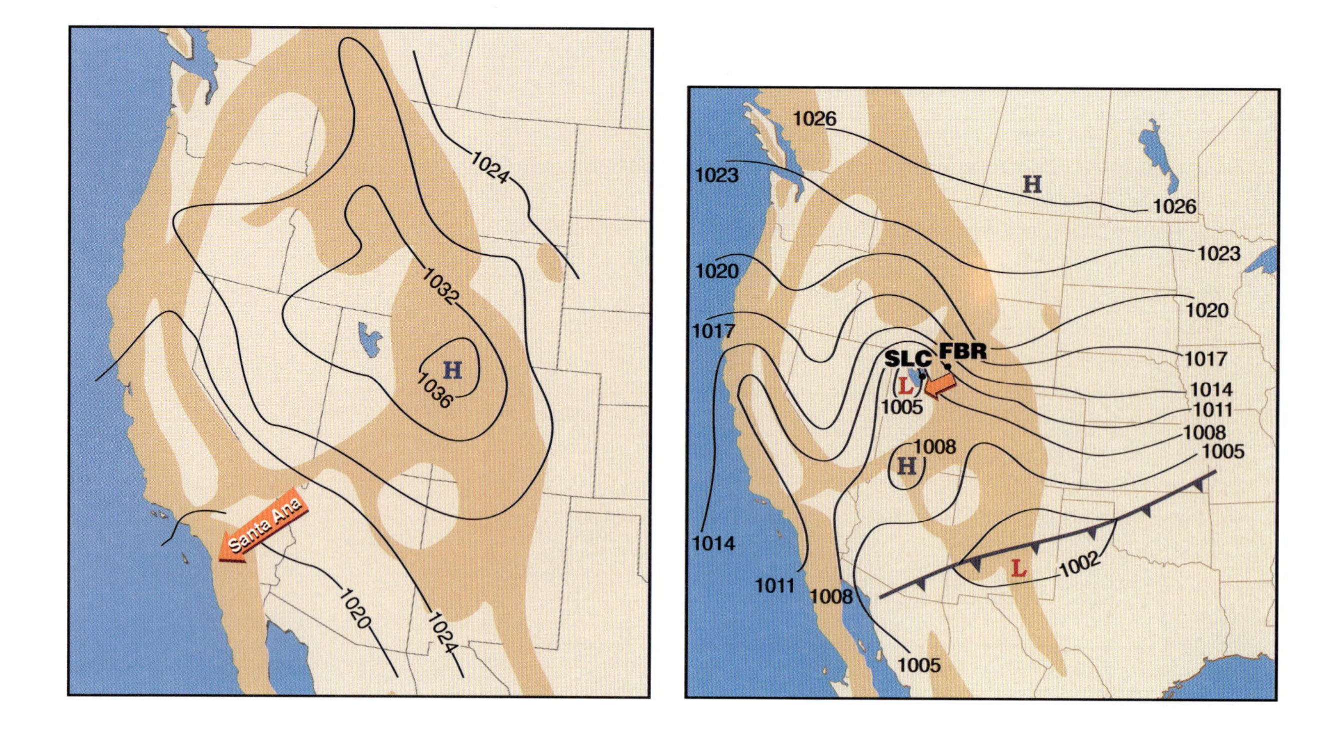

Figure 10.16 (*left*) The Santa Ana winds occur when a strong pressure gradient develops between the Great Basin and the coast of Southern California. (Adapted from Ahrens, 1994)

Figure 10.17 (*right*) This weather chart for 16 May 1952 at 1200 UTC shows the synoptic situation for a Wasatch wind event in which winds reached 95 mph (43 m/s) north of Salt Lake City. Wind strength is proportional to the pressure gradient between Salt Lake City (SLC) and Fort Bridger (FBR). Isobars are sea level pressures in millibars. (Adapted from Williams, 1952)

the Columbia River Gorge, in the Strait of Juan de Fuca, and in and below the Fraser River and Taku River Valleys.

10.3. Flow around Mountains

A flow approaching a mountain barrier tends to go around rather than over the mountains if

- the ridge line is convex on the windward side,
- the mountains are high,
- the barrier is a single isolated peak or a short range,
- the cross-barrier wind component is weak,
- the flow is very stable, or
- the approaching low-level air mass is very shallow.

The Rockies and the Appalachians are so long that flow around them is uncommon. Flow around subranges or isolated peaks occurs most frequently. Examples are found in the Aleutians, the Alaska Range, the Uinta Mountains of Utah, the Olympic Mountains in Washington, and around isolated volcanoes in California, Oregon, Washington, and Alaska.

Flow around mountains can sometimes be detected by watching clouds. On the windward side of the mountain barrier, low clouds are often seen extending, like a scarf, from one edge of the mountain barrier to the other. The clouds may pour over the barrier at its periphery and cascade down the lee slopes. Clouds and precipitation may also form in the lee of the barrier as the air currents that split around the barrier reconverge.

THE ALPINE FOEHN AND THE WESTERN CHINOOK

In the Alps, the characteristics of the foehn and the towns and valleys frequently affected by it are well known. The residents of the Alps are attuned to the foehn, which is a topic of daily discussion and interest. The foehn is valued for the warmth and the partly cloudy skies that it brings, a welcome respite from the cold and foggy conditions that persist for much of the winter on the north side of the Alps generally and on the Swiss Plateau in particular. However, the foehn is also associated with changes in human behavior, health, or psychological outlook and is an active area of research in Alpine medicine. Reports of headaches, low spirits, general ill-being, or depression are common, with some residents reporting that symptoms develop before the onset of strong winds. Various medications are available to treat the general feeling of malaise, described as feeling "foehnig." Schools may close during the foehn, and the effects of the winds can be used as a medical defense in court cases.

In the American West, chinooks are associated with strong, warm winds, although they are also responsible for many warm winter days when winds are not particularly strong and gusty. Some residents complain of headaches during windstorms, but other symptoms are rarely linked to the storms.

10.3.1. *Barrier Jets*

Barrier jets form when a stably stratified, low-level flow approaches a mountain barrier of fairly limited length with an edge or low pass on its left side. The flow, unable to go over the barrier, turns to blow parallel to the longitudinal axis of the barrier. Barrier jets always turn to the left in the Northern Hemisphere as they approach the barrier (figure 10.18). Thus, if a low-level stably stratified air mass approaches a north–south barrier from the west, it is turned to the left and becomes a south wind; a flow approaching from the east is also turned to the left and becomes a north wind.

Barrier jets turn to the left in the Northern Hemisphere as they approach a mountain barrier.

Barrier jets are observed infrequently in North America because of the extreme length of the major mountain barriers, but they have been reported on the west side of the Cascades and Sierra Nevada (figure 10.19), on the east sides of the Rockies and Appalachians, and on the north side of the Brooks Range. Although few observations are available, barrier jets probably develop around the smaller subranges of the Rockies. For example, there is anecdotal evidence that a southerly barrier jet sometimes occurs on the west side of the Colorado Rockies when the prevailing flow is from the west, carrying air around the north (i.e., left) side of the main crest and over the lower lying terrain of southern Wyoming.

Figure 10.18 A barrier jet occurs when winds are turned to the left to flow around the edge of a mountain barrier.

10.3.2. *Flow Splitting and Convergence Zones*

Flow splitting occurs when the length of a mountain barrier is limited and both the right and left sides are open, as is the case with isolated peaks and roughly circular mountain ranges such as the Olympics or the Alps.

Figure 10.19 East–west cross section through a barrier jet that formed on the west side of the Sierra Nevada. Isolines of constant wind speed (*isotachs*) are labeled in m/s and show the high wind speeds at the core of the flow. The southerly flow from the Sierra Nevada barrier jet is thought to be responsible for unusually high precipitation on the east side of the Trinity Alps at the north end of the Sacramento Valley. (Adapted from Parish, 1982)

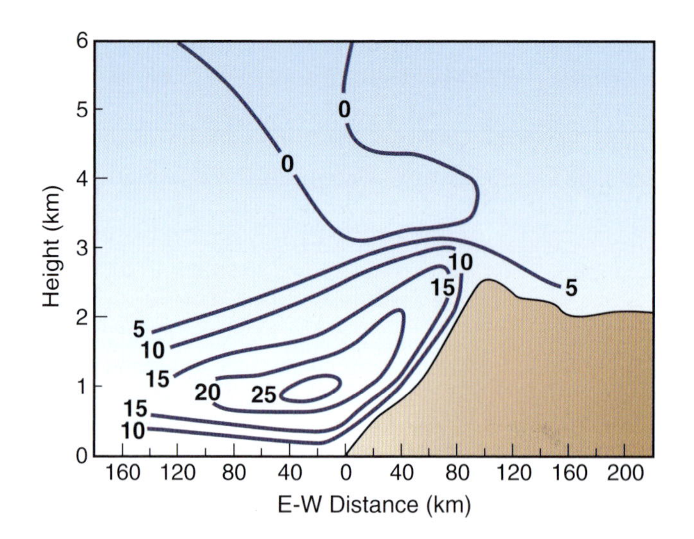

Part of the flow is carried around the left side of the barrier, part around the right side (figure 10.20). The flow around the left side is usually somewhat enhanced by the barrier jet.

The two branches of the split flow converge on the barrier's lee side, causing rising motions above the convergence zone and thus fog, cloudiness, and precipitation. The downwind location of the convergence zone shifts in response to changes in the direction of the approach flow and may also vary diurnally when conditions are influenced by diurnal mountain winds.

In North America, a well-known convergence zone is located in the lee of the Olympic Mountains of western Washington. Onshore flows split around the near-circular Olympic Mountains, flowing north through the Strait of Juan de Fuca and south through the low-lying Willapa Hills. The flows converge on the east side of the Olympics in the Puget Sound area, where low-level moisture is often present. Rising motions above the zone of low-level convergence produce clouds and precipitation. It is difficult to predict the exact location of the Puget Sound convergence zone because it shifts as the onshore wind direction changes, but it often affects operations at SeaTac International Airport south of Seattle. Convergence zones also exist in the lee of the Santa Monica and Santa Ana Mountains of Southern California.

A flow that splits around a barrier converges in the lee of the barrier. Fog, cloudiness, and precipitation may form in the convergence zone.

10.3.3. *Frontal Blockages and Postfrontal Accelerations*

The forward movement of an approaching cold front can temporarily slow as the cold front is forced to ascend a mountain range. As the depth of the cold air mass behind the cold front becomes deeper and deeper, the cold air may begin to invade the valleys, or the front and the cold air behind it may split and flow around the edges of the barrier. If the amount of cold air flowing around the edges of the barrier is small, or if the depth or rate of accumulation of cold air is sufficient, the front will continue to ascend the barrier until it eventually reaches and exceeds the mountain

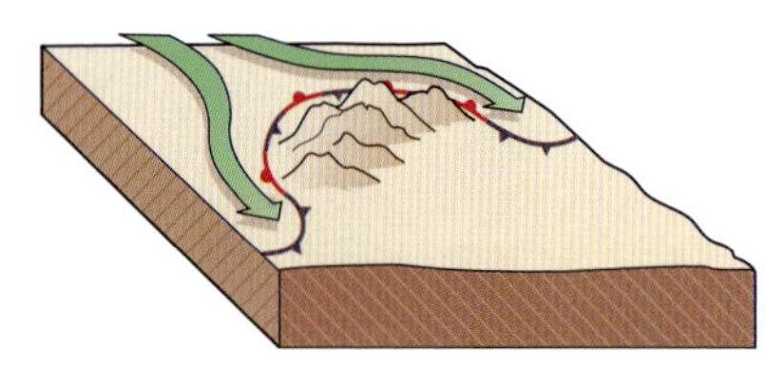

Figure 10.20 A conceptualized view of a cold air mass that approaches the Alps from the northwest and splits to flow around the eastern and western edges of the mountain ridge.

ridge height. The front will then suddenly accelerate as the cold air flows over the mountain ridge and pours down the lee side of the mountain, moving the front past the barrier and exposing the lee slopes to sudden cold air outbreaks (figure 10.21).

A cold front slows as it ascends a mountain barrier and accelerates when the cold air behind it flows over the barrier.

Channeling and distortion of fronts and postfrontal cold air outbreaks frequently occur in the Alps because of the limited horizontal extent and distinct edges of the range (figure 10.22). Similar wind systems are features of the wind *climatology* around the Pyrenees and subranges of the Alps (figure 10.23). Although flow splitting and postfrontal cold air incursions are not common around the long Rocky Mountain and Appalachian ranges, they can occur there when a cold air mass approaches an isolated subrange. A dense network of surface weather stations and radiosonde launch sites in the Alps provides ample documentation of these flows, whereas the low density of weather stations in the United States does not generally provide sufficient data.

10.4. Flows through Gaps, Channels, and Passes

Strong winds are often present in *gaps* (major erosional openings through mountain ranges), in channels between mountain subranges, and in mountain passes. The winds are usually produced by *pressure-driven channeling*, that is, they are caused by strong horizontal pressure gradients across the gap, channel, or pass. The pressure gradient may be im-

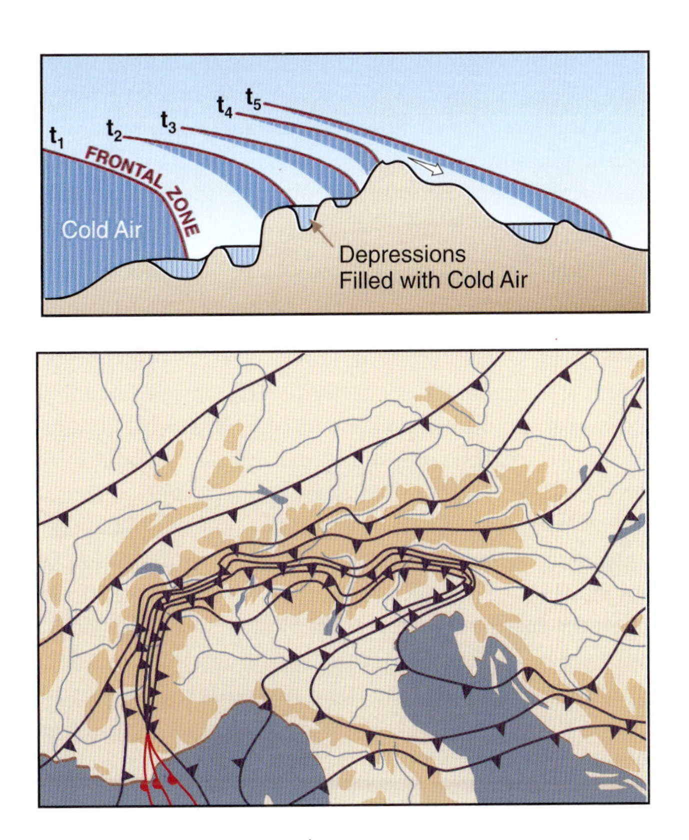

Figure 10.21 (*top*) Fronts are retarded as they move up mountain barriers but accelerate once the barrier is surmounted. The symbols t_1, t_2, t_3, t_4, and t_5) indicate regular time intervals and illustrate the slowing of the front as it climbs the barrier.

Figure 10.22 (*bottom*) Successive 3-hour frontal positions as a cold front approaches and passes the Alps, from 4 March 1982 at 0000 UTC to 5 March 1982 at 1800 UTC. The front slows as it encounters the Alps (dark tan shading). Cold air intrudes into the major Alpine valleys and surges around the eastern and western edges of the Alps. In this case, the stronger surge is around the east side of the Alps. (Adapted from Steinacker, 1987)

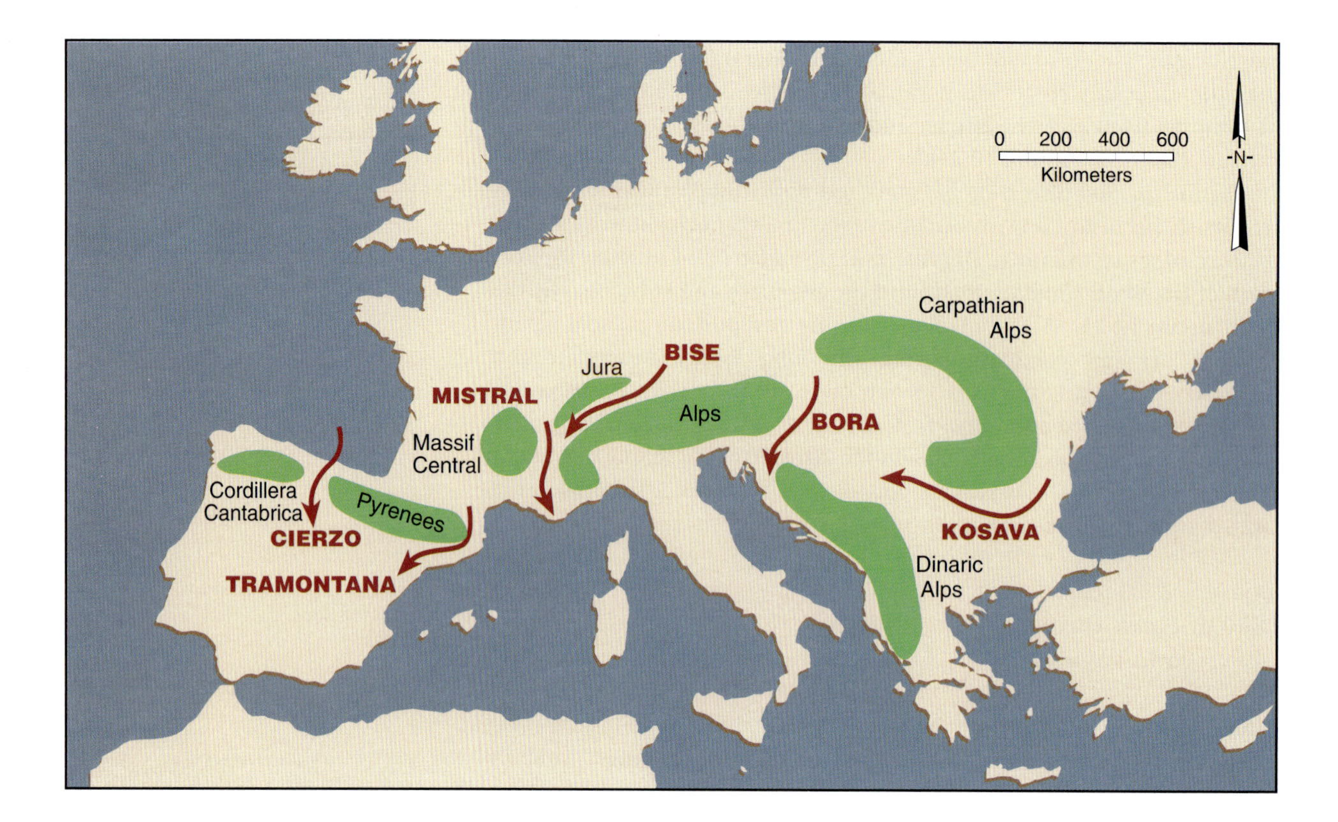

Figure 10.23 Flows between the major mountain barriers in Europe have been given special names. The Bora, flowing between the Alps and the Dinaric Alps, is also a downslope wind descending to the Adriatic coast. (Adapted from Wanner and Furger, 1990)

posed on the terrain by traveling synoptic-scale pressure systems or may result from differences in temperature and density between the air masses on either side of the opening. The differences are usually caused by regional-scale processes, but may also result from smaller scale processes, such as cold thunderstorm outflows. The strongest gap winds occur when synoptic-scale pressure gradients are superimposed on regionally developed pressure gradients. Because the physical processes are similar for gaps, channels, and passes, the phenomenon will be illustrated primarily for gaps.

Strong winds in a gap, channel, or pass are usually pressure-driven, that is, they are caused by a strong pressure gradient across the gap, channel, or pass.

10.4.1. *Flows through Coastal Mountain Ranges*

Regional pressure gradients occur frequently across coastal mountain ranges because of the differing characteristics of marine and continental air masses. In winter, the coldest and densest air is found over the elevated interior of the continent, whereas temperatures on the ocean side are moderated by the open ocean. When the pool of cold air in the interior becomes deep enough, it spills over the lowest altitude gaps, producing strong winds that descend toward the coast (figure 10.24). Further deepening of the cold pool increases the flow through the gap and allows air to flow over low ridges and eventually through higher altitude passes. When low clouds are present, the cloud mass fills the gap or pass and descends like a waterfall into the adjacent valley (figure 10.25). Because major channels or gaps are generally lower in altitude and wider than passes, they carry the strongest flows and largest air volumes. In summer, continental interiors are warmer than the adjacent ocean, and pressure gradients develop, forcing air through the gap from the ocean to the interior. Thus, it is typical for regionally driven pressure gradients in gaps to reverse between winter and summer, with the direction of the flow determined by the direction of the pressure gradient, usually toward the ocean during winter and toward the continental interior during summer.

Perhaps the best known gap wind in the United States develops in the Columbia River Gorge, the main east–west gap through the Cascade Range that connects the Pacific to the interior Columbia Basin (figure 10.26). The pressure difference between stations on the west and east ends of the gorge determines the direction and strength of the winds. In summertime, the Pacific High is off the coast of Oregon and Washington and a low is located over the Columbia Basin, producing an eastward flow that brings marine air up the gorge. The flow is moderately strong and blows opposite to the direction of the river current, producing the dependable up-gorge winds and river waves that are well known to windsurfers. In wintertime, when cold air collects in the Columbia Basin, the pressure gradient is reversed and strong, cold, easterly flows, called Columbia Gorge winds, come down the gorge. The easterly flows undercut warm, precipitating air masses that are being lifted up the western slope of the Cascades, enhancing precipitation and setting the stage for severe ice storms in the lower gorge. Summer and winter winds similar to the gap winds in the Columbia River Gorge also occur through low passes in the Cascade Mountains, including the Snoqualmie, Naches, and Stampede Passes.

Gap winds blow through other major gaps along the West Coast of North America, including the Caracena Strait near San Francisco, the Strait of Juan de Fuca near Seattle, the Fraser Valley near Vancouver, the Stikine Valley near Wrangell, the Taku Straits just south of Juneau, the Copper River Valley just east of Cordova, and the Turnagain Arm south of Anchorage. As in the case in the Columbia River Gorge, the summer

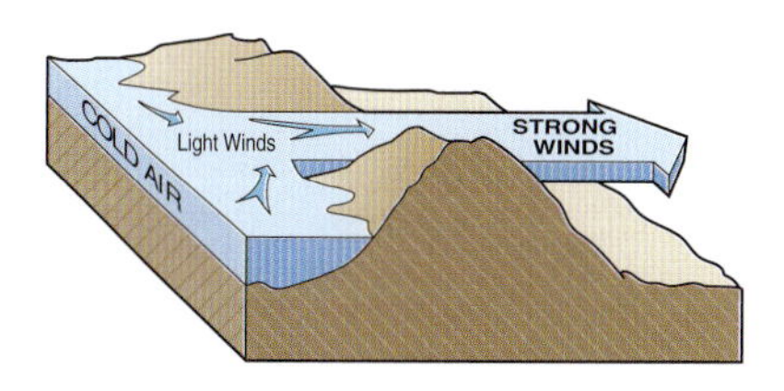

Figure 10.24 Flows can accelerate as they come through mountain passes.

Flows through coastal ranges usually reverse seasonally, flowing toward the ocean during winter and toward the continental interior during summer.

Figure 10.25 A waterfall cloud over Mackinnon Pass (1154 m), a major pass through the Southern Alps of New Zealand on the Milford Track. Clouds are pouring westward across this low pass from the Clinton River Valley into the Arthur River Valley. (Photo © C. Whiteman)

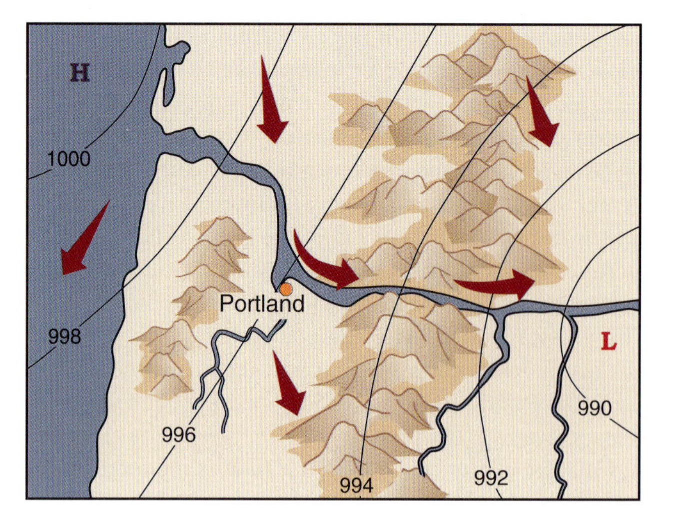

Figure 10.26 Winds are channeled through the Colombia River Gorge. The consistent up-valley flow produces well-developed waves on the Columbia River and wind velocities that are ideal for wind surfing.

inflows through these gaps are relatively weak compared to the strong wintertime outflows.

Wintertime gap winds that blow out onto the Pacific Ocean or onto the waterways of the Inside (or Inland) Passage pose a serious hazard to shipping, especially to slow-moving barges pulled by tugboats. (The Inside Passage is nonetheless the preferred shipping channel because of strong winds and high waves on the open Pacific Ocean.) Of particular concern to barge traffic is the passage between Cross Sound at the north end of Chichagof Island and the anchorage at Cordova, where barge traffic moves onto the open ocean but is also subject to gap winds at the outlet of the Copper River. Winds along the Inside Passage are complicated by channeled winds that frequently blow along the complex of marine channels, canals, or straits, driven by pressure gradients that develop along the length of the channels.

Forecasts of gap winds are usually based on a forecast of pressure differences across the gap. Forecasters monitor surface pressure differences across the gap from surface weather stations and watch the development of synoptic pressure systems. Forecasts of wintertime gap winds along the coast of western Canada and southeast Alaska, for example, rely on the pressure measurements in the cold air mass on the east side of the Coast Range in the Rocky Mountain Trench (e.g., at Whitehorse) and on pressure changes along the coast as low pressure centers or troughs approach the coast from the west. When a low approaches from the west, winds north of the low have an easterly component that strengthens the gap wind flow.

10.4.2. *Flow into Heat Lows*

Confined desert or semiarid basins or plateaus have little soil moisture and vegetation and therefore experience little evaporation and large sensible heat fluxes during daytime. A large sensible heat flux results in a deep, warm convective boundary layer above the surface and relatively low pressures in the basin or on the plateau. A low pressure formed in this way is termed a *heat low*. Horizontal pressure gradients between the heat low and its higher pressure surroundings drive winds that bring air from the surroundings into the heat low, especially at low altitudes where the horizontal pressure gradients are strongest. A heat low contributes to the summertime low pressures over the Columbia Plateau that bring marine air up the Columbia River Gorge. The best known and most strongly developed heat low in the United States forms in summer over the Mojave Desert and the south end of the Great Basin. Winds blow into this low pressure area from the Gulf of California to the south and across passes in the Sierra Nevada to the west and north. Networks of wind machines have been installed on the Tehachapi and Altamont Passes to harness these winds for the production of electrical power.

10.4.3. *The Venturi (or Bernoulli) Effect*

When a valley or other channel has a substantial pressure gradient along its length and a topographic constriction at some point along the channel, air is accelerated through the constriction by the pressure drop across the constriction (figure 10.27). Acceleration through a terrain constriction is called the *Venturi* or *Bernoulli effect*. The flow speed can be roughly estimated by assuming that the mass of the flow is conserved through the channel, so that speed increases when the cross section of the flow narrows and decreases when the cross section widens. When the pressure gradient across the constriction is weak or the constriction provides a near-total blockage, air may pool behind the constriction rather than accelerating through it (section 11.4.2.1).

10.4.4. *Forced Channeling*

Most winds through gaps, channels, and passes are driven by the difference in pressure from one side of the gap to the other. These winds blow across the pressure contours from the area of high pressure to the area of low pressure. Some winds, however, result when strong flows aloft under neutral or unstable conditions are channeled by landform features, such as parallel mountain ranges or valleys. *Forced channeling*, in contrast to pressure-driven channeling, requires the downward transfer of momentum into the channel from winds aloft that are blowing parallel to pressure contours (figure 10.28). Forced channeling often occurs with foehn windstorms and at high altitudes in the mountains, especially at mountain passes. Forced channeling is described in more detail in section 11.3.2.

Forced channeling is the transfer of momentum from winds aloft into a terrain channel.

10.5. Blocking, Cold Air Damming, and Obstruction of Air Masses

A mountain barrier can not only channel an approaching flow over, around, or through gaps and passes in the barrier, it can also block an approaching flow or it can obstruct a shallow, stable air mass that develops on one side of the barrier, channeling it along the foot of the barrier and preventing it from crossing the barrier. Blocking, cold air damming,

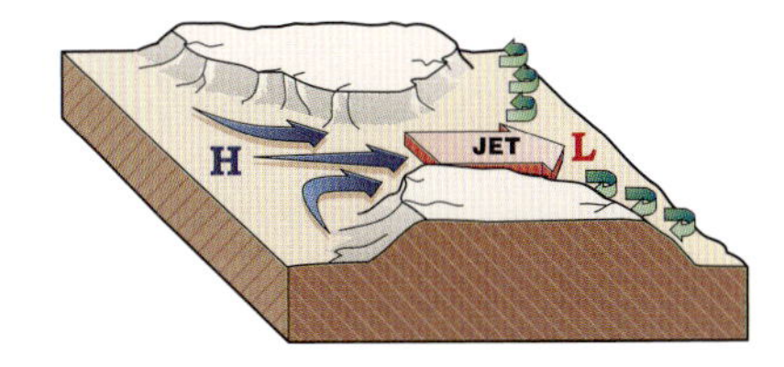

Figure 10.27 The Venturi effect causes a jet to form as winds pass through a terrain constriction and strengthen.

and obstruction all affect stable air masses and occur most frequently in winter.

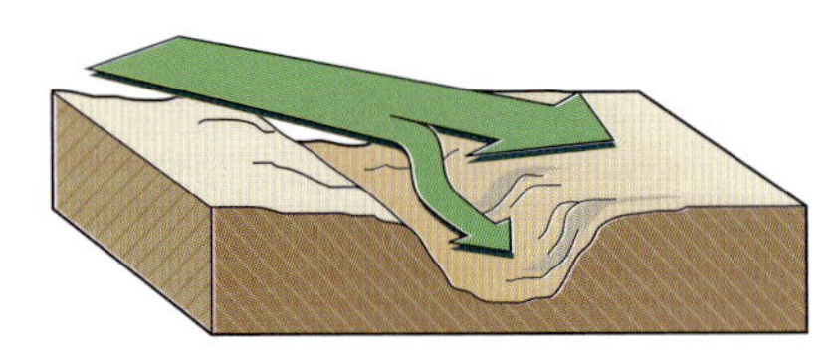

Figure 10.28 Forced channeling occurs when upper winds are brought down into valleys from aloft and turned to flow along the valley's longitudinal axis.

Blocking, cold air damming, and obstruction all affect stable air masses and occur most frequently in winter.

10.5.1. *Blocking*

A low-level, stably stratified flow approaching a mountain barrier can be stopped or blocked at the barrier as dense air builds up over the windward slope of the mountains, causing the formation of a local high pressure (figure 10.29). The resulting mesoscale pressure gradient can counteract the synoptic-scale pressure gradient that drives the approaching flow, sometimes blocking or stalling the approaching air mass. Blocking occurs frequently in winter when cold, stable air masses are common and may persist for extended periods. Smoke and pollution emitted into a blocked layer can be trapped for days in the stalled flow.

The blocked flow upwind of the barrier is usually shallower than the barrier depth. Air above the blocked flow layer may have no difficulty surmounting the barrier and may respond to the effective topography (section 10.1.1), that is, the combination of the orographic barrier and the blocked cold air mass on its upwind side.

The horizontal extent of a blockage upwind of a barrier depends on the height of the barrier, the stability of the approaching flow, and the latitude. The higher the barrier and the more stable the flow, the greater the horizontal extent. As latitude increases, the horizontal extent of the blockage decreases. If an isothermal air mass approaches a 3250-ft-high (1000-m) barrier at 40°N latitude, blocking extends about 125 mi or 200 km upwind of the barrier. The blockage upwind of a mountain half as high (1625 ft. or 500 m) extends about half this distance. The speed of the approaching flow does not affect the horizontal extent of a blockage.

In basins on the windward side of mountain ranges, a blocked air mass can form a more or less horizontally stratified pool of cold air across the entire basin. Blockages occur frequently in winter in the Intermountain Basin west of the Rocky Mountains, with individual episodes persisting for periods as long as 10 days. The strength of the pool (i.e., the temperature difference between the bottom and top of the pool), as well as its depth, vary as cold air drains into the basin from the surrounding mountains and as cold or warm air is advected into the basin by synoptic flows. The strength of the pool increases when warm air is advected above the pool and decreases when cold air is advected above the pool.

The onset and cessation of blocking can be abrupt, occurring over periods as short as one hour and causing sudden changes in dispersion conditions for smoke or air pollution discharges. In winter, blocking is influenced primarily by synoptic conditions, with the pressure gradient, and therefore blocking, increasing when a synoptic ridge passes the barrier and decreasing when a synoptic trough (often a short-wave trough) passes the barrier. If the upwind stable layer is shallow or only weakly stable, blocking ends when daytime convection upwind of the mountain is sufficient to break through the stable layer. The flow in the resulting neutral layer has no difficulty surmounting the mountain barrier. The blockage may reform at night when longwave radiation loss cools the atmosphere.

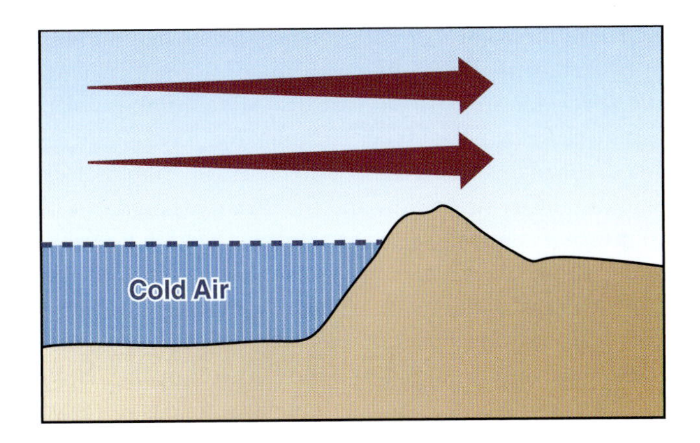

Figure 10.29 Flow blocking produces a layer of stagnant air at low levels upwind of the mountain barrier.

10.5.2. *Cold Air Damming*

Cold air damming is a wintertime phenomenon that occasionally affects the east sides of the Rocky Mountains and Appalachians, resulting in persistent low temperatures, precipitation, and ice storms. A cold, shallow, high pressure center surges southward on the east side of the mountain barrier, a trough or low approaches the mountains from the west, and a pressure gradient is established that initiates an easterly flow over the east side of the barrier. Pressure-driven winds lift the cold air partway up the east slope of the mountains. The air, cooled even further by adiabatic ascent, cannot rise all the way up the barrier and becomes trapped over the east slope. The shallow, wedge-shaped air mass effectively alters the terrain configuration of the barrier to flows that approach it from the east. Warm, moist air to the east of the mountains, also carried westward by the pressure gradient, is less stable than the cold air and can be lifted up and over the cold air dam, causing stratiform or cumuliform cloudiness and precipitation many miles upwind of the barrier. The warm precipitation can turn to snow, freezing rain, or sleet as it falls through the entrenched cold air mass below. The cold air dam becomes more stable when rain or snow falls from above and evaporates into the cold dry air within the dam. The type and location of precipitation along the foothills can be difficult to predict when a cold air dam is present because the position and depth of the cold air mass relative to the mountain barrier can vary with time.

10.5.3. *Obstruction of Air Masses*

In contrast to blocking and cold air damming, which result from the dynamic lifting and cooling of an approaching flow, *obstruction* of an air mass is a static process that can occur on either the upwind or downwind side of a barrier. An air mass develops or moves along one side of a barrier and is unable to surmount the barrier, either because the air mass is too shallow or because the cross-barrier wind component in the air mass is too weak (figure 10.30). Obstruction occurs most frequently along long barriers that prevent cold, stable, very shallow, or dense air masses from flowing around the barrier.

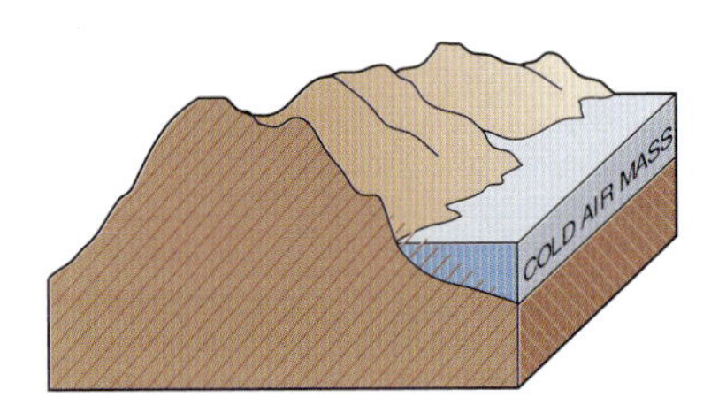

Figure 10.30 Mountain chains make an effective barrier for shallow air masses. The cold air mass here represents a shallow polar air mass on the east side of the Rocky Mountains.

When shallow air masses are common near a long mountain range, the range can be a climatic divide. In North America, the Rocky Mountains and the Appalachians are both climatic divides, confining polar and arctic air masses that form in the higher latitudes and travel south across central Canada to the plains between the two mountain ranges. Because temperatures in the air masses are low and the density is high, the air masses form high pressure centers. Air to the south of the high flows toward the west, forcing the shallow cold air up the Great Plains to the eastern foothills of the Rocky Mountains. Lifting is generally insufficient, however, to carry the shallow dense air over the Rockies. Air to the north of the high flows eastward but is obstructed by the Appalachians. Because the flow on the leading edge of the high pressure area is easterly, the cold shallow air mass hugs the eastern foothills of the Rockies and the Sierra Madre Oriental as it moves south into the Gulf of Mexico and, in some cases, as far south as the Bay of Campeche and the Gulf of Tehuantepec in Central America.

10.6. On the High Plains: The Low-Level Jet

A strong southerly or southwesterly low-level wind, called the low-level jet (LLJ), frequently occurs during nighttime over the Great Plains, which slopes downward from the foothills of the Rocky Mountains to the Mississippi River. This nocturnal flow has a distinctive *jet wind speed profile* (figure 10.31) with maximum speeds often in excess of 35 mph (16 m/s) only 800–1500 ft (250–450 m) above the ground. The LLJ is focused in a narrow band with typical widths of 200–400 miles and extends from the Gulf of Mexico northward into the central Great Plains. The LLJ typically begins at sundown and persists through the night into midmorn-

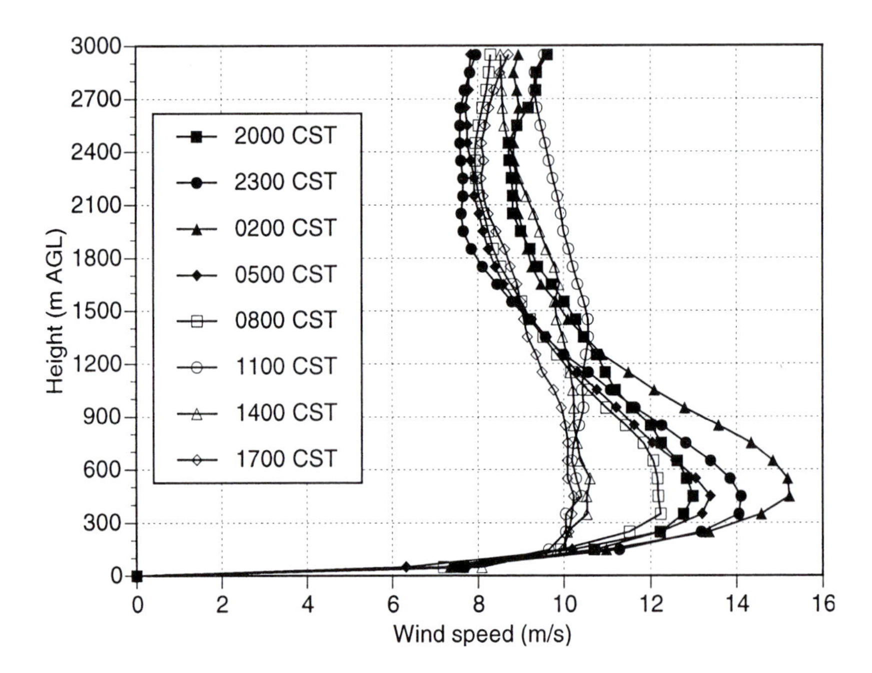

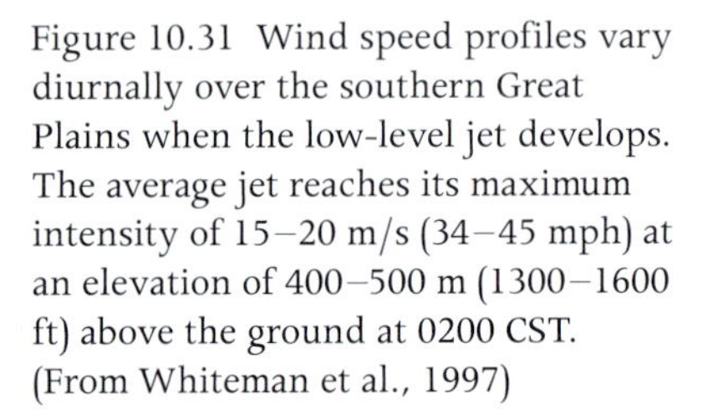

Figure 10.31 Wind speed profiles vary diurnally over the southern Great Plains when the low-level jet develops. The average jet reaches its maximum intensity of 15–20 m/s (34–45 mph) at an elevation of 400–500 m (1300–1600 ft) above the ground at 0200 CST. (From Whiteman et al., 1997)

THE LLJ AND BIRD MIGRATION

Birds and insects use the southerly LLJ to migrate northward in the spring; they often wait for nights when the LLJ is particularly strong or seek flight altitudes that provide the best tailwinds. Many species of birds migrate at night, a fact driven home to meteorologists when a larger network of vertically pointing meteorological radars was installed in the midwest in the early 1990s. This network reported anomalous nighttime winds that were ultimately traced to radar returns from migrating birds. Because birds are better radar scatterers than the atmosphere, the radars measure the ground speed (i.e., the sum of the bird's flight speed, about 10 m/s, and the speed of the jet) of the birds rather than the speed of the air current. In the spring, when the birds fly north with an LLJ tailwind, the radar-reported wind speeds are about 10 m/s too high. In the fall, when the birds fly south into an LLJ head wind, they choose altitudes where the LLJ is relatively weak. Radar-reported wind speeds are about 10 m/s too slow during this time.

ing. Although it can occur in any season, the LLJ is most frequent in the summer half of the year.

A southerly or southwesterly low-level jet occurs frequently during nighttime over the Great Plains.

Although the mechanisms that cause the LLJ to form are still being debated, the correspondence between the spatial extent of the phenomenon and the spatial extent of the Great Plains suggests that topography plays an important role. One theory is that the southeasterly flow around the Bermuda–Azores High over the southern Great Plains is channeled by the slope of the Great Plains and the Rocky Mountains to produce a southerly flow. Another theory suggests that an inclined temperature inversion over the sloping plains provides a thermodynamic driving force. A final factor, independent of topography, that is widely acknowledged as a contributor to the formation of the LLJ is the decoupling of the boundary layer winds from the frictional drag of the earth's surface once surface-based inversions begin to form in the early evening.

The low-level jet can be easily distinguished from the barrier jet (section 10.3.1) that forms east of the Rockies. The LLJ blows from the south and extends far to the east of the Rockies, whereas the barrier jet blows from the north and is closer to the mountain barrier.

11 Diurnal Mountain Winds

Diurnal mountain winds develop over complex topography of all scales, from small hills to large mountain massifs and are characterized by a reversal of wind direction twice per day. As a rule, winds flow upslope, up-valley, and from the plain to the mountain massif during daytime. During nighttime, they flow downslope, down-valley, and from the mountain massif to the plain. Diurnal mountain winds are strongest when skies are clear and winds aloft are weak.

Diurnal mountain winds are produced by horizontal temperature differences that develop daily in complex terrain. The resulting horizontal pressure differences cause winds near the surface of the earth to blow from areas with lower temperatures and higher pressures toward areas with higher temperatures and lower pressures. The circulations are closed by return, or compensatory, flows higher in the atmosphere.

Four wind systems (figure 11.1) comprise the *mountain wind system*, which carries air into a mountain massif at low levels during daytime and out of a mountain massif during nighttime.

- The *slope wind system* (*upslope winds* and *downslope winds*) is driven by horizontal temperature contrasts between the air over the valley sidewalls and the air over the center of the valley.
- The *along-valley wind system* (*up-valley winds* and *down-valley winds*) is driven by horizontal temperature contrasts along a valley's axis or between the air in a valley and the air over the adjacent plain.
- The *cross-valley wind system* results from horizontal temperature differences between the air over one valley sidewall and the air over the opposing sidewall, producing winds that blow perpendicular to the valley axis and toward the more strongly heated sidewall.
- The *mountain–plain wind system* results from horizontal temperature differences between the air over a mountain massif and the air over the surrounding plains, producing large-scale winds that blow up or

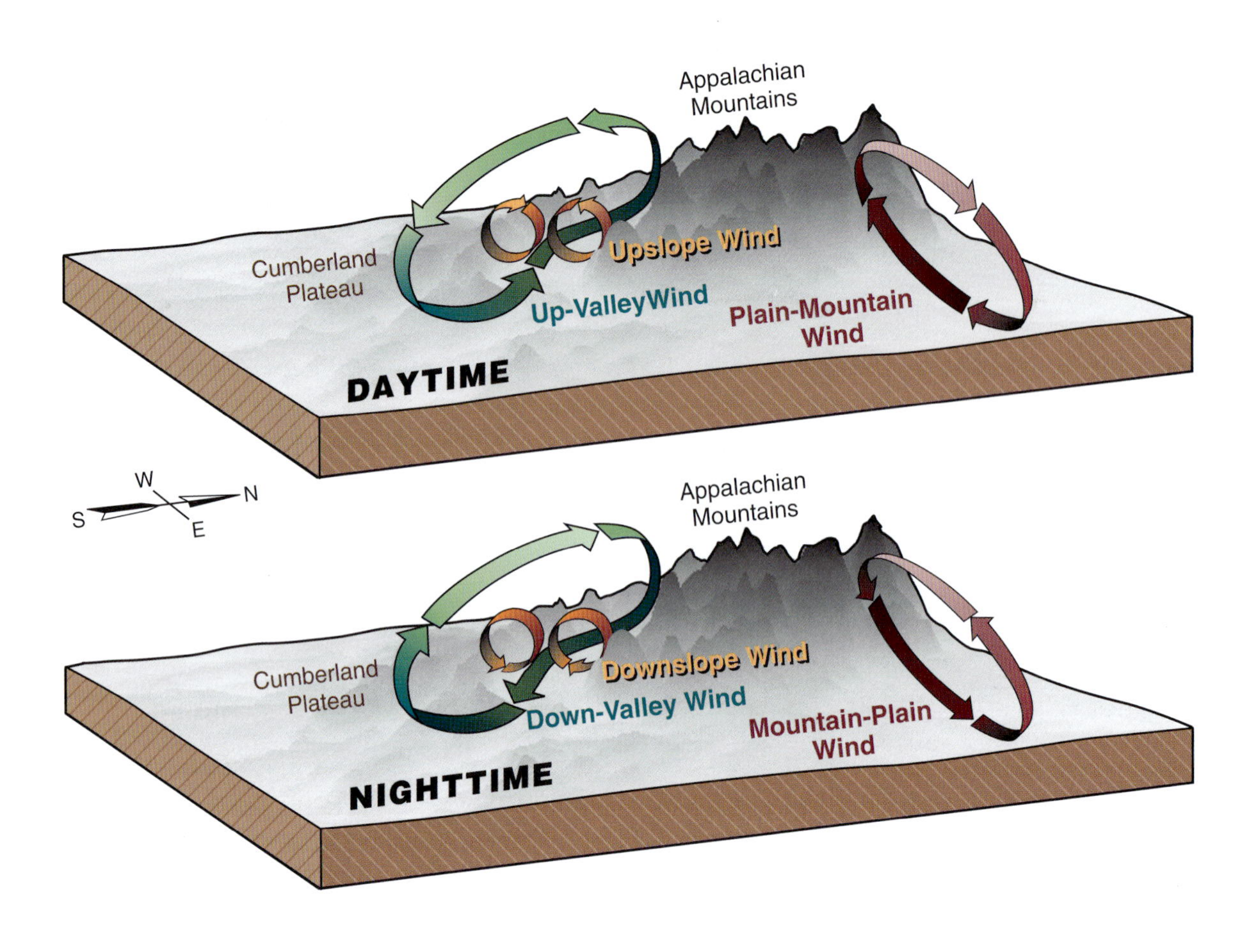

Figure 11.1 Four interacting wind systems are found over mountainous terrain. Three of these are shown schematically in this digital elevation model of the Appalachian Mountains and the Tennessee Valley. The fourth wind system, cross-valley flow (not shown), blows from cooler sidewalls toward warmer sidewalls. (Elevation model provided by J. Fast, Battelle Northwest Laboratories.)

Winds blow up the terrain (upslope, up-valley) when surfaces are heated during daytime and down the terrain (downslope, down-valley) when surfaces are cooled during nighttime.

down the outer slopes of a mountain massif. The mountain–plain circulation and its upper level return flow are not confined by the topography but are carried over deep layers of the atmosphere above the mountain slopes.

11.1. The Daily Cycle of Slope and Along-Valley Winds and Temperature Structure

Because diurnal mountain winds are driven by horizontal temperature differences, the regular evolution of the winds in a given valley is closely tied to the thermal structure of the atmospheric boundary layer within the valley, which is characterized by a diurnal cycle of buildup and breakdown of a temperature inversion. The relationship between winds and temperature structure can best be described by considering the typical diurnal cycle in a representative valley. Figure 11.2 shows the four distinct phases of wind and temperature structure evolution in a valley during the course of a day:

- The evening-transition phase begins when slope winds reverse from upslope to downslope. The downslope winds drain cold air off the sidewalls into the valley, resulting in the build-up of an inversion and causing along-valley winds to begin to reverse direction from up-

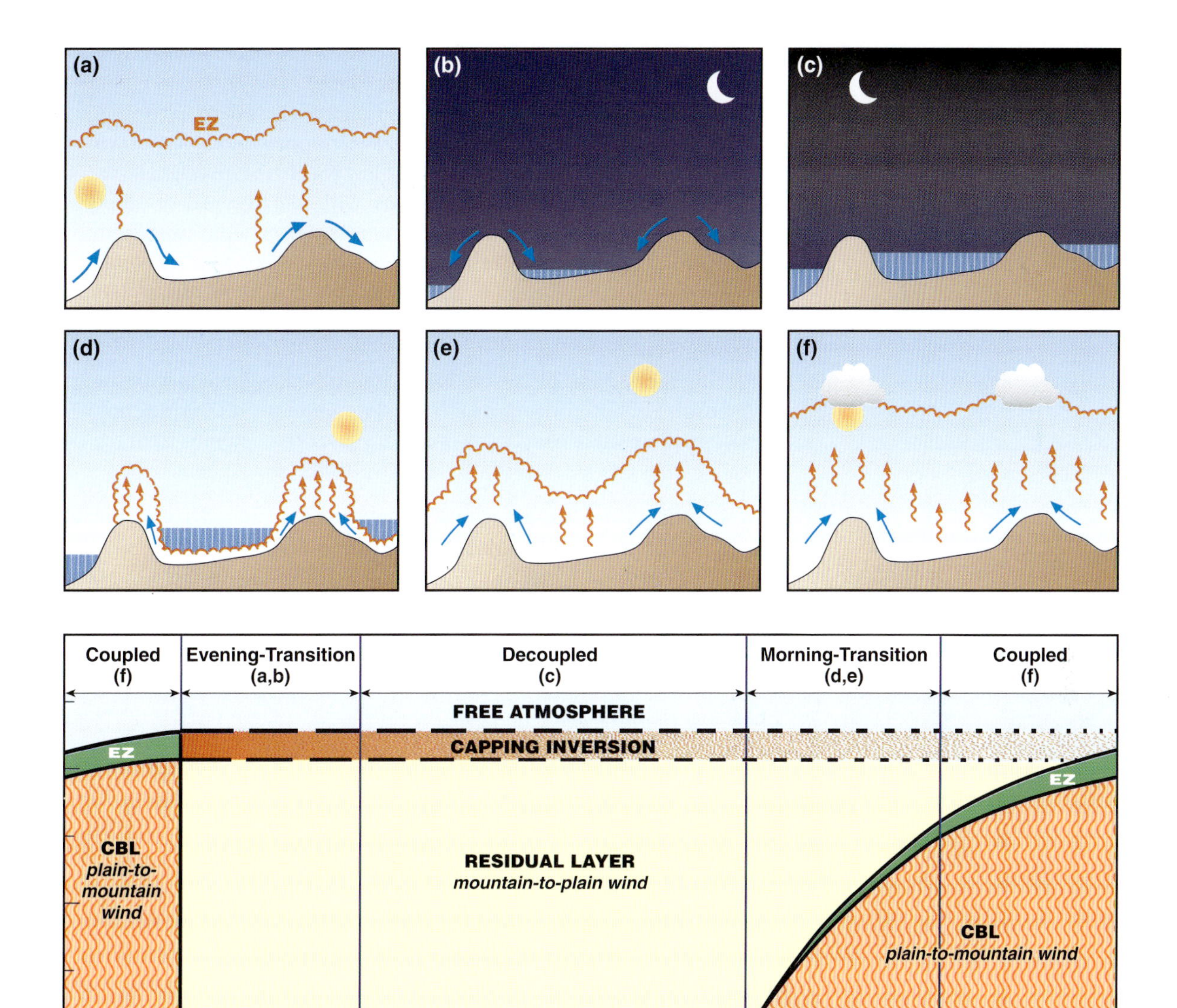

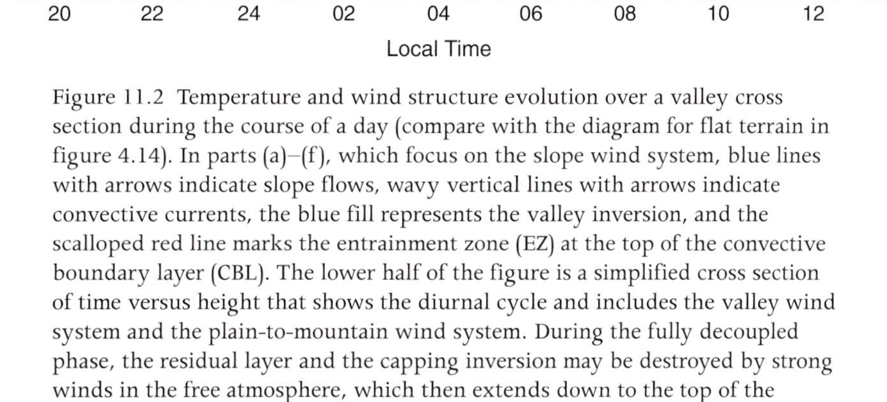

Figure 11.2 Temperature and wind structure evolution over a valley cross section during the course of a day (compare with the diagram for flat terrain in figure 4.14). In parts (a)–(f), which focus on the slope wind system, blue lines with arrows indicate slope flows, wavy vertical lines with arrows indicate convective currents, the blue fill represents the valley inversion, and the scalloped red line marks the entrainment zone (EZ) at the top of the convective boundary layer (CBL). The lower half of the figure is a simplified cross section of time versus height that shows the diurnal cycle and includes the valley wind system and the plain-to-mountain wind system. During the fully decoupled phase, the residual layer and the capping inversion may be destroyed by strong winds in the free atmosphere, which then extends down to the top of the inversion.

valley to down-valley. This transition phase ends when down-valley flows prevail through the depth of the valley.

- During the nighttime phase, the valley atmosphere is decoupled from the atmosphere above ridge-top level. Downslope winds blow down the sidewalls, and down-valley winds blow within the inversion layer.
- The morning-transition phase begins with the reversal of slope winds from downslope to upslope. Convective currents rising from the ground destroy the inversion from below, and along-valley winds reverse from down-valley to up-valley. This transition phase ends when the inversion is destroyed and winds blow up-valley through the depth of the valley.
- During the daytime phase, the valley atmosphere is coupled with the atmosphere above ridge-top level. Upslope and up-valley winds prevail in an unstable convective boundary layer that extends from the valley floor and sidewalls into the above-valley atmosphere.

The evolution of slope and along-valley winds is closely related to the temperature structure in a valley, which is characterized by a nighttime inversion and a daytime convective boundary layer.

These four phases are described in detail subsequently for a typical 24-hour period in a representative 500-m-deep (1600-ft) valley cut into the side of a large mountain range. The slope and along-valley wind systems are emphasized because they have the greatest impact on weather-sensitive activities in mountainous terrain.

11.1.1. *Evening Transition Period*

During the evening transition period, slope and valley winds reverse from upslope to downslope, and a temperature inversion builds up within the valley.

The evening transition phase (figures 11.2a and b) begins in the late afternoon shortly before sunset when the surface energy budget within the valley reverses sign, that is, longwave radiation losses become greater than shortwave radiation gains, causing the surface to cool radiatively. Downward sensible heat flux from the atmosphere to the ground produces a shallow layer of cool, stable air over the valley floor and sidewalls. Because the air above the slopes is colder than the ambient air over the valley center, it slides down the slopes in a shallow layer immediately above the slopes (figure 11.3). The cold air flow follows the *fall line* (i.e., the direction of steepest drop of a slope), and cold air accumulates over the valley center, building up a valley temperature inversion. As the inversion builds, the coldest air is found at the valley floor, but the air over the ridge tops is also cold. The warmest air is generally found at the approximate level of the top of the inversion (figure 11.4) in a *thermal belt* partway up the sidewalls (section 13.3.3.2).

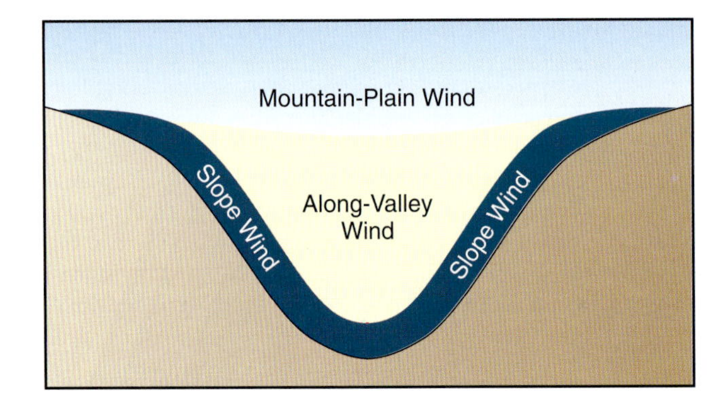

Figure 11.3 Slope winds blow in layers adjacent to the slope, whereas along-valley winds blow over much of the valley cross section. The weak mountain–plain circulation above the valley is often overpowered by strong winds.

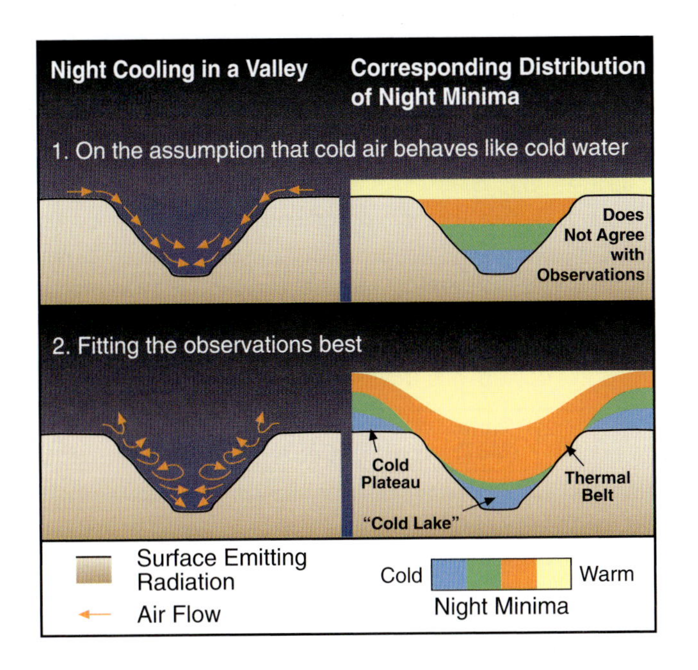

Figure 11.4 A warm zone or thermal belt is typically found on a valley sidewall at night. Colder air is present at the valley floor and on the ridge tops. (Adapted from Geiger et al., 1995)

As the cold air pouring down the slopes converges over the valley center, it rises, cooling the air higher and higher in the valley atmosphere, which in turn causes the inversion to grow deeper and deeper. The temperature inversion grows rapidly on clear nights, often growing to 500 m within 3–5 hours after sunset. Winds within the inversion layer blow down-valley. As the inversion layer deepens, the layer of the down-valley winds also deepens and the winds become stronger. Winds above the inversion continue to blow up-valley. The evening transition period ends when the rapid growth of the inversion slows and down-valley winds prevail through the depth of the fully developed inversion.

The evolution of wind structure during the evening transition period on 15 October 1978 over the center of Colorado's 700-m-deep Eagle Valley is shown in figure 11.5. Arrows indicate wind direction and speed (the longer the arrow, the higher the speed). The first sounding shows the up-valley winds that blew in a deep convective boundary layer during the day. The subsequent soundings show a shallow layer of down-valley winds that increased in depth over time as an inversion grew deeper in the valley. Down-valley wind speeds increased, with peak speeds occurring about midlevel in the valley atmosphere. A thin transition layer, indicated by the two lines crossing the soundings, separates the down-valley winds within the surface-based inversion, which grows in depth after sunset, from the up-valley winds, which continue to blow above the developing temperature inversion.

11.1.2. *Decoupled Period (Nighttime)*

During the night (figure 11.2c), down-valley winds prevail within a fully formed valley temperature inversion. The inversion depth remains essentially constant and is approximately equal to the valley depth. Shallower valleys in climates with strong nighttime cooling often have inversions that extend above the adjacent ridge tops, whereas deeper val-

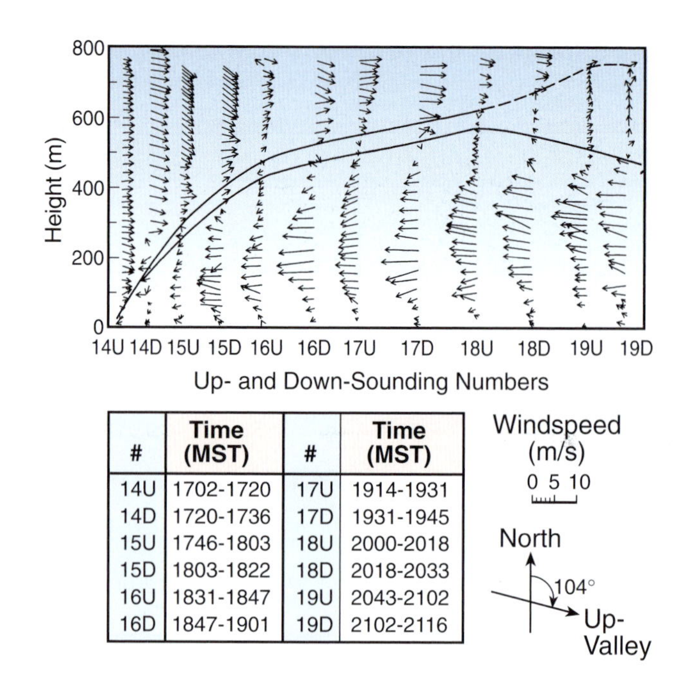

#	Time (MST)	#	Time (MST)
14U	1702-1720	17U	1914-1931
14D	1720-1736	17D	1931-1945
15U	1746-1803	18U	2000-2018
15D	1803-1822	18D	2018-2033
16U	1831-1847	19U	2043-2102
16D	1847-1901	19D	2102-2116

Figure 11.5 Sequential tethered balloon up- and down-soundings of horizontal winds during the evening transition period in Colorado's Eagle Valley. Arrows show wind direction and strength. North and the up-valley direction are shown in the wind legend. The table gives the time intervals corresponding to each of the soundings. The first up-sounding started at 1702 and the last down-sounding ended at 2116 mountain standard time (MST). The lines delineate the transition zone between the down-valley winds within the inversion and the up-valley winds above the inversion. (Adapted from Whiteman, 1986)

leys have inversions that do not completely fill the valley. In moist or cloud-covered valleys, the nighttime inversions are often much shallower than the valley, with the inversion depth often corresponding to the top of a fog, haze, or cloud bank.

The downslope winds continue on the sidewalls. Part of the downslope flow leaves the slopes and flows toward the center of the valley just below the top of the inversion rather than continuing down to the valley floor (figure 11.6). The remaining portion of the downslope flow continues down to the valley floor, but at a slower speed and within a shallower layer.

During this phase, the valley atmosphere is decoupled from the air above the valley. The decoupling can be attributed to two factors. First, the stable temperature structure within the valley suppresses vertical exchange of air between the valley and the free atmosphere. Second, the surrounding ridges shelter the valley atmosphere from light to moderate winds aloft.

During nighttime, down-valley winds prevail within a fully formed valley temperature inversion, and downslope winds continue on the sidewalls.

Above the inversion, the remnant of the daytime convective boundary layer, now called the residual layer, may persist. The turbulence and mixing characteristic of the daytime convective boundary layer have ceased, and the residual layer is bounded at its top by the remnant of the daytime entrainment zone, now called the capping inversion. However, both the residual layer and the capping inversion may be destroyed or carried off by strong winds in the free atmosphere, which then extends all the way down to the top of the inversion.

11.1.3. *Morning Transition Period*

The morning transition period (figure 11.2d and e) begins shortly after sunrise when incoming solar radiation exceeds longwave radiation loss and the surface energy budget reverses on the valley sidewalls. Sensible

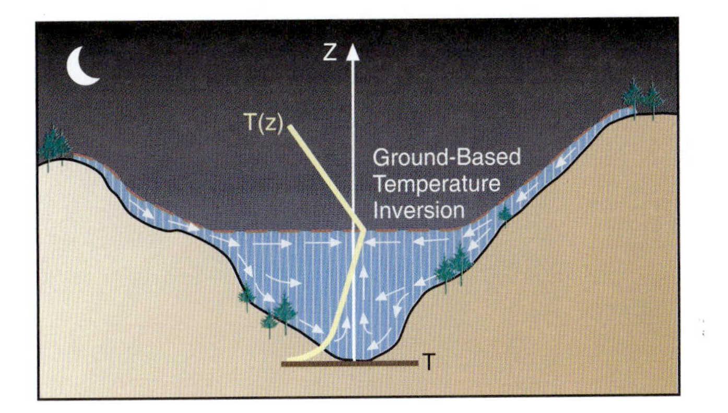

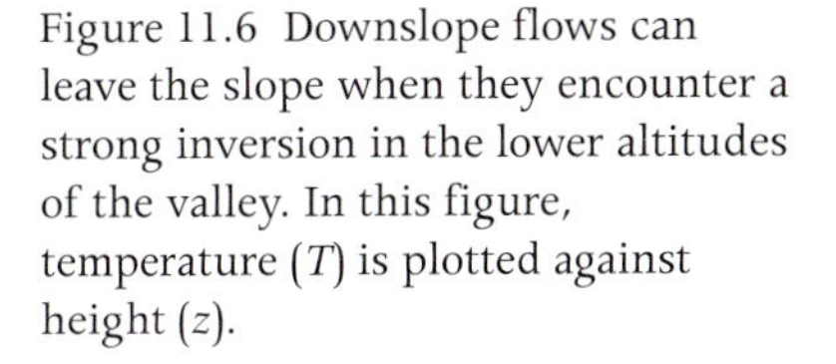
Figure 11.6 Downslope flows can leave the slope when they encounter a strong inversion in the lower altitudes of the valley. In this figure, temperature (T) is plotted against height (z).

heat is transferred from the ground to the air, and the air over the slopes, now warmer than the air over the center of the valley, rises up the slopes, creating upslope flows. Convective currents that rise from the heated valley floor and slopes entrain air from the base of the inversion into a growing convective boundary layer. The upslope flows carry the air up the slopes, removing it from the valley. The remnant of the nighttime inversion, the *stable core*, still contains down-valley winds but is separated from the ground by the growing convective boundary layer. As more and more air is removed from the base of the stable core and carried up the sidewalls, the top of the stable core sinks lower and lower into the valley. The sinking over the valley center produces a distinctive breakup pattern in valley fog or stratus clouds in which the sky clears first in a thin strip along the valley's axis over the valley center (figure 11.7). The subsidence of the stable core compensates for the upslope flows and produces warming, thus effectively distributing the energy received at the valley sidewalls throughout the entire valley atmosphere. The warming of the valley atmosphere eventually causes the along-valley pressure gradient to reverse, and up-valley winds replace the nighttime down-valley winds. The morning transition period ends about $3\frac{1}{2}$ to 5 hours after sunrise when the stable core is finally destroyed and up-valley winds prevail.

During the morning transition period, slope and valley winds reverse from downslope to upslope and from down-valley to up-valley, and the valley temperature inversion is destroyed.

A diagram of the valley atmosphere midway through the morning transition period (figure 11.8) illustrates the wind layers present in the atmosphere as a convective boundary layer grows above the heated surfaces and the stable core shrinks. Winds blow upslope in the convective boundary layer over the sidewalls and up-valley in the shallow convective boundary layer over the valley floor, but down-valley in the stable core. Winds blow from the plain to the mountain in the convective boundary layer over the inclined large-scale slope of the mountain range. The wind direction above these convective boundary layers is determined by synoptic-scale pressure gradients in the free atmosphere.

11.1.4. *Coupled Period (Daytime)*

During daytime, upslope and up-valley winds prevail in a deep convective boundary layer that couples the valley atmosphere with the free atmosphere above the ridge tops.

Between the final destruction of the stable core and the late afternoon or evening reversal of the daytime surface energy budget, the valley atmosphere becomes coupled with the atmosphere above the valley, and upslope and up-valley winds prevail (figure 11.2e and f). Convective cur-

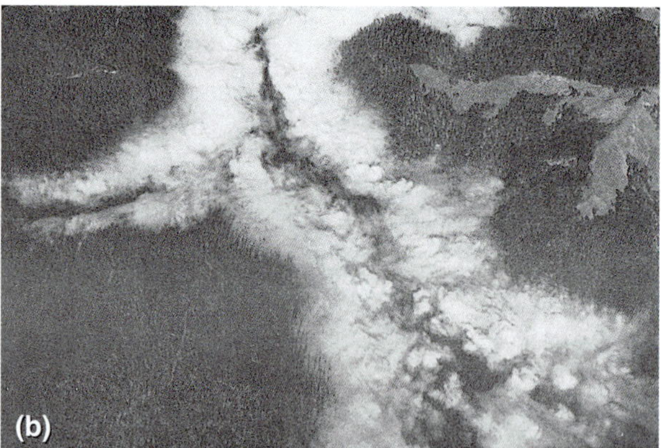

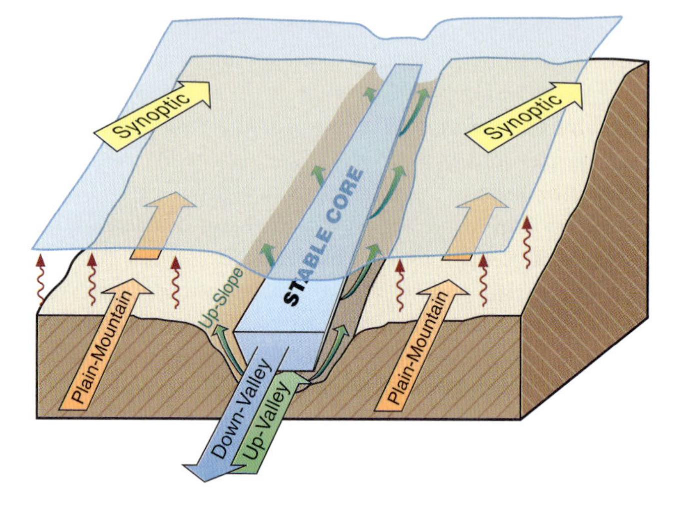

Figure 11.7 (*top*) Breakup of a valley fog layer in the Redwood Creek Valley of northern California, showing the strip of clear air forming over the valley center (a) at 0959 PST and (b) 8 minutes later at 1007 PST. (From Hindman, 1973; photo © E. Hindman)

Figure 11.8 (*bottom*) Valley cross section at midmorning, showing the wind layers during the morning transition period. Each of the wind layers is found within an inclined temperature structure layer or boundary layer. (Adapted from Whiteman, 1982)

rents rise from the valley floor, sidewalls, and ridge tops. The convective boundary layer becomes deeper and deeper as air from the free atmosphere is incorporated into it through the entrainment zone at its top.

Mixing within the convective boundary layer produces fairly uniform wind speeds throughout the layer. Wind speeds increase as the convective boundary layer grows to higher altitudes and encounters stronger winds. Wind speeds usually peak in late afternoon when the convective boundary layer reaches its maximum depth. If winds aloft are light, diurnal mountain winds continue to blow up the valley and the sidewalls. If winds aloft are moderate to strong, they can overpower the diurnal mountain winds. The daytime period ends with the onset of the evening transition period shortly before sunset, and the cycle begins again.

11.1.5. *Diurnal Progression of Wind Directions on Valley Sidewalls*

A distinctive diurnal progression of wind directions on the valley sidewalls results from the timing of the reversal of the slope winds and the along-valley winds and the tendency for the stronger along-valley winds to overpower the weaker and shallower slope winds. The winds shift from upslope just after sunrise, to up-valley during daytime, to downslope at

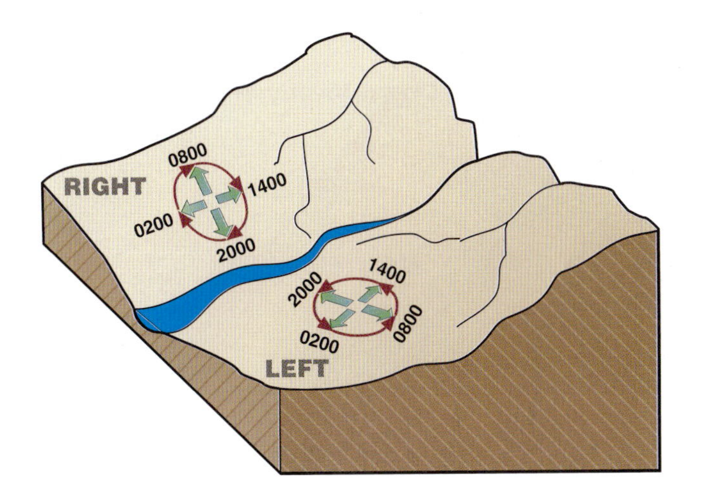

Figure 11.9 The diurnal phasing of slope and along-valley wind systems causes winds to turn clockwise with time on the right sidewall of a valley, but counterclockwise with time on the left sidewall. Numbers indicate local times; green arrows indicate upslope, up-valley, downslope, and down-valley flows. (Adapted from Hawkes, 1947)

sunset, and then to down-valley during the night. On the right sidewall of a valley (as viewed from the head of a valley), the phasing results in a clockwise turning of the winds with time; on the left sidewall, the winds turn counterclockwise with time (figure 11.9).

Whereas slope winds reverse direction near the times of local sunrise and sunset when the surface energy budget reverses and layers of relatively warm or cold air begin to form above the slopes, the reversal of along-valley winds is gradual. At night, cold air must build up at the valley floor before down-valley flows can begin. In the morning, the down-valley flow within the stable core does not reverse until it is warmed sufficiently by compensatory sinking or until a convective boundary layer grows up through it. A larger volume of air, rather than just a thin layer adjacent to the slopes, must be warmed or cooled before the along-valley winds can reverse. The wider or deeper a valley, the greater the volume of air in the valley and the greater the delay in along-valley wind reversal. In Austria's deep Inn Valley, for example, the reversal often occurs 4–6 hours after sunrise or sunset. Because of the shift in sunrise and sunset times (and therefore wind reversal times) with the seasons, the duration of the up-valley wind decreases as winter approaches and the days grow shorter. In fact, winds may blow down-valley all day if the ground is snow covered or the sky overcast. An example of the seasonal variation in along-valley wind reversal times is shown in figure 11.10 for Germany's Loisach Valley.

Winds turn clockwise with time on the right sidewall of a valley (as viewed from the head of the valley) and counterclockwise with time on the left sidewall.

11.2. Modification of Diurnal Mountain Winds by Variations in the Surface Energy Budget

Diurnal winds are well formed in mountainous areas where sensible heat flux is a prominent component of the surface energy budget, resulting in strong heating of the atmosphere during daytime and strong cooling during nighttime. The strongest diurnal wind systems develop in dry, high-elevation climates because of the good exposure to solar radiation during the day and the strong longwave radiative loss during the night. In ad-

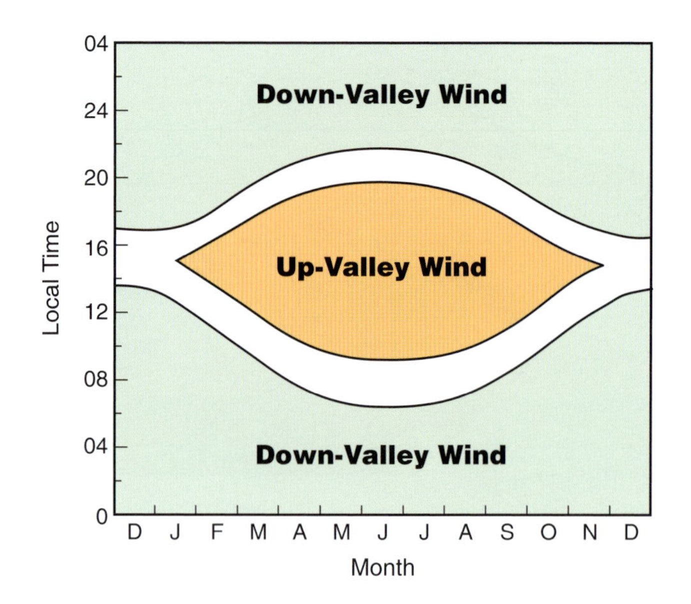

Figure 11.10 Timing of the down-valley (green) and up-valley (orange) wind regimes in the Loisach Valley, Germany, over the course of a year. The transition phase between down-valley and up-valley winds is shown in white. The up-valley wind is absent in December. (Adapted from Reiter et al., 1983)

dition, the latent heat flux, used to evaporate or condense water, is small, making more energy available to heat or cool the atmosphere. Mountains in moister climates have weaker diurnal wind systems because of weaker sensible heat fluxes. In climates where there are seasonal changes in the angle of the sun or in vegetative ground cover, there are also seasonal changes in the strength of diurnal mountain wind systems.

Diurnal winds in a given mountain range are affected in the short term by widespread rainfall, cloud cover, or snowfall. A widespread rainstorm increases soil moisture and thus affects the partitioning of heat in the surface energy budget of the region. Much of the daytime net all-wave radiation is used to evaporate water, leaving little energy for sensible heat flux. Without strong upward sensible heat flux, strong horizontal temperature contrasts do not form and diurnal mountain winds are weak.

Cloud cover reduces daytime and nighttime temperature extremes by reducing incoming solar radiation during daytime and outgoing longwave radiation at night. When cloud cover is present over an entire mountainous region, temperature contrasts that form between different parts of the region tend to be suppressed, weakening diurnal mountain winds.

Diurnal mountain winds are affected by climatic, seasonal, regional, and microclimatic variations in the surface energy budget.

Snow cover decreases the net all-wave radiation at the surface because the high albedo of snow (section 4.4.3) results in a large portion of solar radiation being reflected back into space. Sensible heat flux is usually also reduced. Thus, temperature contrasts that develop during the day are suppressed, and daytime winds over snow-covered surfaces are weak. At night, the high emissivity of the snow (section 4.4.3) enhances long-wave radiation losses, and the snowpack insulates the surface of the snow and the air layer immediately above it from heat stored in the ground. As a result, cooling of the snow surface and the atmosphere just above it is enhanced, and nighttime temperature contrasts and nighttime winds are strengthened.

In addition to climatic, seasonal, and regional variations in the surface energy budget, smaller scale variations influence diurnal mountain winds.

Because mountain topography is complicated, a mosaic of surface energy budget and ecosystem microclimates develops. Slope wind systems are particularly sensitive to microclimatic variations. Along-valley flows are less sensitive because they integrate sensible heat fluxes from all of the microclimates in a valley cross section. The energy budget at a given location is affected by the location's exposure to solar radiation during the day and to longwave radiation loss during the night and by the ground cover (snow or vegetation).

The total receipt of solar radiation and the times of initiation of upslope and downslope flows are affected by the times of sunrise and sunset, which vary significantly in mountainous terrain. The times of sunrise and sunset are determined by the *azimuth angle* and the *inclination angle* of the slope at a given location, by the sun's changing position in the sky during the course of the day and year, and by shadows cast by surrounding terrain (appendix E). Local sunrise time variability on 14 February in Montana's Marshall Creek Canyon, as calculated from a topographic shading model, is presented in figure 11.11. This examples shows that west-facing slopes, especially steep slopes that are shaded by ridges, have a very late sunrise. The initiation of upslope flows is significantly delayed on these slopes.

At night, when the energy budget of a slope is driven by the net loss of longwave radiation (i.e., outgoing longwave radiation exceeds incoming or downward longwave radiation), slope flows are affected by spatial variations in downward longwave radiation. Plateaus or ridge tops receive only small amounts of downward radiation from the cold, radiating sky. Locations lower in the valley, however, receive radiation not only from the portion of the sky that is visible, but also from the warm sidewalls above, which produce more radiation than the sky that they block. Thus, upper slopes and ridge tops experience a larger net longwave loss than the valley floor and lower slopes per unit area of surface and produce cold air

Figure 11.11 Shadows in an area of complicated topography affect the times of local sunrise: (a) varying times of local sunrise in the Montana's Marshall Creek Valley, and (b) topography.

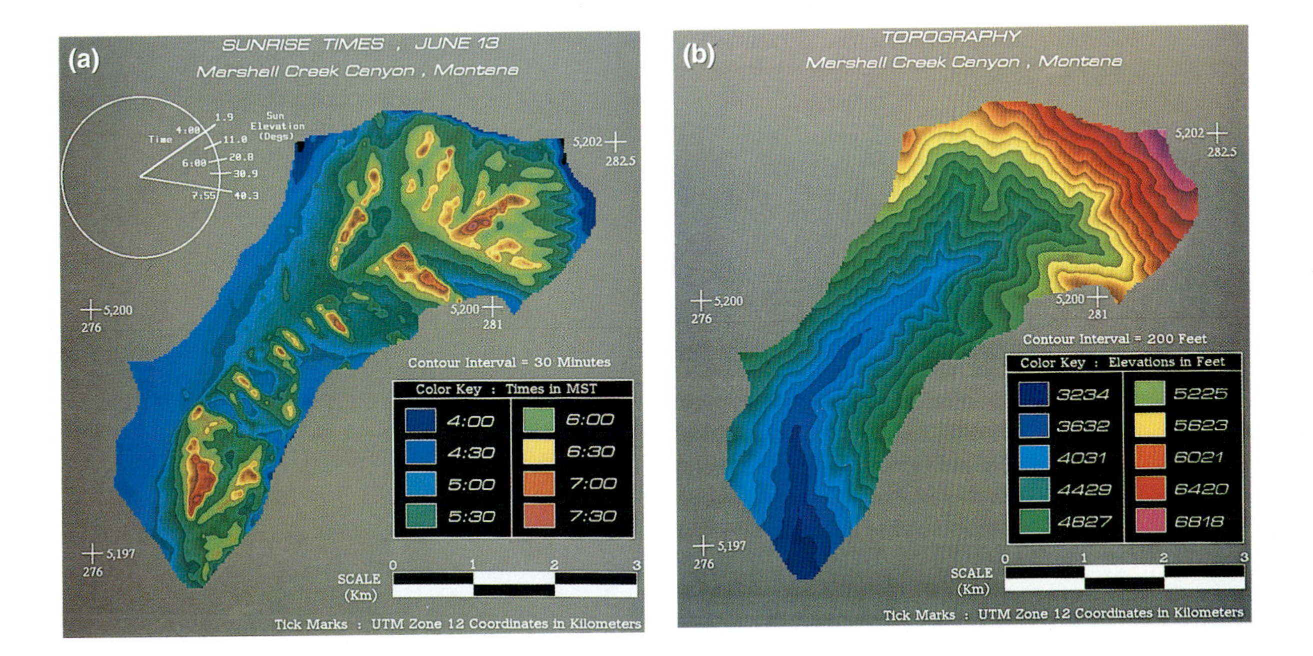

Figure 11.12 (*top*) Upslope flows can occur above the forest canopy in snowy valleys at the same time that downslope flows are occurring on the snowy slopes below the canopy. (Adapted from Hawkes, 1947)

Figure 11.13 (*bottom*) The upslope flow suddenly decreases in strength when a cloud passes in front of the sun.

that drains into the valley. Valleys that drain elevated plateaus and ridges therefore generally experience strong downslope and down-valley flows unless strong synoptic winds carry away the cold air on the upper slopes and ridge tops. Spatial variations in upward longwave radiation complicate the computation of net longwave loss. These upward longwave losses depend on the temperature and emissivity of the surfaces.

Microscale changes in ground cover can produce local variations in wind direction and strength. For example, upslope flows may prevail over evergreen forests, whereas shallow downslope flows are present over the surrounding snow-covered slopes. An evergreen forest is a good absorber of sunlight, assuming that snow has been removed from the branches of trees by the wind. The layer of air above the forest is heated by the upward sensible heat flux from the trees, producing upslope flows above the forest *canopy*. Nearby snow-covered slopes, in contrast, can be cooled for much (or all) of the day, producing downslope flows that blow during the day as well as at night (figure 11.12).

Very short-term variations in the surface energy budget can cause slope flows to blow intermittently at a given site. If, for example, a small cumulus cloud drifts overhead obscuring the sun, the upslope flow over the part of the mountain slope in the shadow of the cloud will suddenly decelerate or stop, while the flow over the rest of the slope continues (figure 11.13).

11.3. Disturbances of the Daily Cycle by Larger Scale Flows

Although temperature contrasts develop daily within complex terrain, they do not always result in a well-formed *diurnal mountain circulation*. The daily cycle of mountain wind systems illustrated in figure 11.2, although typical, can be disturbed by interference from larger scale winds.

Larger scale flows nearly always influence the smaller scale diurnal mountain wind systems. The extent of the influence depends on atmospheric stability and the strength and direction of the winds above the valley relative to the valley axis. The valley atmosphere is completely decoupled from the large-scale flows only when winds aloft are weak or when a deep, strong inversion in the valley limits vertical exchange between the valley atmosphere and the upper air. Winds within the valley are then purely thermally driven flows, blowing up-valley during daytime and down-valley during nighttime.

The extent to which diurnal mountain winds are influenced by larger scale flows depend on atmosphere stability and the strength and direction of the winds above the valley relative to the valley axis.

Disturbances of the mountain wind system by larger scale flows can modify the winds or can prevent them from developing altogether. The frequencies of along-valley winds in Colorado's Brush Creek Valley and in California's Anderson Creek Valley are shown in figures 11.14 and 11.15. In Anderson Creek Valley, a relatively dry valley in the Coast Range that is subject to intrusions of marine air from the Pacific Ocean, an along-valley wind system develops on fewer than 20% of the days in February and March and on as many as 85% of the days in August, averaging 44% for the 10-month period shown. Semiarid Brush Creek Valley in the central Rocky Mountains has along-valley wind circulations on more than 40% of the days in all months and on more than 60% of the days during the warmer months.

11.3.1. *Sudden Strong Wind Break-Ins*

When strong winds from above the mountain crests suddenly break into a valley atmosphere, they can overpower the diurnal mountain winds, causing abrupt changes in wind direction within the valley and sudden increases in wind speed at the ground.

The most common cause of strong wind break-ins is the recoupling of the valley atmosphere to the above-mountain winds. As the convective boundary layer grows deeper and deeper and extends above the valley, strong winds from aloft can be mixed downward through the valley atmosphere. A break-in can occur any time during the daytime phase when the convective boundary layer reaches a layer of strong winds aloft. However,

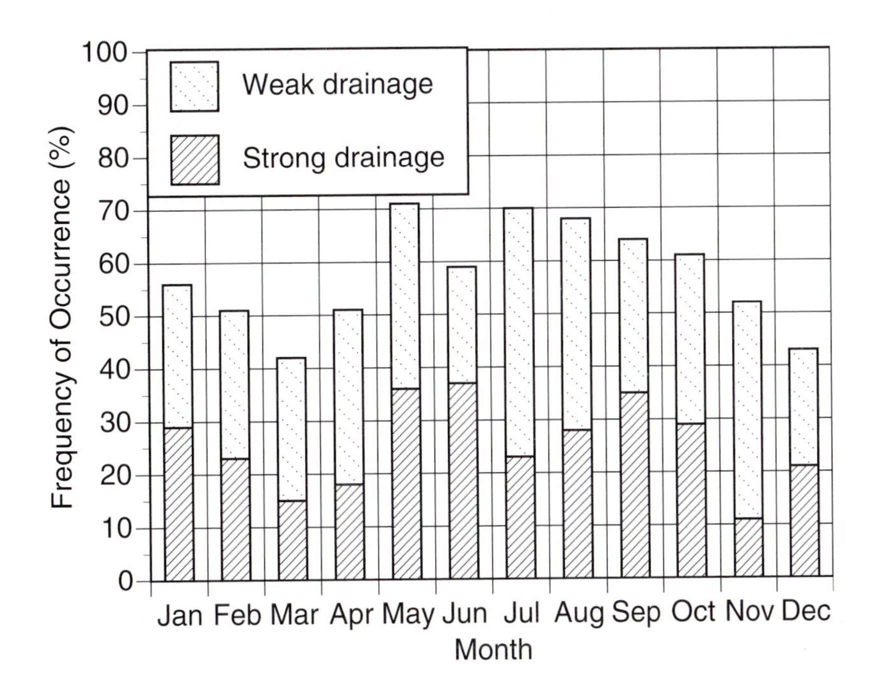

Figure 11.14 Frequency of occurrence of days with diurnally reversing along-valley wind systems in Colorado's Brush Creek Valley, as determined for the period from August 1982 to January 1985. (Adapted from Gudicksen, 1989)

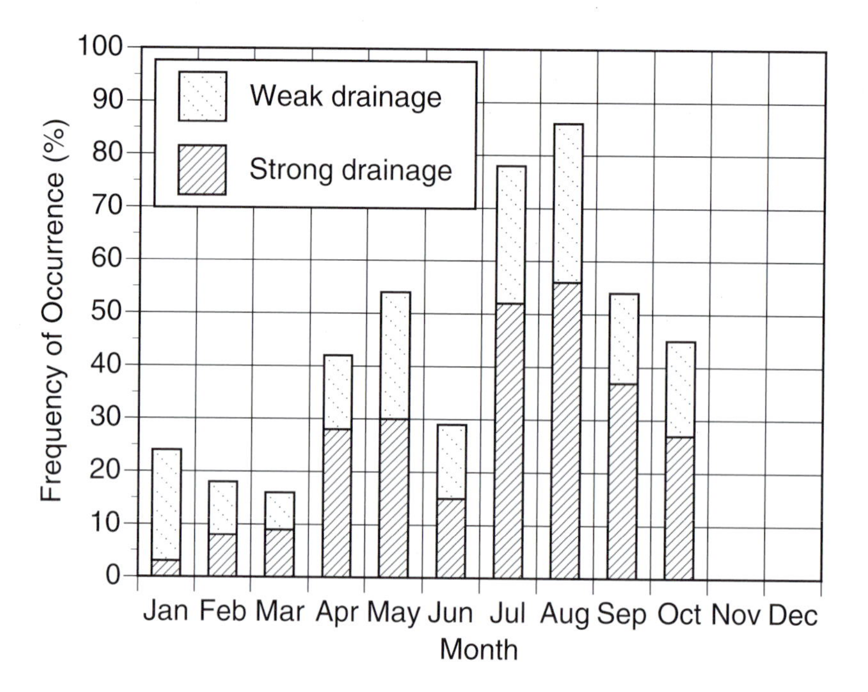

Figure 11.15 Frequency of occurrence of days with diurnally reversing along-valley wind systems in California's Anderson Creek Valley for the months January through October of 1980. (Adapted from Gudicksen and Walton, 1981)

if upper level winds are light, the diurnal wind systems within the valley are not disrupted. Break-ins are common during recoupling in dry continental mountainous areas, such as the western United States, where deep convective boundary layers develop and the winds aloft are often strong.

Break-ins can also be caused by strong vertical wind speed shear above the valley. If winds aloft are significantly stronger than winds within the valley, turbulence in the strong winds can erode the inversion layer that protects the valley, exposing the ground to the strong winds. In a moderately deep valley with a medium to strong inversion, winds at ridgetop level must be stronger than about 6 or 7 m/s (13–16 mph) for the shear to interrupt the development of the diurnal mountain wind systems. Additionally, strong shifting winds associated with fronts (section 6.2) or the intrusion of sea breeze or lake breeze fronts into complex terrain areas (section 11.7.1) can cause strong wind break-ins.

Mountain waves generated by upwind topography, whether produced locally by the valley's upwind ridgeline or imposed over multiple valleys on the lee side of a major mountain barrier, can bring strong winds to the valley floor. The waves are most intense, and therefore the probability of strong wind break-ins highest, when the above-mountain wind is strong and perpendicular to the topographical barrier. The wind strength required varies with atmospheric stability and topography, but, as a general rule, winds must exceed 6 or 7 m/s (13–16 mph) before significant waves form.

The sudden break-in of strong winds into a valley can be caused by the recoupling of the valley atmosphere to the above-mountain winds, by vertical wind speed shear above the valley, by frontal passages, by mountain waves, or by thunderstorm downdrafts.

Strong wind break-ins are not always caused by the downward transport of stronger synoptic winds from aloft. Thunderstorm downdrafts can also penetrate into valleys and interrupt diurnal mountain winds. The strongest downdrafts, and thus the highest probability of strong wind break-ins, occur when the descending air is cooled by evaporation from rain shafts (section 8.7.2).

There are a number of indicators of strong winds aloft, including wind-blown dust, snow, fog, or smoke on ridges, the rapid movement of clouds, the shearing apart of aircraft contrails, and the development of lenticulars, banner clouds, and billows (section 7.1.4). Indicators that are less easily observed include the late formation of a nighttime temperature inversion, an unusually weak down-valley wind system, and a weak or shallow inversion. Strong winds on the upper, unprotected sidewalls of a valley may also indicate strong winds aloft. If the winds above the mountains have an up-valley component, the winds on the upper sidewalls may blow up-valley during the night rather than down-valley as expected.

11.3.2. *Channeling of Synoptic Flows*

Forced channeling or pressure-driven channeling of larger scale winds can interrupt the regular diurnal cycle of mountain winds, causing changes in wind speed and/or wind direction in a valley. Forced channeling (section 10.4.4) is most likely to occur and is strongest in a neutral or unstable atmosphere when strong winds aloft blow directly along a valley's axis. Because the atmosphere tends to be stable at night and neutral or unstable during the day, forced channeling occurs most frequently in daytime. Forced channeling often begins in late morning after the nocturnal temperature inversion has been destroyed and results in an abrupt increase in wind speed and gustiness within the valley. Although forced channeling is strongest when winds aloft blow parallel to the valley axis (i.e., when the pressure gradient aloft is weak in the along-valley direction), winds that blow at an oblique angle to the valley's axis can also be channeled. The flow is channeled in the direction, either up- or down-valley, that requires the smallest directional change of the upper winds. Wind directions within the valley shift suddenly by 180° when the wind direction aloft shifts across a line perpendicular to the valley axis. Winds that result from forced channeling are more likely to affect high terrain near passes and gaps than low-lying areas within a valley.

Pressure-driven channeling (section 10.4) interrupts diurnal mountain winds when the synoptic-scale pressure gradient is stronger than the local, thermally produced pressure gradient that drives along-valley winds. In contrast to forced channeling, pressure-driven channeling is strongest when the along-valley pressure gradient is strongest (figure 11.16a and b) and is weakest when the along-valley pressure gradient is weakest (figure 11.16c). Winds resulting from pressure-driven channeling always blow from the high pressure end of the valley to the low pressure end. Thus, the wind direction does not depend on the time of day, as it does for diurnal mountain winds, but rather on the location of synoptic highs and lows relative to the valley. Pressure-driven channeling is often seen in shallow valleys or in moist, cloudy climates where the diurnal mountain winds are weak and therefore easily overpowered. Because pressure-driven channeling is produced by the large-scale pressure gradients associated with slow-moving low and high pressure systems, changes in along-valley wind speed are gradual rather than abrupt, as is the case with forced channeling. Wind directions within the valley may change by 180° if the upper winds shift across a line parallel to the valley axis. Shifts in wind direction are usually preceded by a gradual decrease in along-valley wind speed. Winds driven by pressure-driven channeling affect the entire valley.

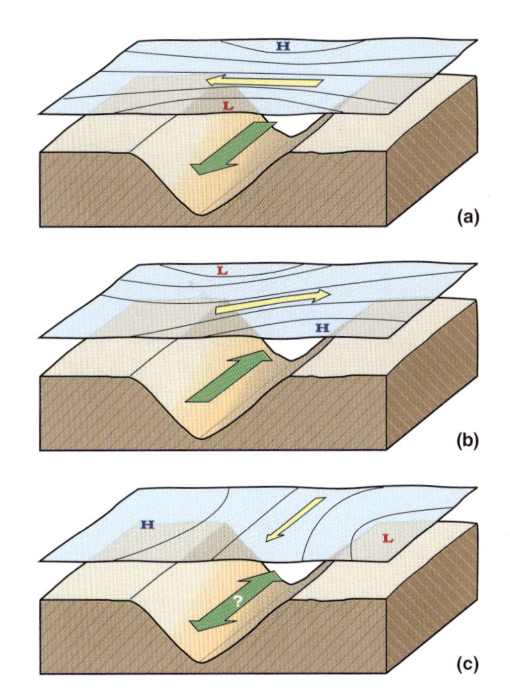

Figure 11.16 In pressure-driven channeling, winds blow along the valley axis from the high pressure end of the valley to the lower pressure end, while winds aloft, in conformance with the Buys-Ballot rule, blow parallel to the isobars with low pressure to the left (Northern Hemisphere): In (a) the wind blows down the valley, in (b) The wind blows up the valley and in (c) The along-valley pressure gradient is negligible so that no pressure-driven channeling occurs.

Forced channeling is strongest when the pressure gradient aloft is weak in the along-valley direction. Pressure-driven channeling is strongest when the along-valley pressure gradient is strongest in the along-valley direction.

Under forced channeling, wind directions within a valley change suddenly by 180° when the wind direction aloft shifts across a line perpendicular to the valley axis. Under pressure-driven channeling, the wind direction changes by 180° if the upper winds shift across a line parallel to the valley axis.

Figure 11.17 Helical flow trajectories can occur during daytime when a cross-valley wind circulation is imposed on an along-valley wind.

11.4. The Four Components of the Mountain Wind System

It is difficult to observe any one component of the mountain wind system—slope winds, along-valley winds, cross-valley winds, or mountain–plain winds—in its pure form. Not only is the mountain wind system disturbed by larger scale flows, but each component can be affected by the others. For example, both upslope and up-valley flows are difficult to distinguish from background synoptic flows. In addition, slope flows are often overpowered by the somewhat stronger and deeper along-valley flows. Generally weak cross-valley flows also interact with along-valley flows, resulting in a helical circulation parallel to the valley axis (figure 11.17) and strengthened upslope flow over the warmer sidewall. Mountain–plain winds, which are not confined by the topography but blow up or down the surface of a mountain massif, are often influenced by synoptic-scale flows and thus vary more from day to day than winds within a valley. Mountain–plain flows are generally so weak that they are difficult to detect on individual days because of synoptic flow variability. The flow strength and direction must be determined climatologically by analyzing many events. More information is available on slope flows and along-valley flows than on the weaker cross-valley flows, which can nonetheless impact air pollution dispersion, and on mountain–plain flows.

The four components of the mountain wind system (slope winds, along-valley winds, cross-valley winds, and mountain–plain winds) interact with each other and with larger scale flows.

11.4.1. *Slope Wind System*

The slope wind system is a closed circulation driven by horizontal temperature contrasts between the air over a slope and the air at the same level over the center of the valley. The temperature contrasts result from the heating or cooling of an inclined boundary layer over the slope.

Slope flows are typically in the range of 1–5 m/s (2–11 mph). Although the highest daytime temperatures and lowest nighttime temperatures are found at the ground, peak wind speeds occur a few meters above the surface because of frictional drag near the ground. Figure 11.18 illustrates the relationship between wind and temperature structures over a midslope location during upslope and downslope flows.

Both upslope and downslope wind speeds generally increase with distance. Downslope flows on the long slope of the continental ice dome in Antarctica can reach gale force (28–47 knots) on the periphery of the continent. The strongest downslope flows occur around sunset when slopes first go into shadow. The strongest upslope flows occur at midmorning when the temperature contrast between the warm, sunlit slopes and the valley atmosphere is strongest.

Both upslope and downslope flows increase in speed and depth with distance.

The depth of slope winds also varies over time and space. Downslope flows are shallower than upslope flows. The depth of a downslope flow increases with distance down the slope and can be estimated as 5% of the drop in elevation from the ridge top. Thus, at 100 m in elevation below the ridge top, a 5-m-deep downslope flow can be expected. Downslope flow depth tends to decrease during the night when an inversion builds up in the valley. Upslope flows increase in depth with distance up the slope and with time, usually attaining a depth of 50–150 m (150–500 ft) during the first several hours after sunrise.

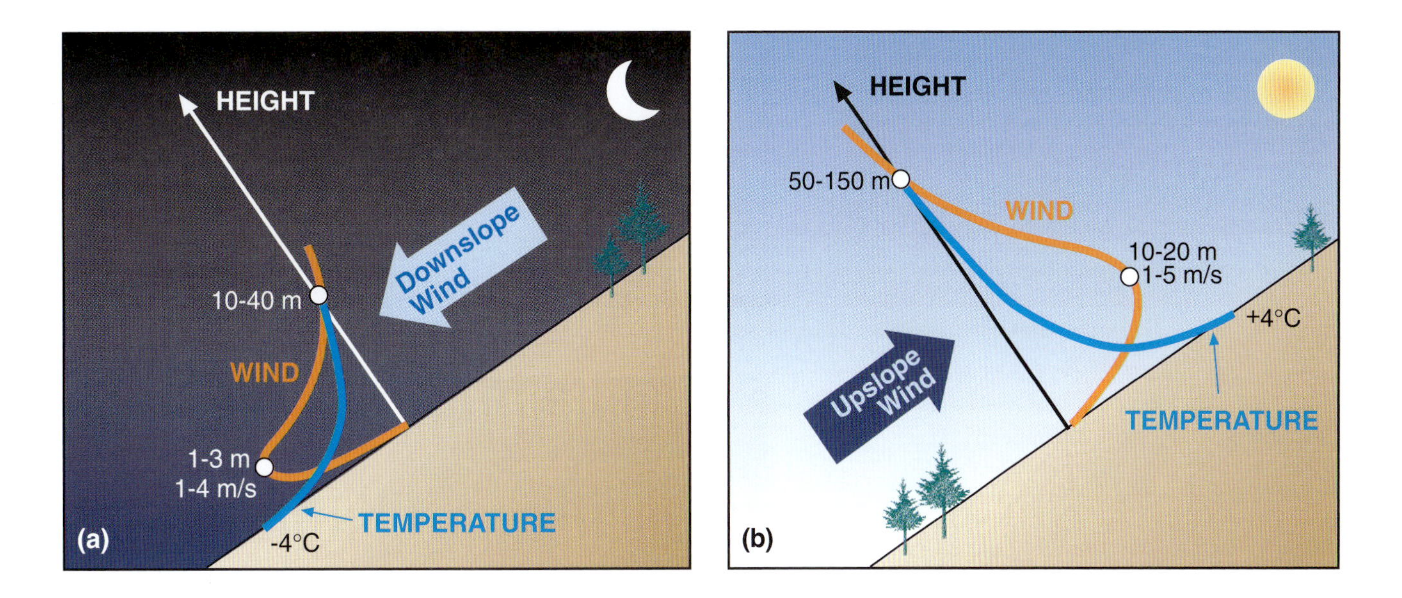

Figure 11.18 Typical profiles of wind and temperature as a function of height above a slope during (a) daytime and (b) nighttime.

The depth of slope flows is affected by the atmospheric stability over the valley center. The stronger the stability (i.e., the greater the resistance to vertical motion in the atmosphere), the shallower the upslope or downslope flows.

Slope flows are particularly sensitive to spatial variations in the surface energy budget (section 11.2) and to small-scale features of the underlying terrain. Downslope flows tend to converge in depressions or gullies on the sidewalls, whereas upslope flows converge over higher ground between depressions or gullies. Slope flows are also subject to mechanical influences, splitting to flow around obstacles such as rocky projections, shrubs, trees, and forest copses.

11.4.2. *Along-Valley Wind System*

Along-valley winds are the lower branch of a closed circulation that moderates horizontal air temperature differences that form within a valley or between a valley and a nearby plain by advecting relatively cold air into an area with relatively warm air. The lower branch of the circulation is illustrated in figure 11.19, which shows two columns of air—one in the valley and one above the plain. The top of both columns is defined by the el-

> COLD AIR AVALANCHES
>
> Mountain climbers bivouacking on the sides of a peak or plateau may observe *cold air avalanches* descending from the peak or plateau. These regular nighttime downslope flow pulsations typically recur at intervals of 5 minutes to 1 hour. They are the result of cold air building up on the peak or plateau until the cold air pool reaches a critical size and cascades down the slopes. They are formed most readily when background or ambient air flows are weak and skies are clear.

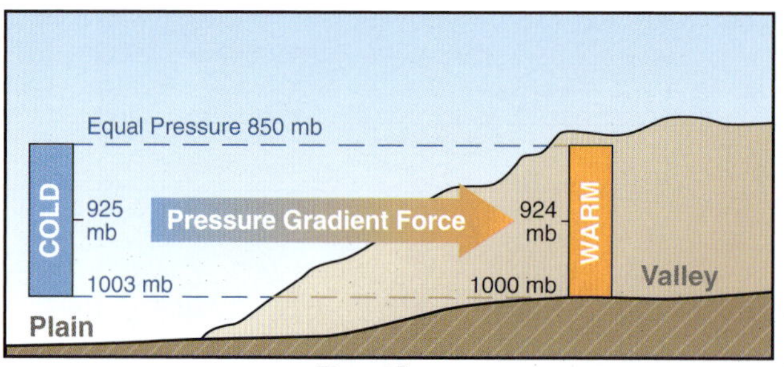

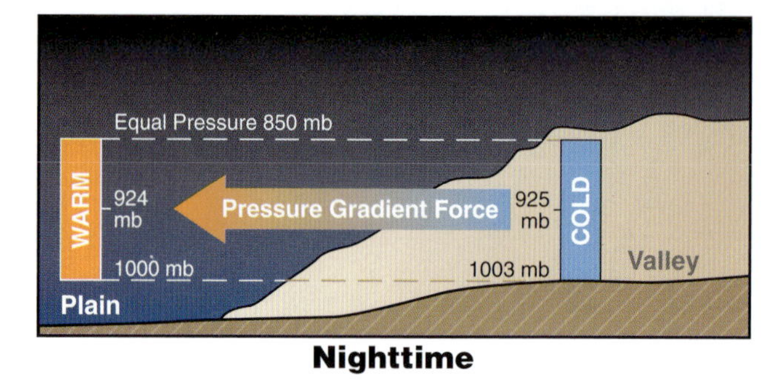

Figure 11.19 Pressure differences develop between a valley and the adjacent plain when the valley atmosphere becomes colder (nighttime) or warmer (daytime) than the atmosphere over the plain. These horizontal pressure differences produce the along-valley wind system. (Adapted from Hawkes, 1947)

evation of the valley ridge tops, and the bottom of both columns is defined by the elevation of the valley floor. Both columns have the same horizontal area at the top. During daytime, the air in the valley warms more rapidly than the air over the plain, reducing the pressure in the valley atmosphere relative to the plain and driving an up-valley wind from the plain into the valley. The air carried into the valley during daytime at low levels is returned to the plain by the upper branch of the circulation. During nighttime, the valley atmosphere cools faster than the plain atmosphere at the same level, producing higher pressure in the valley column and driving a down-valley flow from the valley to the plain. A return current forms above the down-valley flow and carries air back toward the upper end of the valley. The development of a closed circulation requires rising motions (and cooling) above the warm columns and sinking motions (and warming) above the cold columns. These vertical motions thus also play a role in equilibrating horizontal temperature differences within the valley.

The return flow above the valley is rarely observed directly because it is not confined to a channel like the lower branch of the circulation. It is much weaker than the lower branch and weaker than typical prevailing synoptic winds. In some cases, however, the return flow has been observed in the upper altitudes of the valleys below ridge-top level. When confined within valleys, these return circulations are called *antiwinds* because they blow in the direction opposite to the winds in the lower altitudes of the valley. Antiwinds were observed during a series of experiments conducted in valleys extending radially outward from the volcanic cone of Mt. Rainier in Washington, but have otherwise been reported infrequently.

When the compensatory flow of an along-valley wind system is confined within a valley, it is called an antiwind.

Although figure 11.19 illustrates daytime and nighttime pressure gradients for air columns over a valley and an adjacent plain, pressure gradients also occur between air columns located at different points along a valley's axis as the result of along-valley temperature differences. Along-valley pressure gradients drive along-valley winds even in valleys that have no immediately adjacent plain.

11.4.2.1. *Along-Valley Temperature Differences and Cold Air Pools.* Three factors produce along-valley temperature differences:

- Sidewall shape, and thus valley volume, then the changes along the length of a valley (figure 11.20). Sidewalls at the head of a valley are usually convex, enclosing less volume U-shaped or V-shaped side-

INVERSIONS OVER THE PLAINS AND IN COMPLEX TERRAIN AND THE VALLEY VOLUME EFFECT

Temperature inversions in valleys differ significantly from inversions over nearby plains, primarily because valleys undergo stronger diurnal heating and cooling cycles. Valley inversions are usually much deeper but have a somewhat weaker average stability. For example, in clear, undisturbed weather conditions at Denver, Colorado, on the eastern edge of the Rockies and at Grand Junction, Colorado, on the western edge of the Rockies, morning inversions are typically 200 meters (650 ft) deep with average temperature gradients of 30°C/km (16°F/1000 ft). In nearby mountain valleys, however, inversions usually extend through the valley's depth (typically 600 m or 2000 ft) with average temperature gradients of 20°C/km (11°F/1000 ft).

If upper winds are light and skies are clear, temperature inversions develop more regularly from night to night and from season to season in the valleys than over the plains because the daily temperature range in the valleys is greater. The enhanced warming of the valley atmosphere during daytime and the enhanced cooling during nighttime and the resulting larger daily temperature range are primarily due to the *valley volume effect* illustrated in figure I below. Solar radiation entering a valley during daytime and longwave radiation leaving a valley during nighttime are converted to sensible heat that warms or cools the air in the valley. The temperature change produced depends on the volume of the valley. The smaller the volume, the larger the temperature change for an equal heat input or output. The volume of air within a valley is always smaller than the volume of a column of air over the plains that has an equal area at its top because the valley volume is reduced by the solid valley sidewalls. Thus, the same heat gain and loss produces larger temperature changes and a larger temperature range in a valley than in a comparable column of air over the plains.

(*continued*)

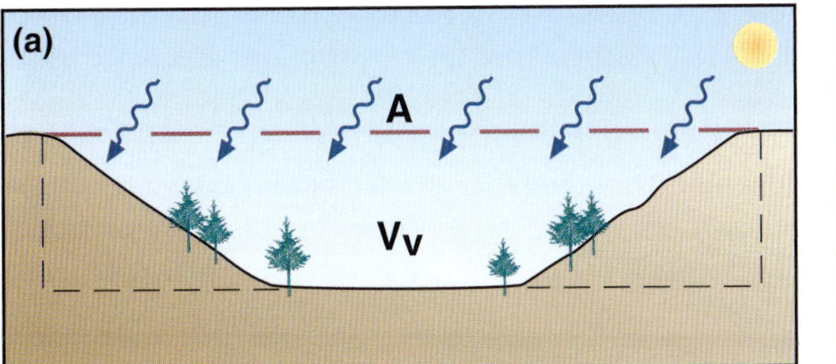

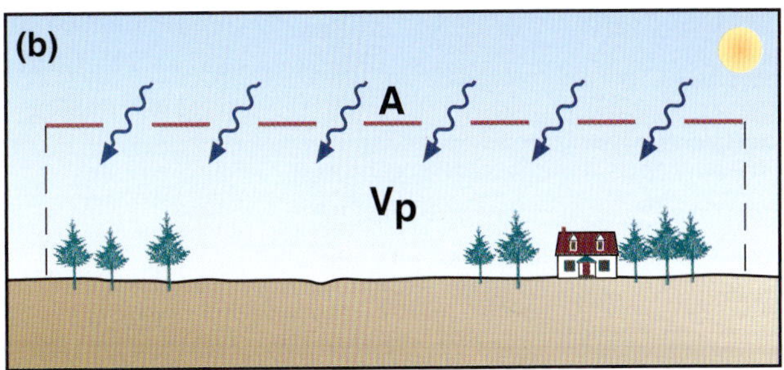

Figure I The air volume in a valley V_v is smaller (a) than the volume V_p in a plains cross section (b) given the same area A at the top of the two cross section.

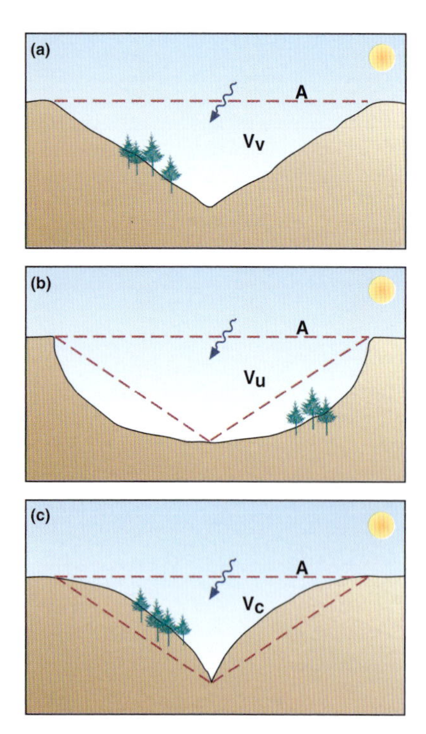

Figure 11.20 The air volume within a valley depends on the shape of the valley cross section. For valleys with the same ridge-to-ridge width, valleys with convex sidewalls (c) enclose less volume than valleys with V-shaped sidewalls (a) or valleys with U-shaped sidewalls (b). The rate of heating (or cooling) of the air within the valley is in direct proportion to this volume. (Adapted from Müller and Whiteman, 1988, and Whiteman, 1990)

Along-valley temperature differences are caused by the valley volume effect and by differences in sensible heat flux.

Terrain constrictions in a valley can result in cold air pools building up behind the constrictions or in air accelerating through the constrictions.

INVERSIONS OVER THE PLAINS AND IN COMPLEX TERRAIN AND THE VALLEY VOLUME EFFECT (*continued*)

The warming and cooling of a column of air in a valley is also enhanced because the air is close to the earth's surface (the valley floor and sidewalls) where most of the heat transfer takes place through radiation, *conduction*, and convection. In contrast, a column of air over the plains at the same elevation is some distance above the surface where these processes transfer heat less efficiently.

Mountain topography not only enhances warming and cooling of the air in a valley, it also protects a valley inversion from ambient winds. The shallower inversions over plains are unprotected and are therefore more easily dispersed by ambient winds.

walls usually found in midvalley. Broad, low inclination concave sidewalls, found near the mouth of a valley enclose an even greater volume. Given the same energy imput or output, the range of temperatures near the mouth of a valley will be more limited than near the head of the valley because a larger volume of air must be heated or cooled. (See the accompanying Point of Interest.)

- Solar or longwave radiation gains or losses vary along the length of a valley. For example, daytime temperatures in a valley segment increase if the albedo of the segment is low. Nighttime temperatures in a valley segment decrease if emissivity is high. Significant radiative differences could occur, for example, among the glaciated head of a valley, rock outcroppings below a glacier, and a forest in the lower valley.
- The rate of conversion of net all-wave radiation to sensible heat flux can vary along the length of a valley. For example, if the upper part of a valley is dry and the lower part is wet, a greater portion of the incoming solar radiation during daytime and longwave radiation during nighttime goes to heat or cool the air in the upper valley than in the lower valley, where much of the available energy is used to evaporate or condense water.

Valleys with terrain constrictions along their length may have segments where the valley volume effect causes a reversal of the nighttime down-valley pressure gradient. This pressure gradient reversal halts the down-valley winds and causes cold air pools to build up behind the constrictions. Calculations of the ratio of valley width to cross-sectional area along a valley's length can be used to estimate the strength of the along-valley winds and to determine whether cold air will stagnate and pool in different valley segments (McKee and O'Neal, 1989). The Mossau and Finkenbach Valleys in Germany's Odenwald region are examples of valleys with multiple terrain constrictions along their lengths (figure 11.21). Cold air pools that form behind the constrictions can cause frosts that damage the valley vineyards. A cold air pool also forms behind a narrow gorge at the lower end of Colorado's Gore Valley west of Vail, where it causes low nighttime temperatures farther up the valley at the Vail Ski Resort and reduces the nighttime dispersion of air pollutants. A city ordinance has been adopted to reduce nighttime particulate emissions from wood-burning fireplaces because of the poor nighttime dispersion conditions in the cold pool (figure 12.11).

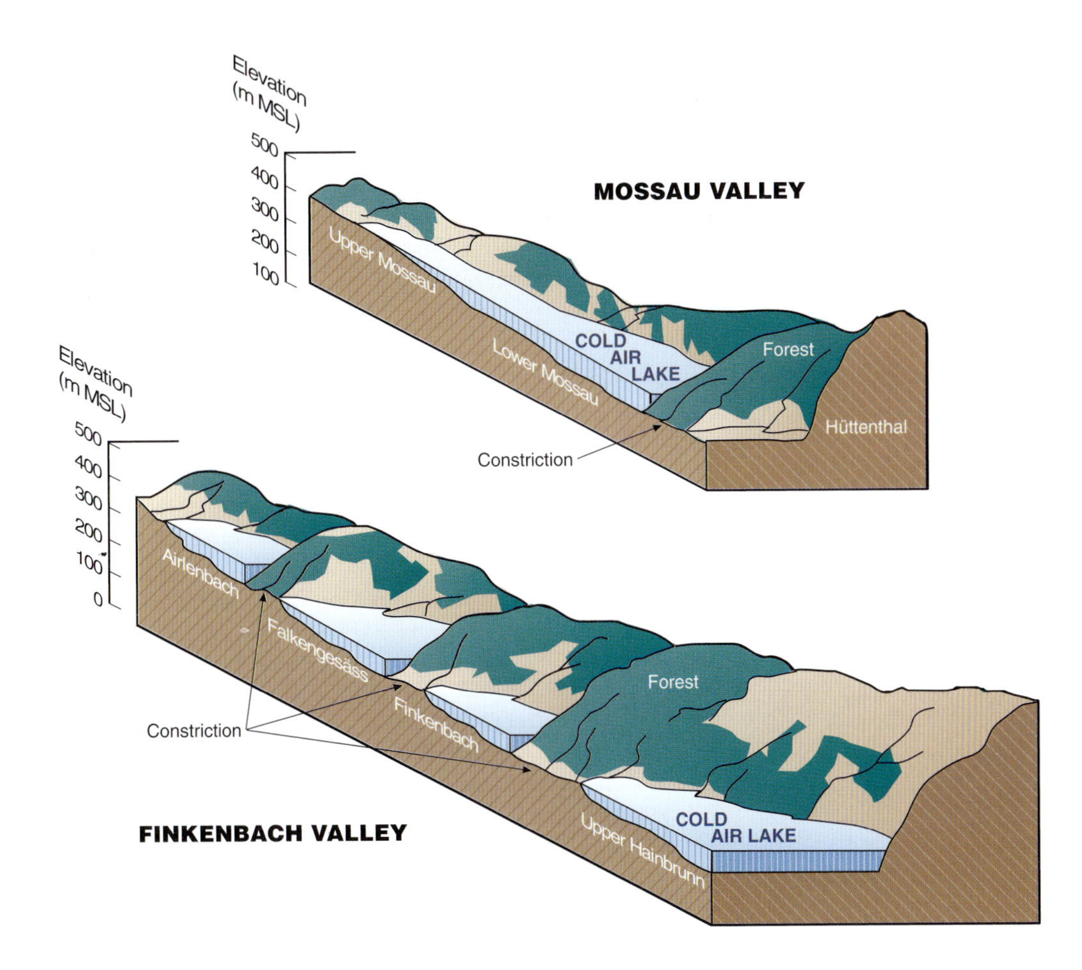

Figure 11.21 Cold air forms pools along the length of Germany's Mossau and Finkenbach Valleys because of the presence of terrain constrictions. (Adapted from Geiger et al., 1995)

Where along-valley pressure gradients are strong, down-valley winds may actually be accelerated through a constriction by the Venturi effect discussed in section 10.4.3. This has been observed, for example, in Colorado's Eagle Valley in a gorge below the confluence with the Gore Valley. Little is known about the topographic and atmospheric characteristics that determine whether cold air will pool behind a constriction or be forced through the constriction at high speed.

11.4.2.2. *Down-Valley Wind Speeds.* Down-valley flows are characterized by a jet wind speed profile. Wind speeds increase with height above the valley floor to a maximum of 3–10 m/s at heights corresponding to 30–60% of the inversion depth. Above this maximum, wind speeds decrease again to the top of the inversion, where a wind reversal is often encountered. Typical strengths and depths of fully developed down-valley winds are illustrated in figure 11.22.

Down-valley flows are characterized by a jet wind speed profile.

The jet wind speed profile is also evident at a valley's exit or a short distance beyond, where down-valley winds often attain their maximum strength with speeds as high as 11–20 m/s or 22–45 mph (figure 11.23). Winds near the ground in the vicinity of these *valley exit jets* are much

Figure 11.22 Typical vertical structure of temperature and down-valley wind near sunrise. The temperature structure sometimes exhibits a near-linear temperature decrease with height as indicated by line (a), but often exhibits a hyperbolic structure, as shown by dashed line (b).

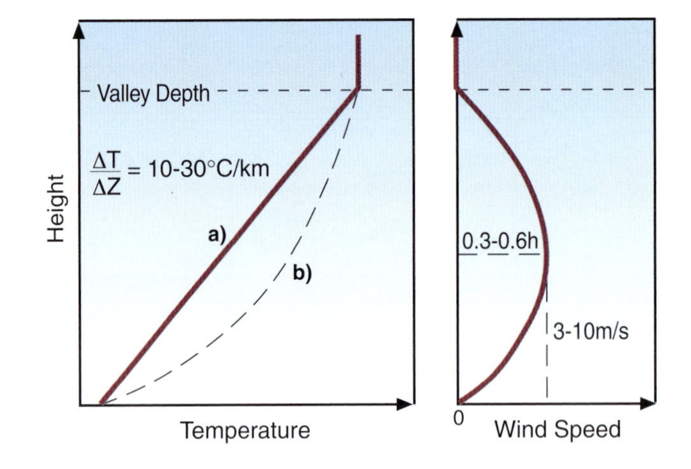

weaker than at the jet axis height, but they are frequently strong enough to cause trees and bushes to sway back and forth.

Nocturnal jet wind speed profiles in Austria's Inn Valley, which ends abruptly at a plain, are shown in figure 11.24. The jet strengthens during

THE MALOJA WIND

Occasionally, observations of wind systems seem to contradict valley wind theory. The best known example is the *Maloja wind* in the upper Engadine Valley of Switzerland near the Maloja Pass. The winds in the upper section of the valley blow up-valley during nighttime and down-valley during daytime because of the unusual topography of the area. The Bergel Valley to the south of the pass is very steep and has ridgelines that extend beyond the pass into the very shallow upper Engadine Valley to the north of the pass, thus effectively extending the valley volume effect and the wind regime of the Bergell Valley into the Engadine Valley. Two valleys draining Mount Rainier in Washington are the only U.S. valleys for which Maloja-type wind systems have been documented (Buettner and Thyer, 1966), but the winds probably occur in other locations as well.

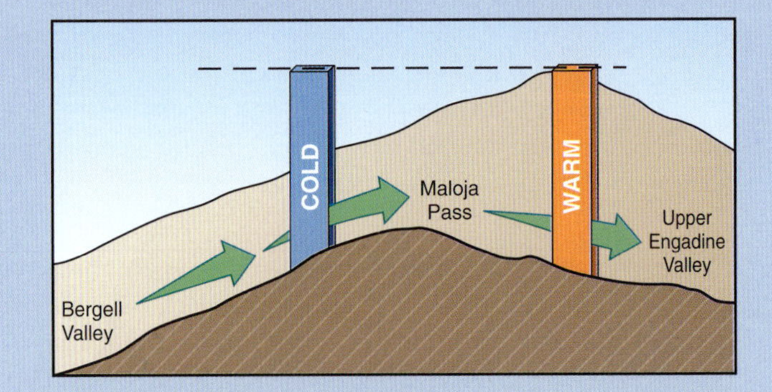

Figure I During daytime, winds from the Bergell Valley come over the Maloja Pass and encroach into the upper part of the Engadine Valley, producing an anomalous daytime down-valley wind called the Maloja wind. This encroachment is thought to be driven by a valley volume effect that produces warmer air in the atmospheric column within the upper Engandine Valley. (adapted from Hawkes, 1947.)

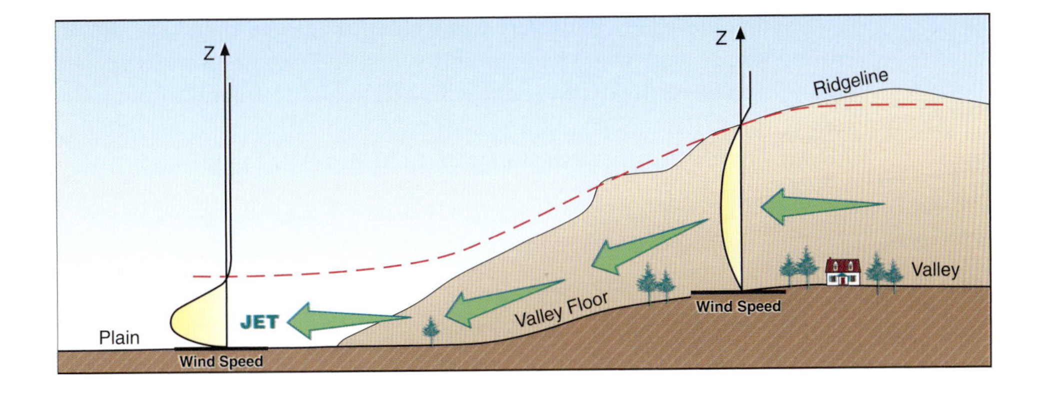

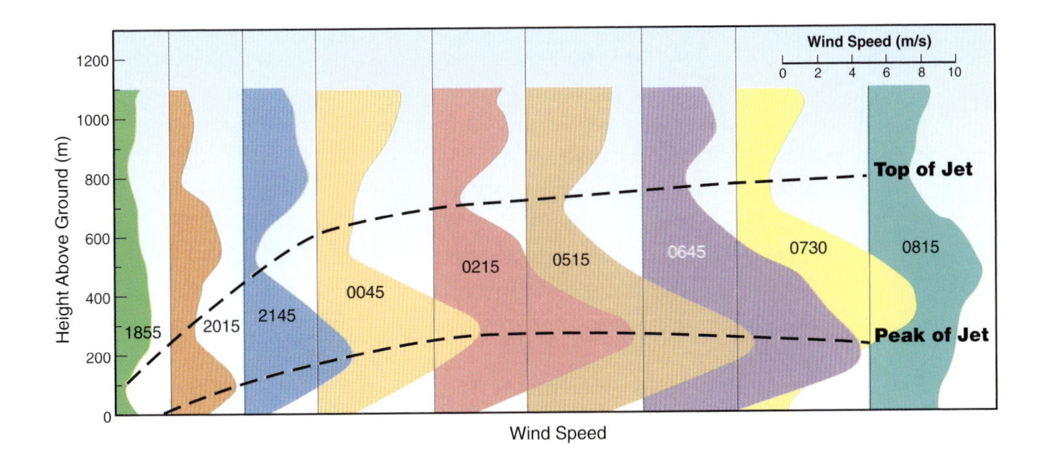

Figure 11.23 Schematic of strong winds issuing from a valley exit. The dashed red line indicates the top of the down-valley flow layer.

Figure 11.24 The development with time of the vertical wind speed profiles on the night of 25–26 March 1982 at Thalreit on the southern German plain less than 10 km (6 mi) from the exit of Austria's Inn Valley. (Adapted from Pamperin and Stilke, 1985)

the night, attaining speeds of 13 m/s (29 mph) at a height of 200 m (660 ft), and can be strong enough to be heard at the ground. Down-valley jet wind speed profiles similar to those observed at the exit of the Inn Valley have been observed at the mouths of other Alpine valleys and in the United States at Paonia, Colorado, where the valley of the North Fork of the Gunnison River fans out into a wide plain, and at the exit of South Boulder Creek, where it flows out into the plains south of Boulder, Colorado. Valley exit jets are probably present in many other U.S. valleys, especially in the drier mountain ranges where conditions are conducive to the development of diurnal mountain wind systems.

Wind speeds are higher near the mouth of a valley for these reasons:

- The winds inside the valley accelerate with distance down the valley.
- The horizontal pressure gradients that drive the wind systems are strongest near the exit where there is an abrupt transition over a short distance between the high pressure in the valley and the lower pressure over the plains.
- Wind speed increases as the cold valley air sinks and fans out over the plain.

These factors have a greater impact on deep valleys that end abruptly at a plain than on valleys that become gradually shallower with down-valley distance.

Daytime up-valley winds are less well-studied than nighttime down-valley winds, but, when background synoptic flows are weak, they are thought to be relatively invariant with height and to extend through the entire valley depth. Maximum speeds vary from valley to valley and day to day but are usually in the range of 3–10 m/s (7–22 mph).

Cross-valley flows blow perpendicular to the valley axis and toward the more strongly heated sidewall.

11.4.3. *Cross-Valley Winds*

Cross-valley winds blow across to the valley axis, when the air above one of the valley sidewalls becomes warmer than the air above the other sidewall. The near-surface flow is toward the more strongly heated sidewall. Compensatory flows in the opposite direction occur at higher altitudes within the valley.

Cross-valley winds are generally weak (2 m/s or less, or 4.5 mph or less). The strength of the flow depends directly on valley width and on the cross-valley temperature gradient, with weaker winds in very wide valleys and in valleys where the temperature differences between the opposing sidewalls are small.

Cross-valley winds are stronger during the day, when the sidewalls are heated unequally by the sun, than during the night, when the outgoing longwave radiation is more evenly distributed across a valley. The strongest cross-valley flows occur early or late in the day in north–south-oriented valleys when one sidewall is in shade while the other sidewall is in full sunlight.

The temperature difference between opposing sidewalls, and therefore the strength of cross-valley winds, is affected by the inclination angles of the sidewalls, by the orientation of the valley axis, and by the season. Steep-walled valleys exhibit larger radiative flux differences between sidewalls than do valleys with shallow sidewalls. Valleys that run north–south have no radiative differences between sidewalls of equal inclination angle at solar noon, but the east-facing sidewalls have much larger radiative loads in morning and the west-facing sidewalls much larger radiative loads in the afternoon. Cross-valley winds thus blow toward the west sidewall in the morning and toward the east sidewall in the afternoon. Valleys that run east–west exhibit large radiative differences in winter between the sunny south-facing sidewall and the opposing north-facing sidewall. These differences peak in midday and are symmetric about noon. In summer, when the sun rises north of east and sets north of west, the north-facing sidewall receives more radiation during a brief period in the early morning and late afternoon. Cross-valley winds in east–west valleys thus flow from north to south during the early morning and late afternoon in summer and from south to north during the remainder of the day. Figure 11.25 illustrates the dependence of insolation differences between sidewalls on season, time of day, sidewall inclination angles, and valley axis orientation. To simplify the illustration, the sidewalls are assigned equal inclination angles and are assumed to have fall lines that are perpendicular to the valley's longitudinal axis.

11.4.4. *Mountain-Plain Wind System*

The mountain–plain wind system is a closed circulation that develops above the slopes of a mountain massif in response to temperature differ-

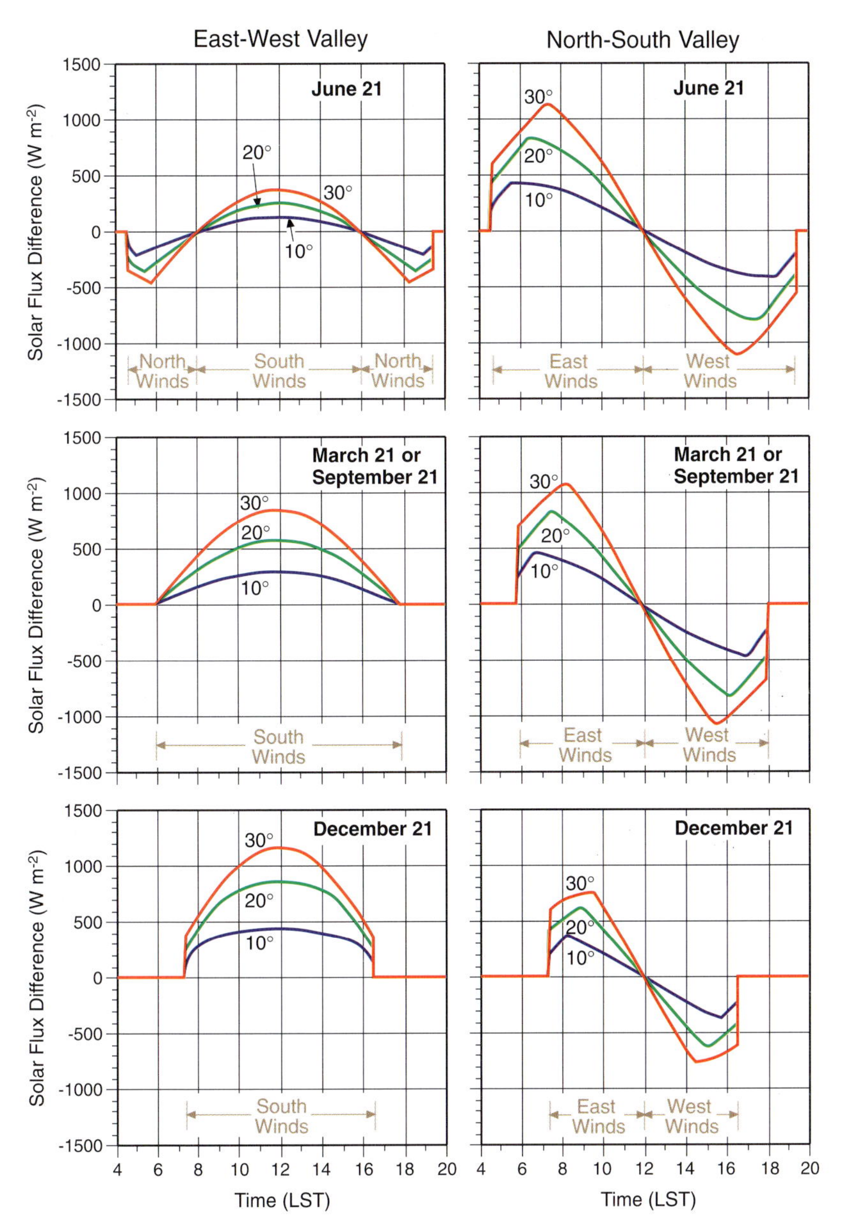

Figure 11.25 Solar radiation differences between the two opposing sidewalls of valleys oriented east–west and north–south at 40°N latitude. Numbers in the figure (10°, 20°, and 30°) are the sidewall inclination angles. The radiation differences are calculated assuming no attenuation of the solar beam by the earth's atmosphere, and thus they overestimate the actual radiation differences in real valleys. The radiative difference is calculated by subtracting the radiative flux on the north- or east-facing sidewall from the flux on the opposing sidewall. Shading of one sidewall by the other is not considered in the computations.

ences between air over the mountains and air over the nearby plains. The temperature difference produces a horizontal pressure difference between the air over the plains and the air in the atmospheric boundary layer that forms over the entire mountainous region. This pressure difference is illustrated using summertime data for the mountains of the

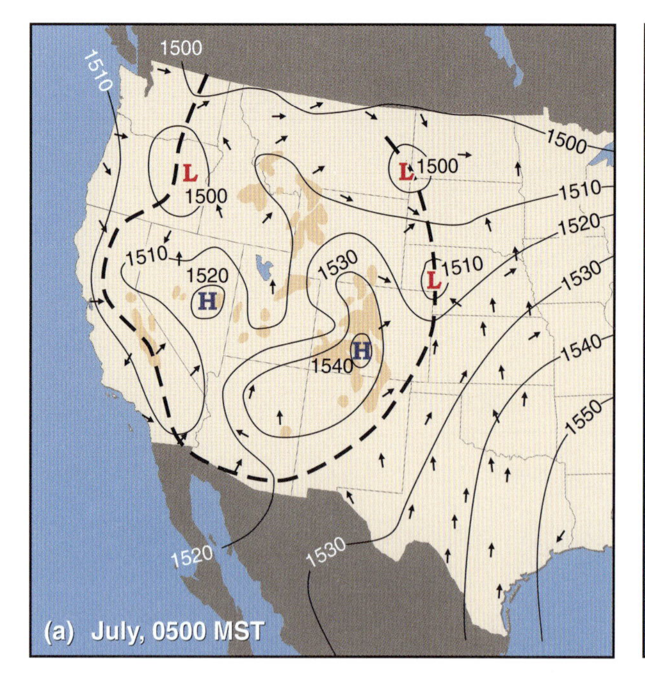

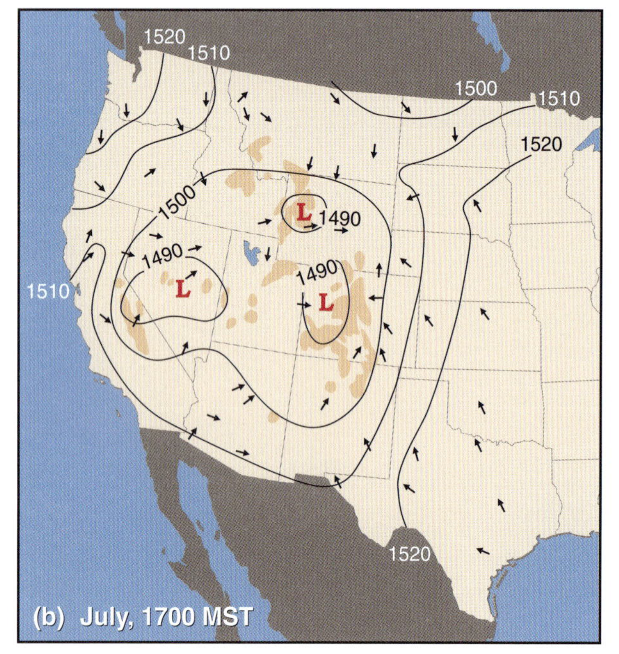

Figure 11.26 Mean height of the 850-mb pressure surface during July at (a) 0500 MST and (b) 1700 MST, as determined from an analysis of 7 years of data. The dashed line in (a) shows the axis of low heights (or, equivalently, pressures) that surrounds the Rocky Mountains. (Adapted from Reiter and Tang, 1984)

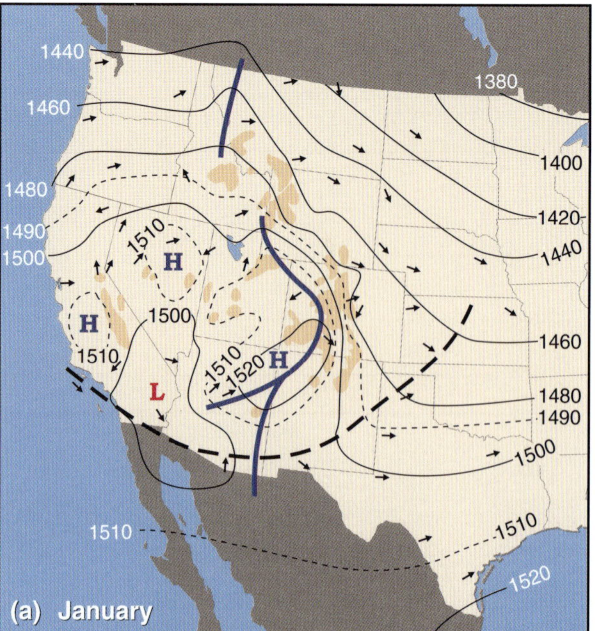

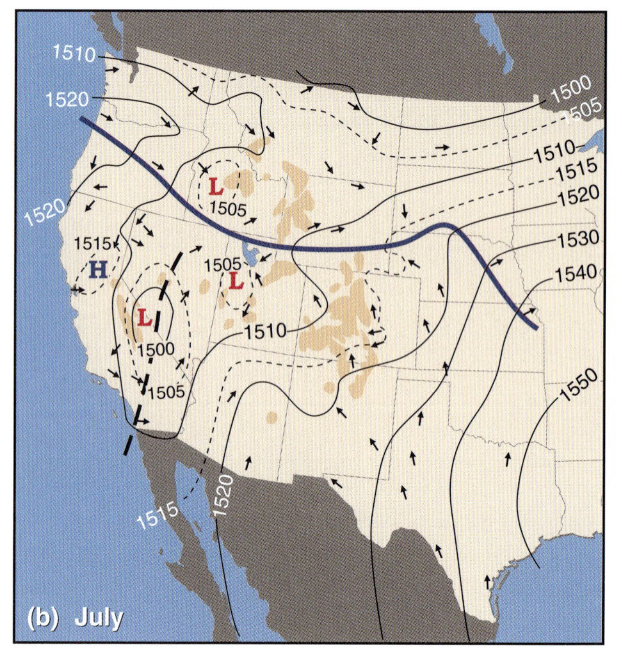

Figure 11.27 Mean height of the 850-mb pressure surface during (a) January and (b) July, as determined from an analysis of 7 years of data. The blue lines indicate axes of high heights (or pressures), and the heavy dashed liens indicate axes of low heights. (Adapted from Tang and Reiter, 1984)

western United States for 0500 MST (figure 11.26a) and for 1700 MST (figure 11.26b). During nighttime, the air over the mountains becomes colder than the air over the plains, producing an area of relatively high pressure over the mountains and causing air to flow from the mountains to the plains. During daytime, the air over the mountains becomes warmer than the air over the plains, producing an area of relatively low pressure over the mountains and winds that blow from the plains to the mountains. Two high pressure centers are located over the western United States at night. One is centered over the highest mountains in Colorado and has a northern extension across the high-elevation Yellowstone Plateau into western Montana and a southern extension over the high plateaus of northern Arizona. The other high pressure center forms over the center of the Great Basin, where cold air drains into the basin from the surrounding mountains.

In addition to the diurnal pressure variations illustrated in figures 11.26a and b, there are also seasonal variations in the pressure differences that develop between mountains and plains. In winter, high pressure develops over the mountains and, in summer, low pressure develops, as shown in figures 11.27a and b. In January, high pressure centers are found over the Four Corners area where Utah, Colorado, New Mexico, and Arizona meet, over California's Central Valley, and over the center of the Great Basin in Nevada. In July, low pressure centers develop east of the Sierra Nevada, over the Snake River Plain in Idaho, and in western Utah. The mountain–plain system is strongest in summer and fall when strong horizontal temperature gradients between the mountains and plains coincide with weak synoptic flows. Wind speeds are highest on the sunny side of the mountain massif where strong upward sensible heat flux enhances temperature and pressure contrasts.

Mountain–plain circulations, which are not confined by the topography, vary diurnally and seasonally.

Mountain–plain flows are generally quite weak, below 2 m/s (4.5 mph), and are easily overpowered by prevailing large-scale flows. They are, nonetheless, significant because the convergence of the upward daytime flows over the mountains produces afternoon clouds and air mass thunderstorms, and the divergent nighttime sinking motions produce late afternoon and evening clearing.

The Southwest Monsoon (section 8.5.3) is an example of a well-formed mountain–plain circulation. In summer, the monsoon is driven by the relatively low pressure that develops over the continent compared to the pressure over the Gulf of California and the Pacific Ocean. It brings moist air into the Rockies and the Great Basin from the Gulf of California.

11.5. Diurnal Mountain Winds in Basins

Basins are valleys that are circular or oval in shape. A typical *diurnal mountain circulation* can develop in a basin if the surrounding mountains are broken by a deep opening that extends to the floor of the basin and is thus similar to the mouth of a valley. If, however, the surrounding mountains are interrupted only by high passes, the basin will experience slope flows, but no along-valley flow will develop. The absence of an along-valley flow affects temperatures, wind speeds, and the buildup of nocturnal inversions.

Along-valley circulations tend to equilibrate valley temperatures with those of the nearby plain, advecting colder air into the valley during daytime when the valley is warming and out of the valley during nighttime when the valley is cooling. In the absence of along-valley flows, basins experience enhanced heating during daytime and enhanced cooling during nighttime. Nighttime cooling is pronounced, as evidenced by the many minimum temperature records that have been set in basins. Record low temperatures have been recorded, for example, at Tincup in Colorado's Taylor Park, in Utah's Logan Sinks, and in Montana's Bighole Basin. The lowest wintertime minimum temperatures in Europe are often found in Austria's Gstettneralm sinkhole. The enhancement of daytime heating is less pronounced because heat escapes through the top of the basin atmosphere as a result of convection.

Winds in a basin are usually light because of the absence of along-valley flows and because the surrounding topography shields the basin atmosphere from synoptic flows. The lowest elevations in a basin are particularly well protected from strong winds aloft by strong temperature inversions that form over the floor of the basin. However, basins with uniform-elevation mountains on their periphery can experience sudden strong winds in the late afternoon or early evening that are not characteristic of a valley atmosphere. Plain-to-mountain winds that suddenly break in to a basin when the surface energy budget reverses in the late afternoon are called plain-to-basin winds. The short-lived plain-to-basin winds pour over the surrounding ridge tops, bringing cooler temperatures to the overheated basin.

The formation of nocturnal temperature inversions is characteristic of the basin atmosphere. Cold air drains off the basin sidewalls at night, collecting in a cold air pool over the basin floor. In summer, the inversion is broken on a daily basis by convection. In winter, the cold pool can persist for days because heating is insufficient to break the inversion. Air pollutants, moisture, fogs, and clouds build up within the layer of cold air. Freezing rain and freezing drizzle occur when rain or drizzle falling into the pool freeze on contact with the cold ground. A wintertime inversion in a basin is destroyed when an approaching low pressure trough advects cold air above the basin, thus destabilizing the basin atmosphere (section 4.3.2). After the trough passes, warm air is advected above the basin, the basin atmosphere stabilizes, and another inversion forms.

Basins tend to have weak along-valley circulations and intense inversions.

11.6. Diurnal Mountain Winds over Plateaus

A small plateau, like a mountaintop, has only a minimal effect on the atmosphere around it because the limited surface area exchanges little heat with the atmosphere and exerts little frictional force on synoptic winds. Conditions on the plateau are thus determined primarily by the synoptic flow. The air above a large plateau, on the other hand, is influenced not only by the synoptic flow but also by a diurnal, thermally driven circulation that is driven by temperature differences that develop between the air over the plateau and the air at the same elevation surrounding the plateau.

Air over a plateau is warmer during the day and colder during the night than the air surrounding the plateau.

During daytime, the surface of the plateau heats the air adjacent to it, and a warm convective boundary layer grows above the plateau. The air within the convective boundary layer is thus warmer than the air at the same elevation surrounding the plateau. The resulting large-scale circulation, called a plain-to-plateau circulation, carries air from the surroundings toward the plateau with a return circulation above the convective boundary layer. The shallow upslope flows on the slopes of the plateau are part of this circulation. During nighttime, the surface of the plateau cools the air above it, and a shallow, stable boundary layer develops over the plateau. The air in this layer is colder than air at the same level over the plain, resulting in a plateau-to-plain circulation that carries air down off the plateau with a return circulation aloft. The downslope flows over the slopes of the plateau are a component of the large-scale circulation.

The plateau-to-plain and plain-to-plateau circulations are generally not strong because the diurnal temperature range of the air on a plateau is limited. During the day, temperatures on the plateau are moderated by the synoptic flow, and at night the coldest air drains off the plateau. In addition, the daytime maximum and nighttime minimum temperatures are not affected by the valley volume effect (section 11.4.2) because the air on a plateau is not confined by sidewalls.

Weak to moderate diurnal mountain winds on plateaus are often overpowered by synoptic flows, which normally strengthen with elevation and therefore have a significant impact on high plateaus. At night, the elevated surface of a plateau is subject to winds that prevail at the approximate elevation of the plateau top. During daytime, strong winds are mixed down from higher elevations above the plateau by the growing convective boundary layer.

11.7. Other Local Thermally Driven Wind Systems

11.7.1. *Sea (or Lake) and Land Breezes*

Sea breezes and *land breezes* are driven by horizontal temperature contrasts that develop between the sea and the adjacent land. Similar breezes that form along the shore of a large lake are called *lake breezes*. The temperature contrasts result from the unequal heating and cooling rates of land and water. If a unit mass of water and a unit mass of land receive the same heat input, the temperature of the water will rise less than the temperature of the land. Further, solar radiation, which is absorbed in a shallow soil layer over land, is distributed through a deeper layer in the ocean. Thus, the temperature of a large body of water changes little between summer and winter and between day and night. In contrast, the temperature of a land surface rises rapidly during daytime, drops rapidly during nighttime, and fluctuates with the seasons.

Given the same heat input, a unit mass of air undergoes greater temperature changes than an equal unit mass of either land or water. The temperature of the atmosphere therefore adjusts quickly to the temperature of the underlying surface, whether land or water. The horizontal gradi-

The temperature contrasts that drive sea and land breezes result from differences in the heating and cooling rates of land and water.

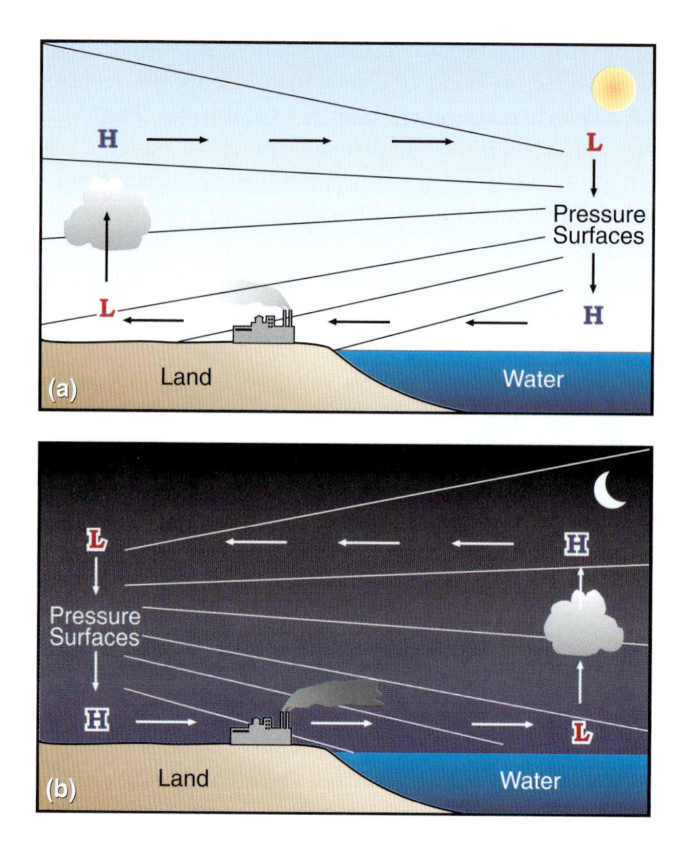

Figure 11.28 Sea and land breezes, like all diurnal circulations, are closed. During the day (a), there is a surface low pressure over the land and a surface high pressure over the ocean. Air rises above the low, causing a high pressure area to form above it. A low pressure area forms aloft as air sinks above the surface high. At night (b), the circulation is reversed. The sloping solid lines indicate pressure surfaces. (Adapted from Ahrens, 1994)

ents in temperature and pressure that develop across coastlines drive the land and sea breezes, with stronger gradients producing stronger winds. When the atmosphere is warmer over the land than over the sea, a low pressure forms over the land and a sea breeze flows from the sea to the land. When the air over the sea is warmer than the air over the land, a land breeze blows from land to sea. Sea breezes predominate during the day, land breezes at night. Both sea and land breezes are closed circulations, with a compensatory flow higher in the atmosphere carrying air in the opposite direction (figure 11.28).

The strength of sea and land breezes varies with the season and the time of day. In middle latitudes, sea–land temperature differences, and therefore sea and land breezes, are strongest in spring and summer. In summer, sea breezes are usually established by midmorning, after the land has warmed sufficiently, and reach peak speeds in the afternoon, when the temperature contrasts are strongest. Temperature contrasts between sea and land are usually weaker during nighttime, resulting in land breezes that are weaker.

The leading edge of the cool moist air off the ocean or off a large lake is called a *sea breeze front* or *lake breeze front* and marks abrupt temperature differences between the cool air that originated over the water and the warm air over the land. When the circulation is strong, the front can be 10–150 km (5–90 miles) inland by midafternoon. The location of the front can be determined visually when the cool air behind the front is hazy or when rising motions near the front, caused by the convergence of the incoming sea air and the air over the land, produce clouds. Convergence is somewhat enhanced over the shoreline by the difference in roughness between the sea and land surfaces (section 10.1.1). If the air ahead of the sea breeze front is unstable, lifting motions up the frontal surface may initiate thunderstorms.

Convergence along a sea or lake breeze front produces clouds and can be enhanced by differences in roughness between sea and land surfaces.

Where there are mountains along a coastline, sea and land breezes are subject to the influence of diurnal mountain winds. Sea breezes can be channeled through valleys or passes, which increases wind speeds, or they can be channeled around isolated mountains. A *sea breeze convergence zone* often forms on the inland side of an isolated coastal mountain range, where the air that has split to flow around the mountain range reconverges. The moist air is forced to rise, producing clouds and rain. The Santa Monica and Santa Ana Mountains of Southern California produce well-known sea breeze convergence zones.

11.7.2. Glacier Winds

Glacier winds are downslope winds over glaciers or snowfields. The winds are driven by the temperature contrast between the air in a shallow layer over the glacier that has been cooled by a downward flux of sensible heat to the snow surface and the warmer air at the same altitude farther down the valley. The cold air flows down the glacier fall line, reaching peak speeds when the temperature contrasts are highest. The strength and depth of a glacier wind depend on the length of the air trajectories over the glacier and on the shape of the outflow terrain. Peak wind speeds on small- to medium-size valley glaciers in midlatitudes are typically around 3 m/s (7 mph) at a height of 2–3 m (7–10 ft) above the glacier surface, although the glacier wind extends through depths of tens of meters. In Antarctica, glacier winds can have 1500-mile-long (2400-km-long) trajectories, producing windspeeds that can exceed 50 mph (22 m/s) in the coastal areas. On the Antarctic coast at Cape Denison, where glacier winds are nearly always present, winds average 43 mph (19 m/s). The trajectory of Antarctic glacier winds is sufficiently long that the Coriolis force (section 5.1.5) turns the winds significantly to the left of the fall line as they descend the continental ice dome.

In the summer, glacier winds generally blow continuously both day and night. During the day, up-valley winds blow over the top of the gla-

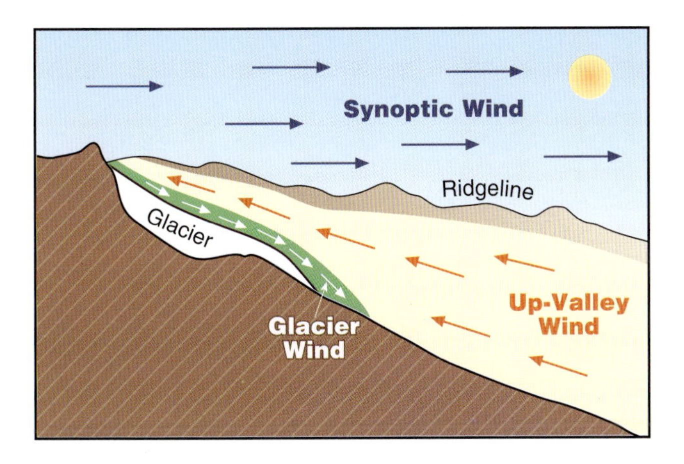

Figure 11.29 Longitudinal section through a valley glacier illustrating the glacier, up-valley, and synoptic winds during daytime. (Adapted from Geiger et al., 1995)

cier winds. The glacier winds disappear some distance below the glacier terminus when the upward heat flux from the glacial outwash plain destroys the cold air layer from below. During the night, down-valley winds blow over the top of the glacier winds, though the two are hard to distinguish because they blow in the same direction. The glacier wind is shown in longitudinal section in figure 11.29.

Because the temperature of a snow surface is always near or below freezing, a temperature inversion is present above a snow surface whenever the air temperature is above freezing. The inversion can be strong, with temperatures increasing by 5–10°C (9–18°F) in the first 2 meters (6 ft) above the surface of the glacier. This stable temperature structure produces a downward flux of heat from the atmosphere to the underlying ice surface, which can result in prodigious melting on summer days. The amount of meltwater at the terminus of a glacier exhibits a strong diurnal cycle with especially high volumes in the late afternoon of warm, sunny days.

11.7.3. *Fire Winds*

Large fires produce their own thermally driven winds. The horizontal temperature and pressure gradients at the edge of the column of air heated by the fire are extreme in comparison to those that drive normal diurnal mountain winds. The resulting *fire winds*, which blow toward the fire column, can be very strong.

Part IV

SELECTED APPLICATIONS OF MOUNTAIN METEOROLOGY

Air Pollution Dispersion

12

12.1. Classification and Regulation of Air Pollutants

Air pollutants are harmful airborne substances (solids, liquids, or gases) that, when present in high-enough concentrations, threaten human health or welfare, harm vegetation, animals, or structures, or affect visibility. Visibility alone is not, however, a reliable indicator of the presence of pollutants. A visible plume of condensed water vapor from an industrial cooling tower decreases solar radiation and increases the frequency of fog and icy road conditions near the cooling tower, but it is not an air pollution plume because it is composed entirely of water. In contrast, an industrial pollutant plume may be nearly invisible after the gross particulate matter has been removed by pollution control equipment, but it may still contain large quantities of pollutant gases.

12.1.1. *Types and Sources of Air Pollutants*

Air pollutants can come from either natural or *anthropogenic* (human) sources. The distinction between the two categories is not always clear. Natural emissions include ash and dust from volcanoes, certain highly volatile chemicals from forests, *aeroallergens* such as ragweed pollen, wind-entrained dust from natural land surfaces, and smoke and ash from wildfires. Wind-entrained dust can, however, come from roadways or land surfaces that have been disturbed by man, some aeroallergens come from plant species introduced to a new habitat by man, and many fires are *prescribed fires*—natural or man-made fires (whether accidental or deliberate) that are allowed to burn in order to meet forest or land management objectives.

Pollutants can be emitted directly into the atmosphere (*primary pollutants*) or produced in the atmosphere (*secondary pollutants*) as a result of chemical or physical transformations of primary pollutants when exposed to other components of air, including other pollutants or water vapor. Examples of transformations include the clumping or coagulation of small particulates into larger particles and the conversion of sulfur dioxide gas emitted from coal-fired power plants to particulate sulfates under humid conditions or to *acid rain* droplets if clouds are present. Some secondary pollutants, such as *photochemical smog* or *ozone*, result from photochemical reactions, that is, chemical reactions that occur only in the presence of solar radiation.

Pollutants may come from *point*, *area*, or *line sources*; the emissions may be continuous or intermittent; and the source strength may be variable or constant. Sources may be at fixed locations or may be mobile; they may be ground-based or elevated. For example, a power plant may produce near-constant continuous anthropogenic emissions from a fixed, elevated point source. A heavily traveled roadway is a fixed, ground-based line source with variable emissions as the traffic volume changes through the day. Individual cars on the roadway are mobile sources. An evergreen forest in the Smoky Mountains on a hot summer day is a continuous ground-based natural area emissions source for terpenes—natural, low-boiling-point compounds that reduce summertime visibility in parts of the Appalachians.

Hundreds of different species of gases and aerosols can be introduced into the atmosphere as pollutants. Most pollutants fall in seven broad pollutant classes: carbon monoxide, oxides of nitrogen, ozone, lead, particulate matter (dust, smoke, and other aerosols), oxides of sulfur, and volatile organic compounds. An annual publication by the Environmental Protection Agency (Environmental Protection Agency, 1995) lists sources for these pollutant classes by industry and state, and provides a complete list of *Nonattainment Areas* in the United States that do not meet various *National Ambient Air Quality Standards*.

12.1.2. *The Clean Air Act*

The U.S. *Clean Air Act* (CAA) of 1970 was promulgated to protect and enhance the quality of the nation's air and to protect public health and welfare. The act empowered the federal government to set national ambient air quality and emission standards and gave the states the primary responsibility for air quality management. The CAA was revised in 1977 and again in 1990 to include stricter federal emission standards for automobiles and industrial installations, as well as incentives for industry to lower emissions of pollutants that lead to acid rain. The states carry out their air quality management responsibilities under the CAA by writing and following *State Implementation Plans* (SIPs) to achieve federal and state standards. The states have the authority to impose their own, stricter, state regulations when they feel that the federal regulations are inappropriate or insufficient to maintain or improve their air quality. The CAA requires the states to identify areas that do not meet national air quality standards, to designate those areas as Nonattainment Areas, and to make plans and take specific actions, specified in the SIPs, to bring the Nonattainment Areas into compliance.

The CAA also provides for the *Prevention of Significant Deterioration* (PSD) of air quality in areas of the country where air quality is above the standards. The CAA program designates three classes of PSD areas, with *Class I Areas* having the most pristine air and receiving the greatest protection under the Clean Air Act. Mandatory Class I Areas include all international parks, national wilderness areas, and national memorial parks over 5,000 acres and national parks over 6,000 acres that were in existence in 1977. In addition, states and Native American tribes can designate additional Class I Areas. PSD regulations in the Class II and III Areas are less strict than those of the Class I Areas. Under the PSD program, applications must be submitted to the Environmental Protection Agency (EPA) or the state for a permit to construct any major new pollution sources or to make major modifications to existing sources. The permit is issued only if the new emissions will not exceed, or contribute to the exceeding of, designated maximum allowable pollution increments in the PSD areas. Major new stationary sources are not allowed to degrade air quality by more than a small increment.

The Clean Air Act and its subsequent amendments are the key regulations that protect the nation's air quality.

The Clean Air Act Amendments of 1977 set a national goal of remedying existing impairments to visibility and preventing future visibility impairment from man-made pollution sources at 158 national parks and wilderness areas across the United States. Under this *Visibility Protection Program*, states and tribes must demonstrate to the EPA in their State and Tribal Implementation Plans that they are making progress in achieving this national goal. The first regulations under the Visibility Protection Program were established for the states in 1980 and for the tribes in 1990. These regulations focus on the small group of air pollution sources to which visibility impairment at a Class I Area can be "reasonably attributed," and they require these sources to adopt the best available retrofit technology to reduce their emissions. Additional regulations are currently being developed to deal with widespread, regionally homogeneous haze that arises from multiple sources and impacts visibility over large areas. Haze is caused by fine dust, smoke, or salt particles that diminish visibility and subdue atmospheric color and contrast.

12.1.3. *National Ambient Air Quality Standards*

National Ambient Air Quality Standards (NAAQS) were designed to limit the ambient concentrations of certain pollutants. *Primary ambient air quality standards* were established to protect human health, whereas *secondary ambient air quality standards* were established for selected pollutants to protect human welfare, including effects on vegetation, visibility, and structures. Some of the concentration standards are imposed over different averaging times or exceedance intervals because human health and welfare are affected by both acute and chronic exposures to pollutants. Violations can thus result from a variety of meteorological conditions, both local and synoptic, long term and short term. The 1996 NAAQS for six major pollutant categories are listed in table 12.1.

National Ambient Air Quality Standards limit the ambient concentrations of different categories of air pollutants.

12.1.4. *Emission Standards*

The EPA's emission standards limit the atmospheric emissions of primary pollutants to ensure that ambient concentrations are below the levels specified in the National Ambient Air Quality Standards. Standards have

Table 12.1 National Ambient Air Quality Standards

Pollutant	Primary: Type of Average	Primary: Standard Level Concentration[a]	Secondary: Type of Average	Secondary: Standard Level Concentration[a]
Carbon monoxide (CO)	8-hour[b]	9 ppm (10 μg/m^3)	no secondary standard	
	1-hour[b]	35 ppm (40 μg/m^3)	no secondary standard	
Nitrogen dioxide (NO_2)	annual arithmetic mean	0.053 ppm (100 μg/m^3)	same as primary standard	
Ozone (O_3)	maximum 8-hour average[c]	0.08 ppm (157 μg/m^3)	same as primary standard	
Lead (Pb)	maximum quarterly average	1.5 μg/m^3	same as primary standard	
Particulates less than 10 micrometers in diameter (PM-10)	annual arithmetic mean[d]	50 μg/m^3	same as primary standard	
	24-hour[d]	150 μg/m^3	same as primary standard	
Particulates less than 2.5 micrometers in diameter (PM-2.5)	annual arithmetic mean[e]	15 μg/m^3	same as primary standard	
	24-hour[e]	65 μg/m^3	same as primary standard	
Sulfur dioxide (SO_2)	annual arithmetic mean	0.03 ppm (80 μg/m^3)	3-hour[b]	0.50 ppm (1300 μg/m^3)
	24-hour[b]	0.14 ppm (365 μg/m^3)		

[a]Value in parentheses is an approximately equivalent concentration in micrograms per cubic meter at standard (normal) atmospheric temperature and pressure.

[b]Not to be exceeded more than once per year.

[c]Attainment is based on a 3-year average of the annual fourth-highest daily maximum 8-hour average ozone concentration.

[d]Particles less than 10 micrometers in diameter are the indicator pollutant. The annual standard is attained when the annual arithmetic mean concentration is less than or equal to 50 μg/m^3; the 24-hour standard is attained when the number of days per calendar year above 150 μg/m^3 is less than or equal to 1.

[e]Particles less than 2.5 micrometers in diameter are the indicator pollutant. The annual standard is met when the 3-year average of the spatially averaged annual means is less than or equal to 15 μg/m^3. The 3-year average of the spatially averaged annual means is determined by averaging quarterly means at each monitor to obtain annual mean PM-2.5 concentrations at each monitor, then averaging across all designated monitors, and finally averaging for 3 consecutive years. The 24-hour standard is met when the 3-year average of the 98th percentile values at each monitoring site is less than or equal to 65 μg/m^3. The 98th percentile means the daily value out of one year of monitoring data below which 98 percent of all values in the group fall.

also been established for specific chemicals released in individual industrial operations, including 169 of the most dangerous pollutants, termed *air toxics*. To meet these emission standards, air pollution control equipment may have to be installed at various stages in a manufacturing or industrial process.

Emission standards differ from ambient air quality standards by directly limiting the emissions of certain primary pollutants from industrial and other operations.

According to data collected by the EPA, the implementation of air quality regulations in the CAA since 1970 has greatly improved the nation's air quality by lowering emissions (figure 12.1). The reduction in atmospheric lead concentrations has been especially notable and is attributed to regulations requiring the use of nonleaded gasoline in automobiles. Air pollutant emissions are projected to rise again in the United States within two decades because of the increase in population.

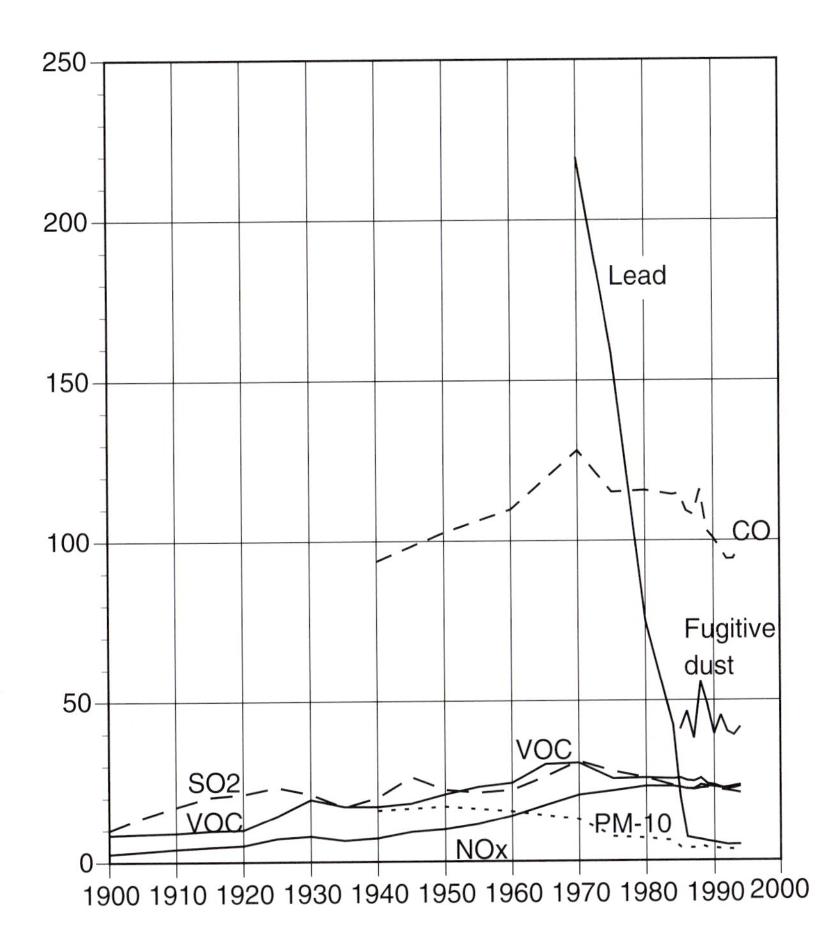

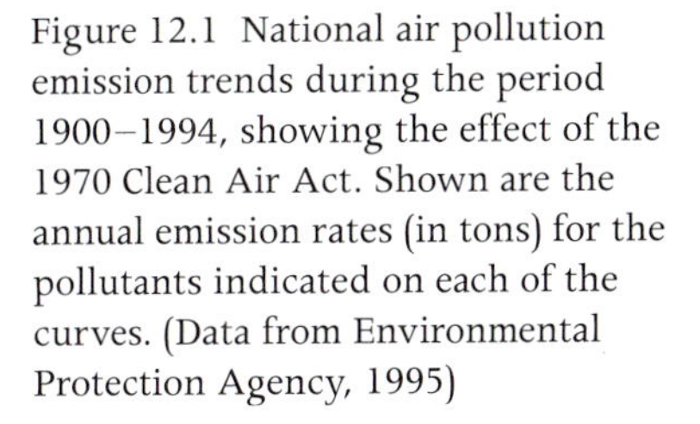
Figure 12.1 National air pollution emission trends during the period 1900–1994, showing the effect of the 1970 Clean Air Act. Shown are the annual emission rates (in tons) for the pollutants indicated on each of the curves. (Data from Environmental Protection Agency, 1995)

12.2. Air Quality Studies and Air Pollution Models

Air quality studies can be designed to monitor ambient air quality, to predict the impact of a proposed new source of emissions, to estimate the effectiveness of proposed corrective actions, or to focus on the physical processes that lead to high concentrations of pollutants. Studies may include an inventory of emission sources, meteorological field measurements to determine typical evolution of wind and temperature structure, climatological analyses of ambient air concentrations of pollutants, upper air wind strengths and directions, inversion frequencies and local wind system characteristics, and computer modeling to simulate air pollution dispersion.

The pollution concentration at a site depends on four factors: the ambient weather, the sources of pollution affecting that site, the atmospheric *dispersion* (i.e., transport and *diffusion*) of emissions between the sources and the site, and the chemical and physical transformations of the emissions during their transport. The *air pollution potential* of a site is the meteorological potential for air pollution problems, considered without regard to the presence or absence of actual pollution sources and is dependent on wind, stability, and *mixing depth* (section 12.4.2). Actual concentrations, of course, also depend on emission rates at sources that affect the site. The determination of "worst case" meteorological or emissions scenarios, that is, the highest possible air pollution potential, is often the first step in selecting an *air quality model* that will predict ambient air pollution concentrations from emission sources.

Air quality models are mathematical, statistical, or physical (e.g., wind tunnel or towing tanks) models that relate air pollution concentrations at various receptor sites to pollutant emissions from one or more sources. The mathematical models stimulate the dispersion of pollutants from an emissions source, including the transport, diffusion, and chemical and physical transformations of pollutants. A number of mathematical models have been developed for use in specific mountainous terrain, but no one model is applicable to air pollution dispersion in the wide range of terrain-forced and diurnal mountain flows found in complex terrain. An appropriate model must accurately simulate the physical processes that are important in the investigation at hand. Model assumptions, approximations, required input, and expected output must be considered before a model can be chosen that accurately simulates the physical processes important in a particular air pollution investigation.

Air quality models are mathematical, statistical, or physical models that relate air pollution concentrations at various receptor sites to pollutant emissions from one or more sources.

Air pollution transport and dispersion models can be classified in a number of ways. The following classification, modified somewhat from one developed by Dr. Robert Meroney (Ekblad and Barry, 1990), includes three *wind field* or pollutant transport models (*diagnostic models*, *perturbation models*, and *full-physics numerical models*) and five types of air quality models (*box models*, *Gaussian plume models*, *phenomenological models*, *particle trajectory models*, and *chemistry models*). The first three models are used to determine wind fields at one or more times. The remaining five models use these wind fields and address the dispersion, chemical transformation, or removal mechanisms of air pollutants during their transport.

12.2.1. *Wind-Field Models*

Diagnostic models produce a wind field over an area by interpolating (under various physical constraints) from a limited number of actual wind observations. The models produce an interpolated or adjusted wind field for one or more specific locations at a single time from observations made at the same time at other locations and thus have no prognostic capability. Because of the many interacting wind systems that affect plume transport and diffusion in mountains, it is difficult to obtain enough data for an optimal analysis, especially during wind transition periods (see, for example, figure 11.8).

Perturbation models use a mathematical technique to build a wind field from a series of solutions to a simplified set of equations that describes atmospheric motions. The total wind field is determined by summing the wind fields produced mathematically by individual wind perturbations associated with terrain shape, surface roughness, stability, and other effects. An assumption is made that only small oscillations in speed and direction are possible about a mean flow state. These models are appropriate only for weak terrain slopes and then only when the temperature and wind profiles follow certain constraints, when wind fields change slowly with time, and when heating or cooling rates are weak.

Full-physics numerical models calculate wind fields and other meteorological fields such as temperature or humidity from the basic conservation equations that describe atmospheric flows; they can simulate all types of atmospheric phenomena, including both terrain-forced and diurnal mountain wind systems. These models, which run on powerful computers, obtain solutions to systems of equations that describe atmospheric flows. To obtain solutions, the three-dimensional calculation grid must first be initialized with values of all basic meteorological variables. Values of the variables on the edge of the calculation grid must also be updated at every model time step using data obtained from some other source (usually other larger scale operational models). Difficulties arise in assigning appropriate initial and boundary conditions and in describing small-scale processes (e.g., turbulence) mathematically. Data assimilation techniques have been developed to "nudge" the model toward solutions that more or less match independent observational data.

12.2.2. *Air Quality Models*

Box models calculate air pollution concentrations by assuming that pollutants are emitted into a box through which they are immediately and uniformly dispersed. The floor and sides of the box are usually defined to coincide with the floor and sidewalls of a valley, with the upper surface representing the top of a mixed layer or some fraction of the height of a surface-based inversion. A series of boxes can be used to stimulate adjacent valley segments having different geometric characteristics. The model allows pollutants to transfer from one box to the next and for clean polluted air to be added or removed from the boxes to stimulate flow up or down valley tributaries. In certain applications, the box is allowed to be transported by the wind (a "moving box" model) or expanded as along-valley winds carry pollutants farther down (or up) a valley. Box models are simple to understand and operate, and they are useful for making initial estimates of air pollution concentrations in valleys. However, the key assumption in the model (that pollutants are immediately and uniformly dispersed throughout the box) is often unrealistic, particularly in stable atmospheres.

Gaussian plume models assume that a plume has its highest concentration at its horizontal and vertical midline directly downwind of the emission source. Concentrations are assumed to decrease toward the outer edge of the plume such that a graph of pollution concentrations on both horizontal and vertical cross sections has a distinctive Gaussian or bell-shaped curve. The rate of diffusion from the plume's midline is specified from the findings of experiments conducted over flat homogeneous terrain and is sometimes adjusted upward in an ad hoc manner to account for enhanced diffusion in mountainous terrain. The Gaussian model assumes straight line winds that do not change speed or direction over time. It therefore cannot handle temporal or spatial variations, nor can it handle any unusual enhancements to diffusion caused by secondary circulations (as from valley tributaries). Modified versions of the model, called *Gaussian puff models* and *Gaussian plume segment models*, have been developed to overcome some of these limitations, but temporally and spatially resolved wind data from field experiments or meteorological models are required to drive them.

Phenomenological models estimate air pollution concentrations resulting from individual phenomena, such as plumes drifting into a hillside (*plume impingement*, sections 12.6.2 and 12.6.3.4), nighttime plume dispersion in confined valleys, or downward mixing of elevated plumes to the ground by convection during daytime (*fumigation*, section 12.6.3.5). The accuracy of many phenomenological models in mountainous terrain is unknown because few have been rigorously tested against field data.

Particle trajectory models are used with wind field models to determine concentrations of nonreactive air pollutants downwind of a source. These models simulate the release of large numbers of particles (up to hundreds of thousands) from sources located within the model's geographical simulation area or *domain*. The paths of the particles through this domain vary as the predicted three-dimensional wind fields change over time. At the end of the simulation, statistical information about the particles' final positions is used to characterize the dispersion. Some particle trajectory models include models that can enhance the effects of turbulence on the particle trajectories.

Chemistry models are used with wind field models to account not only for diffusion of the transported material during transport, but also for chemical and photochemical transformations of mixtures of reactive pollutants, which are characteristic of urban atmospheres. The most detailed chemical models simulate chemical transformations in vapors, liquids, or on aerosol or other solid surfaces, as well as the removal of pollutants from an air mass by rain or by dry deposition on surfaces.

12.3. Wind Speed and Air Pollution Concentrations

12.3.1. *Dilution and Diffusion of Air Pollutants*

Wind speed determines the extent to which a pollutant is diluted at the emissions source (figure 12.2). For example, suppose that sulfur dioxide is being emitted continuously from a stack at the rate of 1 gram per second. If the wind speed at stack height is 1 m/s, one gram of pollutant will be distributed over a 1-m-long horizontal column of air because one meter of air will travel past the stack while the gram of sulfur dioxide is being emitted. If the wind is blowing at 5 m/s, however, the gram of sulfur dioxide will be distributed over a 5-m-long column, and concentrations (grams per cubic meter) will be approximately five times more dilute.

The initial dilution of a pollutant plume depends on the wind speed at emission height.

Strong winds also produce shear-induced turbulent eddies that mix the pollution with clean air and diffuse it about the plume centerline. The rate of diffusion of the plume depends on the intensity of the turbulence and the size of the turbulent eddies relative to the plume (figure 12.3).

Strong winds decrease plume centerline concentrations by increasing plume dilution at the stack and by producing additional or stronger turbulent eddies that diffuse the plume. The decrease of plume rise as the

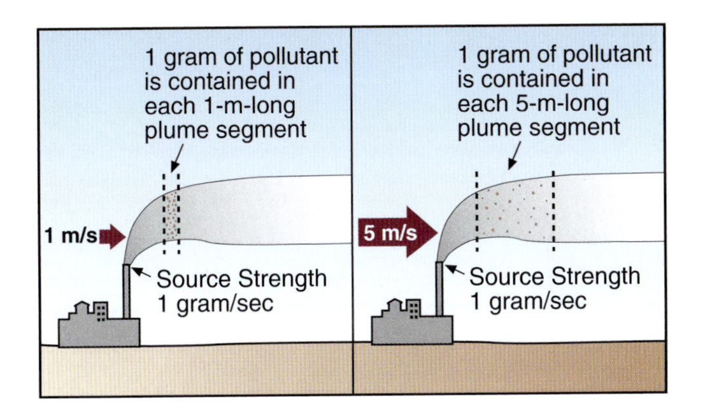

Figure 12.2 Dilution of the plume is enhanced by strong wind speeds at plume height.

plume is bent over in strong winds is a compensating effect that tends to increase ground level concentrations downwind of the stack.

12.3.2. *Fugitive Dust*

Strong winds are generally associated with lower pollutant concentrations because of dilution and enhanced diffusion as a result of turbulence. *Fugitive dust* emissions and smoke emissions from fires, however, usually increase rapidly with wind speed. Fugitive dust is all dust that is not emitted from definable point sources, such as smokestacks. Industrial fugitive dust can be eroded by the wind from storage piles and unpaved roads or kicked up from paved plant roadways by vehicles. Nonindustrial fugitive dust includes roadway dust from both paved and unpaved roads and dust from agriculture, construction, and fires. In 1994, nearly 98% (42 million tons) of the national emissions of dust particles with diameters below 10 micrometers came from nonindustrial sources (Environmental Protection Agency, 1995). The rate of fugitive dust emissions is determined primarily by wind speed but is also affected by surface moisture, ground cover, and particle size.

Fugitive dust emissions increase with wind speed.

Strong winds transport not only dust, but also spores, seeds, pollen, bacteria, and other aeroallergens. Although large dust particles settle out quickly, other smaller airborne materials have lower settling speeds. The smallest particles, some of which are harmful, can remain suspended in the atmosphere for many days.

12.4. Stability, Inversions, and Mixing Depth

When the atmosphere is stable, vertical motions are suppressed and plume diffusion is reduced. When surface-based or elevated temperature inversions form, pollutants are trapped within or below them.

12.4.1. *Surface-Based Inversions*

Surface-based inversions form at night and in winter when the ground loses heat by longwave radiation. Pollutants emitted within the inversion

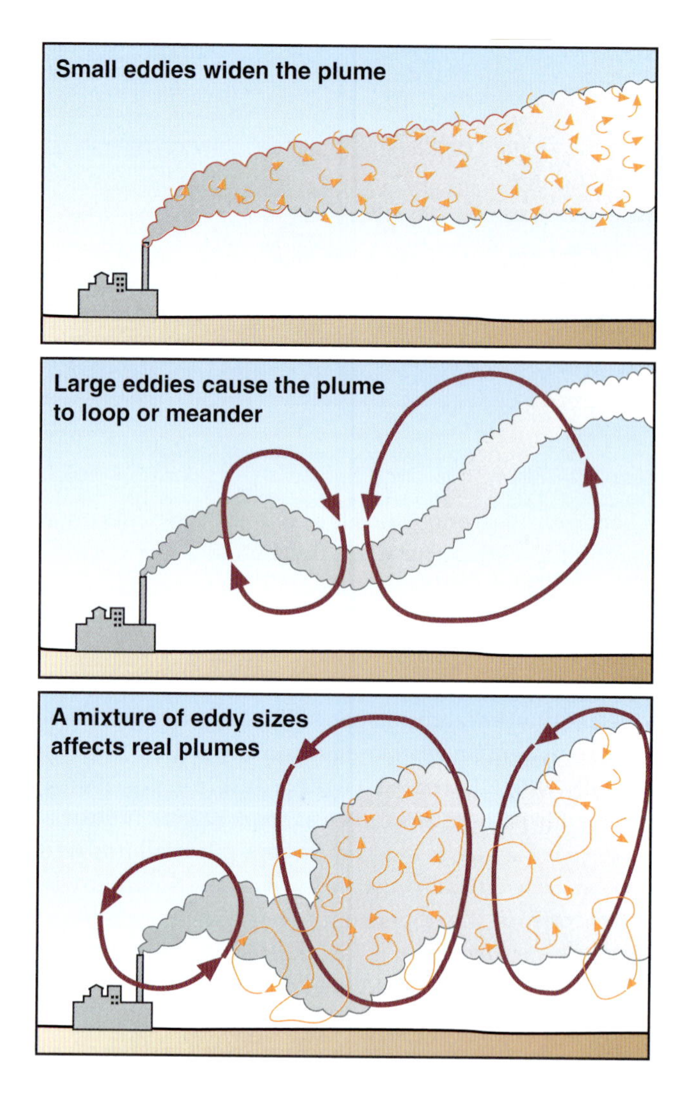

Figure 12.3 The rate of diffusion of a plume depends on the size of the turbulent eddies relative to the plume dimensions. Small eddies diffuse the plume about its centerline; large eddies transport the plume in large looping motions. Eddies that are the same size as the plume produce the most rapid diffusion.

layer, either at the ground or from an elevated source, are trapped within the inversion. Because wind speeds and turbulence are usually low within an inversion, pollution concentrations near the source can be high. If the pollution is emitted at the ground, ground level concentrations will be high. If the pollution is emitted from a stack into a surface-based inversion, the poor vertical dispersion at stack height will leave the pollution plume elevated, resulting in low concentrations at ground level (figure 12.4).

A surface temperature inversion suppresses the vertical dispersion of pollutants released into the inversion layer.

Pollutants from an elevated source can occur in high concentrations at ground level if the pollutant plume is carried toward a hillside that rises above the plume elevation, if the plume is caught in a lee eddy behind an obstruction, or if the plume is suddenly carried to the ground (i.e., the ground is fumigated) when a growing *unstable boundary layer* reaches plume height. If the horizontal and vertical diffusion of the plume through the inversion layer is gradual, ground level concentrations will not be as high because the pollutants are mixed through a greater volume of air before reaching the ground.

A plume from a stack may rise through a shallow inversion layer if the

OWENS VALLEY DUST

Water draining eastward from the Sierra Nevada into California's Owens Valley was diverted to Los Angeles nearly 80 years ago to accommodate the rapid growth of the city. This diversion dried up Owens Lake, a 100-square-mile lake south of Bishop, California, and created one of the largest sources of dust in the United States. About 100 times per year, the winds on the east side of the Sierra Nevada are strong enough to lift dust from the dry, salt-encrusted lake bed and to distribute this dust up and down the Owens Valley from Bishop to San Bernardino and to points east of the valley. Visibility is impaired as the large dust plumes scatter a fine, alkaline grit containing potentially toxic arsenic and selenium salts. Severe dust storms can kick up tens of thousands of tons of dust during periods as short as 12 hours The blowing dust scours away the dry lake surface, which is quickly replaced by a new surface that forms as briny water seeps upward. The new surface feeds the next dust storm and the cycle continues.

stack is high enough, if the effluent is emitted from the stack with a high velocity, or if the effluent is hot and buoyant. When this occurs, local pollution is minimal, but the pollutant source can degrade regional air quality, contributing to *regional haze* and regional acid deposition and causing high pollution background concentrations some distance downwind from the stack.

12.4.2. *Elevated Inversions*

Elevated inversions, like surface inversions, resist vertical motions. Pollutants emitted below an elevated inversion are trapped below it and cannot mix with air (whether polluted or clean) above the inversion layer. They can, however, become well mixed below the elevated inversion by convective currents that rise from the heated ground. The depth of the mixing during daytime is determined by the depth of the growing convective boundary layer. If the mixing depth is low (i.e., the convective

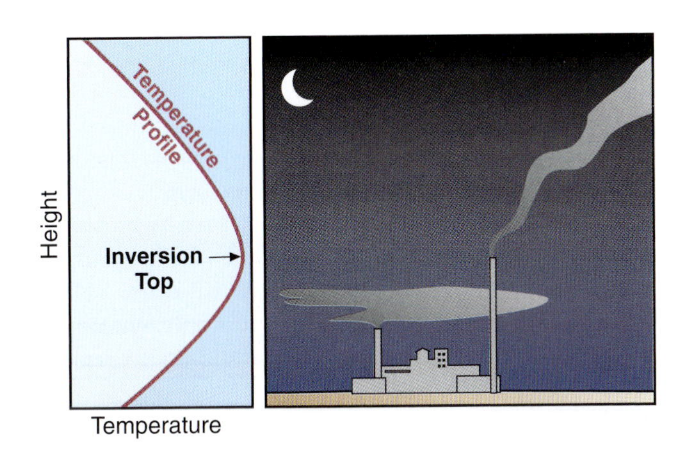

Figure 12.4 A surface-based temperature inversion limits the vertical dispersion of pollutants released within it. For shallow inversions, if the smokestack is high enough or if the effluent is warm enough or has a high exit speed, the plume may penetrate the inversion and be dispersed aloft.

boundary layer is shallow), pollution concentrations can be high (figure 12.5). (During nighttime, when convection is absent, the mixing depth is defined as the depth of the layer of air adjacent to the ground where vertical mixing is accomplished by turbulence produced by vertical shear of the horizontal winds. Mixing depths are typically below several hundred meters at night unless shears are very high).

An elevated inversion caps convective currents that rise from the heated ground and establishes an upper limit to the vertical mixing of pollutants.

A *subsidence inversion* is an elevated inversion layer that forms when sinking motions occur above a surface high pressure area. This inversion sometimes has serious consequences for air pollution. The high pressure areas that produce subsidence inversions generally have low wind speeds, which limit dilution, and clear skies with abundant solar radiation, which increase the photochemical production of secondary pollutants such as ozone. Some of the worst air pollution conditions in the eastern United States are caused by elevated subsidence inversions occurring in connection with summertime high pressure systems that drift slowly eastward across the central and eastern United States. Also, subsidence inversions associated with the Bermuda–Azores High, which can persist for prolonged periods over the northeast coast, produce major ozone episodes in the corridor from Washington to Boston.

Subsidence inversions originate at upper levels of the atmosphere, but can descend, especially in areas subject to semipermanent high pressure centers in marine or coastal areas where convective boundary layers are weak, to become surface-based inversions. These are typical, for example, in the Los Angeles Basin. Here, the subsidence inversion associated with the Pacific High (section 1.4) contributes to the production of photochemical smog.

A *frontal inversion* is an elevated inversion that forms at the top of the cold air behind a cold front or ahead of a warm front, providing a transition between the cold air below and the warm air that flows over it. Although a frontal inversion traps pollutants below it, concentrations are generally not particularly high because winds are strong, wind directions are changeable, and frontal rain and drizzle remove pollutants from the atmosphere. The fronts also move relatively rapidly, decreasing the duration of pollution at a given location.

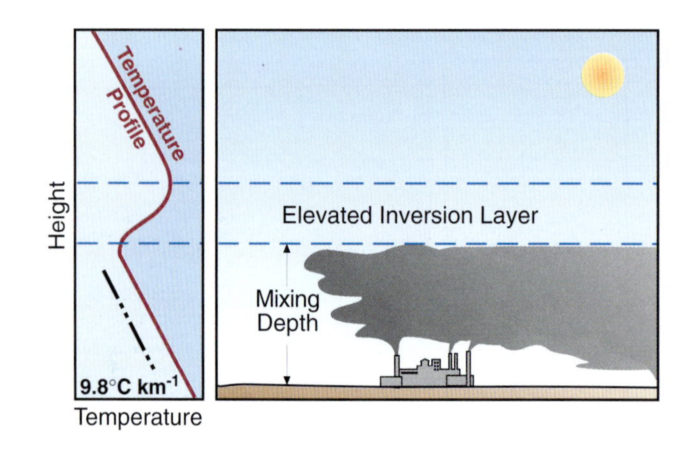

Figure 12.5 An elevated inversion traps pollutants that are emitted below its base. Concentrations can be especially high when the subsidence inversion is near the ground, so that the mixing depth is low.

12.4.3. *Stability and Plume Behavior*

The behavior of a visible plume provides information about atmospheric stability (figure 12.6). Conversely, atmospheric stability, as determined from a vertical temperature sounding, can be used to predict the behavior of an elevated plume by applying the following four principles:

- An elevated stable layer will cap vertical dispersion.
- Stable conditions will suppress vertical dispersion relative to horizontal dispersion.
- Unstable conditions will enhance vertical dispersion relative to horizontal dispersion.
- The more unstable the atmosphere, the greater the dispersion.

Normal diurnal changes in stability cause a succession of typical plume forms during a 24-hour period, as illustrated in figure 12.6. A *fanning*

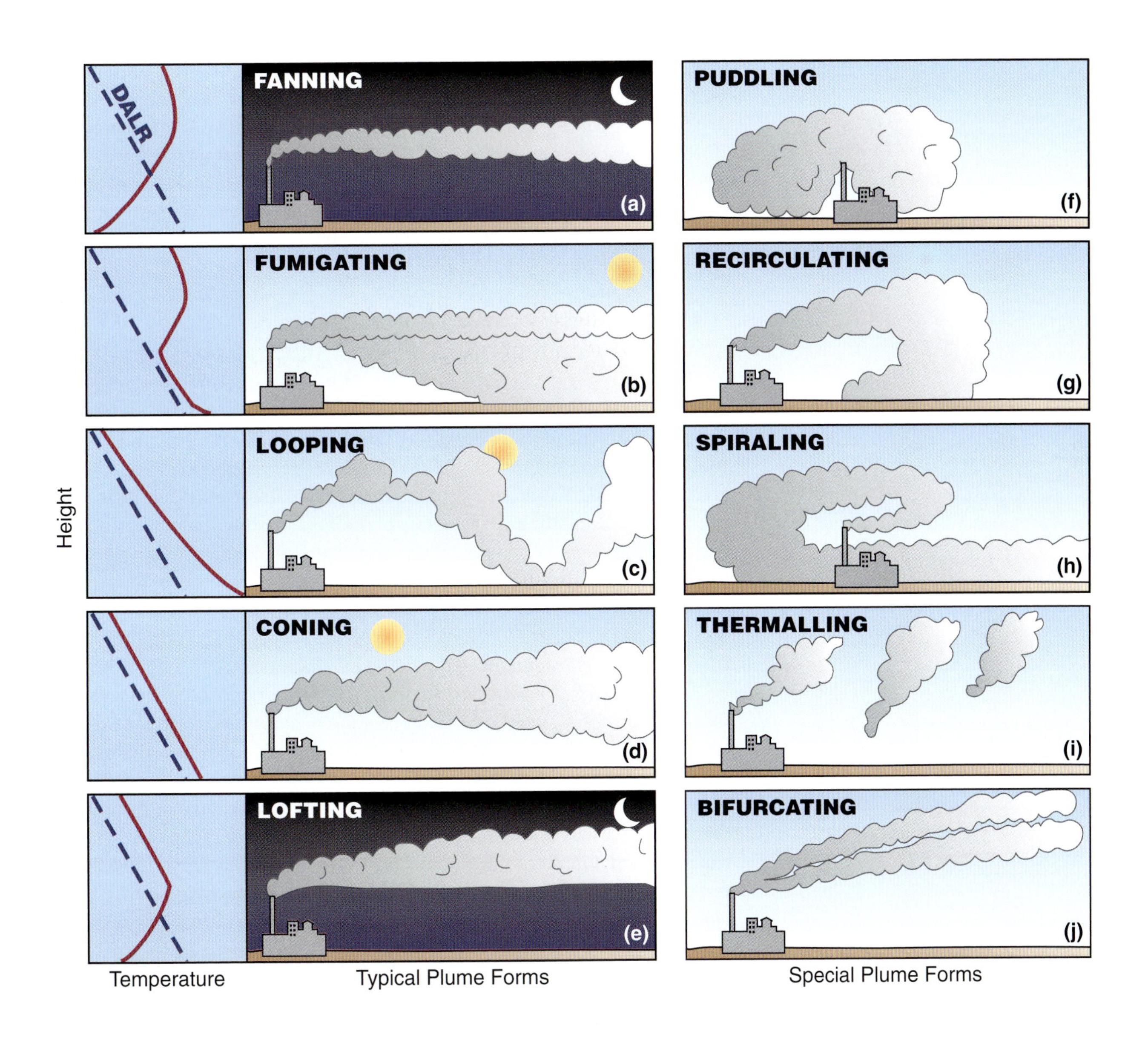

Figure 12.6 Diurnal changes in stability cause a succession of typical plume forms that are described in the accompanying text. The forms of diurnal plumes include (a) fanning, (b) fumigating, (c) looping, (d) coning, and (e) lofting. A variety of special plume forms (f–j) are also shown in the diagram, but without accompanying temperature profiles. (Adapted from Bierly and Hewson, 1962, and Angle and Sakiyama, 1991)

The form of an elevated plume provides information about atmospheric stability.

plume (figure 12.6a) indicates a deep stable (generally nighttime) atmosphere. Horizontal dispersion is greater than vertical dispersion, causing the plume to fan out in the horizontal and meander about at a fixed height. A *fumigating* plume (figure 12.6b) indicates a growing convective boundary layer below the elevated remnant of a nocturnal surface-based inversion. In middle to late morning, the convective boundary layer that grows upward from the heated ground reaches the fanning plume that is in the elevated remnant of the inversion and suddenly mixes the plume downward, fumigating the ground below. A *looping* plume (figure 12.6c) indicates an unstable atmosphere, which normally occurs with surface heating in the afternoon. Vertical dispersion is enhanced relative to horizontal dispersion, causing the plume to undergo vertical oscillations as it is alternately affected by convective plumes that rise from the heated ground and areas of sinking motion between the convective plumes. A *coning* plume (figure 12.6d) indicates a neutral atmosphere. The vertical and horizontal dispersion rates are similar, causing the plume to form a cone with a near-circular vertical cross section and with its vertex at the stack. Neutral atmospheres tend to occur when clouds are present or winds are strong. A *lofting* plume (figure 12.6e) indicates a stable boundary layer topped by a neutral layer. This type of stability is often present in the early evening when a surface-based inversion starts to form. The plume, emitted above the stable boundary layer, will disperse upward through the neutral layer, but it cannot penetrate downward into the stable atmosphere below.

12.5. Synoptic Weather Categories and Air Pollution Dispersion

As large-scale weather systems (low and high pressure centers and fronts) travel around the globe, they affect wind speed and stability and therefore air pollution dispersion. Migratory low pressure centers are associated with strong winds and rising motions that enhance air pollution dispersion by destabilizing the atmosphere and producing precipitation. High pressure centers, on the other hand, are associated with light winds, clear skies, and sinking motions that lead to subsidence inversions, nocturnal surface-based inversions, and photochemical reactions in the atmosphere. Fronts are associated with high wind speeds and variable wind directions and thus enhanced air pollution dispersion.

Atmospheric dispersion varies with the location of the pollutant source relative to fronts and pressure centers.

Five synoptic categories or dispersion environments have been identified, each with a specific location relative to synoptic weather systems and each with wind and stability conditions that can either enhance or suppress the dispersion of air pollutants emitted near the ground. Thus, information from a synoptic weather chart indicates when conditions are favorable or unfavorable for pollution dispersion at a given site. When the pollutant source can be started at will (smoke from prescribed fires, for example), emissions can be scheduled to take advantage of the best synoptic dispersion conditions. Positions of the five categories relative to fronts and high and low pressure centers are shown in figure 12.7. Categories, 1, 2, and 3 have the worst conditions for dispersion of pollution from local sources, category 5 the best. Figure 12.8 shows an ap-

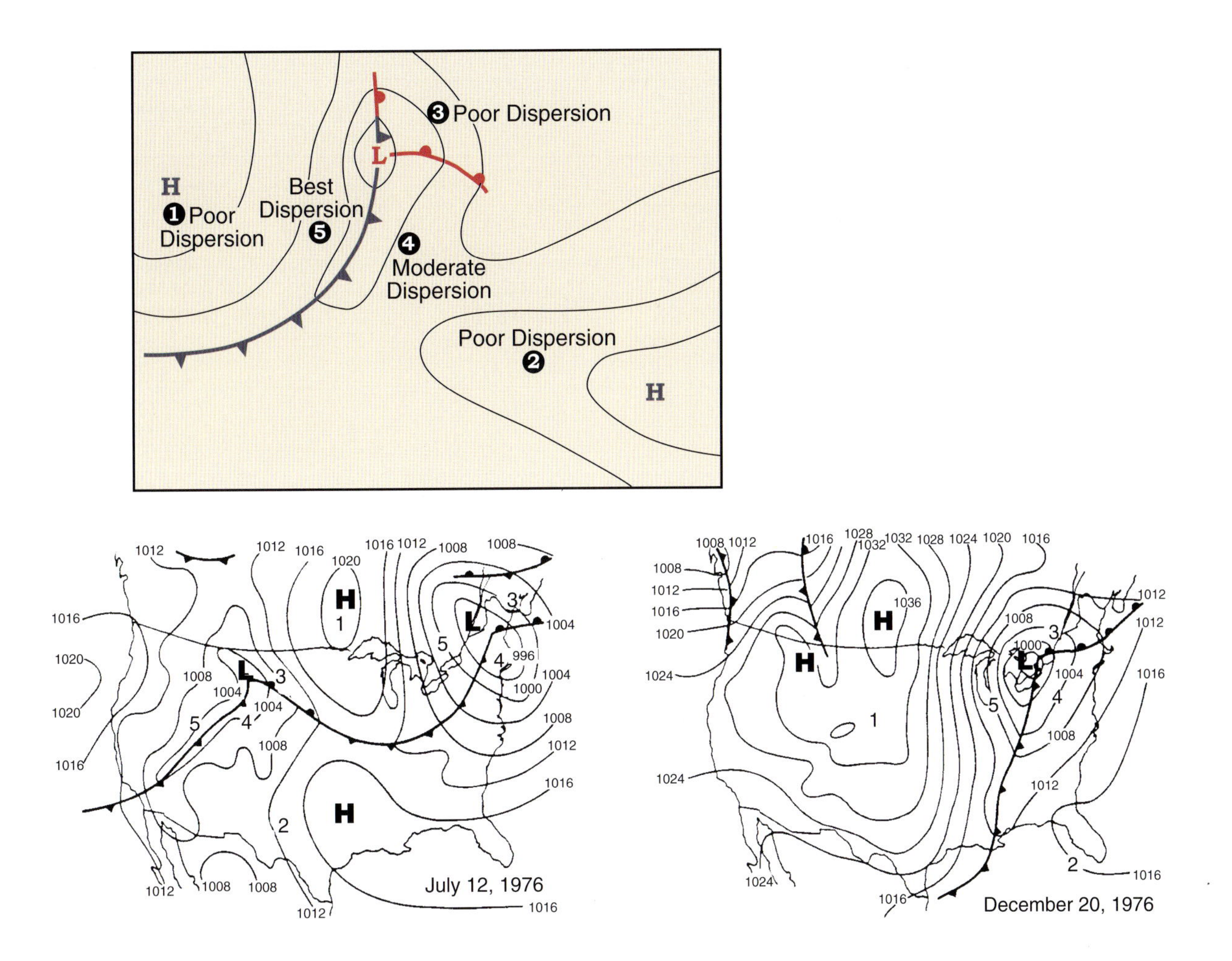

plication of the categorization scheme to summer and winter weather charts.

A list of meteorological characteristics of synoptic categories 1 through 5 that are particularly relevant for air pollution dispersion is given in table 12.2. These characteristics include wind speed, stability, and mixing depth. Also included are air mass type (section 6.1), which affects stability; cloud cover (section 9.3.3) and cloud type, which affect insolation and thus the growth of daytime mixed layers; proximity to fronts (section 6.2) and lows (section 5.1.5), which affects wind speed; and large-scale vertical motion in highs and lows (section 5.1), which affects stability, cloud cover, and the formation of elevated inversions.

Categories 1, 2 and 3 are associated with poor dispersion from local sources. Categories 4 and 5 are associated with moderate to excellent dispersion conditions, primarily as a result of strong winds and weak stability.

Category 1. Regions designated as category 1 (the polar high pressure area behind a cold front) are of particular importance because polar highs affect major U.S. mountain ranges in all seasons. In the summer, the high pressure centers drift slowly eastward across the country and often remain stagnant over a location for days, allowing regional haze and high ozone concentrations to build up. In the winter, high pressure centers

Figure 12.7 (*top*) Dispersion from near-ground pollutant sources is dependent on location relative to fronts and pressure centers. (Adapted from Pielke et al., 1991)

Figure 12.8 (*bottom*) Application of the synoptic categorization scheme to actual summer and winter surface weather charts. (From Pielke et al., 1991)

Table 12.2 Summary of the Dispersion Associated with Moving Synoptic-Scale Weather Systems

	Synoptic Category 1	Synoptic Category 2	Synoptic Category 3	Synoptic Category 4	Synoptic Category 5
Location	Under a polar high	In the vicinity and west of a subtropical ridge	Ahead of a warm front	Ahead of a cold front or behind a warm front	Behind a cold front
Air mass (see figure 6.1 for abbreviations)	Continental polar (cP) or continental arctic (cA) air	Maritime tropical air (mT)	Maritime air overrides continental air (mT/cP, mT/cA, or mP/cA)	Maritime tropical air (mT)	Continental polar (cP) or continental arctic (cA) air
Cloud cover	Clear except tendency for nighttime fog; middle- and/or high-level clouds over and west of the polar high center	Scattered fair weather cumulus in daytime; clear at night, except near convection	Mostly cloudy to cloudy	Clear to partly cloudy, except near squall lines	Clear to scattered or broken shallow convective clouds
Surface winds	Light and variable winds	Light southeast to southwest winds	Light to moderate southeast to east-northeast surface winds	Brisk southwest surface winds	Strong northeast to west surface winds
Stability	Weakly to moderately unstable during day and very stable during night. Stable if snow cover is present or sun angle is low during day	Moderately to strongly unstable during day and moderate to strongly stable during night	Stably stratified day and night	Moderately unstable during day and moderately stable during night	Near-neutral day and night
Vertical motion	Synoptic descent	Synoptic descent, becomes stronger as ridge axis is approached	Synoptic ascent	Weakening synoptic descent as the cold front approaches	Synoptic ascent
Inversion and the atmospheric boundary layer (ABL)	Synoptic subsidence inversion and/or warm advection aloft create an inversion which caps the ABL	Synoptic subsidence inversion caps the ABL	Frontal inversion caps the ABL	Weak synoptic subsidence inversion caps the ABL	Deep ABL

Adapted from Pielke et al., 1991.

travel slowly southward or southeastward and are shallow and stable, bringing clear skies and strong insolation, which promote the development of a daytime convective boundary layer. However, the growth of the boundary layer is limited by a subsidence inversion, trapping pollutants in the well-mixed but shallow convective boundary layer and resulting in regional haze. Clear skies and strong insolation also lead to high daytime ozone concentrations, which promote photochemical reactions. At night, clear skies inhibit air pollution dispersion by strengthening the development of strong temperature inversions. (In the winter, when insolation is weak or when snow covers the ground, inversions can persist both day and night.) Within the inversion, weak winds and high stability limit the dispersion of pollutants, cause the buildup of haze, and lead to fog if surface moisture is present. A high pressure center also brings weak synoptic winds. Thus, local wind systems such as mountain–valley or land–sea breezes often govern the dispersion of air pollutants. These diurnal circulations may be confined below subsidence inversions or, in mountainous terrain during winter or at night, within the topography, thus trapping air pollutants in a small volume of air.

Category 2. Although category 2 regions (the maritime tropical air in subtropical high pressure areas) have dispersion characteristics similar to those under polar highs (category 1), subtropical highs are often located too far south to affect major U.S. mountain ranges in winter. In summer, however, ozone concentrations can be high over much of the East Coast and the Appalachian Mountains because of the influence of the Bermuda–Azores High, and the Pacific High can extend as far north as southeast Alaska and can affect the Coast Range.

Category 3. Dispersion in category 3 regions (the polar or arctic air mass ahead of a warm front) is also poor, but the warm frontal area is distinctly smaller than the high pressure regions in categories 1 and 2. Maritime tropical air flowing over the polar or arctic air mass produces extensive cloudiness to the north of the warm front. This cloudiness limits insolation and therefore the growth of a convective boundary layer. The shallower convective boundary layer produces a smaller dilution and thus higher air pollution concentrations. Precipitation associated with the cloudiness, however, may remove air pollution from the cold air mass. The cold air mass itself is stably stratified, contains light to moderate winds from the southeast, and is capped by a warm frontal inversion. These characteristics limit both the vertical and horizontal dispersion of air pollutants.

Category 4. Dispersion conditions in regions designated as category 4 (ahead of a cold front or behind a warm front, i.e., in the warm sector) are moderate, as brisk southwest winds carry maritime tropical air in advance of an approaching cold front. During daytime, the maritime tropical air mass is moderately unstable because of the low-level influx of warm air and strong insolation, and air pollutants are well dispersed. The deep, daytime mixed layer is capped by a subsidence inversion, but the inversion weakens or disappears as the front approaches. At night, however, the clear or partly cloudy skies allow the ground to lose heat by longwave radiation, resulting in the formation of nocturnal temperature inversions, which trap pollutants near the ground.

Category 5. Regions designated as category 5 (the continental polar or arctic air behind a cold front) experience the best dispersion conditions. The sky is cloudy at the cold front, but the air behind the front is generally clear or contains shallow cumulus clouds, allowing a deep convective boundary layer to develop. Neutral stability in the cold air and synoptic rising motions both night and day allow pollutants to disperse well vertically. Strong cold winds from the north further enhance dispersion conditions.

12.6. Mountainous Terrain and Atmospheric Dispersion

Atmospheric dispersion of pollutants is more complicated over complex terrain than over homogeneous terrain because of the terrain-forced (chapter 10) and diurnal mountain (chapter 11) flows that develop in the mountains and interact on a wide range of time and space scales. Atmospheric conditions in complex terrain and the terrain itself can either enhance or suppress air pollution dispersion relative to the surrounding plains. Table 12.3 lists the characteristics of complex terrain meteorology that can lower pollution concentrations (enhance dispersion) and those that can raise pollution concentrations (decrease dispersion).

The high altitude of mountainous areas increases the emissions of some pollutants. Automobiles and other combustion sources produce more carbon monoxide and particulates at high altitudes because lower oxygen levels decrease the efficiency of combustion. Particulate production is also increased by emissions from the furnaces and fireplaces that are used for space heating at the lower temperatures found at high altitudes and by road dust from the sanding of snow-covered roads. The production of ozone may increase because of intense solar radiation and the trapping of air masses in confined valleys or basins. Rain acidity may also increase in mountainous areas where frequent cloudiness and precipitation enhance removal of oxides of nitrogen and sulfur.

12.6.1. *Effect of Mountains on Regional and Hemispheric Pollution*

The release of natural and anthropogenic pollution in different parts of the world can affect large areas of the earth's surface over long periods of time and may even alter the earth's climate. Clouds of Saharan dust cross the Atlantic Ocean from northern Africa to the southeastern United States. Emissions from major volcanic eruptions (e.g., from Mount Pinatubo, El Chicón, or Mount St. Helens) circle the earth and remain in the stratosphere for months or years. Smoke plumes from fires in the rain forests of South America or Indonesia are transported long distances, and industrial emissions from many parts of the Northern Hemisphere produce haze in the Arctic. Fallout from the April 1986 Russian nuclear accident at Chernobyl was tracked around the globe. Chlorofluorocarbons

Table 12.3 Effects of Mountain Terrain on Pollution Concentrations

Concentrations decrease because . . .

- wind speeds are generally higher on exposed mountainsides than over lower-lying terrain.
- the roughness of the terrain and forest canopies increases turbulent dispersion.
- precipitation increases with elevation, causing pollutants to be washed out.
- the diurnally forced winds over sloping valley and sidewall surfaces rarely drop to zero for prolonged periods.

Concentrations increase because . . .

- pollutants are carried along the same path day after day when channeled by terrain.
- wind speeds in valleys and basins are often reduced by the sheltering of surrounding ridges.
- plumes may directly impinge on higher elevation areas.
- temperature inversions form frequently in low-lying valleys and basins and persist longer than over flat terrain.
- valley walls limit plume spread and the volume of air through which the plume can be dispersed.
- surface friction produced by terrain roughness reduces wind speeds.
- fumigations occur over the same terrain day after day.
- sinking motions during inversion breakup bring pollutants closer to the ground.

or CFCs that were used as propellants in spray cans contribute to the destruction of the stratospheric ozone layer that shields the earth's surface from harmful high-energy ultraviolet radiation from the sun. (Production of CFCs was banned in the United States after 1995.)

Because mountains influence the general circulation of the earth's atmosphere (section 5.2.1), they play a role in many hemispheric and regional air pollution problems, the most important of which are *acid precipitation* and regional visibility impairments.

12.6.1.1. *Acid Precipitation.* Acidity is measured on a pH scale from 0 to 14. A neutral solution has a pH of 7, values greater than 7 are alkaline, and values less than 7 are acidic. The pH scale is logarithmic, so a decrease of 1 pH represents a tenfold increase in acidity. Normal rainfall is naturally somewhat acidic, with pH values in the range of 5.0–5.6, because atmospheric carbon dioxide, a minor atmospheric constituent (section 3.2.1), dissolves in raindrops, producing a weak solution of carbonic acid. Acid rain or acid precipitation (in the form of rain droplets, snowflakes, or even fog droplets that are deposited on trees and other surface structures) has pH values below this normal range.

Normal rainfall is slightly acidic with pH values between 5 and 5.6. Acid precipitation has pH values below 5.

Acids are produced in the atmosphere when oxides of nitrogen and sulfur, released during combustion in industrial operations and in vehicles, interact with water in clouds, dew, or frost. The precipitation of such acids may occur at long distances from the emissions source, especially if the source is elevated. For example, acid precipitation in a large part of eastern North America, with pH values between 4 and 4.5, is attributed largely to high sulfur dioxide emissions in the industrialized Ohio Valley. Acid precipitation was studied intensively in the United States in the 1980s, and the results of the studies were used in the 1990 amendments to the Clean Air Act to mandate substantial reductions in nitrogen and sulfur oxide emissions.

Mountain ecosystems tend to have poorly buffered soil that is unable to neutralize acids introduced by precipitation or deposition. Because of the poorly buffered soil, acids can change soil chemistry, damaging forests and making them more susceptible to disease, insect pests, and drought. The general increase in precipitation with elevation (section 8.5.1) increases acid precipitation at the higher elevations. In addition, acidic particulates can be deposited directly when cap clouds (section 7.1.4.5) form on the ridge tops and persist for hours, exposing mountaintops to heavy concentrations of pollutants. The clouds convert the oxides of sulfur and nitrogen in the continuous stream of air that flows through them to acidic droplets, which are deposited on the ground, vegetation, or structures. Over time, soil acidity builds up, damaging the mountaintop ecosystems (figure 12.9).

Because of poorly buffered soils, mountain ecosystems are especially sensitive to acid precipitation.

Mountain watersheds formed in granitic or quartzitic rocks or soils have streams and lakes that are dilute in dissolved minerals, almost like distilled water, and are therefore poorly buffered aquatic ecosysytems. Acid precipitation can readily change water chemistry in these ecosystems, causing loss or modification of aquatic fauna and flora. Examples of sensitive, high-elevation lake watersheds are found in the Sierra Nevada and in the mountains of Montana, Idaho, Wyoming, and Colorado.

Figure 12.9 Forest damage to red spruce thought to be exacerbated by acid fog near the summit of Mt. Mitchell, North Carolina (Photo © C. Whiteman)

12.6.1.2. *Regional Visibility Impairments.* Visibility varies widely over the United States (figure 12.10) as a result of both natural and anthropogenic factors. Visibility is impaired by scattering and absorption of light by suspended particles and gases. Haze particles, which typically have diameters less than about 0.2 micrometers, become larger as they absorb water even at humidities below 100%. This swelling of haze particles in humid conditions (*damp haze*) is largely responsible for the lower visibility in the eastern United States and along the West Coast. In the drier climate of the interior western United States, damp haze is less of a problem, but *dry haze* nonetheless produces significant restrictions on visibility.

Anthropogenic visibility impairments are produced on a regional scale when there are multiple air pollution sources and, like the individual plume forms in figure 12.6, are indicators of atmospheric stability. *Plume blight* occurs near pollutant sources where multiple smoke, dust, or colored gas plumes can be distinguished as individual plumes. *Layered haze* develops when air pollution from multiple point, line, or area sources is transported long distances, producing distinguishable layers of discoloration. Plume blight and layered haze occur within deep, long-lived stable layers. Both are most common in winter and, ironically, are most easily seen in areas of the country having good visibility (e.g., Denver's "brown cloud"). Regional haze indicates the presence of an unstable or neutral atmospheric boundary layer capped by an elevated temperature inversion. Pollution from many sources becomes well mixed in the boundary layer, forming a widespread homogeneous haze layer that reduces visibility in every direction. The depth of the haze layer is the mixing depth. In the drier regions of the mountainous West, the daytime surface energy budget is characterized by large sensible heat fluxes that produce deep daytime convective boundary layers or mixed layers (sometimes to altitudes above 5000 m or 16,000 ft above mean sea level). Regional haze layers tend to be deeper in the western than in the eastern United States. Regional haze in all parts of the country is most noticeable from an elevated vantage point (an airplane or a mountainside) above the layer of haze. Examples of visibility impairment are shown in Figures 12.11–12.14.

Regional visibility impairments include dry and damp haze, plume blight, layered haze, and regional haze.

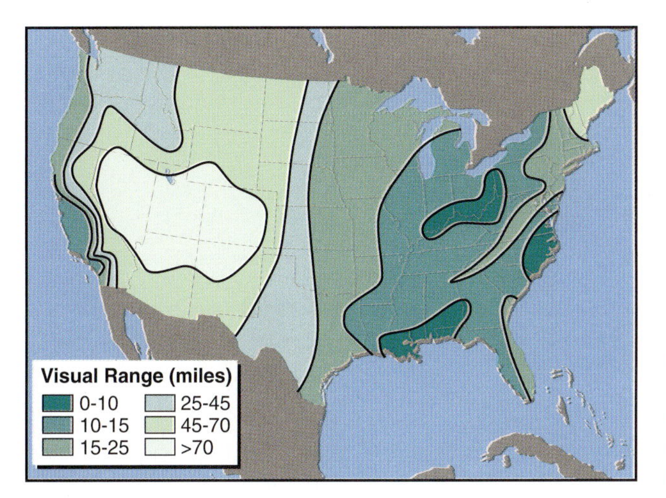

Figure 12.10 Visual range in miles (1 mi = 1.6 km) as obtained from airport observations in nonurban areas of the United States. Visual range represents the farthest distance from which a large black object can be seen against the horizon. (Adapted from Trijonis and Shapland, 1978)

Figure 12.11 Particulate pollution at Vail, Colorado, on a winter morning in the late 1970s, comes mostly from wood-burning stoves and fireplaces, restaurant grills, and roadway dust. Some individual plumes can be identified, but the effect is one of plume blight, as pollutants from the various sources build up overnight in the nocturnal valley temperature inversion. (Photo © C. Whiteman)

Figure 12.12 Layered haze at Boulder, Colorado, as viewed looking eastward from Green Mountain on a fall day. A pollutant plume from the Valmont Power Plant meanders northward in the stable air above a well-mixed layer of regional haze. The pollutants in the regional haze layer have become mixed throughout the depth of the growing daytime convective boundary layer. (Photo © C. Whiteman)

Figure 12.13 A regional haze is present below a subsidence inversion, as seen in December looking eastward from 17,000 ft (5180 m) on Mexico's Mt. Popocatepetl. (Photo © C. Whiteman)

12.6.2. *Pollution Dispersion in Terrain-Forced Flows*

Terrain-forced flows are generally characterized by high wind speeds and turbulence and are therefore associated with enhanced dispersion and reduced air pollutant concentrations. Because wind speeds usually increase with elevation, exposed mountainsides experience higher winds, more dilution, and more mechanically generated turbulence than adjoining plains. Thus, most of the terrain-forced flows discussed in chapter 10 (mountain and lee wave flows, damaging downslope windstorms, foehn or chinook winds, barrier jets, post-frontal wind accelerations around mountains, channeling of strong winds through passes and gaps, and low-level jets) alleviate air pollution problems. In some instances, however, terrain-forced flows can cause higher ground-level air pollution concentrations, either through decreases in diffusion or through trajectory changes that bring the plumes closer to the ground or repeatedly along the same pathways. Ground-level concentrations of air pollutants can increase (figure 12.15):

- when winds are strong enough to stir up dust,
- when terrain-forced flows carry an air pollution plume into a mountainside (plume impingement),
- when converging flows bring plumes from different sources together,
- when wind speeds drop (in flow separations and wakes),
- when a stable air mass stagnates either upwind or downwind of a mountain barrier, or
- when pollutants are channeled repeatedly over the same terrain.

Even in these cases, however, the increase is usually offset to some extent by compensating factors that are also associated with the terrain-forced flows.

Terrain-forced flows, which tend to be strong and turbulent, are usually associated with good pollution dispersion.

Terrain-forced flows can cause plume impingement at the top of a mountain barrier or on the mountainsides. A plume upwind of a mountain barrier will be transported over the barrier or around it, depending

Figure 12.14 Photographs of Canyonlands National Park, Utah, on three clear July days at approximately the same times of day, illustrating visibility variations in regional haze. The figures represent (a) the best 10%, (b) the average, and (c) the worst 10% of visual air quality days. (National Park Service photos)

on the height of the plume relative to the dividing streamline height (figure 10.1). If the plume originates above the dividing streamline height (section 10.1), it is carried over the barrier. In a stable atmosphere, a plume that originates below the dividing streamline height is carried around the barrier. Both the flow over the barrier and the flow around the barrier speed up as they move past the barrier, as indicated by the convergence of (i.e., decreasing distance between) the streamlines, which brings the plume closer to the ground and may increase ground-level concentrations. Because the speedup of the flow and the descent of the plume counterbalance each other to some extent, the impact of plume impingement on ground-level concentrations is difficult to predict on mountainsides or mountaintops.

Plumes can also directly impact exposed hillsides when the prevailing flow blows a plume toward higher terrain. This is often a temporary condition that occurs when the wind shifts across the hillside, although a continuous impingement may occur when a flow toward a hillside is carried up or down the slope by the diurnal slope wind system. Mechanical turbulence at the slope can decrease these concentrations somewhat compared to impingement without turbulence. Models are available to predict concentrations at the slope under these conditions, but they have not yet been adequately tested with observational data.

A convergence zone (section 10.3.2) on the lee side of an isolated barrier may receive pollutants from both branches of the converging flows.

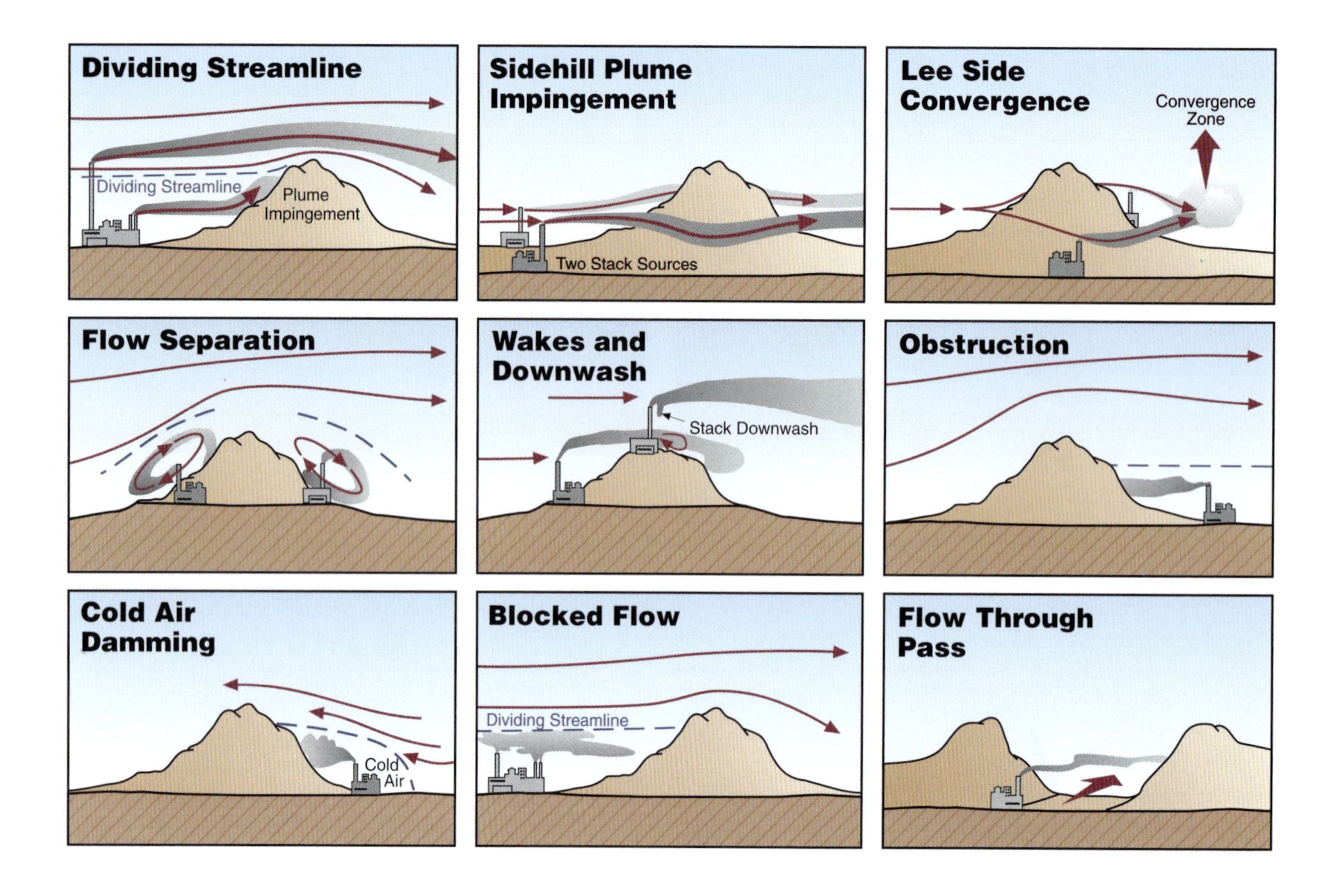

Figure 12.15 Terrain-forced flow phenomena that can produce poor dispersion conditions.

Concentrations in the convergence zone are not, however, additive; rather, mixing produces concentrations intermediate between those in the two branches. High concentrations in the convergence zone are exacerbated by local emissions into the weak winds at the convergence zone. The low-level flow convergence causes rising motions over the convergence zone that produce clouds and precipitation. Clouds and precipitation can decrease concentrations by removing air pollutants through snowfall or rainfall, but in-cloud chemical reactions can increase concentrations of secondary pollutants. Local concentrations also vary as shifts in the ambient wind direction cause shifts in the location of the convergence zone. When the flow approaching a mountain barrier is strong and the underlying terrain steep, flow separations can form on the windward and leeward sides of a mountain barrier (section 10.1). Air pollution concentrations at the ground increase because the dilution of air pollution plumes captured within the flow separations is inhibited by lower wind speeds and because the pollutants are recirculated in the separation eddies.

Wakes generated in the lee of obstacles cause wind speeds to drop and turbulence to increase. The lower wind speeds reduce dilution of air pollutants emitted within or entrained into the wake, thus increasing concentrations. The turbulence enhances the diffusion of air pollutants, but can also cause abrupt increases in concentrations at the ground by suddenly carrying the plume downward. Plumes emitted from smokestacks and buildings can be subject to *downwash,* or the downward transport into the wake that forms downwind of buildings, resulting in an increase in pollution concentrations very near the source. Engineering design standards have been formulated to ensure that smokestack heights and stack exit velocities are high enough to minimize downwash in the lee of the stack.

Mountain chains obstruct the cross-mountain transport of shallow, dense air masses, causing them to stagnate either upwind or downwind of the barrier. For example, continental arctic (cA) and polar (cP) air masses that move southward into the Great Plains from Canada are obstructed by the Rockies to the west and can stagnate for days upwind of the barrier. The diffusion of air pollution emitted into the air mass depends on the characteristics of the air mass, which vary with position relative to the high and low pressure centers and fronts (section 12.5). The most serious local air pollution problems in these air masses occur in weak flows near a high pressure center where the stable air mass is capped by a subsidence inversion. If cold air damming occurs (section 10.5.2), winds within the cold air mass move the mass up the eastern slope of the Rockies but are too weak to carry it over the barrier. The slope of the barrier is then exposed to pollutants trapped within the air mass, but the diffusion of pollutants emitted into the air mass is influenced by the weak thermal flows that originate in and on the mountains.

A similar situation exists when a stable air mass is blocked upwind of a mountain barrier (section 10.5.1). Under these stagnant conditions, pollutants can accumulate over several days. Ground-level concentrations are usually highest in the immediate vicinity of near-surface pollutant sources. Weak diurnal mountain wind systems may, however, modulate the pollution concentrations as daytime and nighttime flows vary in speed and direction. If pollutant plumes are emitted from smokestacks higher in the blocked flow layer, they are often transported within the stable air mass and are spread more horizontally than vertically by light, meandering winds associated with diurnal wind systems. The plumes accumulate and recirculate over time, increasing ambient pollutant concentrations within the stable air mass. The highest ground-level concentrations of pollutants emitted higher in the blocked flow layer occur during daytime when the growing convective boundary layer reaches the plume height, producing a sudden fumigation at the ground below the plume. These fumigations occur at about the same time at all sites below the plume, even if they are widely distributed across a blocked flow region. Relief from high pollutant concentrations comes when the flow is unblocked. This unblocking may occur rather suddenly over periods as short as several hours.

When an approach flow is channeled through passes or gaps in a mountain barrier, wind speeds increase through the terrain constriction. Although the higher wind speeds enhance the dilution of pollutants within the flow, the channeling causes the pollutants to be carried repeatedly over the same narrow pathway, thus increasing the exposure to pollutants over time.

12.6.3. *Pollution Dispersion in Diurnal Mountain Flows*

High air pollution concentrations within a mountainous region are usually associated with local diurnal mountain winds rather than with terrain-forced flows and can be attributed in part to the low speeds (several meters per second) of along-valley winds. Because wind strength determines the extent to which pollutants are diluted at their source, weak along-valley winds can result in high pollutant concentrations within the plume. Measurements of midvalley flow strengths are therefore critical to all air pollution studies of valleys. Measurements must be taken through the depth of the valley atmosphere and not just at the ground because wind directions can vary with height by 180° during the morning and evening transition periods (sections 11.1.3 and 11.1.1). A vertical wind profile in a valley often shows a jet structure, with the strongest nighttime winds at midlevels in the inversion layer, with weaker winds both above and below. A plume emitted at the level of the jet wind maximum experiences the best initial dilution.

Diurnal mountain winds, which are generally weak, local in extent, and associated with stable layers in the atmosphere, do not dilute pollution concentrations as effectively as terrain-forced flows.

The strength of along-valley flows varies significantly from valley to valley and from one section of a valley to another, depending on the valley volume effect (section 11.4.2). Flow strength governs the initial dilution of air pollution plumes (section 12.3.1 and figure 12.2). For example, down-valley winds of 7 m/s in Colorado's Eagle Valley produce high initial dilutions, whereas 1 m/s winds at the same level at nearby Vail, Colorado, in the Gore River Valley produce low dilutions. Valleys or valley sections with convex sidewalls that drain plateaus generally experience higher winds than valleys or valley sections with concave sidewalls and broad floors (section 11.4.2). Thus, wind speeds and plume dilutions are often larger near the head of a valley where the valley volume is small relative to the horizontal distance from ridge top to ridge top than in the lower part of a valley where the valley volume is larger relative to the distance between ridge tops.

Along-valley flow strength is the most critical factor in air pollution dispersion in valleys.

The length of a valley also affects air pollution concentrations. Pollutants emitted into a very long valley are carried down the valley by nighttime down-valley winds, but they may never reach the mouth of the valley. When the winds reverse during the morning transition period, the pollutants are carried back up the valley. Pollutants emitted into a relatively short valley may be carried out onto the adjoining plain by the nighttime down-valley winds. After sunrise, the plume breaks up as a convective boundary layer grows above the plain, and up-valley flows, which do not extend very far beyond the valley mouth, carry only a portion of the pollution back into the valley.

Pollution concentrations within a valley are affected by the air drained into the valley by nighttime flows from tributaries. If clean air drains from the tributaries into the main valley, pollution concentrations in the main valley decrease. If nighttime flows bring polluted air from a tributary into the valley, concentrations increase in the main valley below the tributary. When the wind direction reverse after sunrise, the pollution emitted in a tributary valley is carried primarily up the main valley, although up-valley flows in tributaries along the plume path may pick up a portion of the pollution. Pollution concentrations in the tributary valley are thus lowered, while concentrations in the main valley are raised.

Diurnal mountain winds can cause high long-term (annual and 24-hour) and high short-term (1-hour) average pollutant concentrations. The highest long-term average concentrations occur where pollutants are consistently carried along pathways defined by the slope and along-valley winds and where air stagnates in pools behind terrain constructions in a valley or basin. The highest short-term average concentrations are usually associated with strong downslope flows during the evening transition period that carry pollutants into low-lying areas, with plume impingement on mountainsides and with fumigations during the morning transition period when elevated pollutant plumes are mixed downward through the growing convective boundary layer over the valley floor and sidewalls.

The highest long-term average pollution concentrations occur where pollutants are consistently carried along the same pathways or where air stagnates in pools behind terrain constrictions.

12.6.3.1. *Defined Plume Pathways.* When air pollution is emitted over flat terrain, winds can carry the plume in any direction (figure 12.16). When air pollution is emitted in a valley, the plume is carried along a pathway defined by the slope and along-valley winds, unless winds aloft are strong enough to overpower the local diurnal mountain winds. If the pollution source is on the valley floor, the plume is carried up the valley during the day and down the valley during the night. If the source is on a sidewall, the plume is carried up the sidewall and valley during the day and down the sidewall and valley during the night. Diffusion causes a plume to spread during its transport. For example, a plume emitted at night over the center of a uniformly wide valley diffuses horizontally (and also vertically) as it is carried down the valley, bringing the pollution closer and closer to the valley sidewalls at plume height. The sidewall concentrations are greatest some distance from the source, and concentrations decrease with more distance as the plume is dispersed more uniformly across the valley's width.

12.6.3.2. *Pooling.* Terrain constrictions along the length of a valley interrupt along-valley flows and cause cold air to stagnate in pools behind the constrictions (section 11.4.2.1). Pollutants emitted within these cold-air pools build up during the time of stagnation, whereas pollutants emitted above the pools may be well dispersed. Valleys are designated as *trappers* or *drainers* to indicate their susceptibility to cold-air pooling. The trap-

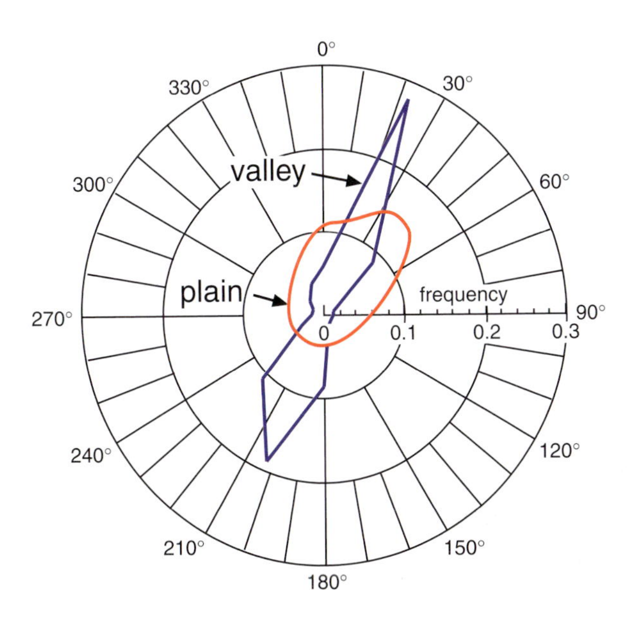

Figure 12.16 A polar plot of the relative frequency of different wind directions is called a *wind rose*. Wind roses for valleys are bidirectional, showing that winds usually blow along the valley's axis, either in the up-valley or down-valley direction. In contrast, wind roses for homogeneous nonmountainous terrain tend to be more circular, indicating that winds are more widely distributed over all directions.

pers can often be identified from topographic maps by looking for effective terrain constructions or by observing weak winds or low minimum temperatures within segments of a valley. Vertical wind profile measurements can also be used to classify valleys into these two categories and to determine the depth of the pool and the optimum elevation for plume dispersion.

Stagnation of cold air and the build-up of pollution concentrations is also a problem in basins, where dispersion is generally poorer than in valleys. Nighttime cooling is stronger in basins, resulting in deeper or more intense temperature inversions. In addition, air tends to stagnate in basins, especially when synoptic winds above the valley are weak. During winter, when persistent cold pools or inversions form in basins, pollutants may accumulate and recirculate within the basin over several days. Daytime fumigations can occur nonetheless if the unstable boundary layer grows upward to the plume-carrying height. The stagnations are eventually broken by traveling synoptic-scale disturbances, resulting in the sudden ventilation of the basin.

Serious photochemical smog episodes occur in populated mountainous areas of the United States when ozone and its chemical precursors (hydrocarbons and oxides of nitrogen) are trapped in valleys or basins under sunny skies. The Los Angeles Basin is well known for its smog, but photochemical smog occurs in other major population centers of the West, including Phoenix, Salt Lake City, and Denver.

12.6.3.3. *Peak Downslope Flows.* Pollution concentrations are often highest on valley floors during a brief (1–3 hours) period in the late afternoon or early evening when the main valley atmosphere is near-neutral because peak downslope flows occur that carry pollutants emitted on the sidewalls to the valley floor and because dilution of pollutant concentrations is reduced as the along-valley flows slow and reverse. The continuity and strength of drainage flows in the late afternoon and early evening can cause severe smoke management problems from prescribed and natural fires that are burning on elevated terrain overlooking sensitive valley locations. Burning operations, where they can be controlled, should be ended before these drainage flows begin, particularly if smoke is likely to drift onto highways.

High pollutant concentrations on valley floors often occur in the late afternoon or early evening when along-valley winds slow and reverse and when downslope flows carry air from the sidewalls down to the valley floor.

Later in the evening, a valley temperature inversion forms, growing deeper through the night. The high stability in the inversion can cause downslope flows from the upper slopes to separate from the slope and ride over the top of the inversion. Pollutants emitted on the sidewalls are thus carried only partway down the slope before leaving the slope and flowing toward the valley center, increasing concentrations at elevations above the valley floor. This type of flow has been documented in Colorado's Brush Creek Valley by observing the behavior of oil fog plumes generated on one of the valley's upper sidewalls (figure 12.17). A series of photographs taken in the early morning hours of 18 September 1984 show the diffusion of the oil fog (illuminated by searchlight) that was emitted into a shallow downslope flow on the mesalike ridge top above the valley at the edge of a box canyon tributary. The upper half of the valley atmosphere was nearly isothermal, with a strong inversion in the lowest 150 m above the valley floor. The smoke cascaded over the near-

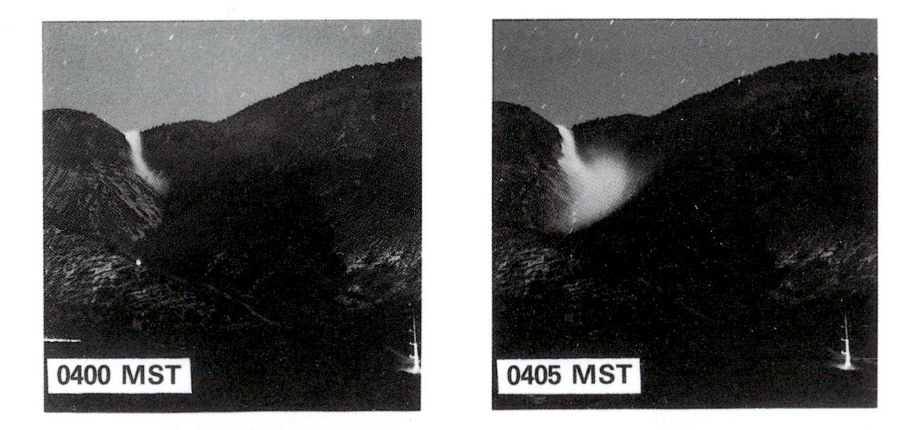

Figure 12.17 Sequence of nighttime photographs showing the movement of a smoke plume released continuously at the top of a box canyon tributary to Colorado's Brush Creek Valley. (From Whiteman, 1990. Photographs by J. M. Thorp and M. M. Orgill, Pacific Northwest National Laboratory, reproduced by permission)

vertical cliffs of the box canyon, came further down the valley sidewall, and then left the slope to travel out over the main valley within the temperature inversion.

This variation in pollution concentration is familiar to residents in the Geysers area of California. A "rotten egg" smell can be detected on the valley floor in the late afternoon and evening when hydrogen sulfide emitted on the sidewalls from geothermal wellheads is transported to the valley floor by downslope winds. Later in the evening, as the valley inversion grows in depth, the downslope flows leave the slopes and carry the emissions aloft toward the valley center, where they remain during the night as elevated plumes within the surface-based valley inversion. The plumes are fumigated to the valley floor the next morning when convective boundary layers grow upward from the heated ground.

12.6.3.4. *Plume Impingement*. Plume impingements occur when a plume emitted below ridge height is carried in a direction more or less perpendicular to the ridgeline to impact directly onto the mountain slope. Impingements can also occur when a wind shift causes an elevated plume to sweep across a slope. Impingements normally occur on the hillsides of isolated mountain barriers rather than in valleys, because plumes in valleys are usually transported parallel to the ridgeline by along-valley winds. Peak concentrations at the slope are generally somewhat lower than the centerline concentration in the plume just before it reaches the slope, because the impinging plume is diffused by shear-induced turbulence in the slope flows and, during daytime, by fumigation. During daytime, the impinging plume can be carried up the slopes by upslope flows. At night, the plume can be entrained into downslope flows and carried down the slopes. Impingements may occur within valleys when strong cross-valley flows penetrate into the valley from above and overpower the local diurnal wind systems. They may also occur briefly during the morning and evening transition periods if the cross-valley flows are strong when the along-valley winds are slowing and reversing.

12.6.3.5. *Fumigations* Fumigations are produced at the ground when a convective boundary layer grows upward into a plume that is trapped in an elevated stable layer. The plume is suddenly brought down to the ground by the convective currents, producing high surface concentrations (i.e., fumigating the ground). Fumigations occur on the floor and sidewalls of valleys in middle to late morning when convective boundary layers grow upward from the valley floor and sidewalls and reach elevated plumes that were carried down-valley in the nighttime stable layer. The convective boundary layer typically reaches the height of the elevated plume within 2 to 5 hours after sunrise. One of the distinctive characteristics of fumigations is their day-to-day regularity in producing sudden increases in surface concentrations all along the length of the valley at about the same time of day.

Fumigations produce sudden increases in surface concentrations in middle to late morning along the length of a valley.

In valleys, elevated plumes are usually transported down the valley during nighttime. After sunrise, however, one sidewall may be heated more than the other, producing a cross-valley flow toward the heated sidewall. The elevated plume then drifts toward the sunny sidewall during its continued down-valley transport and is fumigated onto the sunny slope by the convective boundary layer that develops there. An upslope flow occurs in the boundary layer, so that the pollutants are carried up the slope after fumigation. In valleys where sinking motions are strong in the stable core, the plume sinks closer to the valley floor before the fumigation begins, and the growth of the convective boundary layers can be retarded by the strong subsidence. These two factors together produce even higher concentrations on the valley floor or slopes.

Fumigations have been observed in both wide and narrow valleys. They were first identified in the wide Columbia River Valley near Trail, British Columbia (Hewson and Gill, 1944), and have also been observed in Colorado's narrow Brush Creek Valley. An air motion tracer experiment in the Brush Creek Valley used a chemical tracer as a proxy elevated plume and observed the dispersion of the plume during the nighttime and morning transition periods. Fumigations were responsible for producing the highest concentrations at the surface. The dispersion of the plume is illustrated schematically in figure 12.18. The northwest–southeast oriented valley drains to the southeast. Because of the valley's orientation, the northeast-facing sidewall (on the left in the figure) faces the sun early in the morning and is heated more than the opposite sidewall. This heating produces a cross-valley flow in the stable core that causes the plume to drift toward the heated sidewall as it is carried down the valley. The pollution in the plume fumigates the sidewall as the con-

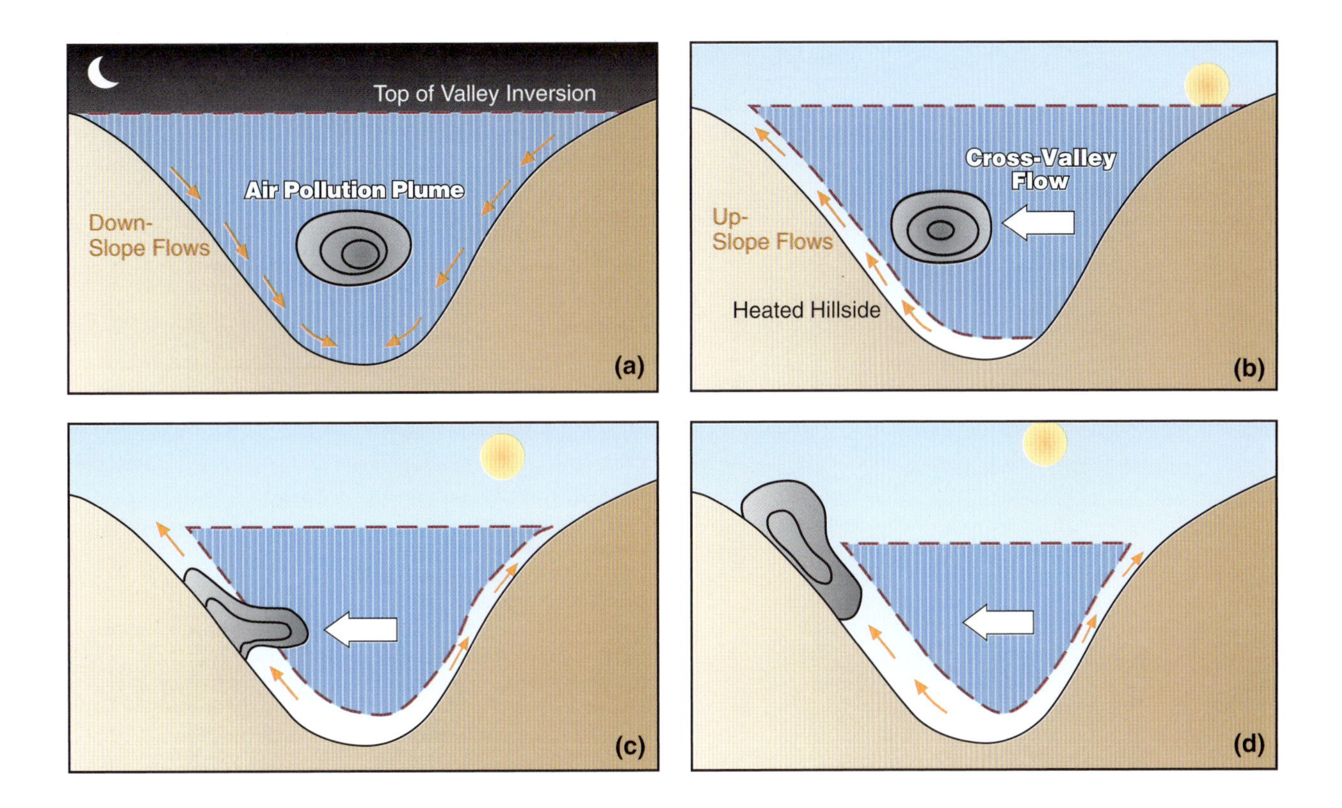

Figure 12.18 Schematic depiction of fumigations in Colorado's Brush Creek Valley: (a)–(d) represent successive times from just before sunrise to midmorning. The blue-colored region is the stable core. (Adapted from Bader and Whiteman, 1989)

vective boundary layer grows upward into the plume. The pollution is then removed from the valley by upslope flows. Mesoscale numerical model simulations suggest that the cross-valley flow is much weaker in this valley in winter when insolation falls more equally on the two sidewalls.

12.7. Assessing Air Pollution Potential in Mountain Terrain

Because topography and meteorological conditions vary significantly from one valley to the next, the air pollution potential of an individual valley in which a major new pollutant source is planned should be carefully assessed by an air pollution specialist trained in mountain meteorology. Assessments are made using meteorological models, air pollution models, air motion tracer experiments, or analyses of existing data. These studies are supported by engineering information on air pollution source characteristics and by meteorological field measurements taken in different seasons to determine the typical evolution of the wind and temperature structure in the valley. Climatological analyses of relevant meteorological data available for the valley and its surroundings (e.g., upper air wind strengths and directions, inversion frequencies, and local wind system characteristics) are helpful, as are on-site inspections (supplemented by topographical maps) of the valley's topography. Different observa-

Table 12.4 Factors to Consider in Determining Air Pollution Potential of an Area of Complex Terrain

Terrain (from maps and/or site visit)	Orientation of valley to sun Shape of valley cross section Length of valley Location and size of tributary valleys Exposure of the valley (elevation, passes, cols, ridges of different heights) to upper winds Effects of unequal solar heating on cross-valley flows
Pollution source	Source height and diameter Point, line, or area source configuration (e.g., tailing piles, roads) Effluent velocity and temperature Emissions composition and rate Diurnal changes in source strength Plume height relative to slope flow depth, valley inversion depth, and ridges
Meteorology (from soundings and surface observations)	Along-valley wind system strength and depth Wind system reversal time Strength of cross-valley flows Valley inversion strength and depth Atmospheric stability Potential for diurnal pollutant recirculation (depends on transport speed, transport time, and valley length) Channeling (forced or pressure-driven) of the synoptic flow by topography Exposure to larger scale thermally forced mountain flows
Climate	Seasonal and diurnal climate of upper wind directions and speeds, stability, local wind systems Seasonal insolation and net all-wave radiation calculations Effects of climate on surface energy budget factors (e.g., soil moisture, snow cover, vegetation cover, cloud cover) that influence wind system development and temperature structure characteristics and evolution
Dispersion	Dilution of pollutants in the main valley due to inflows from tributaries Existence of terrain constrictions that will trap pollutants in valley segments Enhancement of dispersion by terrain or forest canopy roughness

tional and modeling approaches have been used for different locations. A better understanding of the physical processes that affect air pollution dispersion in mountainous areas and sufficient observational data are needed to develop and test complex terrain models. Important points to consider when assessing air pollution potential in complex terrain are listed in table 12.4.

13 Fire Weather and Smoke Management

CARL J. GORSKI
ALLEN FARNSWORTH

> Long before humans arrived in North America, there was fire. It came with the first lightning strike and will remain forever. Unlike earthquakes, tornadoes, and wind, fire is a disturbance that depends upon complex physical, chemical, and biological relationships. Wildland fire is inherently neither good nor bad, but it is the most powerful natural force that people have learned to use. As an inevitable natural force, it is sometimes unpredictable and potentially destructive and, along with human activities, has shaped ecosystems throughout time.
>
> Philpot and Schechter (1995)

Wildland fires consume large areas of forest and grasslands every year. Fires are described in terms of *fire behavior*, which includes rate of spread and *fire intensity*. A fire that spreads rapidly burns less of the available fuel per square unit of area than a fire that moves slowly and allows the flaming front a longer *residence time*. A fire with flames that reach only two feet above the ground produces less heat and is less destructive than an intense fire that *crowns*, that is, has long flames and burns at the top (i.e., *crown*) of the forest canopy (figure 13.1).

Fire suppression activities are initiated when a wildfire threatens people, property, or natural areas that need protection. These activities include dropping water or chemicals on a fire and establishing a *fire line* around the fire. A fire line is a zone along a fire's edge where there is little or no fuel available to the fire. Roads, cliffs, rivers, and lakes can be

Carl J. Gorski is Evaluations and Operations Meteorologist, NOAA National Weather Service, Salt Lake City, Utah. Allen Farnsworth is a Prescribed Fire Specialist, U.S. Forest Service, Flagstaff, Arizona.

Figure 13.1 Spruce and subalpine fir crowning on the Blackwell Fire in the Payette National Forest, August 1994. (Photo © A. Farnsworth)

part of a fire line, or land can be cleared by firefighters. *Backfires* may be set within the fire line to burn toward the fire, widening the fire line and reducing the likelihood of the fire spreading beyond it (figures 13.2 and 13.3). Fires can cross a fire line if the intensity is high or if *spotting* occurs, that is, if the wind carries burning material (*firebrands*) beyond the fire and across the fire line (figure 13.4).

A wildland fire can be very destructive, but it can also be beneficial and may be used by land resource managers to accomplish specific ecological objectives. For example, smaller fires can reduce the danger of a large catastrophic fire by burning off underbrush. Fire can also be used to prepare land for planting, to control the spread of disease or insect infestations, to benefit plant species that are dependent on fire, to influence plant succession, or to alter the nutrients in the soil. When a fire is used to manage land resources, it is called a prescribed fire. Prescribed fires can be either natural fires that are allowed to burn under specified conditions

Figure 13.2 (*top*) Backfiring in light grass north of Kingman, Arizona, in July 1994. (Photo © A. Farnsworth)

Figure 13.3 (*bottom*) Backfiring in heavy fuels on the Payette National Forest north of McCall, Idaho, in August 1994. (Photo © A. Farnsworth)

Figure 13.4 *Spot fire* located over 0.5 mile (0.85 km) from the main fire beginning to crown in ponderosa pine. (Photo © A. Farnsworth)

or fires that are ignited by resource managers. The desired effects of a prescribed fire and the conditions under which the fire can be ignited or allowed to continue to burn are specified in a written fire prescription. The smoke produced by a prescribed fire must be managed so that air quality standards are not compromised and a reduction of visibility, especially along highways and in urban areas, is avoided. In the case of a natural prescribed fire, *smoke management* may require that the fire be suppressed. A prescribed fire, whether natural or management-ignited, that burns beyond the conditions of the prescription is considered to be a wildfire.

13.1. The Fire Environment

There are three components of the fire environment: fuel, topography, and weather. These three components interact (figure 13.5), and the sum of their interactions determines fire behavior. The characteristics of each component directly affect fuel flammability, rate and amount of heat release during fuel combustion, and fire spread.

Wildland fuels are any organic material, living or dead, that can ignite and burn. Fire behavior is affected by the fuel's size, shape, compactness, horizontal continuity, vertical arrangement, *fuel loading* (the weight of fuel per unit area), and chemical content. The amount of wildland fuel available for burning depends on fuel moisture, which, in turn, depends directly on past and present atmospheric humidity and precipitation. Different fuels respond to changes in humidity and precipitation at different rates.

Topography is the most constant over time of the three major components making up the fire environment, but it may vary considerably over space. Elevation, aspect, steepness of slopes, and landform characteristics (such as valleys, slopes, and ridges) must all be considered when assessing the influence of topography on weather and wildland fires. Topography is linked to spatial variations in climate, which play a major

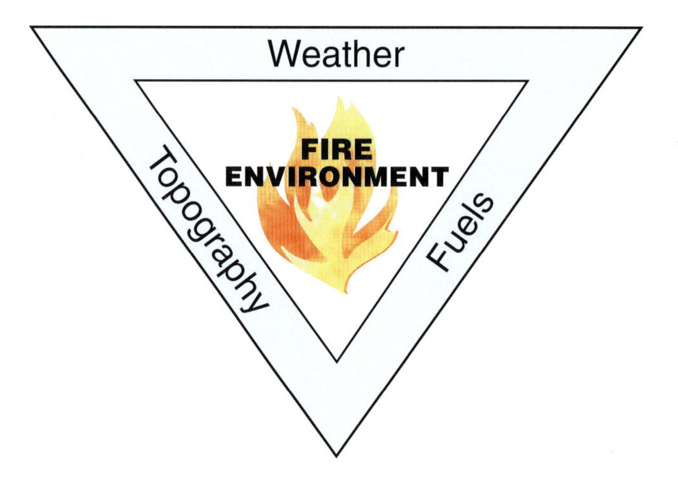

Figure 13.5 The fire environment triangle.

role in determining the types of wildland vegetation found at a given site, and to temporal and spatial variations in weather, which affect the flammability of vegetation.

Weather conditions directly affect fire behavior and significantly influence smoke production and dispersion. Weather conditions can result in the ignition of fires by lightning from thunderstorms and the rapid spread of fires as a result of strong winds. On the other hand, an increase in humidity or precipitation can slow or extinguish fires. Of the three fire environment components, weather is the most variable over time and is the most difficult for the resource manager to interpret and predict. Resource managers and firefighters conducting fire suppression or using fire as a management tool must monitor the weather at all times to make safe and effective firefighting decisions. The importance of monitoring weather and predicting the resultant fire behavior cannot be overstressed. It is one of the 10 Fire Orders (table 13.1) all firefighters must obey: "Recognize current weather conditions and obtain forecasts." The Fire Orders were written by a task force studying ways to prevent firefighting fatalities and were later expanded to include 18 Watch Out Situations (table 13.2). The risk involved in firefighting can be significantly reduced if wildland fire managers pay attention to the ever-changing weather conditions that impact fire behavior and follow the 10 Fire Orders and the 18 Watch Out Situations.

Fuel characteristics, topography, and weather conditions determine fire behavior.

Table 13.1 Fire Orders

Fight fire aggressively but provide for safety first.
Initiate all action based on current and expected fire behavior.
Recognize current weather conditions and obtain forecasts.
Ensure instructions are given and understood.
Obtain current information on fire status.
Remain in communication with crew members, your supervisor, and adjoining forces.
Determine safety zones and escape routes.
Establish lookouts in potentially hazardous situations.
Retain control at all times.
Stay alert, keep calm, think clearly, act decisively.

From National Wildfire Coordinating Group, 1989.

Table 13.2 "Watch Out Situations"

1.	Fire not scouted and sized up
2.	In country not seen in daylight
3.	Safety zones and escape routes not identified
4.	Unfamiliar with weather and local factors influencing fire behavior
5.	Uninformed on strategy, tactics, and hazards
6.	Instructions and assignments not clear
7.	No communication link with crew members/supervisors
8.	Constructing line without safe anchor point
9.	Building fire line downhill with fire below
10.	Attempting frontal assault on fire
11.	Unburned fuel between you and the fire
12.	Cannot see main fire, not in contact with anyone who can
13.	On a hillside where rolling material can ignite fuel below
14.	Weather is getting hotter and drier
15.	Wind increases and/or changes direction
16.	Getting frequent spot fires across line
17.	Terrain and fuels make escape to safety zones difficult
18.	Taking a nap near the fire line

From National Wildfire Coordinating Group, 1989.

13.2. Fuel Moisture Content

The moisture content of natural fuels is the most important variable in determining fire ignition, rate of combustion, and energy output from fires and is almost as important as wind in determining the intensity and the rate of spread of wildfires. When the fuel moisture content is high, fires burn poorly. When the fuel moisture content is low, fires start easily and spread rapidly. The percentage moisture content of fuels can be calculated from the following formula:

$$\text{fuel moisture content} = \frac{\text{field weight} - \text{oven dry weight}}{\text{oven dry weight}} \cdot 100\%$$

Fuel moisture content is directly tied to past and current meteorological conditions. The moisture content of dead fuels in the fire season is typically in the range 1.5–30%. Live fuels have moisture contents in the range of 35% to well over 200% (Schroeder and Buck, 1970). Most fuel complexes contain a combination of dead and live fuels. *Live fuel moisture* typically varies seasonally rather than on a daily basis, as the plant goes through various phenological changes. Live fuels respond to changes in atmospheric moisture by increasing or decreasing *evapotranspiration* rates through their living tissue. However, during long periods of drought, soil moisture cannot replenish moisture at the tissue level as fast as it is being given up by evapotranspiration, and the moisture content of live fuels decreases. In contrast, *dead fuel moisture* varies daily because dead fuels constantly exchange moisture with the surrounding air. There is a net gain in dead fuel moisture during periods of rain or high humidity and a net loss when the air is dry. Dead fuel moisture content is carefully monitored throughout the fire season and is a key component in the calculations for determining a *fire danger* rating as specified by the

Fuel moisture content greatly influences the rate of combustion. The moisture content of dead fuels responds to changes in ambient humidity.

National Fire Danger Rating System (Deeming et al., 1978). This system, which is used by all federal and most state resource managers in the United States, synthesizes information about all three components of the fire environment to estimate the potential for fire.

Dead fuels are divided into four size classes based on the diameter of the fuel. Each class has a characteristic reaction time (or *time lag*) to changes in atmospheric moisture, resulting in the designations 1-hour, 10-hour, 100-hour, and 1000-hour fuels (table 13.3). One-hour and 10-hour fuels have a short time lag, that is, they respond quickly to changes in atmospheric moisture. These fuels, which include grasses, needles, and twigs, ignite easily, carry most of the fire, and tend to flare up during the day and die down at night in response to diurnal humidity changes. In contrast, 100-hour and 1000-hour fuels are larger diameter fuels that respond more slowly to humidity changes. These larger fuels show very little diurnal variation in moisture content but are responsive to seasonal changes. The four size classes are used by the National Fire Danger Rating System and *fire behavior models*, which are used to predict the characteristics of a given fire.

The time lag of a fuel indicates how quickly the fuel moisture content changes from a uniform starting equilibrium value to approximately 63% of a new uniform equilibrium value, assuming that the environment of the fuel changed instantaneously. (This change is equivalent to $100(1 - 1/e)$, where $e = 2.718$. The value of e is the base of natural logarithms.) Figure 13.6 shows time lag calculations for 10-hour time lag fuels.

13.3. Fire Weather in Complex Terrain

The term *fire weather* refers to weather conditions that influence fire ignition, behavior, or suppression. A number of meteorological parameters impact fire potential and fire behavior, including wind, precipitation duration, past lightning activity, and current minimum and maximum values of temperature and relative humidity. Fire danger ratings for large administrative areas are established daily by resource managers and are based on climatological data and one or more current observations taken from representative sites.

Most fire behavior models are sensitive to topography and require weather inputs matched as closely as possible to the fire site. Inputs to the models generally include slope, aspect, time of day, time of year, amount of direct insolation impacting on fuels, fuel moisture content, wind speed and direction, temperature, and relative humidity. The meteorological parameters that have the greatest impact on fire potential and

Table 13.3 Time Lag Fuel Categories

Type of fuel	Size of fuel elements
1-hour fuel	0 to .25 inch (0 to 6 mm)
10-hour fuel	0.25 to 1.0 inch (6 to 25 mm)
100-hour fuel	1.0 to 3.0 inches (25 to 76 mm)
1000-hour fuel	3.0 to 8.0 inches (76 to 203 mm)

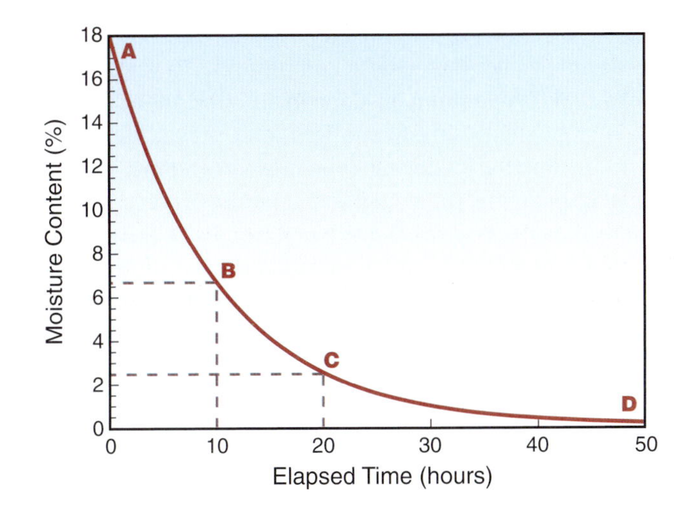

Figure 13.6 Basic time lag concept for 10-hour time lag fuels. At point A, fuel moisture is uniformly 18%. At hour 0, the environment is suddenly changed so that a new equilibrium of 0% moisture content exists. The fuel responds exponentially to this change so that after 10 hours (point B) the fuel moisture falls to 6.6%. At 20 hours (point C) it falls to 2.4%, and at 50 hours (point D) it is nearly at equilibrium. (Adapted from Fosberg, 1977)

fire behavior are atmospheric humidity and temperature, wind, and atmospheric stability.

13.3.1. *Humidity and Temperature*

Low relative humidity is an indicator of high fire danger. Brotak and Reifsnyder (1977a, 1977b) found that 93% of fires examined coincided with a low humidity in the lower atmosphere (dew-point depression > 10°C at 850 mb). Moisture in the atmosphere, whether in the form of water vapor, cloud droplets, or precipitation, is the primary weather parameter that affects fuel moisture content and thus the flammability of wildland fuels. The amount of moisture that fuels can absorb from or release to the air depends largely on relative humidity. Of course, atmospheric moisture is also necessary to produce cloud cover, rainfall, and lightning, which affect fire potential, fire ignition, and fire behavior.

Large fires coincide not only with below normal relative humidity, but also with above normal temperatures (Potter, 1996). High ambient temperatures decrease the moisture content of both live and dead wildland fuels and thus increase their flammability. In addition, high temperatures and direct sunlight preheat the fuels, bringing them closer to their ignition point. Firefighter fatalities have occurred during several fires when record high temperatures were set. For example, on the day of the Mann Gulch Fire in Montana (5 August 1949), a record high temperature was recorded at Helena, Montana (McLean, 1994). All-time-high temperature records were also set in Arizona on 26 June 1990, the day that fatalities occurred on the Dude Fire just below the Mogollon Rim.

Firefighters can see or feel most of the elements of fire weather. However, subtle changes in relative humidity that cannot be felt or seen can have a significant impact on fire behavior. Humidity thresholds for critical fire behavior vary over time and space and are different for different fuel types. Careful records must therefore be maintained for each geographic area. The most important weather instrument for a firefighter in the field is a sling psychrometer (section 3.2.3 and appendix B) or other accurate hygrometer.

The meteorological parameters with the greatest impact on fire potential and fire behavior are atmospheric humidity and temperature, wind, and atmospheric stability.

Record high temperatures and low relative humidity increase fire danger. During critical fire operations, fire crews should monitor the temperature and relative humidity frequently.

13.3.1.1. *The Effects of Aspect and Elevation On Humidity and Temperature.* The receipt of solar radiation in complex terrain varies significantly from slope to slope, depending on slope inclination angle and aspect (appendix E). A slope's exposure to sunlight (and prevailing winds) determines its microclimate, including precipitation, temperature, and humidity, thus influencing the types and amount of vegetation that can flourish there and affecting fire behavior (figure 13.7). Geiger et al. (1995) calculated that under full sunshine in midlatitudes, south slopes are on average about 5°F (3°C) warmer than north slopes because of the aspect angle. The warmer south slopes also experience lower relative humidity and lower soil moisture content. When a fire moves onto a slope with a different aspect, changes in temperature, humidity, and the resultant moisture content of fuels change the fire's intensity and rate of spread.

Elevation directly influences temperature and relative humidity and thus fire behavior. In the well-mixed daytime atmosphere, temperature decreases with height in the boundary layer, whereas the water vapor mixing ratio becomes well-mixed throughout the depth of the boundary layer and thus changes little with height. In this situation, relative humidity increases with height (section 3.2.3). The increase in relative humidity with elevation increases the moisture content of fuels at higher elevations.

At night, when skies are clear, temperature inversions form in valleys,

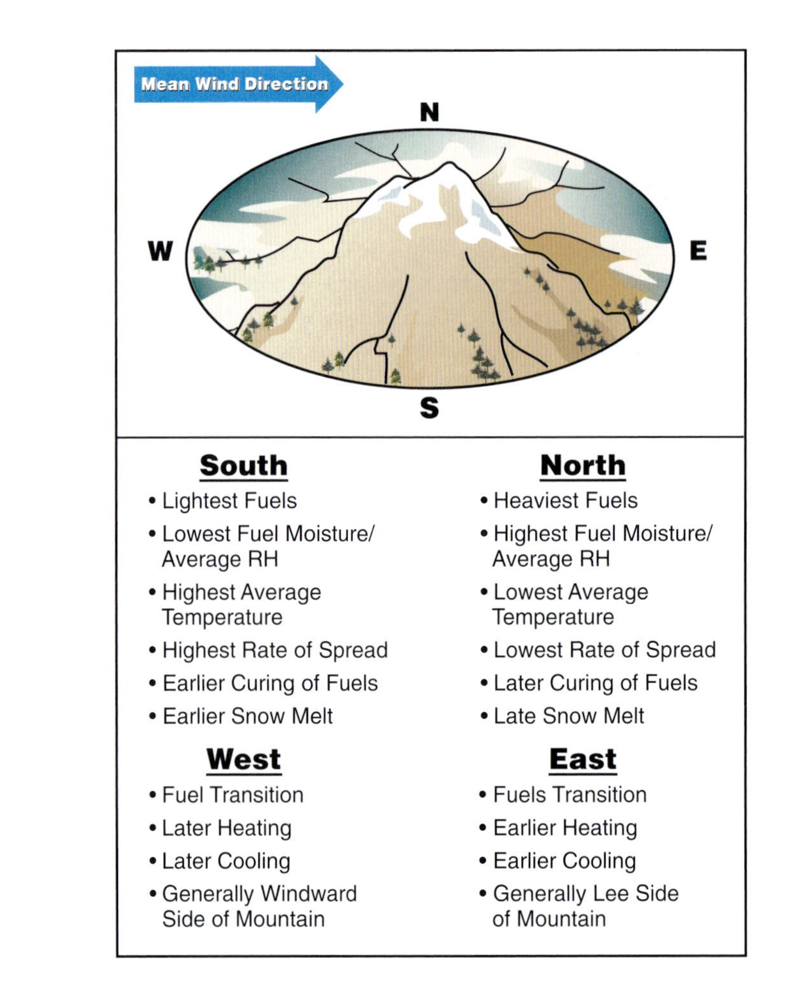

Figure 13.7 The microclimate of a mountain slope depends on its aspect angle, which determines its exposure to sunlight and prevailing winds and thus affects precipitation, temperature, humidity, fuels, and fire behavior. Shown are microclimate factors for midlatitude slopes oriented in the four cardinal directions in the Northern Hemisphere. (Adapted from National Wildfire Coordinating Group, 1994)

causing the relationship between surface temperature, humidity, and elevation to reverse. Cold dense air flows downslope and collects in the valley bottoms, leaving warmer temperature and lower humidity over the upper sidewalls (figure 11.4). As a result, *fine fuels* (dead fuels such as grass, twigs, and needles that are less than $\frac{1}{4}$ inch, or 6 mm, in diameter and react quickly to changes in humidity) at lower elevations have a higher moisture content at night.

The exposure of a slope to solar radiation and prevailing winds affects fire behavior on that slope.

Large diurnal temperature swings (section 11.5) are common in wide valleys and basins in the interior western mountains of the United States during the summer. For example, at the Granite Fire in a high mountain valley in eastern Oregon during August 1986, diurnal temperature variations were routinely in excess of 60°F (33°C). The high frequency of clear skies and low atmospheric moisture creates optimal conditions for maximum daytime insolation and efficient radiative cooling at night. On days when conditions are ideal for maximum heating and the ambient air temperature is above 90°F (32°C), ground temperatures can approach 150°F (66°C). At night when skies are clear and winds are light, radiation-induced nighttime inversions can cause temperatures to increase as much as 25°F (14°C) in the first 250 ft (75 m) above the earth's surface.

In a well-mixed atmosphere (such as might occur on the afternoon of a sunny summer day), observed temperatures at one elevation can be used to estimate temperatures at the same time at higher or lower elevations. The elevation of the observation site is subtracted from the elevation where a temperature estimate is desired. This is multiplied by the dry adiabatic lapse rate of 5.4°F per 1000 ft (9.8°C/km) to obtain a temperature correction, which is then subtracted from the observed temperature value. When the atmosphere is well mixed, the temperature decreases with altitude and relative humidity increases with altitude.

13.3.1.2. *The Effects of Surface Properties On Humidity and Temperature.* Heat exchange processes that occur between the earth's surface and the atmosphere are the primary determinants of the ambient temperature of the air. The temperature of a surface, whether soil, fuels, or water, depends largely on its properties (figure 13.8), including solar reflectivity or albedo, absorptivity, transparency, specific heat, conductivity, and surface moisture. Variations in these properties can result in large spatial differences in air temperatures across a landscape even when insolation is homogeneous. In general, dark, bare soils and forest litter absorb most of the visible wavelength radiation and become hotter than lighter colored soils, which reflect much of the radiation back into space. Bodies of

Figure 13.8 Land surface properties can significantly affect ambient air temperatures. (Adapted from National Wildfire Coordinating Group, 1994)

water are transparent and can absorb more energy through a greater depth than can opaque soils. Organic materials are generally poorer conductors of heat than rocks or wet soils.

Vegetative cover moderates air temperatures by intercepting both incoming and outgoing radiation. Low, sparse vegetation produces little shade and thus has much less effect on the temperature of the underlying surface than tall, dense vegetation. Green foliage, because of respiration, photosynthesis, and evapotranspiration, does not heat up as much as bare ground or surface organic litter.

In complex terrain, surface properties and vegetation can vary over short distances, causing significant differences in heat absorption. If contrasting surface features are large, the differential heating can alter the local thermally driven wind regimes either by enhancing the normal thermal gradients or by producing anomalous gradients. Efficient absorbers and radiators of heat (such as lava beds), heat sinks (such as large bodies of water), or highly reflective dry lakes can enhance or reduce the overall slope/valley wind speeds, delay their onset, or even reverse the typical slope/valley diurnal cycle.

13.3.1.3. *Humidity, Fuel Moisture Content, and Fire Severity.* Relative humidity and the moisture content of 1-hour and 10-hour fuels are good indicators of *fire severity*, that is, the intensity and duration of a fire (table 13.4). When both relative humidity and fuel moisture content are high, there is very little danger of fire ignition. As relative humidity and fuel moisture content decrease, ignition and spotting increase and fires burn more aggressively. When relative humidity drops below 15% and the moisture content of 1-hour and 10-hour fuels is below 5% in the mountainous western United States, *extreme fire behavior* can be expected.

Table 13.4 Fire Severity Related to Relative Humidity and Fuel Moisture

Relative humidity (%)	1-hour fuel moisture (%)	10-hour fuel moisture (%)	Relative ease of ignition and spotting; general burning conditions
>60	>20	>15	Very little ignition; some spotting may occur with winds above 9 mph
46–60	15–19	13–15	Low ignition hazard, campfires become dangerous; glowing brands cause ignition when RH is <50%
41–45	11–14	10–12	Medium ignitability, matches become dangerous; "easy" burning conditions
26–40	8–10	8–9	High ignition hazard, matches always dangerous; occasional crowning, spotting caused by gusty winds; "moderate" burning conditions
15–25	5–7	5–7	Quick ignition, rapid buildup, extensive crowning; an increase in wind causes increased spotting, crowning, loss of control; fire moves up bark of trees igniting aerial fuels; long-distance spotting in pine stands; "dangerous" burning conditions
<15	<5	<5	All sources of ignition dangerous; aggressive burning, spot fires occur often and spread rapidly, extreme fire behavior probable; "critical" burning conditions

High rates of spread, prolific crowning and spotting, the presence of *fire whirls*, and a strong *convection column* above the fire can preclude direct control of the fire.

13.3.2. *Wind*

Wind is the most critical factor affecting fire behavior and is the most important element in wildfire and prescribed fire plans. Characteristics of wind are as follows:

- Wind speed and direction are the primary determinants of rate and direction of fire spread and of the transport of smoke.
- Wind carries away moisture-laden air and thus hastens the drying of wildland fuels.
- Wind aids combustion by increasing the supply of oxygen to the fire.
- Wind increases fire spread by carrying heat and burning embers to new fuels.
- Wind bends flames closer to the unburned fuels, thus preheating fuels ahead of the fire.
- Wind influences the amount of fuel consumed by affecting the residence time of the flaming front of the fire. The stronger the wind, the shorter the residence time and the less fuel is consumed.

Wind is the most critical factor affecting fire behavior and is the most important element in wildfire and prescribed fire plans.

Wind is the most variable weather element that affects the fire environment and thus is the hardest to predict. Wind variability over time and space (especially in complex terrain where topography influences the wind regime) can pose safety and fire control problems, which can result in firefighter fatalities. Wind direction and wind speed must be constantly monitored by all firefighters.

13.3.2.1. *Wind Composition*. The wind field that impacts the fire environment is a composite of synoptic winds, terrain-forced winds, and smaller scale diurnal mountain winds. In the fire community, synoptic winds are called *general winds* or *free air winds*, and diurnal mountain winds are called *local convective winds*. To understand the winds at a specific site, the resource manager must understand the day-to-day contribution of transitory synoptic-scale winds, the influence of topography on the synoptic winds, the diurnal evolution of the local, thermally driven wind regime, and the daily influence of atmospheric stability on the wind patterns.

Synoptic-scale winds aloft are more likely to reach the surface when they are strong (above 10 mph) and blow parallel to a valley and when the atmosphere is unstable. Synoptic-scale winds have a greater effect on upper slopes than lower slopes.

Diurnal mountain winds are best developed when skies are clear and synoptic-scale winds are weak (chapter 11). Up-valley and upslope winds tend to be gusty, whereas down-valley and downslope winds are usually steady. Along-valley winds can overpower the weaker slope winds.

For purposes of fire weather, winds are measured at three different levels of the atmosphere: (1) in the free atmosphere above the frictional influence of the surface, (2) at a standard 20-foot (6.1-m) level above the vegetation, and (3) at midflame height. The direction and speed of winds in the atmosphere are obtained through existing meteorological data, from aircraft observations, by releasing and tracking a weather balloon on-site, or through visual indicators. These ambient winds govern the direction of smoke transport and of potential spotting. The *20-foot wind* is a near-surface wind measured at a standard 20-foot (6.1-m) height above the vegetative surface or the top of the forest canopy. The 20-foot wind is averaged over a 2- to 10-minute period and may therefore be lower than instantaneously measured wind. The 20-foot wind is used for calculating fire danger parameters in the National Fire Danger Rating System and for estimating the distance that firebrands lofted by the fire will be carried by the wind, resulting in spotting. *Midflame winds* are winds measured closer to the ground at the midflame level, which varies from fire to fire and from fuel type to fuel type. The midflame wind speed and direction at the flaming front of a fire are used to calculate fire spread (Rothermel, 1983). The winds that affect fire behavior are shown in figure 13.9.

Wind speeds at the 20-foot level are taken at fixed locations or from portable weather stations that have been deployed during fires or projects. Most midflame wind speeds are estimated from observations taken at eye level, but they can also be estimated from the observed or forecast 20-foot wind speeds. A reduction factor applied to the 20-ft wind speed must be used to account for the increase in surface friction closer to the ground. Adjustment tables have been developed for fire behavior models; these tables take into account differing vegetative regimes, such as open grasslands or densely packed forest stands, and the exposure of fuels on a slope to the ambient wind.

13.3.2.2. *Winds of Most Concern to Firefighters.* Synoptic and local diurnal mountain wind regimes may be considered problem winds to firefighters. If wind speeds are low and the direction is consistent, the resultant fire behavior is usually predictable, and firefighters can develop strategies to execute safe and successful fire suppression activity. If winds become so strong that fire lines are lost and large acreages are rapidly consumed, the safety of firefighters may be threatened and suppression efforts may have little impact on fire spread. Winds that totally dominate the fire environment have been termed *Winds of Most Concern to Firefighters*. These winds fall into the following categories: frontal winds,

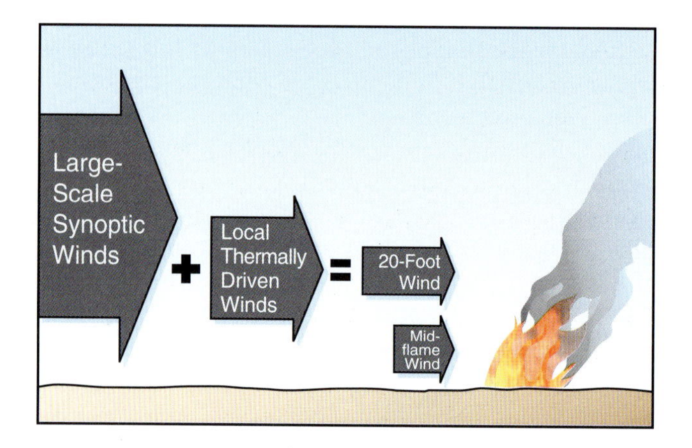

Figure 13.9 Local thermally driven winds combine with larger scale synoptic winds to produce the winds that affect fires. These winds are measured in the field at two levels by firefighters: at 20 feet and at midflame height.

foehn winds, thunderstorm winds, whirlwinds, low-level jets, and inversion-breakup winds.

Frontal winds are associated with frontal passages (section 6.2). The winds are strong and shift direction in a clockwise fashion as the front passes. Frontal winds can cause a fire to abruptly change intensity or direction. Firefighters must be aware of the timing and strength of the frontal passage and anticipate changing wind directions.

Foehn winds are strong, dry winds caused by the compression of air as it flows down the lee side of a mountain barrier (section 10.2). The combination of high wind speeds and low atmospheric moisture can cause high rates of fire spread. When a foehn wind occurs after a long period of dry weather, fire behavior can be extreme (figure 13.10). For example, fires in Southern California that are exposed to Santa Ana winds can be even more difficult to control if they occur in the fall or early winter after a long, dry summer (section 10.2.2).

Winds associated with thunderstorms, whether indrafts or downdrafts, can change both direction and speed suddenly, resulting in sudden changes in the rate and direction of spread of a fire, as well as in the fire's intensity. This unpredictable fire behavior can be a significant hazard to firefighters. Heat rising from a fire can form a convection column strong enough to trigger the development of a thunderstorm, even on an otherwise cloudless day.

Whirlwinds are produced by strong shear behind obstacles or by intense convection in light winds. Shear-produced whirlwinds are often found on lee slopes or behind shallow ridges. Convectively produced whirlwinds called fire whirls are created by the intense horizontal convergence and rising motions that occur in the highly unstable atmosphere at the base of fire columns. There are various scales of whirlwinds, but very few dominate the entire fire environment for any length of time. They can, however, be troublesome to firefighters when they scatter firebrands or cause spotting across fire lines.

Low-level jets exhibit a jet profile (section 10.6) in which maximum winds attain speeds of 25–35 mph (11–15 m/s) at altitudes from 100 feet (30 m) to several thousand feet (a thousand meters) above the ground. They occur most often at night in the central plains of the United States ahead of cold fronts but can occur over any type of terrain. They are usually associated with synoptic-scale events, such as low pressure troughs or the breakdown of an upper level high pressure system.

The normal postsunrise breakup of a nocturnal inversion can result in the sudden downward transport of momentum from winds aloft to the surface. The sudden increase in wind speed and possible change in wind direction can quickly increase the rate of spread or change the direction of spread of a fire, thus threatening the safety of firefighters. Abrupt changes in wind speed caused by inversion breakups are sometimes attributed incorrectly to surfacing low-level jets.

Winds of most concern to firefighters include frontal winds, foehn winds, thunderstorm winds, whirlwinds, low-level jets, and inversion-breakup winds.

Figure 13.10 This sign warns Swiss citizens of the increased fire danger during the foehn. (Translation: "Fire Danger during the Foehn: Smoking and open fires are prohibited during the foehn. Restrictions during foehn absolutely must be observed.")

13.3.2.3. *The Effects of Slope Inclination Angle and Wind On Fire Behavior.* The slope angle and the direction and speed of the prevailing wind determines the surface fire shape and the intensity of a fire. When winds are calm, a fire on flat terrain has a circular shape. When winds are present, the fire is elliptical in shape, elongating as wind speeds increase. The slope inclination angle does not affect the shape of a fire as dramatically as strong winds, but it can have a profound effect on direction and rates of fire spread and on fire size and intensity. Fire moves more quickly up a slope than across flat ground because flames at the head of a fire tilt toward the sloping surface and preheat the upslope fuels through radiation and convection. The preheated fuels then combust at a significantly higher rate.

When winds are calm, the rate of fire spread increases with the slope inclination angle. If there is a wind present, the portion of the wind that aligns up the slope combines with the slope effect for a faster spread rate up the incline. If a portion of the wind is directed down the slope, that portion of the wind is subtracted from the overall slope effect, and the fire spreads more slowly. For example, a fire burning in heavy grass fuels with a steady 7 mph (3.1 m/s) midflame wind speed can spread at a rate of 0.9 mph (0.4 m/s) over flat ground. Another fire, however, under identical meteorological conditions on a 45° slope might spread at the faster rate of 1.6 mph (0.7 m/s). Most fire behavior models use a vector addition or subtraction method to account for wind on slopes to project fire movement in complex terrain. The combined effect of wind and slope inclination angle is shown in figure 13.11.

13.3.2.4. *Fire-Induced Winds.* When rapid combustion occurs, the fire injects large amounts of heat into the atmosphere. The intense vertical convection generated by the fire creates strong horizontal convergence. Winds produced by this convergence can dominate the local wind field, affecting both the intensity and spread rate of the fire. Wind speeds on the fire line caused by convergence have been reported to exceed 40 mph (18 m/s). Indrafts on the Sundance Fire in northern Idaho during the afternoon of 1 September 1967 were estimated at speeds up to 120 mph (54 m/s) by analyzing tree blowdowns (Anderson, 1968). Reports of unusually strong winds near active flaming fronts should be used as an indica-

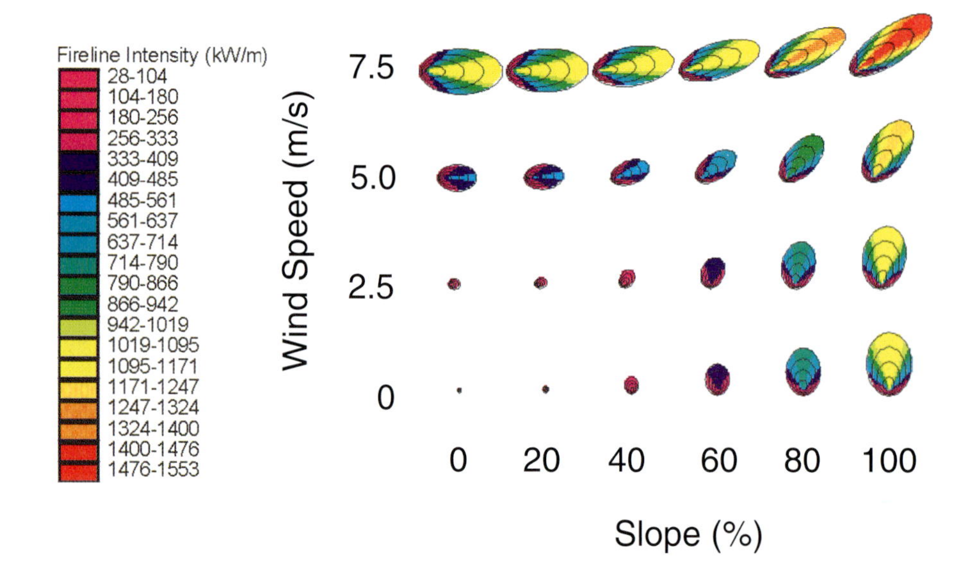

Figure 13.11 Surface fire shapes and intensity patterns depend on slope angle and the direction and speed of the prevailing wind. In this set of model simulations, the wind direction is assumed to blow across the slope from left to right. The combination of slope angles and wind speeds produces the differing fire shapes and intensity patterns shown. (From Finney, 1998)

tor of the fire's intensity but should not be used as input to fire behavior models, which calculate spread rate based on ambient winds and slope angles.

When winds are calm, fire size, spread rate, and intensity increase with slope inclination angle.

13.3.3. *Atmospheric Stability*

Ambient winds and slope angles are the principal determinants of forward spread rates for fires.

Stability is a measure of the resistance of the atmosphere to vertical motion (section 4.3). When the atmosphere is unstable, vertical motions increase, contributing to increased fire activity by:

- allowing convection columns to reach greater heights, producing stronger indrafts and convective updrafts;
- increasing the lofting of firebrands by convective updrafts;
- increasing the occurrence of dust devils and fire whirls; and
- increasing the downward transfer of momentum from winds aloft, thus increasing the potential for gusty surface winds.

Fire danger increases when the atmosphere is unstable.

Atmospheric stability can be determined by observing clouds, smoke plumes, winds, and visibility (section 4.3.1). Firefighters should observe indicators of stability at all times (figures 13.12 and 13.13). When visibility is poor or views are blocked by topography or trees, lookouts should be posted to alert fire crews to the presence of any visual indicators of unstable air.

13.3.3.1. *Atmospheric Instability, Low Humidity, and Large Fires.* Large wildfires are often associated with low surface humidity and with unstable conditions, as indicated by temperature lapse rates that are greater than the 3.5°F per 1000 feet or 6.5°C/km lapse rate of the Standard Atmosphere. An effort has thus been made to develop simple predictive or diagnostic indices for the danger of large wildfires based on routine temperature and humidity soundings. Similar indices have long been used by meteorologists to predict thunderstorm activity. The Lower Atmosphere Stability Index (LASI), also known as the Haines Index (Haines, 1988), is used to predict the likelihood of large fire growth by determining the sum (A + B) of an atmospheric stability index A and a lower atmosphere dryness index B. The stability index is determined from measurements of the temperature difference between two atmospheric levels and the dryness index is determined from measurements of the dew-point

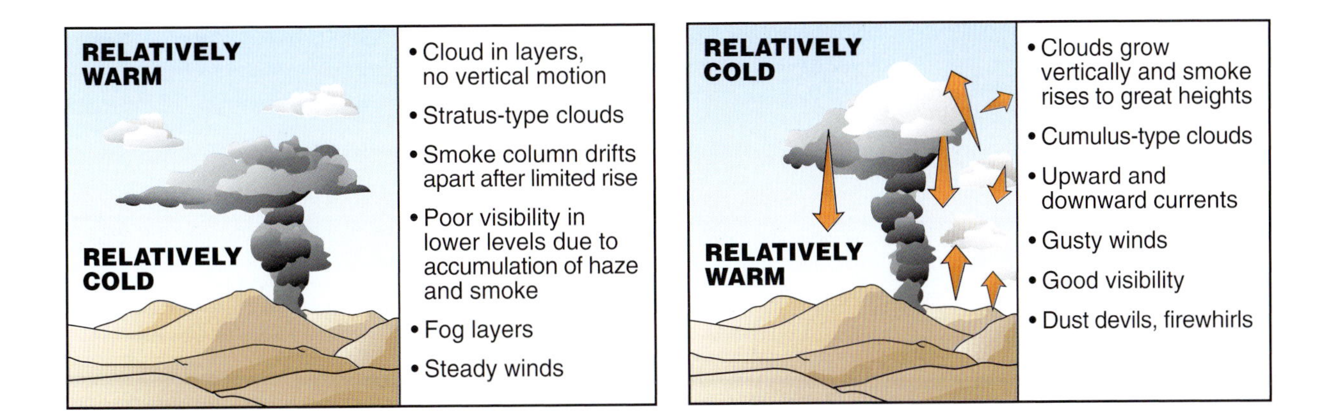

Figure 13.12 (*left*) Indicators of stable air. (Adapted from Schroeder and Buck, 1970)

Figure 13.13 (*right*) Indicators of unstable air. (Adapted from Schroeder and Buck, 1970)

depression (i.e., the difference between the temperature and the dew point, a measure of air moisture content). Because there are large variations in elevation across the United States, the index is calculated for three different pressure ranges (low elevation, 950–850 mb; midelevation, 850–700 mb; high elevation, 700–500 mb). The LASI index for a given elevation range is determined by summing the index values A and B from table 13.5. If the sum is 2 or 3 the potential for large fire growth is very low, if 4 the potential is low, if 5 the potential is moderate, and if 6 the potential is high. The LASI is calculated routinely and appended to most fire weather forecasts. The LASI must be used with caution because it provides no measure of fuel moisture, soil wetness, or other variables on which fire depends, and it can be adversely influenced by erroneous sounding data, nonrepresentative soundings, or rounding errors during calculation.

The Lower Atmosphere Stability Index predicts the potential for fire growth based on atmospheric stability and lower atmospheric dryness. When the LASI reaches 5 or 6, the probability of extreme fire behavior increases significantly.

13.3.3.2. *Inversions.* Low temperature and high relative humidity are found in mountain valleys at night and in winter when radiation inversions form. Under inversion conditions, fuel moisture content is higher, thus decreasing fire spread rates and intensities. When daytime heating breaks up a nocturnal temperature inversion in a valley or basin, firefighters see a marked increase in fire activity because temperatures increase, relative humidity decreases, and winds may increase and suddenly change direction.

Fire activity decreases at night when an inversion is present.

The warmest nighttime air temperatures in valleys are often found in a thermal belt midway up the slope (section 11.1.1). The location and strength of the thermal belt depend on temperature inversion depth and strength, which vary depending on topography, time of night, season, and size of the valley. The thermal belt is characterized by low diurnal variation in temperature and humidity; warmer average temperatures than at the valley floor or ridge tops; lower average humidity than the surrounding slopes; and faster drying following precipitation events. Thermal belts can and often do have a significant effect on fire control efforts. Because there is little or no diurnal fuel moisture recovery, the ther-

Table 13.5 Lower Atmosphere Stability Index Calculations

Elevation	Stability term (A)	Dryness term (B)
Low	temperature at 950 mb − temperature at 850 mb	dew-point depression at 950 mb
	A = 1 when 3°C or less	B = 1 when 5°C or less
	A = 2 when 4–7°C	B = 2 when 6–9°C
	A = 3 when 8°C or more	B = 3 when 10°C or more
Medium	temperature at 850 mb − temperature at 700 mb	dew-point depression at 850 mb
	A = 1 when 5°C or less	B = 1 when 5°C or less
	A = 2 when 6–10°C	B = 2 when 6–12°C
	A = 3 when 11°C or more	B = 3 when 13°C or more
High	temperature at 700 mb − temperature at 500 mb	dew-point depression at 700 mb
	A = 1 when 17°C or less	B = 1 when 14°C or less
	A = 2 when 18–21°C	B = 2 when 15–20°C
	A = 3 when 22°C or more	B = 3 when 21°C or more

[a]The LASI value is given by the sum A + B as determined separately for low, medium, and high elevations. When the LASI value reaches 5 or 6, the probability for extreme fire behavior increases significantly.

mal belt can have the most active fires, which may continue burning throughout the night.

Large diurnal temperature swings under generally undisturbed weather conditions indicate a high likelihood of strong surface-based nocturnal inversions. The time of breakup of nocturnal inversions depends on the time of year, cloudiness, the length and width of the valley, and the strength of the diurnal mountain winds. Breakup time varies from one valley to the next, but it tends to be regular in a given valley and is typically $3\frac{1}{2}$ to 5 hours after sunrise. This regularity is a good forecasting aid. Fire behavior after the breakup of an inversion depends largely on atmospheric stability and can thus vary from day to day. The stability of the valley atmosphere can be determined from aircraft temperature observations or balloon soundings or it can be estimated from visual indicators (figures 13.12 and 13.13 and table 4.5).

Along coasts that have cold ocean currents offshore, fire activity may be inhibited by the high humidity and cooler temperatures associated with *marine air intrusions*. The cool, moist marine air in the *marine inversion* layer intrudes into coastal valleys, varying in depth from a few hundred to several thousand feet. Fires can, however, burn actively in the warm dry conditions above the marine inversion.

Fire activity decreases in the presence of a marine inversion and increases in the presence of a subsidence inversion.

In summer, elevated subsidence inversions are produced by subsiding air in high pressure centers (section 12.4.2). If a high pressure center persists for a period of days, the subsidence inversion slowly lowers, bringing warmer temperatures and drier air to lower altitudes both day and night, thus increasing fire potential over large areas (figure 13.14). The tops of mountain ranges experience the warm dry air first. High pressure centers are usually slow-moving and can dry out fuels to the point that burning conditions become severe. For example, in the ponderosa pine forests of the Southwest, if the subsidence inversion keeps temperatures above 10–13°C (50–55°F) overnight, active burning may be promoted on the following day.

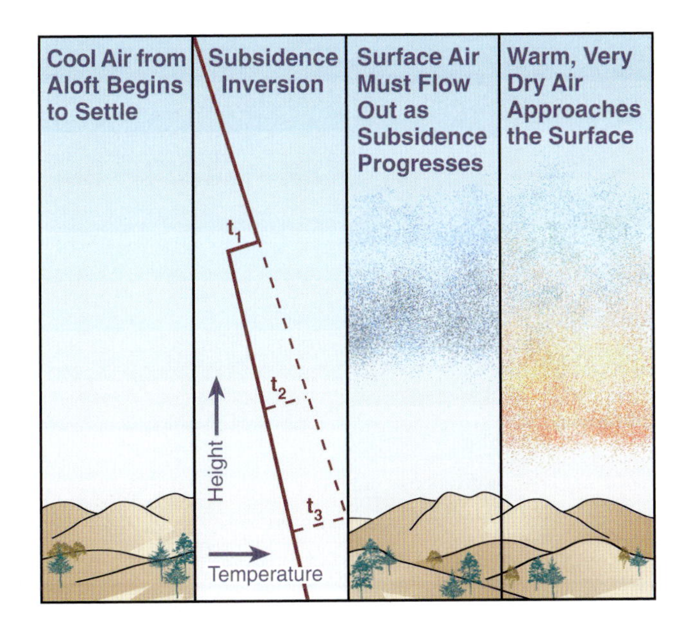

Figure 13.14 Subsidence inversions are produced when air from midtropospheric levels sinks in high pressure centers. The subsidence warms and dries the air, enhancing burning conditions, especially on mountaintops. (Adapted from Schroeder and Buck, 1970)

13.4. Critical Fire Weather

13.4.1. *Fire Seasons*

Fire seasons occur at different times of the year in different regions of the country, depending on seasonal variations in weather. The typical fire season at any given location has numerous hot and dry days, yet wildfires are usually clustered within relatively short periods. These periods are characterized by one or more critical fire weather conditions, including strong wind, very low relative humidity, high temperature, highly unstable air masses, and dry lightning. Table 13.6 summarizes fire seasons for the United States.

Critical fire weather conditions include strong wind, very low relative humidity, high temperature, highly unstable air masses, and dry lightning.

As discussed previously, strong winds can be associated with foehn events, thunderstorms, whirlwinds, frontal passages, low-level jets, and inversion breakups. Synoptic patterns that bring a combination of critical fire weather conditions include

- foehn wind events, which bring not only strong winds, but also drying as a result of compression heating;
- the presence of dry thunderstorms, which produce strong downdrafts and lightning, but little precipitation;
- the passage of dry cold fronts, which bring strong winds but little moisture; and
- the presence of a high pressure center, which is characterized by low humidity and high temperatures. The breakdown of upper high pressure ridges, which results in both extremely dry air and atmospheric instability, is particularly critical for fires. In the zone between the high pressure ridge and the ensuing trough, the pressure falls, the atmosphere becomes unstable and gusty winds can be expected. Fuels are still dry, but clouds and precipitation have not yet formed.

Table 13.6 Fire Seasons

Season	Locations	Comments
East of Rockies		
April–October	North states, Lake states, NE and NW Plains	
March–November	Southeast, west Gulf States, Ohio and Middle Mississippi Valley regions	Lulls during the summer and occasional periods of high fire danger in winter
Yearly	South Plains	
Rockies, Intermountain, and Southwest		
June–October	Northern Rockies, Northern intermountain	
May–October	Central intermountain	
Yearly	Southwest	Minimum in winter, secondary minimum in August due to monsoon
West Coast		
June–September	Pacific Northwest	Occasionally, the season may begin as early as April and end as late as November.
Yearly	California	

13.4.2. *Droughts*

There is a close relationship between the occurrence of large fires and prolonged drought because drought dramatically decreases the moisture content of both live and dead fuels. Droughts can shorten the time when plants are putting on maximum growth and absorbing large amounts of moisture. Long-term drying during the growing season accelerates the curing of live fuels, abnormally lowers the moisture contents of all sizes

CHANGES IN FOREST COMPOSITION CONTRIBUTE TO INCREASE IN SEVERE FIRES

A series of severe and costly fire seasons have occurred in the United States since 1988. In 1988, the greater Yellowstone area became the center of national and international attention as a result of large spectacular fires. In 1991, two tragic fires occurred at the urban-wildland interface near Oakland, California, and Spokane, Washington, within days of each other. In 1994, the southwestern United States ignited with a record fire season, and 14 firefighters lost their lives at the South Canyon Fire near Glenwood Springs, Colorado. In 1996, a record fire season was recorded for the high elevations of eastern Oregon. In these severe fire seasons, multiple smoke plumes from large fires could be viewed from weather satellite imagery, as shown in the figure below.

A number of factors contributed to these large destructive fires, but one factor stands out above the rest: the change in the composition of North American forests since the European settlement of the continent. Massive quantities of forest fuels have built up over the last 50 to 100 years as the natural cycle of frequent fires in forests that are dependent on fire for regeneration has been broken by management activities such as logging, grazing, and fire suppression. In addition to the exclusion of fire, periodic droughts, the expanding urban wildland interface, and the spread of disease and insect infestations to large areas have changed the forests and increased the potential for devastating fires.

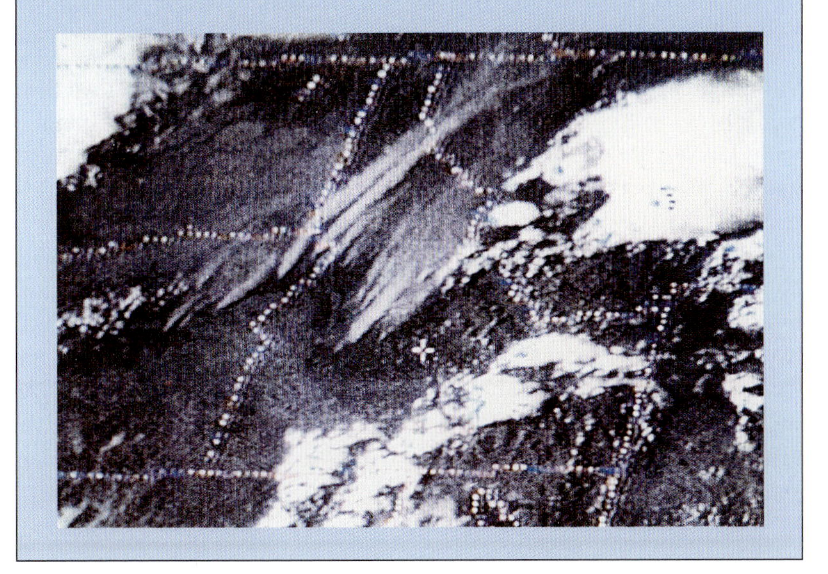

Satellite photo of smoke plumes from multiple fires over eastern Oregon and central Idaho, 29 July 1989. (National Weather Service photo.)

of fuels, and can thus accelerate the onset of a fire season. When the moisture content of living plants (particularly evergreens) falls below 100%, the potential for *crown fires* increases significantly. In chaparral ecosystems, very active fires are likely if the live fuel moisture dips below 60% (Agee, 1993). In the South Canyon Fire near Glenwood Springs, Colorado, failure to recognize the low live moisture content of Gambel oak was a contributing factor in underestimating the fire behavior that led to firefighter fatalities on 6 July 1994.

Drought also increases the loading of dead fuels and dries larger dead fuels (1000-hour) and deep layers of *duff* (organic matter in various stages of decomposition on the forest floor) and organic soils, making them more likely to ignite. The result can be extreme fire behavior, including crowning, high rates of spread, and the release of large amounts of heat.

Two common drought indices used by resource managers to gauge the severity of drought conditions are the *Palmer Drought Severity Index* and the *Keetch–Byrum Drought Index*. The Palmer Drought Severity Index is a water balance equation that tracks water supply (precipitation and stored soil moisture) and water demand (potential evapotranspiration, soil recharge, and runoff). The index is calculated by the National Weather Service using weekly averages of temperature and precipitation from climate stations across the United States. The Keetch–Byrum Drought Index, which is part of the National Fire Danger Rating System, is a calculation of the net effect of evapotranspiration and precipitation in producing a cumulative moisture deficit in deep duff and organic soils. It can be calculated using maximum temperature and 24-hour rainfall values from a single station. The index is used extensively throughout the southeastern United States.

Drought reduces the moisture content of both live and dead fuels and can result in extreme fire behavior.

13.4.3. *Fire Weather Watches and Red Flag Warnings*

The National Weather Service's *Fire Weather Watch* and *Red Flag Warning* programs alert fire managers to developing weather conditions that, when coupled with critically dry wildland fuels, could lead to dangerous fires. These watches and warnings enable fire fighting agencies to manage resources and prepare for appropriate suppression activities to protect life and property (table 13.7). A Fire Weather Watch is issued when Red Flag Warning criteria are expected to be exceeded within 12 to 48 hours. The criteria for the issuance of watches and warnings are developed locally. The criteria may be, for example, dry lightning coupled with low 100-hour or 1000-hour class fuel conditions, high LASI coupled with 10-hour fuel moisture below a certain reading, or wind speeds that are forecast over a threshold value developed for the geographical area. The matrix used in northern California for issuing Red Flag Warnings for foehn winds is shown in table 13.8. It assumes that 1-hour and 10-hour fuel moistures are low, annual grasses are cured, and no wetting rain has fallen in the last 24 hours. Both Fire Weather Watches and Red Flag Warnings are disseminated through National Weather Service and fire agency communication networks.

Table 13.7 Fire Weather Watches and Red Flag Warnings

Fire Weather Watch
- is issued 12 to 48 hours before Red Flag criteria are expected to be met or exceeded
- provides resource managers with lead time to take appropriate action

Red Flag Warning
- is issued when Red Flag criteria are met or exceeded, or if forecasts expect the criteria to be met or exceeded within 24 hours
- requires a high degree of confidence that conditions will develop
- is directly tied to firefighter safety and protection of property

13.4.4. *Extreme Fire Behavior*

During the life span of large catastrophic fires, there are usually periods of extreme fire behavior including

- fires with high rates of spread,
- intense burning as indicated by long flame lengths and high heat releases,
- spread of the fire by long-distance spotting,
- large acreages consumed by crown fires,
- the presence of fire whirls, and
- *firestorms*, raging fires of great intensity that spread rapidly.

Extreme fire behavior implies a level of unpredictability and danger that ordinarily precludes direct attack by firefighters. To better understand and predict extreme fire behavior, firefighters must know when a fire is controlled by its environment (in most cases by winds or slope angles) and when a fire is controlled by its own convective circulation. *Wind-driven fires* are much more common than *plume-dominated fires* (also called *convection-dominated fires*).

Table 13.8 Red Flag Criteria Matrix[a] for Issuing Warnings for Foehn-Aggravated Fires in Northern California

	Sustained 20-foot wind speed			
Relative humidity (RH)	6–11 mph	12–20 mph	21–29 mph	30+ mph
Daytime min RH 29–42% and/or Nighttime max RH 61–80%				Issue a warning
Daytime min RH 19–28% and/or Nighttime max RH 46–60%			Issue a warning	Issue a warning
Daytime min RH 9–18% and/or Nighttime max RH 31–45%		Issue a warning	Issue a warning	Issue a warning
Daytime min RH <9% and/or Nighttime max RH <30%	Issue a warning	Issue a warning	Issue a warning	Issue a warning

[a]This matrix is used locally to guide decisions on issuing Fire Weather Watches and Red Flag Warnings for foehn winds. A Fire Weather Watch is issued if the criteria are expected to be met more than 16 hours from the forecast time; a Red Flag Warning is issued if the criteria are met or are expected to be met within 16 hours of the forecast time. Foehn winds include the north winds that affect the Sacramento Valley and foothills and the east winds that occur on the western slopes of the Sierra Nevada and Coast Range.

13.4.4.1. *Wind-Driven Fires.* When a fire is wind-driven, the direction and speed of spread are determined primarily by ambient wind speed and direction (figure 13.15). The convection column above the fire leans distinctly downwind and does not extend to great heights. Burning embers lofted within the convection column and then deposited ahead of the fire are major contributors to rapid spread rates. Spotting distances are commonly $\frac{1}{8}$–$\frac{1}{4}$ mile (0.2–0.4 km), but during extreme fire conditions, spotting can occur at distances more than one mile from the fire. Spread rates in a wind-driven fire are usually 1–7 miles per hour (1.6–11.2 m/s) but can be faster on steep slopes. A wind-driven fire is elliptical in shape, with the long axis of the ellipse parallel to the wind direction. Although a wind-driven fire is dangerous, wind direction and speed can usually be predicted with a fair degree of accuracy, and firefighters can be prepared for the resulting fire spread. Fire behavior models handle these projections reasonably well (Rothermel, 1983).

13.4.4.2. *Plume-Dominated Fires.* A fire that is convection- or plume-dominated, sometimes referred to as a *blowup fire,* is extremely dangerous because of the possibility of abrupt changes in the fire's spread rate or direction. Firefighters at plume-dominated fires have noted indrafts into the column from every sector of the fire (Goens, 1998). Convection columns are usually well developed, reaching heights of more than 20,000 feet and resembling cumulonimbus clouds. Plume-dominated fires are characterized by extensive vertical development (figure 13.16). Spotting is profuse and occurs in all directions, but generally at distances of less than $\frac{1}{4}$ mile (400 m). Fire whirls often form on the fire perimeter. The fire often spreads in "pulses" (Rothermel, 1991).

The behavior of wind-driven fires can usually be predicted based on wind direction and wind speed.

Plume-dominated fires can produce a cumulus or cumulonimbus cloud, sometimes referred to by firefighters as *pyro-cumulus* or *pyro-cumulonimbus.* The latter can produce powerful downburst outflows (Goens, 1998). The presence of virga (section 8.4) or precipitation from the cell or convection column is often an indicator of a potential downburst. Often a calm develops when the indraft into the fire stops just be-

Figure 13.15 Wind-driven fire on the Coconino National Forest, 18 May 1996. The Horseshoe Fire grew to 8650 acres (3460 hectares) as a result of southwest winds gusting to over 50 mph (22 m/s). (Photo © A. Farnsworth)

fore the outflow from the cell begins (Rothermel, 1991; Goens, 1998). At the Summit Fire in eastern Oregon in August 1996, firefighters reported light rain and thunder accompanying the downburst winds.

Near-record or record temperatures, an unstable atmosphere, and a high LASI are indicators that conditions may be favorable for a plume-dominated fire. Rapid development of the convection column over the fire or nearby thunderstorms indicates a potential for development of a plume-dominated fire. Firefighters on the fire line on plume-dominated fires cannot see all that is developing above them. It is critical that observers stationed far enough away to monitor the convection column or other indicators of unstable conditions advise firefighters on the fire line.

Figure 13.16 Conditions in August 1996 at the Summit Fire on the Umatilla National Forest in eastern Oregon led to plume-dominated fire behavior on two consecutive days. Light rain, thunder, lightning, and strong downburst winds developed from the convection column. (Photo © A. Farnsworth)

Plume-dominated fires can produce a cumulonimbus cloud, which can generate powerful downbursts. Plume-dominated fires can abruptly change direction and rate of spread.

13.5. Prescribed Fire and Smoke Management

Management-ignited prescribed fires used by land resource agencies to meet specific ecological objectives must be carefully planned to ensure that the fires burn under control and that smoke from the fires has minimal impact on air quality and visibility. The written prescription for a fire is based on a site-specific analysis, which includes logistical information about the area to be burned. The area's size and outline, the availability of external fire lines, internal modifications that are necessary within the area to construct fire lines or manipulate the fuels (by piling, for example), and the availability of facilities within the area, such as weather instruments, helicopter pads, water sources, and access lanes, all affect program efficiency and firefighter safety. Meteorological conditions, especially wind speed, wind direction, and atmospheric stability, must be specified in the prescription because they affect not only fire behavior, but also smoke dispersal.

Smoke management combines the use of meteorology, fuel moisture, fuel loading (availability of burnable fuel), suppression, and burning techniques to keep the impact of smoke within acceptable limits. Good air quality is maintained when the atmosphere is unstable and the winds are sufficient to disperse smoke into a large volume of air. The written prescription for a fire must address the following issues:

- the objectives and expected benefits of the fire
- environmental or legal constraints
- smoke management and air quality standards
- the safety of personnel conducting the burn
- the health, safety, and property of others who may be directly affected by the fire
- the potential for off-site impact on visibility and public health
- costs
- possible alternatives

In conducting a prescribed fire, management can control three variables: the location of the fire, the time when the fire is ignited, and the *ignition pattern* or *firing pattern* used (figure 13.17 and table 13.9). The

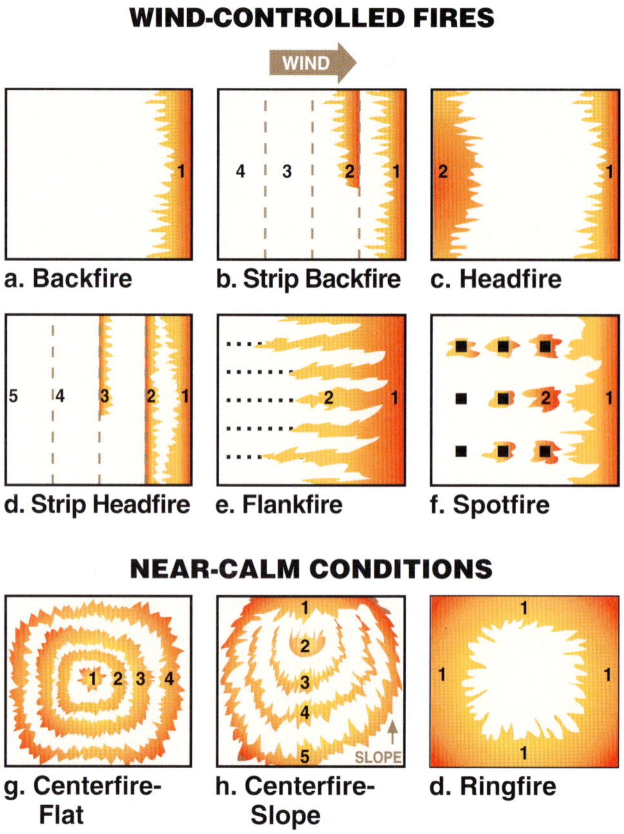

WIND-CONTROLLED FIRES

WIND

a. Backfire
b. Strip Backfire
c. Headfire
d. Strip Headfire
e. Flankfire
f. Spotfire

NEAR-CALM CONDITIONS

g. Centerfire-Flat
h. Centerfire-Slope
d. Ringfire

Figure 13.17 Firing or ignition patterns for prescribed fires. Patterns (a)–(f) are for wind-driven fires, whereas patterns (g)–(i) are used only under low wind speed conditions. The numbers indicate the order of ignition, and the dashed lines indicate internal fire lines. (Adapted from Barney et al., 1984)

fire must be ignited within a time frame determined by the ecological objectives of the burn. Within this time frame, a specific time is selected based on weather and fuel conditions. A firing pattern is then chosen to produce a fire with a rate of spread and an intensity that will have the desired effect on vegetation and minimize smoke production.

13.5.1. *Air Quality and Safety Concerns*

Smoke from a prescribed fire (figure 13.18) must be managed to meet all federal, state, and local air quality regulations (chapter 12). Prescriptions for management-ignited prescribed fires in mountainous terrain must consider the three components of the fire environment: the amount of fuel that will be burned, which impacts smoke production, the influence of topography on smoke trajectories, and the effect of weather on fuels and smoke.

Smoke from a prescribed fire must be managed to meet all federal, state, and local air quality regulations.

Prescriptions must specifically address the impact of emissions on ambient air pollution and Nonattainment Areas, EPA-designated Class I Areas, and smoke sensitive areas (chapter 12). Agencies can use small test fires coupled with visual observations to ensure that the smoke plume does not affect smoke-sensitive areas. These observations should be included in the documentation of the fire. It is also important to anticipate public concerns. Public acceptance of an active prescribed fire program is contingent upon effective smoke management.

FIRING PATTERNS

The firing ignition patterns shown in figure 13.17 are defined (Barney et al., 1984) as follows:

- A *backfire* is ignited across the leeward edge of an area and burns into the wind. It produces the lowest intensity fire and requires the most time to burn an area. It is usually applied as a strip backfire in which the fire is set along internal transverse fire lines.
- A *flankfire* is ignited in parallel lines leading to windward from the leeward edge. It burns an area more rapidly and at higher intensity than a backfire. (Backfires and flankfires are often used to secure the leeward edges of an area to be burned before other firing patterns are ignited to ensure against fire escapement.)
- A *headfire* is ignited along a line perpendicular to the wind direction and runs to leeward. It is usually applied as a strip headfire as a precaution against wind change. It covers an area more rapidly and at higher intensity than the preceding two patterns.
- A *centerfire*-flat is ignited at a chosen spot on flat terrain and then in concentric circles, a spiral, or other appropriate shape. This type of fire results in strong central convection, inward spread, and high fire intensity. The firing pattern is modified on slopes to account for the faster speed of fire propagation up the slope (centerfire-slope). A *ringfire* is a variant of the centerfire.
- A *spotfire* is ignited by firing equally spaced spots in rapid succession or using simultaneous ignition. The spacing is chosen so that the fires draw together rather than making individual runs.

Individual firing patterns are chosen to meet differing management objectives and have the advantages and disadvantages listed in table 13.9.

Smoke impacts not only air quality, but also operational safety and firefighter health. Careful consideration should be given when establishing the locations of incident command posts and air operation bases (figure 13.19) to ensure the posts are usable, even if overnight inversions trap smoke from the fire. Reduced visibility due to smoke is a hazard for both fire and civilian vehicular traffic. Smoke from one prescribed fire can impact an adjoining prescribed fire operation, possibly causing delays or cancellation.

13.5.2. *Meteorological Conditions and Smoke Management Strategies*

Most management-ignited prescribed fires have two phases. The active phase is usually completed several hours after ignition; the smoldering phase can last for several days or longer, depending on how thoroughly fuels were consumed during the active phase. Prescribed natural fires may cycle repeatedly through these two phases. Smoldering combustion generates three to five times more smoke than active burning (Pyne et al., 1996). Meteorological conditions, particularly wind speed and direction

Table 13.9 Advantages and Disadvantages of Different Firing Patterns for Prescribed Fires

Technique	Where used	How Done	Advantages	Disadvantages
Backfire	Under tree canopy, in heavy fuels near fire lines	Backfire from downwind line; may build additional lines and backfire from each line	Slow, low intensity, low scorch, low spotting potential	Expensive, smoke stays near ground, the long time required may allow wind shift
Flankfire	Clear-cuts, brush fields, light fuels under canopy	(1) Backfire from downwind line until safe line created (2) Several burners progress into wind and adjust their speed to give desired flame	Flame size between that of backfire and headfire, moderate cost, can modify from near backfire to flankfire	Susceptible to wind veering; good coordination among crew necessary
Headfire	Large areas, brush fields, clear-cuts, under stands with light fuels	(1) Backfire from downwind line until safe line created (2) Light headfire	Rapid, inexpensive, good smoke dispersion	High intensity, high spotting potential
Strip headfire	As for headfire, also partial cuts with light slash under tree canopies	(1) Backfire from downwind line until safe line created (2) Start headfire at given distance upwind (3) Continue with successive strips of width to give desired flames	Relatively rapid, intensity adjusted by strip widths, flexible, moderate cost	Need access within area; under stands having three or more strips burning at one time may cause high intensity fire interaction
Centerfire	Clear-cuts, brush fields	(1) For centerfire, light center first (2) Ring is then lighted to draw to center; often done electrically or aerially	Very rapid, best smoke dispersion, very high intensity, fire drawn to center away from surrounding vegetation and fuels	May develop dangerous convection currents; may develop long-distance spotting; may require large crew
Spotfire	Large areas, brush fields, clear-cuts, partial cuts with light slash, under tree canopies; fixed wing aircraft or helicopters may be used	(1) Backfire from downwind line until safe line is created (2) Start spots at given distances upwind (3) Adjust spot to give desired flames	Relatively rapid, intensity adjusted by spot spacing, can get variable effects from headfires and flankfires, moderate cost	Access within area necessary if not done aerially

From Martin and Dell, 1978.

and mixing depth, affect the dispersion and transport of smoke produced during both phases.

There are three strategies for minimizing the impact of smoke from a prescribed fire: avoidance, dilution, and emission reduction. Avoidance and dilution strategies depend almost entirely on meteorological conditions, specifically wind speed and direction and mixing depth. Emissions can be reduced by combining favorable weather conditions with ignition techniques.

Figure 13.18 (*top*) Convective plumes from prescribed fires in the Coast Range of western Oregon. (Oregon Department of Forestry photo, provided by Mike Ziolko)

Figure 13.19 (*bottom*) Fire managers should consider the potential impacts of smoke when locating incident command posts and helicopter bases. (Photo © A. Farnsworth)

Avoidance practices minimize smoke impact by scheduling fires for times when winds carry smoke away from sensitive areas, such as highways or built-up areas. Winds affect the trajectory of smoke from a fire during both active burning and smoldering. During the active burning phase of a large prescribed fire, a convection column may be generated. The height of the column depends on the intensity of the fire. If the column is higher than topographical features that may trap smoke, the wind profile through the height of the column determines the smoke's trajectory. If the prescribed fire burns through sparse fuels, or if the moisture content of fuels is high, the fire may not produce sufficient heat energy to generate a convection column. In this case, surface winds play the largest role in dispersing smoke. Surface winds are also the primary determinant for the smoke trajectory during the smoldering phase when the convection column dissipates and smoke remains near the ground, generally following topographic features. If the smoldering phase lasts through the night, downslope and down-valley winds transport the smoke. These winds are usually light, but they can carry smoke great dis-

Prescribed natural fires may cycle repeatedly through active burning and smoldering phases. More smoke is generated by smoldering than by active burning.

tances. Visibility at the surface can be significantly reduced and the hazard for vehicular traffic significantly increased as a result of higher nighttime humidity, the hygroscopic properties of the aerosols, and the concentration of smoke in the shallow downslope and down-valley layers. Burning should be avoided when stable layers are present because smoke could be trapped in a small volume of air. Normal diurnal variations in stability in complex terrain (strong convection during the day, the formation of inversions at night) can be incorporated into an avoidance strategy to minimize the impact of smoke.

Winds affect the trajectory of smoke from a fire during both active burning and smoldering.

The dilution of smoke concentrations depends on the stability of the lower atmosphere. Maximum dilution is ensured if management-ignited fires are scheduled so that the greatest smoke production occurs when the atmosphere is deep and unstable. Afternoon periods when strong heating creates deep mixed layers and the lowest fuel moisture are generally favorable for burning. Nighttime and early morning periods are generally unsuitable unless cloud cover or strong winds prevent the formation of nocturnal inversions. (In addition, burning conditions are poor during the night and in the early morning because of moderately high near-surface humidity.) Weak inversions or shallow stable layers can be overcome if fire intensity is high. An intense fire results in rapid fuel consumption and significant buoyancy within the convection column above the fire, which can carry the smoke beyond the top of the mixed layer. Buoyancy weakens during the later stages of the prescribed fire when combustion rates decrease.

Maximum dilution is ensured if management-ignited fires are scheduled so that the greatest smoke production occurs when the atmosphere is deep and unstable.

Emissions can be reduced by timing ignition to take advantage of meteorological conditions and by using ignition techniques that reduce smoke production (section 13.5). Fires can be ignited during the warmest and driest time of day, and rapid firing methods can produce a fire that consumes large acreages in a short time period. In addition, the fire can be forced to spread downslope, using either the wind or firing patterns, so that it consumes more fuel during the active phase, leaving less fuel to smolder. Burning under light surface wind conditions or burning on level terrain can also increase fuel consumption by maintaining slow spread rates. Smoke production is significantly reduced if the prescribed fire is in a location that has been previously burned (figure 13.20).

Figure 13.20 Prescribed burning on the Coconino National Forest, Arizona, November 1995. This area was also burned under prescription in the fall of 1989 and 1992. The quantity and duration of smoke was greatly reduced on the second and third prescribed burns. (Photo © A. Farnsworth)

Guidelines for implementing the three strategies to reduce the impact of smoke from prescribed fires are given in table 13.10.

13.5.3. *Smoke Management Forecasts*

Smoke dispersal forecasts provide fire managers with meteorological information pertinent to maintaining air quality standards. The most common parameters used to assess dispersion potential are mixing depth and the average wind in the mixed layer, called the *transport wind*. Mixing depth, the depth of the convective boundary layer during daytime, and the depth of shear-induced mixing during nighttime, was discussed previously in sections 4.4.4 and 12.4.2. The greater the mixing depth and the stronger the transport wind, the better the smoke dispersal.

There are a number of simple schemes to forecast local mixing depths. The Miller–Holzworth method (Holzworth, 1972) is illustrated in figure 13.21. Afternoon mixing depth is estimated by extrapolating the fore-

Table 13.10 Guidelines for Reduction of Smoke Impacts from Prescribed Fires

Define the burn's objectives.
Obtain and use weather and smoke management forecasts.
Don't burn during pollution alerts or stagnant conditions.
Know and comply with air pollution control regulations.
Burn when conditions are good for rapid dispersion.
Use caution when near or upwind of smoke-sensitive areas.
Use caution when smoke-sensitive areas are down the drainage path.
Estimate the amount and concentration of smoke you expect to generate.
Notify your local fire control office, nearby residents, and adjacent landowners.
Use test fires to confirm smoke behavior.
Use backing fires when possible.
Burn during the middle of the day.
Consider burning in small blocks.
Do not ignite organic soils.
Be very cautious of nighttime burning.
Anticipate down-drainage smoke drift.
Mop up along roads.
Have an emergency plan.

From Wade, 1989.

cast afternoon surface temperature upward using the dry *adiabat* until it intersects the morning temperature sounding. The height from the surface to the intersection is the estimated afternoon mixing depth. This method assumes that there are no major air mass changes from morning to afternoon and that cold or warm air advection are negligible. It is therefore an appropriate forecast tool only during periods of static weather patterns. Once a forecast mixing depth is obtained using the Miller–Holzworth method, routine wind forecasts from the National Weather Service can be used to estimate the transport wind in the mixing layer.

Some state air regulatory agencies use a *ventilation index*, which is the product of the mixing depth and the transport wind speed, in determining whether to issue burning permits:

$$\text{ventilation index (m}^2\text{/s)} = \text{mixing depth (m)} \times \text{transport wind speed (m/s)}$$

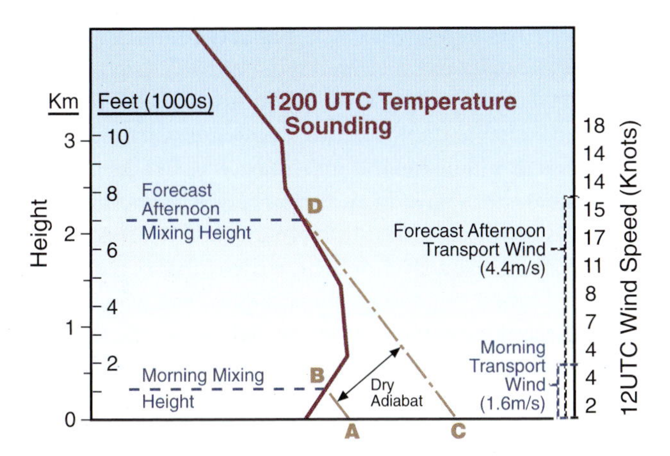

Figure 13.21 Miller–Holzworth method for forecasting the afternoon mixing depth.

The most common parameters used to assess dispersion potential are mixing depth and transport wind.

Mixing depth and wind guidelines have also been established by some land management agencies for prescribed fire operations. For example, the guidelines for burning in mixed grass prairies in the northern Great Plains call for mixing depths of 2000 feet (610 m) or greater, surface inversions breaking before noon, and surface winds of 5–20 mph (2–9 m/s).

Dispersion models (section 12.2) can also be used to forecast smoke dispersal. Wildland smoke dispersion models have been developed specifically to make initial estimates of smoke dispersal for planning prescribed fires, addressing not only dispersion and transport of smoke, but also smoke emissions from the burning of wildland fuels (Breyfogle and Ferguson, 1996).

13.6. Monitoring Fire Weather and Smoke Dispersion Parameters

Weather observations are essential for predicting wildland fire behavior and for monitoring prescribed burns. The fire manager determines when and what weather data are needed to accomplish the task at hand safely and efficiently.

The first step in a good monitoring program is to study past fire weather and fire behavior for the geographic area of interest and to establish a climatological and historical reference (including frequency of fires, time since the last fire, maps of past fires, and weather records) for fire management decisions. Climatological data and historical records of fires are also helpful in making long-range forecasts of fire weather and fire behavior.

The second step is to collect data on-site during the fire to build a detailed data base for predicting fire behavior. The location at which observations are made should be representative of the fire but not influenced by it. Observations should be as current as possible, and the more observations from the same site, the better. On-site weather observations for wildland fires should include the state of the weather, dry bulb temperature, wet bulb temperature, eye-level or 20-foot wind speed, and direction and remarks that note any significant past weather or weather observed at a distance.

A good fire weather monitoring program includes climatological data and data collected on site during the fire.

On-site data may include observations from fire lookout towers and guard stations or from manual fire weather stations. Manual data from fuel sticks (sets of wooden dowels that mimic the humidity uptake of 10 hour fuels) are especially useful. Fixed wing or rotor aircraft can provide vertical temperature profiles over the fire area to determine inversion heights and atmospheric stability, estimate winds aloft, and provide visual observations of plume height and trajectory. In addition, a number of new instruments that have been developed over the past decade are available at a reasonable cost to monitor and record weather, to sample fuels, and to document fire behavior. These instruments take much of the guesswork out of predicting fire weather and fire behavior. The large volume of data collected must be managed so that the decision maker is not overwhelmed, and precise records must be kept, not only to document the fire, but also to provide data for the development of more effective means of managing future fire projects.

Modern communication networks have greatly expanded the information available to decision makers in the field and allow information to be relayed quickly to dispatch and weather forecast offices. Cell phones, laptop computers, the Internet, faxes, and the NOAA Weather Radio all provide quick and up-to-date weather information that can increase the margin of safety for firefighters. (Although NOAA radio does not broadcast fire weather forecasts, it does provide supplemental weather information that can be useful to field operations.)

Assessments and observations taken for smoke management should include the following:

- the timing of the previous day's inversion breakup (if an inversion is present)
- estimates of inversion height (to be made before ignition)
- observations of convection column development, height, and direction of plume (to be made during ignition)
- observations of visibility both under and above inversions
- assessment of potential impact on Class I Areas and visibility under the current or forecast weather regime
- identification of areas vulnerable to downwind smoke impacts on visibility, especially highways

13.6.1. *Manual Field Measurements*

The most common field instrumentation carried by firefighters to monitor the weather and to obtain accurate field readings is the *belt weather kit*, which includes a wind meter and a sling psychrometer (figure 13.22). A number of tools are available in forestry supply catalogs to supplement the belt weather kit, including digital maximum and minimum hygrothermometers that have a constant display of temperature and relative humidity, as well as electronic moisture meters for measuring fuel moisture contents. Some of these moisture meters, however, are inaccurate at moisture readings lower than 12%.

Figure 13.22 A standard-issue belt weather kit contains a Dwyer wind meter, a sling psychrometer, a container for distilled water, psychometric tables, and a compass. (Photo © A. Farnsworth)

Figure 13.23 Releasing a pilot balloon to determine the speed and direction of winds that affect a fire. (Photo © A. Farnsworth)

Other useful items include:

- a pair of compact binoculars for observing fire behavior, terrain, fuels, and weather features;
- global positioning system units for plotting field locations and fire shapes;
- a clinometer for determining the steepness of slopes and the heights of trees; and
- a small 35-mm camera or video camera for documenting fire weather and fire behavior for future reference or training purposes.

13.6.2. *Pilot Balloons*

Small latex *pilot balloons*, or *pibals*, are often released and tracked visually near wildfires or prescribed fires (figure 13.23). As the balloons ascend, they provide a vertical profile of the wind structure above the release site. In remote locations far from the National Weather Service's network of upper air stations or wind profilers, pilot balloon information is key in assessing winds aloft that could impact fire behavior or smoke trajectories.

A 10-gram pilot balloon is used to measure wind profiles. The balloon is inflated with helium until it just lifts an object weighing 44 grams. If the balloon is then tied off and released (without the 44-g weight), it will ascend at a rate of about 500 feet per minute (2.5 m/s). By tracking the balloon and measuring the elevation and azimuth angles for each minute of flight, wind speed and direction profiles can be computed. These measurements are usually taken with a theodolite, and the data are processed by a computer program. For a prescribed fire where trained theodolite operators are unavailable, balloons can be released and tracked visually to estimate winds aloft before the fire is ignited. The pilot balloon can provide approximate smoke drift information during marginal "go" and "no-go" days or on prescribed fires close to sensitive areas, such as Class I (section 12.1.2) or urban areas. Each balloon release should be recorded and included in the documentation of the prescribed burn.

A pilot balloon can also be used to measure the height of a smoke plume. The balloon is released downwind and away from the main convection column, and its ascent is timed until it disappears into the smoke. The plume height is calculated by multiplying the number of minutes by the ascent rate of about 500 feet per minute (2.5 m/s). This information should also be included in the documentation of the fire.

13.6.3. *On-Site Upper Air Soundings*

In some cases, detailed upper air soundings are needed for wildfire or prescribed burns. Field-deployable instruments that provide higher resolution sampling of the upper atmosphere at remote sites include sodars, aircraft sondes, and minisonde systems. Sodars, also called acoustic sounders, are remote sensing instruments that measure wind speed and wind direction profiles by emitting soundwaves that are scattered back from the atmosphere at heights below 3000 feet (about 1000 meters). Sodars have been used operationally on fires in the southwestern United

States (Svetz and Barnett, 1997). Aircraft sondes use a global positioning system and an instrument package affixed to the wing of an aircraft. Temperature, dew point, and wind profiles are sampled up to the operating limits of the aircraft. Minisonde systems use an inexpensive disposable balloon instrument package that radios temperature and humidity information to the ground as it ascends. The balloon is tracked and winds are calculated using the same method described for pibals (section 13.6.2).

13.6.4. *Remote Automatic Weather Stations*

Land management agencies have developed a substantial network of Remote Automatic Weather Stations (section 9.3.1) that provide hourly weather and fuels information. Figure 13.24 shows a RAWS unit, and figure 13.25 shows the distribution of RAWS across the country. The RAWS data can be accessed through the USFS Weather Information Management System or through the Western Region Climate Center in Reno, Nevada.

Several different types of portable weather stations with sensors mounted on short towers or masts (like the RAWS network stations) can be deployed at temporary sites to gather and transmit data via radio, cellular phone, or satellite broadcast directly to fire managers. These stations are often erected to collect data during prescribed fires or during persistent wildfires.

Figure 13.24 Oak Creek, Arizona, Remote Automatic Weather Station (RAWS) platform. (Photo © A. Farnsworth)

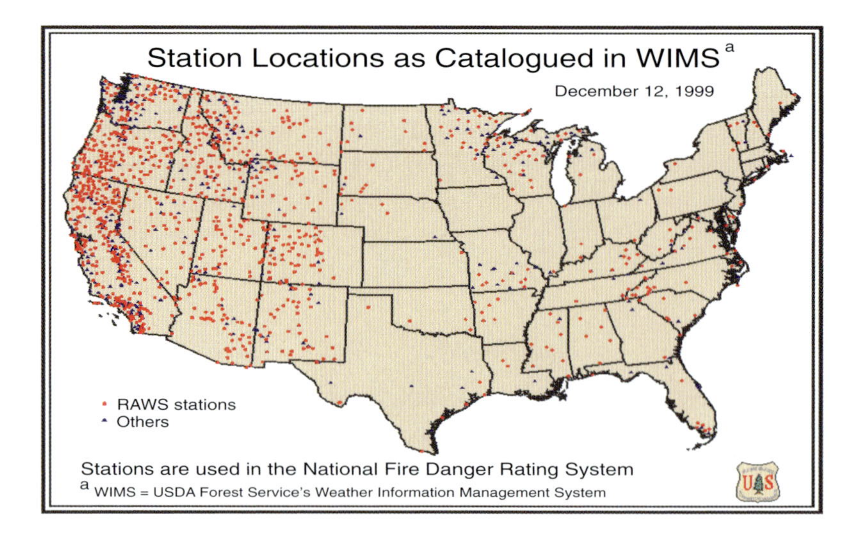

Figure 13.25 Remote Automatic Weather Stations are distributed widely across the country and are used for fire weather and other applications.

13.6.5. *Lightning Detection Networks*

Since the 1970s, lightning strokes have been detected remotely from networks of sferics-sensing instruments (section 8.7.3). These networks have replaced the practice of flying aircraft behind thunderstorms to detect lightning. Near real-time lightning data are very important to agencies responsible for responding to wildland fires. In the United States, approximately 10,000 fires are started annually by lightning. In many parts of the western states and Alaska, lightning is the major cause of wildfires.

13.6.6. *Doppler Weather Radar*

The National Weather Service operates a network of Doppler weather radar sites (figure 13.26) that monitor the atmosphere 24 hours a day. The sensitive radars in the NEXRAD Network are used to analyze internal storm motions, track wind motions, estimate rainfall intensities, and even detect convection columns from nearby wildfires. Data from the radar network are used to update fire managers on the current status of thunderstorm development, outflow boundaries, and gust fronts from distant unseen thunderstorms and to provide information on areas of heavy rainfall accumulations. They are also used to warn land management agencies of impending flash floods where logging operations and campgrounds are vulnerable to high water. Beam blockage by surrounding mountains limits radar coverage in the western United States, especially at the lowest elevations. For example, radar coverage of the intermountain west at the 4000-ft level above the radar sites is generally less than 50%. The fractional coverage improves to about 70% at levels 10,000 ft above the radar sites. Coverage over the flatter eastern United States is nearly complete.

13.6.7. *Satellite Remote Sensing*

Sophisticated remote sensing technology has become available to land management agencies in recent years. Remote sensing by satellite has ma-

Figure 13.26 National Weather Service Doppler radar on Virginia Peak, 47 miles (76 km) northeast of Reno, Nevada, at 8299 ft. (2530 m) elevation. The radar antenna is shown. Electronic preprocessing of the radar data is handled on site, and the data are then transmitted to the forecast office, where they are displayed and processed further. (NWS Western Region photo)

tured to the point that fuel inventories and the moisture status of the living plant cover can be estimated. A satellite-based comparison of the emissivity of the vegetation in the near infrared and the visible red spectrums provides information about the viability of the green living material in plants (*greenness*). Observations of vegetative greenness can assess the impact of drought on living plants when fuels have cured and provide an estimate of broad area fire potential. The Normalized Difference Vegetation Index (NDVI) is one of the most common derived vegetation greenness indices from remotely sensed data. Analyzing current and historical NDVI values can be a tool to compare real-time vegetative conditions with those of past historical drought years.

Both geostationary and polar orbiting satellites have the capability to detect fires in remote locations. The GOES (Geostationary Operational Environmental Satellite) geostationary and NOAA (National Oceanic and Atmospheric Administration) polar orbiting satellites offer high resolution and high sensitivity both in the visible and near infrared parts of the spectrum. GOES visible images with 1-km resolution are available every 15 minutes and permit meteorologists to locate and track smoke plumes from large fires. The GOES near infrared (3.9 micrometers) imagery permits detection of individual hot spots created by active burning. The detectable size depends on the heat signature of the fire. Extremely hot fires as small as 2 acres (0.8 hectare) can be detected under ideal conditions, and fires larger than 100 acres (40 hectares) can be determined routinely (Weaver and Purdom, 1995).

13.6.8. *Obtaining Fire Weather Forecasts*

The National Weather Service's Fire Weather Program provides a complete suite of forecast products that are used to assess general fire danger and produce site-specific forecasts for fire behavior. Routine fire weather forecasts are generally produced twice daily, once in the morning and again in the afternoon during the fire season. These are narrative forecasts that contain a weather discussion and forecasts for two days with extended outlook. Transport and stability information for smoke dispersal may also be included. Information on 24-hr weather trends used to calculate National Fire Danger Rating indices are available in the afternoon through the USFS's Weather Information Management System and through the Internet homepages of most National Weather Service forecast offices.

Site-specific forecasts are made by request. The resource manager must first take a representative observation at or near the fire site (state of weather, sky cover, temperature, relative humidity, eye-level winds) at a location where the observation is not influenced by the fire. If unique weather elements must be forecast, a current observation of that element should be given to the forecaster to provide a calibration point for the forecast. A map of the project or wildfire, if it can be transmitted to the forecaster, provides valuable information that may improve the forecasts. The meteorologist makes a forecast that takes into account the elevation, aspect, and other topographic or mesoscale influences at the fire site. The forecast usually concentrates on the next 12-hour period, but may also include-longer range outlooks. The forecast is relayed back to the fire site, where it may be accepted by on-site personnel for input into a fire behavior forecast or a go/no-go decision on a prescribed burn or spray operation. Sometimes a meteorologist is requested to be on site during dangerous wildfires or large prescribed burns. The meteorologist then sets up weather observation networks and directly provides site-specific forecasts to the resource manager or fire personnel.

Routine fire weather forecasts are generally produced twice daily, once in the morning and again in the afternoon, during the fire season. Site-specific forecasts are made on a request basis.

14

Aerial Spraying

HAROLD W. THISTLE
JOHN W. BARRY

Aircraft are used in a number of resource management operations, including fire suppression, seeding and fertilizing operations, and the application of pesticides to agricultural, forest, and rangelands (figure 14.1). The objectives of any aerial application are to apply the material, either liquid or solid, to the target area safely, efficaciously, and economically and to avoid *drift*, that is, off-target displacement of the agents. Barry (1993) is a general reference for aerial spraying of forests. Picot and Kristmanson (1997) provide an overview of all aspects of this topic. Bache and Johnstone (1993) give a detailed description of spray meteorology. The emphasis in this chapter is on the role of meteorology in the aerial application of liquid pest control agents to manage plant, fungal, and animal pests in mountainous forested areas.

The effectiveness of a spray operation depends on the timing of the operation relative to phenological conditions, the characteristics of the forest canopy or rangeland being targeted, the spray formulation, pilot skills and attitude, the aircraft type and spray equipment used, and weather conditions. Pest control agents are regulated by federal, state, and local agencies. Restrictions on the use of agents are specified on the product label and may include weather conditions.

Weather is one of many factors that affect the effectiveness of a spray operation. Other factors include phenological conditions, forest canopy characteristics, the spray formulation, the spray equipment, aircraft type, and pilot skills and attitude.

Drift reduces the efficacy of a spray operation and can have unintended and undesirable impacts on nontarget species, residences, and public areas near the target area. Although there is a driftable component in every spray operation, the drift potential is generally greater for liquids than for solids because the size of liquid droplets becomes smaller after release into the atmosphere, depending on the volatility of

Harold W. Thistle is with the U.S. Forest Service, Missoula Technology and Development Center, Missoula, Montana. John W. Barry is retired from the U.S. Forest Service in Davis, California.

Figure 14.1 An example of a forest aerial spraying operation. In this case, a biological insecticide is being sprayed from a Bell JetRanger helicopter to control a gypsy moth infestation in Utah in 1993. (USDA Forest Service photo)

the substance itself, the aircraft and spray equipment used, and the meteorological conditions at the time of spraying. The smaller the droplets, the greater the potential for drift. Weather conditions have a significant impact on drift because wind speed and direction, temperature, humidity, and atmospheric stability affect the transport, diffusion, evaporation, settling, and deposition of both solid particles and liquid droplets.

The collection of meteorological data and the use of professional weather forecasts are thus an integral part of a spraying operation. Using the meteorological principles discussed in previous chapters, the project manager can evaluate the information available and integrate it into the operational plan and field operation.

It is difficult to make general statements about the most effective way to achieve a desired coverage or to minimize drift because aerial spray operations in complex terrain vary markedly in objective and approach. For example, a program could involve the application of herbicide with a desired droplet diameter of 700 micrometers to achieve rapid deposition. A mosquito control operation, on the other hand, may have a desired droplet diameter of 70 micrometers to achieve drift so the droplets will encounter mosquitoes in flight. In certain applications, solid particulate or granular material is used, whereas in others, highly volatile liquids may be required. The project manager must evaluate the objectives of the operation and all of the conditions that affect spray operation effectiveness.

A TYPICAL SPRAY OPERATION SCENARIO

Spraying operations in forested areas are typically conducted in the early morning during the growing season when wind speeds are low, humidity is high, and temperatures are relatively cool. A typical schedule for a spray operation (times are local) follows.

1700 hours: Area weather forecasts for 72 hours, 24 hours, and 12 hours and site-specific weather forecasts for 12 hours are provided to the project manager. Longer lead-time forecasts allow the manager to anticipate antecedent conditions and to begin planning for the next operational phase.

1730 hours: A weather briefing is provided for the spray team, and the project manager makes the go/no-go decision.

0200–0500 hours: Weather observers, spray deposition monitors, and aircraft crews are deployed.

0600–1000 hours: Spray operations are conducted.

1000–1400 hours: Field work is completed, the operation is critiqued, and preparations are made for the next day's operations.

14.1. Overview of Aerial Spraying

Forest aerial spraying is conducted to meet a variety of management goals, including pest or vegetation control, fertilization, defoliation, and control of diseases. The problem is evaluated initially by entomologists, pathologists, silviculturalists, foresters, or other specialists in field investigations. A control strategy is then developed and a schedule is established that takes phenological considerations into account and defines the weather conditions required for the operation.

Aerial spraying is used to address a wide variety of forest management goals, including pest or vegetation control, fertilization, defoliation, and control of diseases.

14.1.1. *Phenological and Meteorological Time Constraints*

Whether applying seed, fertilizer, or pest control agents, the life cycle of the plants or insects being targeted places time constraints on the operation and can limit spraying to a short period of time. For example, the western spruce budworm larva is susceptible to the control agent after it has emerged from the bud. If the larva is still in the bud, it will not be exposed to the pest control agent.

Meteorological conditions also influence the timing of a spray operation. For example, seeds or fertilizer may require precipitation within a certain time period before or after the application. Heavy precipitation after spraying of a pest control agent, however, can dilute the agent or wash it off the foliage.

14.1.2. *Materials Sprayed*

Pest control agents applied by aircraft are usually chemical (organic or inorganic) or biological (e.g., microorganisms or invertebrates) agents formulated to be applied dry or in a liquid carrier, typically water. Liquid sprays are applied in amounts of 0.2 to 40 liters per hectare (0.02 to 4.3 gal per acre), whereas dry materials are applied at 0.2 to 4 kg per hectare (0.2 to 3.6 lbs per acre). Ultralow volume sprays (less than 5 liters/hectare or 0.5 gal/acre) containing smaller droplets with high concentrations of the active ingredient are preferred when applying insecticides in forestry because they are more efficacious and more economical to apply, even though they are more susceptible to drift. Aqueous herbicide mixes are usually applied in larger droplets that are less prone to drift. Some herbicides are now being formulated in dry particles to be applied in low volumes.

All liquids and liquid–solid mixtures have a *volatile fraction* that indicates how much of the material could evaporate. Droplets released in a spray operation become smaller through evaporation as they move through the atmosphere. Evaporation is enhanced under dry atmospheric conditions. Mineral oil carriers are sometimes used when water evaporation is expected to be a problem but oil carriers have the potential for more distant drift. There are commercially available evaporation inhibitors. These materials generally work as surfactants, forming a surface layer of less volatile material on the droplet surface. Evaporative inhibitors have met with mixed success and must be investigated and tested before being used in a given setting. Although dry particles do not evaporate, drift can be a problem during dry applications because the fine particulate associated with the dry materials is driftable and may be toxic or allergenic.

14.1.3. *Aircraft and Spray Equipment*

Both rotary-wing aircraft (helicopters) and fixed-wing aircraft are used in aerial spray operations. Helicopters are usually preferred when the target area is in complex terrain because of the greater maneuverability required when the canopy or ground is uneven. The helicopter flight paths typically follow contours in these settings. When fixed-wing aircraft are used in valleys, they are usually flown in quasi-horizontal flights over the sidewalls parallel to the valley axis, rather than up or down sidewalls, for reasons of performance, safety, and efficiency of coverage. Figure 14.2 shows several examples of fixed-wing and rotary-wing aircraft used commonly in forest spraying operations.

Helicopters are usually preferred in complex terrain because of the greater maneuverability required when the canopy or ground is uneven.

The aircraft designed for liquid spraying are fitted with pressurized tanks, booms, and nozzles. Liquid pest control agents are sprayed under hydraulic pressure through nozzles with restrictive orifices or through rotary nozzles capable of fine atomization. Droplet size can be controlled by adjusting the pressure and the size of the orifices, the type of nozzle, and the angle of the nozzle. Generally, the smaller the orifice and the higher the pressure, the smaller the atomized droplets. Air speed across the orifice, which varies with aircraft speed, aircraft altitude, and orifice orientation relative to the airframe, reduces droplet size. For rotary atomizers, the higher the rotation rate, the smaller the atomized drops. There are many different types of aerial application systems for dispersal of dry and liquid (water or oil carrier) pest control agents, seed, and fertilizer tank mixes.

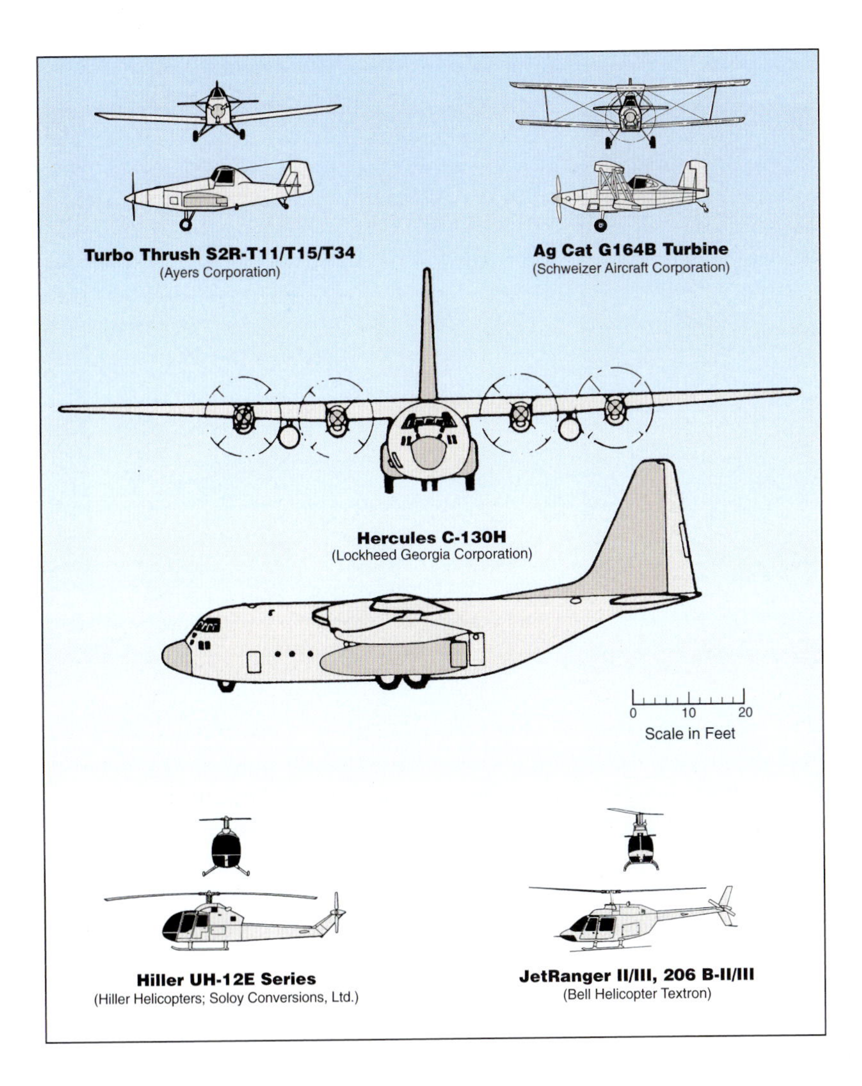

Figure 14.2 Fixed- and rotary-wing aircraft used frequently in forest spraying projects. (Adapted from Hardy, 1987)

14.1.4. *Geographic Information and Positioning Technology*

Treatment areas are chosen on the basis of field surveys and are indicated on maps. If the prescription includes use of aerial spraying, a spray strategy is established for the treatment areas, termed *spray blocks*. Aerial sprays are generally applied to forestlands and rangelands from 15 to 50 meters above the canopy. The height of the spray aircraft above the canopy influences the width of the swath and spray displacement.

The development of Geographic Information Systems (GIS) and Global Positioning Systems (GPS) over the last decade has greatly improved the quality of aircraft navigation and, thus, the uniformity of coverage of spray blocks. The first of these systems, GIS, allows digital mapping data to be quickly displayed, overlain, and analyzed by computer (figure 14.3). This provides a quasi-three-dimensional view of the spray block that is very helpful in identifying the spray block in the field and in planning the flight strategy for the field spraying operation. GIS is also used as a base on which to display the actual spray swaths flown (as determined with GPS).

The second system, GPS, is a navigation or position location system that

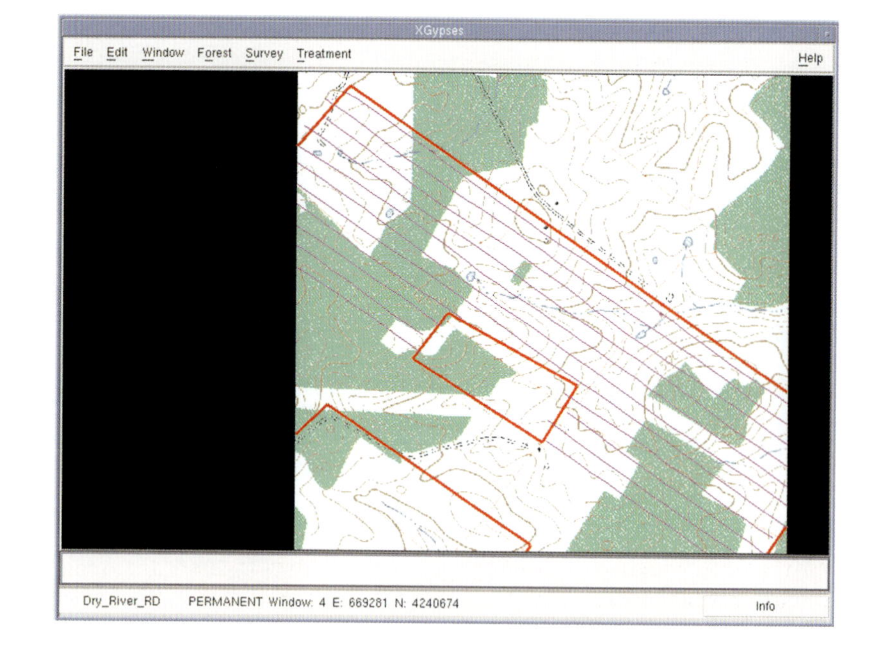

Figure 14.3 Spray block (delineated by red border with internal exclusion area) overlain onto a USGS topographic map of western Virginia. The actual flight lines (blue lines) were generated in the cockpit based on GPS input. This map is in the GypsES modeling system (Ghent et al., 1996). The spray block position can be transferred electronically to an in-cockpit GPS spray navigation system. Actual flight lines and spray lines can then be downloaded back from the aircraft to GypsES to evaluate the spraying. (USDA Forest Service photo)

uses signals transmitted from a constellation of polar orbiting earth satellites to determine position on the surface of the earth or in the earth's atmosphere. The satellite constellation is managed by the U.S. Department of Defense (DOD). DOD intentionally introduces an error into the signal available to civilian receivers. This error can be determined and removed in a differencing process known as differential GPS (DGPS), and the position of the spray aircraft can be determined in near real time. The positional accuracy available using DGPS is within 2 m (6 ft) of the actual position horizontally. This is approximately two orders of magnitude better than accuracies available with previous systems. Positional updates can be obtained at high frequency, making the technology especially suited for aircraft navigation. Errors in the vertical are larger but the use of these systems as altimeters is also possible and largely dependent on the application. With GPS, aircraft guidance has become much more accurate, spray aircraft can be monitored in real time, data on time, position, spray status (on–off), volume sprayed, etc., can be recorded for mission evaluation, and weather can be distinguished from other application variables. These data are especially important for contract evaluation and legal challenges.

GPS has improved the guidance of spray aircraft and monitoring of spray operations.

The GPS receiver must receive an unobstructed signal from four of the GPS satellites and have either a line of sight to a transmitter that broadcasts differential corrections or access to satellite broadcasts of differential corrections. Obstructions in complex terrain can block the GPS signals. However, since full deployment of the satellite constellation, all but the most severe terrain, where aerial spraying is less feasible, is covered. Satellite broadcasts of differential corrections can also be obscured by terrain. Tests in the central Appalachians indicate that signal loss due to terrain obstruction is comparable to loss due to steep turns made in aerial spraying. The blockage of correction signals is dependent on both the severity of the terrain and latitude, and it can be roughly estimated geometrically for any situation.

14.2. Meteorological Factors that Affect Aerial Spraying Operations

Droplets in the atmosphere are affected by wind fields, relative humidity, temperature, atmospheric stability, and precipitation. These parameters influence the rate at which droplets evaporate and the direction in which they are carried and are therefore an important consideration in planning a safe and environmentally sound aerial spraying project.

Droplets in the atmosphere are affected by wind fields, relative humidity, temperature, atmospheric stability, and precipitation.

14.2.1. *Wind Fields and Droplet Trajectories*

The trajectory of an individual spray droplet released from an aircraft is determined by its settling velocity, which is dependent on its size and density, and by two superimposed wind fields: the wake vortices produced by the spray aircraft and the ambient winds. The smaller the droplet, the lower the settling velocity and the greater the influence of the two wind fields.

The trajectory of a spray droplet is determined by its settling velocity, which is dependent on its size and density, the wake vortices produced by the spray aircraft, and the ambient winds.

Calculations of spray drift and droplet deposition must take many additional factors into account, including nozzle type, aircraft speed, humidity, stability, and surface roughness. Two critical considerations are that spray equipment does not produce uniform droplets but rather droplets in a range of sizes and that the droplet size spectrum changes during transport as the droplets evaporate (section 14.2.2).

14.2.1.1. *Droplet Settling Velocities.* A droplet falling through still air accelerates but quickly (often within the time it takes to fall 2 m or 6 ft) approaches a steady settling velocity that represents a balance between the gravitational acceleration of the droplet and the drag exerted on the droplet by the air. The settling velocity V_s can be estimated accurately for droplets below about 70 micrometers in diameter by the Stokes formula:

$$V_s = \frac{g\rho d^2}{18\eta}$$

where g is gravitational acceleration (9.8 m s^{-2}), η is air viscosity (1.72×10^{-5} kg m^{-1} s^{-1}), ρ is droplet density (1000 kg m^{-3} for water), and d is droplet diameter (m). Droplets larger than about 70 micrometers fall somewhat more slowly than predicted by the Stokes formula primarily because they become flattened fore and aft and thus present a greater frontal area (and therefore more drag) than a sphere of the same volume. Settling velocities for various size droplets are presented in table 14.1. A different method of estimating the settling velocity of water droplets over a broader range of sizes is given, along with equations, in appendix A.

Because the atomized material is composed of droplets in a range of sizes, an accurate description of the sprayed material should include the number of droplets in different size categories. However, this information can be difficult to determine. In practice, the droplets are often described in terms of their volume median diameter (VMD), which is a statistical measure of the median droplet size. Fifty percent of the spray volume is composed of droplets with diameters less than the VMD, and 50% is com-

Table 14.1 Settling Velocity of Water Droplets of Various Sizes

Droplet diameter (micrometers)	Settling velocity ($m\ s^{-1}$)
20	0.012
40	0.047
60	0.102
80	0.175
100	0.270
120	0.355
140	0.445
160	0.536
180	0.625
200	0.705
250	0.940
300	1.150
350	1.200
400	1.630
500	2.080

From Quantick, 1985.

posed of droplets larger than the VMD. Where possible, measures of the distribution of droplet sizes and particularly a measure of the number of small and thus driftable droplets should be determined.

14.2.1.2. *Aircraft Wake Vortices.* A photograph of smoke from a ground source entrained into a vortex generated by the wing of a spray aircraft is shown in figure 14.4. The vortices are large, organized, swirling motions that are centered near the wing tips. The paired vortices rotate toward each other and descend toward the ground. This motion can enhance deposition of the spray from low-flying aircraft.

The wake from a rotary-wing aircraft (helicopter) is a combination of a hover wake and a vortex pair. When the aircraft is hovering, the hover wake affects the column of air above and below the aircraft. The column

Figure 14.4 Smoke from a fixed source on the ground aids in visualizing the vortex created by a passing aircraft. This vortex is generated by the wing of the aircraft. (National Aeronautical and Space Administration photo)

of air influenced by the wake below the rotor expands laterally as it approaches the ground. A helicopter hovering above an area to be sprayed shakes the leaves, branches, and stems of the vegetative canopy, greatly increasing the penetration of spray droplets into the canopy. Once the helicopter begins forward flight, a pair of vortices forms behind the helicopter to the left and right of the flight path. As the helicopter speed increases, the wake vortex pair approximates more and more the vortex pair that forms behind a fixed-wing aircraft.

Downdrafts in aircraft wing vortices enhance the deposition of spray from low-flying aircraft.

The position of the spray nozzles relative to the wake vortex centers is critical to the initial motion of the spray droplets (figure 14.5). Droplets released outboard of the vortex center are sprayed into an updraft. If the droplets sprayed into the updraft have settling velocities less than the updraft velocity, they are carried upward rather than downward toward the targeted canopy. As the aircraft moves away and the vortex energy is dissipated, these elevated droplets are more likely to drift and miss the intended target. Nozzle position may actually be mandated in certain situations by regulation. Detailed computer simulation of the aircraft wake provides insight into this subject.

The vortices decay with time following aircraft passage, with the decay typically lasting about one minute. As the vortices decay, the spray droplet cloud begins to be influenced more by droplet settling speeds and ambient meteorology than by the vortices.

In the case of low-flying aircraft used in applications over flat cropland, the wake vortex pair interacts strongly with the surface, and the bulk of the spray is deposited during this interaction. In forestry applications over complex terrain, the aircraft typically is much higher and deposition to the canopy occurs well away from the influence of the aircraft wake.

The *swath width* (the width over which the desired deposition is achieved) is largely a function of release height for a given aircraft/spray system configuration. The spray spreads laterally, initially because of the wake vortex pair and subsequently because of atmospheric turbulence.

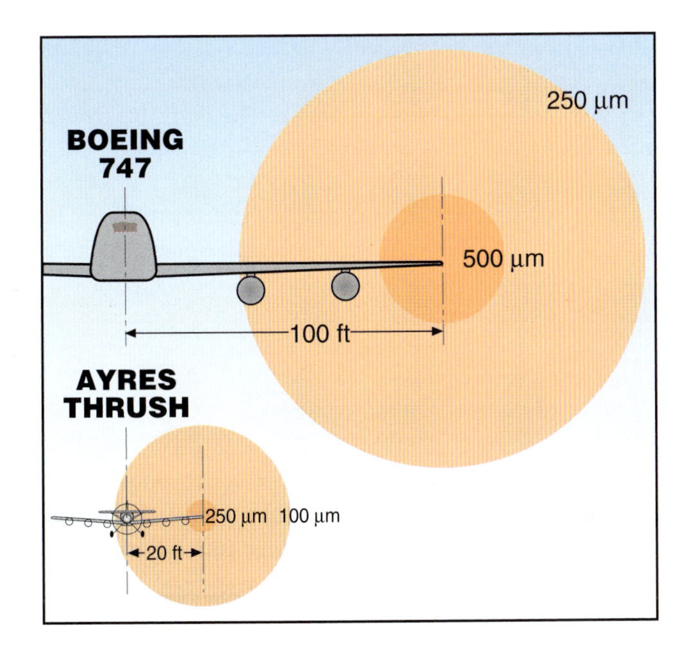

Figure 14.5 Influence of aircraft wingtip vortices on the motion of different size droplets. Within the marked circles, the motions of droplets that are smaller than the sizes indicated will be dominated by the wingtip vortex wind field. (Adapted from Bilanin et al., 1989)

Off-target drift also increases with increasing release height as the transit time of the spray in the atmosphere increases and, in the case of volatile materials, the driftable fraction increases through evaporation. Thus, there is an operational trade-off between drift and swath width.

14.2.1.3. *Effects of the Mean Ambient Wind Field.* The mean ambient wind direction determines the mean direction spray travels beyond the immediate wake of the application aircraft. The ambient wind speed during a spray operation determines the average angle of fall of the spray material. The higher the wind speed, the shallower the fall trajectory and the farther droplets land from the release point. When droplets are sprayed into an ambient wind, the distance S from the aircraft's flight path to the deposition point depends on the strength of the wind U and the length of time that the droplets are exposed to the wind, which depends in turn on the release height H and the settling velocity V_s, following the formula

$$S = \frac{HU}{V_s}$$

This equation yields a gross estimate of the length of the trajectory but ignores the aircraft wake, turbulence, and other physical factors. It is most accurate for very large droplets (>1000 micrometers).

Wind direction determines the direction that spray travels beyond the immediate wake of the application aircraft. Wind speed determines the average angle of fall of the spray material.

Spray operations are generally terminated when wind speeds exceed 5 m/s or 11 mph. At high wind speeds, a larger percentage of the mass of applied material is carried off target as larger droplets are displaced by the wind. Flight paths can be offset to compensate for steady winds consistently from the same direction, but spray deposition is generally harder to control in higher winds. Over complex terrain, higher winds may increase aviation hazards more significantly than over flat ground because of mechanically generated eddies near terrain obstacles (section 10.1). However, in application over extensive flat or rolling forestland, winds over 5 m/s (11 mph) may be preferred because mechanical turbulence near the treetops can aid in mixing the spray material into forest canopies. Usage of certain pest control agents in strong winds may be precluded by label restrictions.

Spray operations are generally terminated when wind speeds exceed 5 m/s or 11 mph.

14.2.1.4. *Effects of Atmospheric Turbulence.* The paths of droplets in the atmosphere away from the aircraft wake are affected not only by the mean wind, but also by turbulent motions within the mean flows (figure 14.6). Turbulence levels depend on the amount of wind shear above the surface and the atmospheric stability. An unstable atmosphere is characterized by the presence of rising plumes or convective bubbles from the heated ground. A stable atmosphere suppresses turbulence. The intensity of turbulence typically varies during the day in response to changes in atmospheric stability (section 12.4).

Turbulence intensity depends on wind shear and atmospheric stability, which varies diurnally.

For volatile droplets, turbulence increases evaporation, decreases droplet size, and increases drift. The droplet trajectory usually lengthens with increasing turbulence as the droplet moves through the turbulent vortices, although some droplets are brought to the surface more quickly if caught in a downward motion. The change in trajectory with turbulence is a function of turbulence and drop size and can be calculated statistically. As the droplet moves through the turbulent eddies, droplet ve-

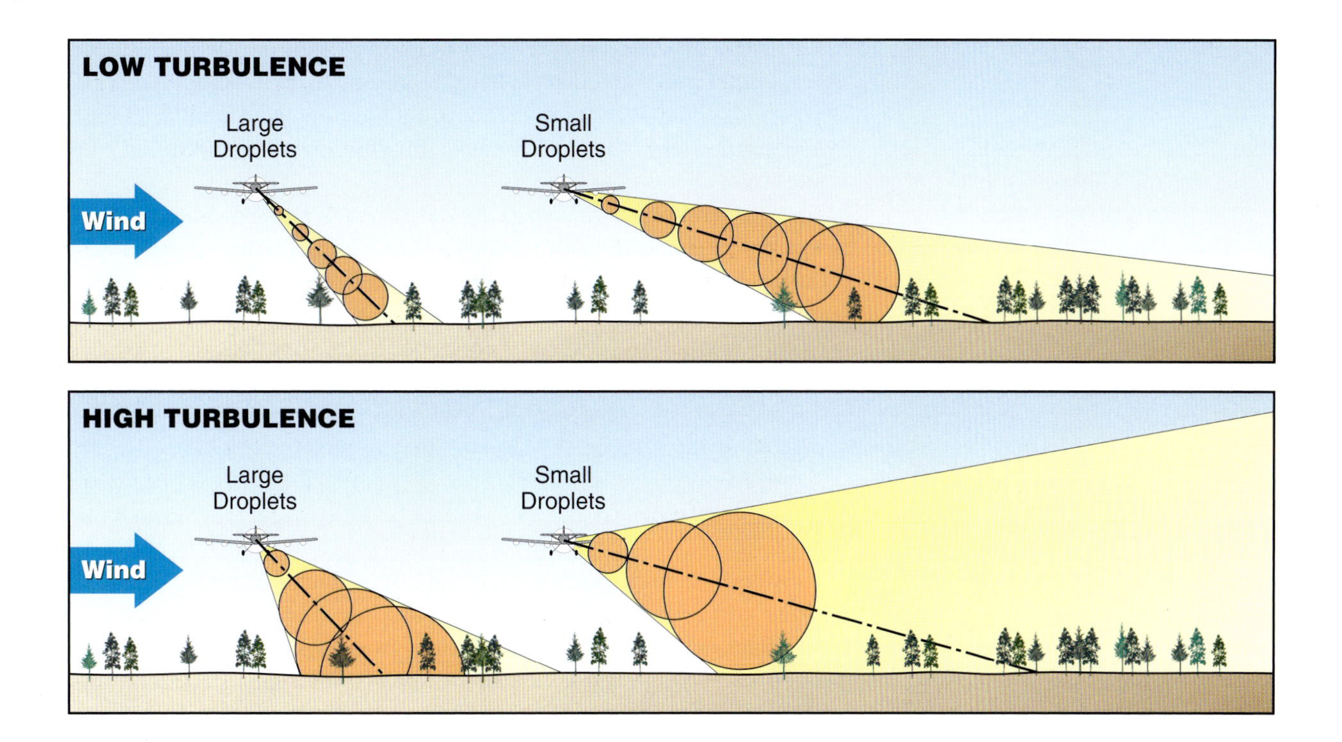

Figure 14.6 Droplet paths for large and small droplets, illustrating the greater dispersion of the droplets about their mean path in strong turbulence. (Adapted from Quantick, 1985)

locity constantly increases and decreases because of inertial forces as the droplet encounters eddies of differing velocity (and the eddies themselves decay due to shear and air viscosity). The droplet is surrounded by a thin layer of air, which is higher in humidity than its surroundings because of evaporation from the droplet. As the droplet accelerates and decelerates, this layer is stripped away. This lowers the humidity of the air near the droplet surface, thus increasing evaporation. Turbulence can also aid in canopy penetration, as mentioned previously.

14.2.2. *Humidity and Temperature*

Ambient humidity influences droplet evaporation. Evaporation reduces droplet size and increases droplet susceptibility to drift. The lower the humidity, the greater the rate of evaporation. Smaller droplets are especially subject to evaporation. Spray droplets may lose their volatile fraction within seconds of their release, depending on the relative humidity. Droplet size also influences evaporation, with smaller droplets evaporating more quickly than larger droplets because

- the ratio of surface area to volume increases as the drops get smaller, causing smaller droplets to lose a larger percentage of their volume more rapidly;
- the smaller radius of curvature of small droplets allows molecules to escape from the droplet surface more easily; and
- small droplets settle more slowly and are thus exposed to evaporative losses for a longer period of time than larger droplets.

The lower the humidity, the greater the rate of evaporation, the smaller the droplets, and the greater the droplet susceptibility to drift.

As shown in section 3.2.3, relative humidity varies inversely with temperature, so that low relative humidity often occurs during the warmest times of the day and high relative humidity often occurs near sunrise when temperatures are lowest. Thus, to minimize drift, spray operations are often conducted in the early morning hours when humidity is high, evaporation is low, and strong convection associated with strongly unstable atmospheres is not a factor.

Spray operations are often conducted in the early morning hours when humidity is high and evaporation is low.

Spray operations may have to be halted when ambient humidity falls below the level specified on the label of the product being sprayed. Even if the product label does not specify a humidity limit, spraying should be stopped when it is clear that the spray is evaporating before reaching the ground. For volatile agents, long residence times after deposition are best achieved by spraying during periods of higher relative humidity. Long residence times are necessary because most pest control agents must be contacted or ingested by the targeted pest to be effective.

14.2.3. *Atmospheric Stability*

Atmospheric stability (section 4.3) is a critical factor affecting spray deposition and drift. Its effect on materials sprayed from aircraft is similar to its effect on the dispersion of plumes from continuous point sources (section 12.4.3), and the same physical principles apply. In general, dispersion of fine spray droplets is suppressed in stable atmospheric layers and enhanced in unstable layers. Stable conditions are usually encountered early in the morning when the vertical temperature profile is characterized by a near-surface temperature inversion with low mean wind speeds and higher relative humidity. Vertical dispersion of material sprayed into this stable layer is suppressed. If the droplets are small, they may hang in the air, with little downward movement and are thus susceptible to drift at even very low wind speeds.

Dispersion of fine spray droplets is suppressed in stable atmospheric layers and enhanced in unstable layers.

As a general rule, spraying in strong inversions should be avoided. The strong stability within the early morning temperature inversion suppresses not only vertical dispersion of spray materials but horizontal dispersion as well. Droplets that hang in the air thus remain in a narrow, highly concentrated plume. Humidity may also be high, suppressing droplet evaporation. When the convective boundary layer begins to grow upward into the inversion, the droplet plume can be fumigated, bringing this relatively highly concentrated material to the ground. General winds, local winds (for example, downslope or drainage winds), or turbulence associated with the growing convective boundary layer may carry material off target. The driftable fraction can be lessened by increasing droplet size. Droplet transit time can be lessened by lowering release height.

Material sprayed into a less stable, neutral, or unstable layer above the surface-based inversion is dispersed primarily upward and laterally, with only the heavier droplets falling through the inversion and depositing on the target canopy below. Material that remains in the dry, turbulent air above the inversion is subject to longer range drift because winds are strong and droplet size is reduced by evaporation.

Unstable conditions develop when the ground is heated by insolation after sunrise and convective currents begin to rise from the surface, mixing and heating a layer of air above the ground. This layer of heated air, capped by the remnants of the nocturnal inversion, becomes progressively deeper until the inversion layer is eventually destroyed (section

11.1.3). Material sprayed into this layer is carried by convective eddies. The eddies increase in size as the layer deepens, causing the plume to mix through a deep atmospheric layer (figure 12.3), making it increasingly more difficult to get the spray material to deposit reliably in the canopy. Spraying is often halted before midmorning as the convective motions strengthen. Under these conditions, material can be carried aloft and transported a long distance, though off-target effects are usually negligible because the material is widely dispersed.

Spraying in or above strong surface-based inversions or during strong convection should be avoided.

Neutral conditions develop at any time of the day or night and in all seasons when there is extensive cloud cover or when the atmosphere is well mixed by strong winds. Although neutral conditions associated with strong winds are not suitable for spraying, neutral conditions associated with overcast skies and lower wind speeds provide the best spraying conditions in some applications. Extensive cloud cover reduces insolation at the surface (and thus the growth of convective boundary layers), reduces daytime air temperatures, and maintains high humidity. Wind directions tend to remain relatively constant. Mechanical turbulence disperses the droplets into the canopy, and small droplets subject to drift are not as concentrated as they would be in more stable conditions. Overcast skies can be forecast fairly reliably and can persist for long periods, allowing spraying to continue throughout the daylight hours, rather than stopping in midmorning when convection gets too strong.

Neutral conditions, which are associated with overcast skies, and low wind speeds provide the best spraying conditions in some applications.

14.2.4. *Precipitation*

Aerial spray operations are generally halted when precipitation occurs. Precipitation

- can reduce the effectiveness of the spray operation by washing the spray material off the targeted foliage surfaces;
- may cause unintended environmental problems by washing the material into nearby streams and ponds;
- can complicate monitoring and measuring efforts and can cause general working conditions to deteriorate; and
- can reduce visibility and thus pose an aviation hazard.

Aerial spray operations are generally halted when precipitation occurs.

14.3. Spray Deposition

The bulk of the spray, after release from the aircraft, is carried into the canopy by the processes discussed in section 14.2.1. Entry of the spray droplets into the canopy, however, does not ensure that the droplets are deposited there. The deposition process depends on biological factors, physical factors, equipment, and meteorological conditions. The biological factors include foliage element size, orientation, shape (needle or flat leaf), surface texture (smooth or hairy), and leaf surface moisture. The density of the canopy is also important, because the likelihood that droplets will deposit increases as the number of foliage elements encountered increases. The *Leaf Area Index* is the foliage area per area of ground surface. LAI is a useful measure of canopy density. One way to measure foliage density is to drop a plumb bob from above the canopy and count how many foliage elements the line touches between the canopy top and the ground. This number can be converted to Leaf Area Index. This method is most appropriate for relatively closed deciduous canopies.

The fraction of droplets approaching a targeted surface that actually deposit on that surface is denoted by the term *collection efficiency*. Collection efficiency is dependent on the characteristics of the targeted surface and droplet size. Envision a droplet falling into a canopy in light winds. Larger droplets with greater settling velocities tend to impact and deposit on the first foliage element they encounter. Smaller droplets with lower settling speeds are more likely to be carried around foliage elements in the small-scale mean and turbulent flows, thus decreasing the collection efficiency. A large droplet (say, 300 micrometers in diameter) might have a collection efficiency of 0.8 (80%) for an encounter with a single leaf, so that the chance of failing to deposit on the leaf is 0.2 (20%). If the droplet is expected to encounter four leaves as it falls through the canopy, then the chance that it will deposit on one of the four leaves is $(1 - 0.2^4)$ or 0.9984 (99.84%). In contrast, a small droplet (say, 30 micrometers in diameter) might have a collection efficiency of 0.2 (20%), so that the chance of failing to deposit is 0.8 (80%). If this droplet encounters four leaves, the probability that it will be deposited is $(1 - 0.8^4)$ or 0.5904 (59.04%) after four encounters.

The larger droplets in the droplet size distribution are more likely to deposit on the first canopy element encountered (usually at the canopy top). Small droplets, however, are less likely to deposit on the first foliage element they encounter and are more likely to move farther into the canopy before depositing. This differing behavior of large and small droplets can be used to select droplet sizes to be sprayed into a canopy to target a specific canopy layer.

Spray falls vertically only when wind is absent and when the spray is beyond the influence of the aircraft wake. Even a light wind causes droplets to descend at an angle, thus increasing the length of their path through the canopy. The longer the transit path, the greater the number of foliage elements encountered. A strong wind may impart impaction energy to the droplet that gives it the inertia to overcome canopy element boundary layer effects. However, the presence of wind does not necessarily increase collection efficiency. In strong winds, some canopy elements minimize their drag by changing orientation (streamlining), which decreases the projected surface area to the oncoming droplet. Fluttering and other canopy movements also complicate overall canopy collection.

14.4. Additional Considerations in Complex Terrain

The transport, diffusion, and deposition of aerial sprays is more complicated over complex terrain than over homogeneous terrain because of the terrain-forced flows produced by the interaction of prevailing winds and the topography (chapter 10) and because of the diurnal flows that develop regularly in mountainous areas (chapter 11).

14.4.1. *Diurnal Mountain Flows*

Diurnal mountain flows (chapter 11) produce regular changes in wind direction and speed, turbulence, and atmospheric stability. Thus, the four phases of the cycle (the evening transition, nighttime period, morning transition, and daytime period) have distinctly different meteorological characteristics that must be considered when planning aerial spray operations. Safety and work schedules must also be considered, and they generally limit spray operations to daytime hours. Aerial spraying in forests is usually conducted during the brief morning transition period, as it represents the best compromise of meteorological and operational factors.

Aerial spraying in forests is usually conducted during the morning transition period.

During the morning transition period, the valley atmosphere gradually becomes coupled with the atmosphere above the ridge tops, and stability and winds change rapidly. Early in this phase, the deep valley temperature inversion that is characteristic of the nighttime period is still present. Spray droplets that are too small to fall entirely through the inversion to deposit on the forest canopy can be carried long distances downslope and down-valley with little vertical dispersion. Off-target drift can therefore be quite concentrated. Spray conditions improve later in the morning transition period when sunlight causes shallow convective boundary layers to form over the slopes and humidity remains relatively high. Operational considerations usually dictate that spraying begin at daylight. The near-neutral to unstable convective boundary layer contains weak upslope winds and is capped by an elevated stable core. Material sprayed into the convective boundary layer is easily transported down into the canopy by the aircraft's wake vortices. Small droplets become harder to control as the convective boundary layer becomes deeper and the stable core shrinks. The end of the morning transition period is defined by the destruction of the stable core. With the onset of the daytime period, vigorous deep convection and the downward transport of stronger winds from aloft make it difficult to deposit material reliably in the canopy. Spraying is often halted at the end of the morning transition period anyway because of low humidity.

Spraying is usually halted near the end of the morning transition period when convection, turbulence, and upslope flows become strong over the canopy and when humidity decreases, increasing droplet evaporation.

Spray operations conducted during the morning transition period should include careful monitoring of changes in wind direction. Any drift is transported by the slope and along-valley wind systems. Early in the period, spray drifts downslope and down-valley. Later, when sunlight warms the slopes, the downslope flows reverse to upslope. The reversal of the down-valley wind, however, occurs at different times in the different layers of the vertical temperature structure (figure 11.8), reversing to up-valley much earlier in the shallow convective boundary layer over the valley floor than in the elevated stable core. Spray aircraft flights should be scheduled so that drift is in a desired direction (either up or down the slopes or valley) and sensitive areas are protected.

The sometimes complex pattern of shading and insolation intensity in valleys during the morning transition period can complicate spraying operations, but can also identify locations in the valley that have the best spraying conditions at different times during the morning transition period. Intense insolation on the sidewall facing the sun, for example, produces stronger convection and earlier inversion destruction above that sidewall. This sidewall should be sprayed early, before convection becomes vigorous. Ridges should also be sprayed early, because they are above the strong valley inversion and will develop vigorous convection early (especially the side of the ridge facing the sun). However, there are many areas of the valley that, because of their slope and orientation, receive only weak insolation during the morning transition period. These slopes can be sprayed later in the morning transition period until the time of final destruction of the stable core.

The daytime period is characterized by deep turbulent convection, which often mixes strong winds down from above the valley atmosphere. Temperatures are higher than in the morning and relative humidities are lower, causing an increase in driftable fraction as a result of droplet evaporation. Small drops may go up instead of down, adversely affecting the continuity of spray coverage in the canopy and increasing drift as spray material is carried by upslope and up-valley flows. The vigorous convection also makes flying conditions difficult.

During the evening transition period, the valley atmosphere is gradually decoupled from the atmosphere above the ridge tops. A temperature inversion grows from the valley floor and slopes, and wind directions shift from upslope to downslope and from up-valley to down-valley. If spraying is begun too early during this period, the same problems are encountered as during the daytime. Later in this phase, smaller spray droplets that cannot settle through the surface-based inversion are subject to drift.

During the nighttime period, the strong nighttime inversion limits vertical dispersion of material sprayed within the inversion. The high stability in the deep temperature inversion causes fanning (section 12.4.3) of the sprayed material and down-valley drift of undeposited material in a concentrated plume. These problems persist into the early part of the morning transition phase.

14.4.2. *Terrain-forced Flows*

Terrain-forced flows typically have less impact on aerial spraying operations than diurnal mountain winds because spraying operations are conducted when wind speeds are low. Terrain-forced flows are generally not strongly developed when winds are below 4 m/s (9 mph). When the prevailing winds exceed 7–10 m/s (16–22 mph), however, wakes from terrain obstacles and turbulence or downdrafts from rotors and lee waves can develop, posing a hazard to spray aircraft and significantly affecting the transport and dispersion of airborne particles, as summarized in figure 12.15. Special cloud types and other indicators of strong winds and possible aircraft flight hazards were described in section 7.1.4. It is important to reiterate that pesticide label restrictions may preclude spraying when winds are strong.

14.5. Collection of Meteorological Data

Monitoring weather conditions and obtaining meteorological support are integral parts of the overall spray project. The type of data collected and the frequency with which it is collected are determined by the objectives of the data collection program, which can include any or all of the following:

- to assist the operational manager in planning the application
- to assist the on-site manager in making informed application decisions in real time
- to document the occurrence or nonoccurrence of pest control agent deposition in the forest canopy
- to document the occurrence of nonoccurrence of pest control agent drift and the direction of drift
- to understand the application outcome in order to improve the effectiveness of future spray operations

Data used in go/no-go spray decisions must be accessible to the operational staff and should come from the instrument in a clear, ready-to-use, understandable form. If the data are to be used in post-spray analysis, data storage considerations must be addressed. Common meteorological instruments and data collection systems are described in many excellent texts (appendix C).

For an effective and environmentally safe spray operation, the following meteorological data are required from representative locations: wind speed, wind direction, temperature, humidity, an indicator of stability (solar or net all-wave radiation, temperature at two levels, or a turbulence indicator, such as the standard deviation of the wind direction), and measurements of recent precipitation (and forecasts of imminent precipitation). In addition, pressure measurements may be useful as an indicator of approaching synoptic weather systems (section 5.1.1).

14.5.1. *Siting of Instruments*

Most meteorological data used in spray operations are collected by recording instruments, or data loggers, placed in meteorological shelters or on short towers (figure 14.7). Tethered or free-flying balloons or remote sensing devices (for example, radars, sodars, lidars) are used to make observations above the ground. The aviation hazard posed by instrumented towers and tethered balloons must be considered, especially in the confined air space of mountain valleys. Federal Aviation Administration approval must be obtained if towers exceed 50 ft (15 m) in height, if moored balloon diameters exceed 6 feet or gas volumes exceed 115 ft^3, or if the balloon is to be used within 5 nautical miles of an airport. Other restrictions on moored balloons in flight near clouds, in low-visibility conditions, or at elevations exceeding 500 feet require special permission as detailed in FAA regulations. The meteorological instruments must be set up in locations that are representative of the spray site. They must also be sampled at a frequency that allows means and variances of the quantities to be calculated on appropriate time scales for the planned analyses. A professional meteorologist should assist in selecting suitable locations and *sampling frequencies* in complex terrain.

On-site operational and meteorological data may be provided by one or more trained on-site observers who report back to the operational manager by radio. These observers are deployed to safe locations where they can observe the spray aircraft and determine whether the sprayed mate-

Figure 14.7 Example of a portable weather station deployed on an aerial application project. This station measures two levels of wind, temperature, and humidity and also includes a net radiometer. The station is operated by battery power, with batteries charged during daytime by solar panels. The data from this station are recorded on-site and also transmitted to a base station through a radio link. (USDA Forest Service photo)

rial is being properly deposited in the canopy. When equipped with portable meteorological instruments, they can report meteorological observations that supplement those obtained from data loggers.

A number of anemometers and wind vanes are available with different response characteristics. Some are rugged for long-term, unattended operation in all types of weather. These, however, are less sensitive than the more delicate, research-grade equipment that can record weak or rapidly changing winds.

Where multiple anemometers are exposed at different heights on a tower, vertical spacing is an important issue. Because winds increase logarithmically with height above an unobstructed surface (appendix A, eqn. A.11), anemometers can be spaced logarithmically with height (i.e., close together near the surface and farther apart with altitude). Winds within a forest canopy are usually weak and do not vary much with height, although the variability depends somewhat on canopy density. Above the canopy, a logarithmic profile begins with winds increasing more rapidly with height. Anemometers should not be placed in locations where they are exposed to nonrepresentative winds, such as over ridge tops, where winds speed up, or behind obstacles, where winds slow down (section 10.1).

Thermometers are generally colocated with anemometers. They must be shielded from direct solar radiation, isolated from nearby heat sources, and aspirated (i.e., ventilated by fans or other devices) to avoid local heat build-up.

Humidity sensors should be sited away from local moisture sources and should be shielded from solar radiation. Humidity sensors are also often aspirated.

Radiometers are usually exposed to an unobstructed sky, although occasionally researchers position net radiometers in the canopy to measure local components of the radiation budget (section 4.4.2). Because net radiometers have both an upward and downward view, it is important to site them over a uniform surface with representative color, roughness, and moisture. Radiometers provide information on surface heating, stability, sky cover, and illumination time.

The exposure of a pressure sensor is generally not critical, although it should be protected from excessive temperature variations and from insects that might obstruct the pressure port. Pressure sensors must be calibrated occasionally to maintain accuracies of 1 or 2 mb.

Vertical soundings of wind, temperature, and humidity are often necessary to gain a better understanding of the effects of atmospheric structure on spray deposition and drift. Tethered or free-flying balloons should be located so that they provide representative vertical soundings at the spray site or on the spray transport path. Tethered balloons are recoverable and do not require tracking equipment. The tether, however, can pose a hazard to aircraft.

Remote sensing instruments such as sodars can be very useful tools on spray projects. This type of instrument can yield meteorological information near the aircraft with none of the dangers associated with towers or tethered balloons in the aircraft flight path.

14.5.2. *Sampling Duration*

Meteorological sampling usually begins two hours before spraying starts and continues for one hour after spraying stops to provide a record in case conditions before or after the actual spraying influence the spray operation. Observations may be taken for a few days before the project to assist the forecasters in developing daily trends. Sampling may be required for a longer period after spraying stops to monitor the persistence or recirculation of spray material in the atmosphere. Because meteorological sampling is done in conjunction with material monitoring, samples must be taken until the spray cloud passes the sampler farthest from the target area.

14.5.3. *Sampling Frequency*

A wide range of sampling frequencies, denoted in *Hertz* (Hz), the number of samples per second, can be selected depending on the objectives of the data collection program, the characteristics of the sensors, the characteristics of the data loggers and data storage devices, the variability of the parameter being monitored, and cost constraints. A high sampling frequency provides the most complete data set but also increases the cost of instrumentation, data loggers, data processing, and analyses. Data collected at a high frequency can be averaged to provide low-frequency information. High-frequency information, however, cannot be derived from low-frequency data.

The response time or *time constant* of the sensor must be considered when selecting the sampling frequency. The time constant is the time it takes for an instrument to achieve 63.2% of full response to a step change in the parameter being measured. For example, if the temperature instantly changed from 0°C to 10°C, the time constant of a thermometer measuring the change would be the time required for the thermometer to reach 6.32°C. A sensor with a long time constant has a slow response to changes in the parameter being sampled. Sampling slow response sensors at a high rate produces data that are influenced by sensor characteristics rather than by actual atmospheric changes.

The nature of the actual signal from the transducer is also a consideration when determining sampling frequency. For instance, a common method of measuring wind speed is to attach a magnet to a set of cups that spin at a speed proportional to the wind moving past them. As the magnet turns, an induced voltage is measured, which changes sign relative to its position in the magnetic field. Thus, one cycle is produced for every turn of the anemometer cups. Typically this type of sensor has a counter that electronically stores one count for every cycle. The signal from the counter changes in discrete increments. If the cups turn one revolution for every 6 m of wind passage and the wind is blowing at 1 m/s, the counter will only increase once every 6 seconds. If the counter is sampled at 10 Hz, 60 samples will be collected without providing any new information.

Since 1980, logging rates for commercially available data loggers have increased by at least three orders of magnitude. Where 10 kHz analog to digital (A/D) conversion boards were considered fast in 1980, 1000 kHz is possible today. (The total quoted rate must be divided by the number of sensors to be sampled, so that a 1000 kHz data logger will sample at 10 kHz per sensor over 100 sensors or channels). Modern data recording equipment is thus capable of sampling rates that are suitable for nearly all routine meteorological sampling tasks.

The most common limitation encountered with modern data collection systems is the size of the data storage device (e.g., hard disk or magnetic tape). The data storage capacity required is determined by multiplying the duration of sampling by the sampling frequency summed over all the sampling channels. A 1-megabyte storage device can only store 1000 seconds of 1-byte data when data are sampled at 1 kHz. Storage capacity is typically expressed in bytes, so calculating the number of bytes per sample allows calculation of storage needs, though it is wise to collect a sample data set during testing to verify assumptions regarding storage needs.

Atmospheric turbulence data are sometimes collected at rates up to 100 Hz, whereas some upper air data are collected only twice per day (i.e., 0.00002 Hz). Typical sampling frequencies for meteorological data in support of aerial spraying fall between these two extremes. Turbulent flux sensors (usually used only in research programs), such as sonic anemometers, hot-wire anemometers, and single-turn thermocouples, are generally sampled at 10 Hz. Cup anemometers and wind vanes are often sampled at 1 Hz. This sampling rate provides detailed data for meteorological analyses that may not be necessary in a strictly operational environment. Net radiometers, thermistors, and relative humidity sensors are usually sampled at 0.1 Hz. Most upper air measurements are made at lower frequencies (0.01 Hz or less). Pressure has been considered a low-frequency measurement, though recent technological advances allow higher frequency measurements when necessary in research.

14.6. Computer Modeling

Computer simulation of aerial spraying can be used to plan spray operations, analyze performance after operations are completed, and present information to the public. Meteorological data collected on-site before, during, and after the spray operation are critical input to computer models. Because the quality of the simulations is dependent on the quality of data provided, it is important to consider modeling objectives when planning the collection of meteorological data.

Models are extremely useful in the planning of spray operations because they allow the project manager to test the effect of dozens of variables, including equipment and properties of the spray material, as well as meteorological parameters, to determine optimal conditions for an effective and environmentally safe operation.

Computer models are used for planning spray operations because they allow the testing of dozens of variables to determine optimal conditions for an effective and environmentally safe operation.

Generally, three types of models are currently widely used in pesticide dispersion and drift applications. The first is a Gaussian approach that follows an established, conventional method used in regulatory air pollution modeling. It is most appropriate for gaseous diffusion. The second focuses on the machine energy, both of the pressurized system and of the release vehicle wake. It uses simple meteorological transition and transport models. The third is a physical approach often based on detailed fluid dynamics (the Navier–Stokes equations) applied in the atmosphere.

The first type of model assumes that the wind direction determines the plume centerline and that the crosswind distribution is a Gaussian or bell-shaped curve (see Turner, 1969, for a discussion of Gaussian techniques). Gaussian models are mass conservative and steady state. The plume is narrower near the source and becomes wider downwind. The maximum airborne concentration is higher near the source and decreases downwind. A crosswind slice of equal thickness at any distance downwind will integrate to the same mass if no depletion of the plume material is considered. The rate of plume spread is determined by dispersion coefficients that vary with stability; these rates also can be varied to consider cases of very large surface roughness such as tall buildings. This type of model is not of much interest to researchers because it is effectively a statistical model with the downwind and crosswind distribution of pollutant mass (expressed as concentration) input a priori in the model. From a theoretical standpoint, there is not much new to be learned using this approach.

Terrain greatly complicates dispersion modeling in general. The Gaussian approach was developed and validated over uniform, flat terrain. Though many schemes have been developed to use Gaussian models in complex terrain, two primary assumptions are violated. The first assumption is that single point meteorology can be used to reasonably describe meteorology over some large area. In actuality, abrupt spatial changes in steepness, orientation, and roughness can cause meteorological conditions to vary dramatically over small areas. The second assumption is that the droplet plume shape can be described generically. The droplet plume "shape" is often controlled by the terrain itself and can be as variable as the "shape" of the terrain.

The second type of model focuses on the wake of the disseminating aircraft, using the weight of the airplane and the wingspan to calculate the strength and location of wake vortices that entrain the spray droplets. The droplets then travel in the atmosphere in these swirling vortices until the vortical energy is dissipated either by the surface or by ambient wind and turbulence (section 14.2.1.2). This type of model focuses on the droplets and considers ambient temperature and humidity in a droplet evaporation algorithm. These models explicitly consider droplet settling velocity and use the ambient wind and turbulence to calculate vortical decay and to transport the droplet after the machine energy has dissipated. The models commonly used in agricultural spraying are greatly simplified, but are nonetheless used in some research applications. This type of model must be coupled with another approach to describe transport of material that is still aloft after the wake vortices have decayed. Deposition onto sloping surfaces by the wake vortices can be simply addressed. However, transport of material that remains aloft after the wake vortices have dissipated requires a different approach and is difficult to calculate because of the spatial variability of the meteorology and of the terrain itself, as discussed previously.

The final class of models generally use the Navier–Stokes equations to describe the atmospheric dynamics. The Navier–Stokes equations describe motion in a turbulent fluid and attempt to give a four-dimensional representation (three spatial dimensions and time). The equations use a velocity vector, pressure, fluid (air) viscosity, and a stability term to calculate how the fluid moves and changes with time. The equations cannot be solved without making assumptions about the flow that impart uncertainty to the solution. These assumptions and approaches to this problem are areas of ongoing research. The explicit physics in these models makes them suitable for explicit consideration of terrain, though real terrain data must be available and the accuracy of the droplet transport predictions is related to the accuracy of the wind field and meteorological calculations and/or measurements.

The FSCBG model (Teske et al., 1993) is a FORTRAN-based model developed with the support of the USDA Forest Service that simulates aerial spraying in uniform terrain. It is available to the public, has a reasonably friendly user interface, and can be run on a modern personal computer. Based on a Gaussian dispersion model and an aircraft wake droplet trajectory model, FSCBG contains a large library of aircraft, nozzle types, and drop size distributions. This model and derivative models have recently been adopted for use in pesticide regulatory applications, and graphical outputs from the models are included in environmental assessments and environmental impact statements. An example of output from the model is shown in figure 14.8.

Comprehensive models are not yet available to fully describe spray deposition and drift in complex terrain. The USDA Forest Service has done considerable work in the modeling of pesticide droplet transport in complex terrain, and some simple models are just becoming available that describe key physical processes that affect the dispersion of aerial spray in

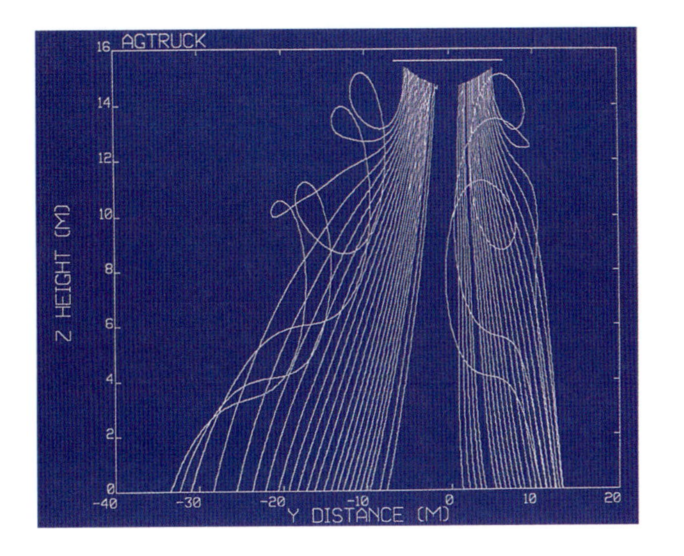

Figure 14.8 Simulation of droplet trajectories in the wake of an AgTruck spray aircraft as modeled by the FSCBG model. Modeling of aerial application allows planning and analysis not otherwise possible. (USDA Forest Service photo, provided by R. B. Ekblad)

idealized valleys. These models include some of the processes that drive the diurnal cycle of winds and temperature inversion structure in valley terrain, as described in section 11.1.

14.7. Integration of Meteorological Information into Operations

14.7.1. *Meteorology and the Operational Plan*

Every spray project should be conducted according to an operational plan. The plan should include project objectives, safety, job descriptions and responsibilities, an organizational chart, and a public involvement plan, as well as the details of the operation itself, including communications, decision standards, monitoring, aircraft and application equipment, application methodology, and emergency response.

Weather considerations should be integrated throughout the operational plan. The plan should discuss and emphasize spray team training, weather forecasting, weather observations, meteorological instrumentation and data acquisition, the importance of weather input in operational decisions, the evaluation of operational effectiveness, and the dissemination of information to the public.

Weather considerations should be integrated throughout the operational plan.

14.7.2. *Project Personnel and Meteorological Training*

The project manager should have meteorological training (provided by this manual or other written resources, classroom instruction, and/or field experience) that will enable the manager to evaluate the impact of weather conditions on the aerial spray operation and to make informed operational decisions. In fact, all members of the spray team need an overview of how weather conditions affect spray operations and how meteorological data are collected, analyzed, and incorporated into the decision-making process. Team members assigned to take ground observations must be trained in the use of meteorological instruments and in the taking and reporting of field observations; project personnel assigned to weather observation and meteorological data collection may also have other responsibilities, for example, spray deposition monitoring.

Some aerial spray field organizations include a spray strategist, who has expertise in aerial spraying, dispersion models, spray dispersion, spray drift, and complex terrain meteorology. The spray strategist coordinates the efforts of weather observers and professional forecasters and in this capacity is a key advisor to the project manager.

14.7.3. *Weather Forecast Support*

Good forecast support is critical to a safe, effective spray project. Arrangements with forecast sources for information on obtaining professional weather forecasts (section 9.4) should be made early in the planning stages. Seasonal demand for forecasting can be high during the fire season.

Meteorologists who support spray projects should be given a thorough briefing on the project objectives, scheduling, and methods, as well as on safety and environmental concerns. To provide effective forecasts for the operational site, the forecaster must be familiar with the topography, altitude, vegetative cover, exposure to the sun, and seasonal weather characteristics of the site. Local meteorological data collected specifically for the project should be made available to the forecaster. Although it is desirable to have a meteorologist on site to assist the project manager, the forecaster is usually at a nearby airport or fire weather center and is seldom in the field with the spray team.

A professional weather forecaster is often assigned to provide both area and project site-specific forecasts. The project manager generally makes go/no-go decisions based on 7-day, 3-day, and 1-day general-area weather forecasts. During the spray operation, area forecasts for 72 hours, 24 hours, and 12 hours are provided, as are 12-hour site-specific forecasts. Forecasts are updated as needed.

14.7.4. *Evaluation of Spray Operations*

Computer modeling is a valuable tool for evaluating the effectiveness of a completed operation and for making realistic estimates of both spray deposition and drift. Recent improvements in aircraft navigation (section 14.1.4) and recording of the swath paths flown by spray aircraft have the potential for improving these postspray simulations in the near future.

Every spray operation should conclude with a critique of the operation. The effectiveness of meteorological data collection, weather observations, and forecasts are among the aspects evaluated. Project records, including weather data, model outputs, aircraft flight paths, recommendations, safety and accident reports, and public relations information, should be archived as reference material for future operations and for use in the event of legal challenges (e.g., spray drift litigation).

14.7.5. *Meteorological Information and Public Relations*

Communication with the public, especially landowners near the target area, and with the news media is an important aspect of spray operations. Concerns about the substances being sprayed, the environmental impact of the operation, and the possibility of drift are best addressed early in the project by providing information about the project objectives, such as the use of herbicides to prepare sites for restoring habitat, controlling exotic weeds, reducing the potential for landslides and flooding, or the use of insecticides to reduce tree mortality in a critical watershed, historic area, or recreational area.

Given that all aerial spray projects are dependent on weather, the public should be aware of how weather influences the project and how the project manager uses weather information to make safe and effective field decisions. Graphical output from computer models, such as FSCBG, driven by a range of weather scenarios can be used to communicate to the public that project managers have planned carefully to mitigate drift and to reduce the potential for off-target impacts.

References

Agee, J. K., 1993. *Fire Ecology of Pacific Northwest Forests*. Island Press, Washington, DC.

Ahrens, C. D., 1994. *Meteorology Today: An Introduction to Weather, Climate and the Environment*, 5th ed. West Publishing Co., St. Paul, Minnesota.

Anderson, H. E., 1968. Sundance Fire: An Analysis of Fire Phenomena. USDA Forest Service Research Paper INT-56. (Intermountain Research Station, 324 25th Street, Ogden UT 84401)

Angle, R. P., and S. K. Sakiyama, 1991. *Plume Dispersion in Alberta*. Standards and Approvals Division, Alberta Environment, Edmonton, Alberta.

Bache, D. H., and D. R. Johnstone. 1993. *Microclimate and Spray Dispersion*. Ellis Horwood Series in Environmental Management, Science and Technology. Ellis Horwood, New York.

Bader, D. C., and C. D. Whiteman, 1989. Numerical simulation of cross-valley plume dispersion during the morning transition period. *J. Appl. Meteor.*, 28: 652–664.

Barney, R. J., G. R. Fahnestock, W. G. Herbolsheimer, R. K. Miller, C. B. Phillips, and J. Pierovich, 1984. Fire Management. In: K. F. Wenger (ed.), *Forestry Handbook*, 2nd ed. Wiley-Interscience, New York, 189–251.

Barry, J. W., 1993. Aerial Application to Forests. In: G. A. Matthews and E. C. Hislop (eds.), *Application Technology for Crop Protection*, CAB International, Wallingford, Oxon, United Kingdom, 241–273.

Beran, D. W., 1967. Large amplitude lee waves and chinook winds. *J. Appl. Meteor.*, 6: 865–877.

Bérenger, M., and N. Gerbier, 1956. Les mouvements ondulatoires à St. Aubansur-Durance (Basses-Alpes); première campagne d'études et de mesures. January 1956. Monographie No. 4 de la Météorologie Nationale. Direction de la Météorologie Nationale, Paris.

Berndt, H. W., and B. W. Fowler, 1969. Rime and hoarfrost in upper-slope forests of eastern Washington. *J. Forestry*, 67: 92–95.

Best, A. C., 1950. Empirical formulae for the terminal velocity of water droplets falling through the atmosphere. *Quart. J. Roy. Meteor. Soc.*, 76: 302–311.

Bierly, E. W., and E. W. Hewson, 1962. Some restrictive meteorological conditions to be considered in the design of stacks. *J. Appl. Meteor.*, 1: 383–390.

Bilanin, A. J., M. E. Teske, J. W. Barry, and R. B. Ekblad, 1989. AGDISP: The aircraft spray dispersion model, code development and experimental validation. *Trans. Am. Soc. Agric. Eng.* 32: 327–334.

Breyfogle, S., and S. A. Ferguson, 1996. User Assessment of Smoke-Dispersal Models for Wildland Biomass Burning. PNW-GTR-379. (Available from USDA Forest Service, Pacific Northwest Research Station, 333 S.W. First Ave, P.O. Box 3890, Portland, OR 97208)

Brotak, E. A., and W. E. Reifsnyder, 1977a. Predicting major wildfire occurrence. *Fire Management Notes*, 38: 5–8. (Available from USDA Forest Service, Fire and Aviation Management, 201 14th Street SW, P.O. Box 96090, Washington, DC 20090-6090)

Brotak, E. A., and W. E. Reifsnyder, 1977b. An investigation of the synoptic situations associated with major wildland fires. *J. Appl. Meteor.*, 16: 867–870.

Buettner, K. J. K., and N. Thyer, 1966. Valley winds in the Mount Rainier Area. *Arch. Meteor. Geophys. Bioklim.*, B14: 125–147.

Burrows, A. T., 1901. Yearbook of the U.S. Department of Agriculture. U.S. Department of Agriculture, Washington, DC.

Byers, H. R., and R. R. Braham, Jr., 1949. *The Thunderstorm*. U.S. Weather Bureau, Washington, DC.

Carney, T. Q., A. J. Bedard, Jr., J. M. Brown, J. McGinley, T. Lindholm, and M. J. Kraus, 1996. *Hazardous Mountain Winds and Their Visual Indicators*. NOAA Handbook, National Oceanic and Atmospheric Administration, Environmental Research Laboratories, Boulder, Colorado.

Collier, M., 1990. *An Introduction to the Geology of Death Valley*. Death Valley Natural History Association, Death Valley, California.

Crawford, K. C., and H. R. Hudson, 1973. The diurnal wind variation in the lowest 1500 ft in central Oklahoma: June 1966–May 1967. *J. Appl. Meteor.*, 12: 127–132.

Deeming, J. E., R. E. Burgan, and J. D. Cohen, 1978. The National Fire-Danger Rating System; USDA Forest Service General Technical Report INT-39. (Available from USDA Forest Service, Intermountain Research Station, 324 25th Street, Ogden, UT 84401)

Doesken, N. J., and A. Judson, 1996. *The Snow Booklet—A Guide to the Science, Climatology and Measurement of Snow in the United States*. Colorado Climate Center, Department of Atmospheric Science, Colorado State University, Fort Collins, Colorado.

Ekblad, R. B., and J. W. Barry, 1990. Selection and Verification of Complex Terrain Wind Flow Model for Spray Transport—Briefing Paper and Progress Report. USDA Forest Service Report 5E52P29-WIND, July 1990. Technology and Development Program, Missoula, Montana.

Environmental Protection Agency, 1995. National Air Pollutant Emission Trends, 1900–1994. EPA-454/R-92-011. October 1995. Office of Air Quality Planning and Standards, Research Triangle Park, North Carolina.

Etling, D., 1989. On atmospheric vortex streets in the wakes of large islands. *Meteor. Atmos. Phys.*, 41: 157–164.

Finney, M. A., 1998. FARSITE: Fire Area Simulator—Model Development and Evaluation. RMRS-RP-4. March 1998. USDA Forest Service Rocky Mountain Research Station. (Available from Intermountain Fire Sciences Laboratory, P.O. Box 8089, Missoula, MT 59807)

Fosberg, M. A., 1977. Forecasting the 10 hour Time Lag Fuel Moisture. USDA Forest Service Research Paper RM-187. (Rocky Mountain Research Station, Fort Collins, CO)

Garnier, B. J., and A. Ohmura, 1968. A method of calculating the direct shortwave radiation income of slopes. *J. Appl. Meteor.*, 7: 796–800.

Geiger, R., R. H. Aron, and P. Todhunter, 1995. *The Climate Near the Ground*, 5th ed. Friedr. Vieweg & Sohn Verlagsgesellschaft mbH, Braunschweig/Wiesbaden.

Ghent, J. H., D. B. Twardus, S. J. Thomas, and M. E. Teske, 1996. GYPSES: The Gypsy Moth Decision Support System. ASAE Paper #AA96-005, Amer. Soc. of Agric. Eng., 2950 Niles Rd., St. Joseph, Michigan, 49085-9659.

Glenn, C. L., 1961. The chinook. *Weatherwise*, 14: 175–182.

Goens, D. W., 1998. Weather and fire behavior factors related to the 1990 Dude Fire near Paysen, Arizona. *Preprints, 2nd Conference on Fire and Forest Meteorology*, Phoenix, Arizona. American Meteorological Society, Boston, Massachusetts, 153–158.

Goyer, R., 1996. A change in the weather. *Flying*, July: 74–77.

Graydon, D., and K. Hanson (eds.), 1997: *Mountaineering: The Freedom of the Hills*, 6th rev. ed. The Mountaineers, Seattle.

Gudiksen, P. H., 1989. Categorization of nocturnal drainage flows within the Brush Creek Valley and the variability of sigma theta in complex terrain. *J. Appl. Meteor.*, 28: 489–495.

Gudiksen, P. H., and J. J. Walton, 1981. Categorization of nocturnal drainage flows in the Anderson Creek Valley. *Preprints, 2nd Conference on Mountain Meteorology*, Steamboat Springs, Colorado. American Meteorological Society, Boston, Massachusetts, 218–221.

Haines, D. A., 1988. A lower atmospheric severity index for wildland fire. *National Weather Digest*, 13: 23–27. (Available from National Weather Association Publications, 6704 Wolke Court, Montgomery, AL 36116-2134)

Hardy, C. E., 1987. Aerial Application Equipment. Report 3400-Forest Pest Management. April 1987. USDA Forest Service, Equipment Development Center, Missoula, Montana.

Hawkes, H. B., 1947. Mountain and valley winds with special reference to the diurnal mountain winds of the Great Salt Lake region. Ph.D. dissertation, Ohio State University.

Hewson, E. W., and G. C. Gill, 1944. Meteorological investigations in Columbia River Valley near Trail, B.C. *Bur. Mines Bull.*, 453: 23–228.

Hindman, E. E., 1973. Air currents in a mountain valley deduced from the breakup of a stratus deck. *Mon. Wea. Rev.*, 101: 195–200.

Hoffman, M. S., 1987. *The World Almanac and Book of Facts 1988*. World Almanac, New York.

Holle, R. L., and R. E. López, 1993. Overview of real-time lightning detection systems and their meteorological uses. NOAA Technical Memorandum ERL NSSL-102, National Severe Storms Laboratory, Norman, Oklahoma.

Holzworth, G. C., 1972. Mixing Heights, Wind Speeds and Potential for Urban Air Pollution throughout the Contiguous United States. Office of Air Programs Pub. No. AP-101. U.S. Environmental Protection Agency, Washington, DC.

Justus, C. G., 1985. Wind Energy. In: D. D. Houghton (ed.), *Handbook of Applied Meteorology*. American Meteorological Society, Boston, 915–944.

López, R. E., R. L. Holle, A. I. Watson, and J. Skindlov, 1997. Spatial and temporal distributions of lightning over Arizona from a power utility perspective. *J. Appl. Meteor.*, 36: 825–831.

Martin, R. E., and J. D. Dell, 1978. Planning for Prescribed Burns in the Inland Northwest. General Technical Report PNW-76, Pacific Northwest Forest and Range Experiment Station, Portland, Oregon.

Math, F. A., 1934. Battle of the chinook wind at Havre, Montana. *Mon. Wea. Rev.*, 62: 54–57.

McCullough, E. C., 1968. Total daily radiant energy available extraterrestrially as a harmonic series in the

day of the year. *Arch. Meteor. Geophys. Bioclimatol., Ser. B*, 16: 129–143.
McKee, T. B., and R. D. O'Neal, 1989. The role of valley geometry and energy budget in the formation of nocturnal valley winds. *J. Appl. Meteor.*, 28: 445–456.
McLean, N., 1994. *Young Men and Fire*. University of Chicago Press, Chicago.
Meyer, S. J., and K. G. Hubbard, 1992. Nonfederal automated weather stations in the United States and Canada: A preliminary survey. *Bull. Amer. Meteor. Soc.*, 73: 449–457.
Müller, H., and C. D. Whiteman, 1988. Breakup of a nocturnal temperature inversion in the Dischma Valley during DISKUS. *J. Climate Appl. Meteor.*, 27: 188–194.
National Weather Service, 1995. *Federal Meteorological Handbook No. 1: Surface Weather Observations and Reports*, 5th ed. December 1995. National Oceanic and Atmospheric Administration, Washington, DC.
National Wildfire Coordinating Group, 1989. *Fireline Handbook*. NWCG Handbook 3, PMS410-1, NFES #0065, Washington, DC. (Available from National Interagency Fire Center, Attn Supply, 3905 Vista Ave, Boise, ID 83705)
National Wildfire Coordinating Group, 1994. Intermediate Wildland Fire Behavior S-290 Student Workbook, Publication NFES #2378. (Available from National Interagency Fire Center, Attn Supply, 3905 Vista Ave, Boise, ID 83705)
Nuss, W. A. 1986. Observations of a mountain tornado. *Mon. Wea. Rev.*, 114: 233–237.
Orgill, M. M., 1981. ASCOT: A Planning Guide for Future Studies. PNL-3656, ASCOT/80/4. Pacific Northwest Laboratory, Richland, Washington.
Orville, R. E., 1994. Cloud-to-ground lightning flash characteristics in the contiguous United States: 1989–1991. *J. Geophys. Res.*, 99(D5): 10833–10841.
Pamperin, H., and G. Stilke, 1985. Nächtliche Grenzschicht und LLJ im Alpenvorland nahe dem Inntalausgang. [Nocturnal boundary layer and low level jet near the Inn Valley exit] *Meteor. Rundsch.*, 38: 145–156.
Parish, T. R., 1982. Barrier winds along the Sierra Nevada Mountains. *J. Appl. Meteor.*, 21: 925–930.
Peterson, A. E., 1962. Lightning hazards to mountaineers. *Amer. Alpine J.*, 13: 143–154.
Philpot, C., and C. Schechter (eds.), 1995. Federal Wildland Fire Management Policy and Program Review. Final Report, December 18, 1995. (Available from National Interagency Fire Center, External Affairs Office, 3833 South Development Ave., Boise, ID 83705-5354)
Picot, J. J. C., and D. D. Kristmanson, 1997. *Forestry Pesticide Aerial Spraying: Spray Droplet Generation, Dispersion and Deposition*. Environmental Science and Technology Library, Kluwer Academic Publishers, Boston, Massachusetts.
Pielke, R. A., R. A. Stocker, R. W. Arritt, and R. T. McNider, 1991. A procedure to estimate worst-case air quality in complex terrain. *Environ. International*, 17: 559–574.
Potter, B. E., 1996. Atmospheric properties associated with large wildfires. *International J. Wildland Fire*, 6: 71–76.
Putnam, W. L., 1991. *The Worst Weather on Earth: A History of the Mount Washington Observatory*. American Alpine Club, New York.
Pyne, S. J., P. L. Andrews, and R. D. Laven, 1996. *Introduction to Wildland Fire*, 2nd ed. John Wiley and Sons, Inc., New York.
Quantick, H. R., 1985. *Aviation in Crop Protection, Pollution and Insect Control*. Collins Professional and Technical Books, London.
Reifsnyder, W. E., 1980. *Weathering the Wilderness: The Sierra Club Guide to Practical Meteorology*. Sierra Club Books, San Francisco.
Reiter, E. R., and M. Tang, 1984. Plateau effects on diurnal circulation patterns. *Mon. Wea. Rev.* 112: 638–651.
Reiter, R., H. Müller, R. Sladkovic, and K. Munzert, 1983. Aerologische Untersuchungen der tagesperiodischen Gebirgswinde unter besonderer Berücksichtigung des Windfeldes im Talquerschnitt. [Aerological investigations of diurnal mountain winds with special consideration of wind fields in the valley cross section] *Meteor. Rundsch.*, 36: 225–242.
Riehl, H., 1965. *Introduction to the Atmosphere*. McGraw-Hill, New York.
Rosenthal, J., 1972. Point Mugu Forecasters Handbook. PMR-TP-72-1. Pacific Missile Range, Point Mugu, California.
Rothermel, R. C., 1983. How to Predict the Spread and Intensity of Forest and Range Fires. Intermountain Research Station. General Technical Report INT-143. (Available from USDA Forest Service, Intermountain Research Station, 324 25th Street, Ogden, UT 84401)
Rothermel, R. C., 1991. Predicting Behavior and Size of Crown Fires in the Northern Rocky Mountains. Research Paper INT-438, USDA Intermountain Research Station. (Available from USDA Forest Service, Intermountain Research Station, 324 25th Street, Ogden, UT 84401)
Schaaf, C. L. B., J. Wurman, and R. M. Banta, 1986. Satellite Climatology of Thunderstorm Initiation Sites in the Rocky Mountains of Colorado and Northern New Mexico. AFGL-TR-86-0075, Environmental Research Papers, no. 952. Air Force Geophysics Laboratory, Hanscomb AFB, Massachusetts.
Schroeder, M. J., and C. C. Buck, 1970. *Fire Weather*. USDA Agriculture Handbook 360, Publication NFES 1174, PMS 425-1. (Available from National Interagency Fire Center, Attn Supply, 3905 Vista Ave, Boise, ID 83705)
Slade, D. (ed.), 1968. *Meteorology and Atomic Energy 1968*. U.S. Atomic Energy Commission, Division of Technical Information, Oak Ridge, Tennessee.
Steinacker, R., 1987. Orographie und Fronten [Orography and Fronts]. *Wetter und Leben*, 39: 65–70.
Stensrud, D. J., R. L. Gall, S. L. Mullen, and K. W. Howard, 1995. Model climatology of the Mexican monsoon. *J. Climate*, 8: 1775–1794.
Stull, R. B., 1995. *Meteorology Today for Scientists and*

Engineers: A Technical Companion Book. West Publishing Co., Minneapolis-St. Paul., Minnesota.

Svetz, F., and A. Barnett, 1997. SODAR and decision making during the Fork Fire. *Fire Management Notes*, 57(2): 28–31. (Available from USDA Forest Service, Fire and Aviation Management, 201 14th Street SW, P.O. Box 96090, Washington, DC 20090-6090)

Tang, M., and E. R. Reiter, 1984. Plateau monsoons of the Northern Hemisphere: A comparison between North America and Tibet. *Mon. Wea. Rev.*, 112: 617–637.

Teske, M. E., J. F. Bowers, J. E. Rafferty, and J. W. Barry, 1993. FSCBG: An aerial spray dispersion model for predicting the fate of released material behind aircraft. *Environmental Toxicology and Chemistry*, 12: 453–464.

Trijonis, J. C., and D. Shapland, 1978. Existing visibility levels in the U.S. Report to the Environmental Protection Agency, Grant 802815. Technology Services Corporation, Research Triangle Park, North Carolina.

Tucker, D. F., 1997. Surface mesonets of the Western United States. *Bull. Amer. Meteor. Soc.*, 78: 1485–1495.

Turner, D. B., 1969. Workbook of Atmospheric Dispersion Estimates. Public Health Service Environmental Health Series Pub. no. 999-AP-26. National Air Pollution Control Administration, Cincinnati, Ohio.

Wade, D. D., 1989. A Guide for Prescribed Fire in Southern Forests. USDA Forest Service Technical Publication R8-TP11, Publication NFES 2108. (Available from National Interagency Fire Center, Attn Supply, 3905 Vista Ave, Boise, ID 83705)

Wallace, J. M., 1975. Diurnal variations in precipitation and thunderstorm frequency over the conterminous United States. *Mon. Wea. Rev.*, 103: 406–419.

Wallace, J. M., and P. V. Hobbs, 1977. *Atmospheric Science: An Introductory Survey*. Academic Press, San Diego.

Wanner, H., and M. Furger, 1990. The Bise— limatology of a regional wind north of the Alps. *Meteor. Atmos. Phys.*, 43: 105–115.

Weaver, J. F., and J. F. Purdom, 1995. Observing forest fires with the GOES-8, 3.9 μm imaging channel. *Weather and Forecasting*, 10: 803–808.

Whiteman, C. D., 1982. Breakup of temperature inversions in deep mountain valleys: Part I. Observations. *J. Appl. Meteor.*, 21: 270–289.

Whiteman, C. D., 1986. Temperature inversion buildup in Colorado's Eagle Valley. *Meteorol. Atmos. Phys.*, 35: 220–226.

Whiteman, C. D., 1990. Observations of thermally developed wind systems in mountainous terrain. In: Blumen, W. (ed.), Atmospheric Processes Over Complex Terrain. Amer. Meteor. Soc., Boston, Massachusetts, *Meteor. Monogr.*, 23(no. 45): 5–42.

Whiteman, C. D., and K. J. Allwine, 1986. Extraterrestrial solar radiation on inclined surfaces. *Environ. Software*, 1: 164–169.

Whiteman, C. D., X. Bian, and S. Zhong, 1997. Low-level jet climatology from enhanced rawinsonde observations at a site in the southern Great Plains. *J. Appl. Meteor.*, 36: 1363–1376.

Whiteman, C. D., and J. G. Whiteman, 1974: An Historical Climatology of Damaging Downslope Windstorms at Boulder, Colorado. NOAA Tech Report ERL 336-APCL 35, NOAA Environmental Research Laboratories, Boulder, Colorado.

Williams, P., Jr., 1952. Wasatch winds of northwest Utah. *Weatherwise*, December: 130–132.

World Meteorological Organization, 1987. *International Cloud Atlas, Vol. II*. WMO, Geneva, Switzerland.

Zishka, K. M., and P. J. Smith, 1980. The climatology of cyclones and anticyclones over North America and surrounding ocean environs for January and July, 1950–77. *Mon. Wea. Rev.*, 108: 387–401.

Appendixes

A. Formulas

Humidity Variables

If temperature and pressure measurements are available, a number of secondary humidity variables can be computed from the measured values of relative humidity. The choice of humidity variable is a practical matter, and conversions among humidity variables are frequently made.

Relative humidity as a percentage is given by

$$RH = 100\frac{e}{e_s}, \tag{A.1}$$

where e is the actual water *vapor pressure* and e_s is the water vapor pressure that the air would have if it were saturated at its current temperature. The maximum or saturation vapor pressure is the vapor pressure that occurs when the air just becomes saturated. It is a function of temperature only and is defined by the Clausius–Clapeyron equation

$$e_s = e_0 \exp\left[\frac{L}{R_v}\left(\frac{1}{T_0} - \frac{1}{T}\right)\right] \tag{A.2}$$

where $e_0 = 6.11$ mb is the saturation vapor pressure at $T_0 = 273$ K, $R_v = 461$ J K^{-1} kg^{-1} is the gas constant for water vapor, and T is absolute temperature in Kelvin (absolute temperature can be calculated by adding 273.16 to the temperature in degrees Celsius). The variation of saturation vapor pressure with temperature is shown in figure A.1. L is the latent

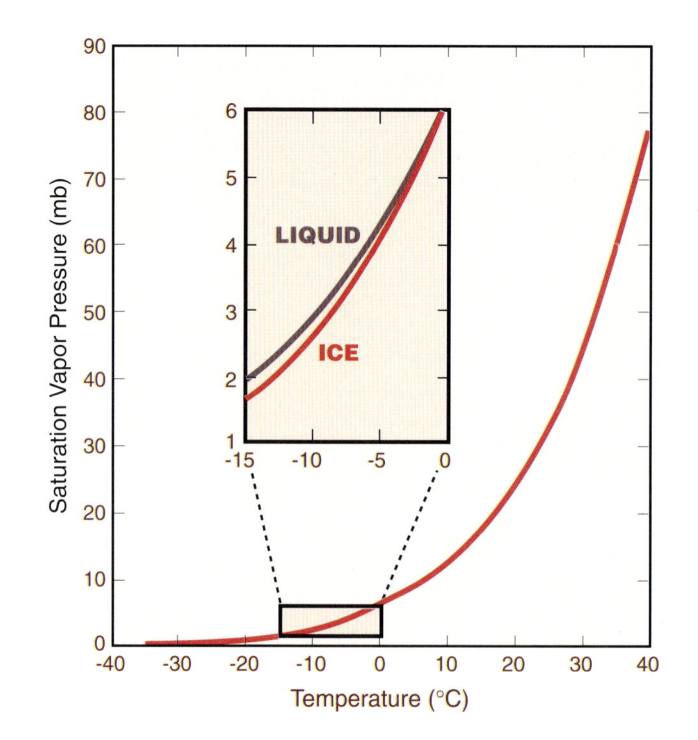

Figure A.1 Saturation vapor pressure over liquid water as a function of air temperature. The inset figure shows that the saturation vapor pressure over ice is lower than over water.

heat, or the heat released or absorbed per unit mass when water changes from one phase (liquid, solid, or vapor) to another. The saturation vapor pressure with respect to an ice surface is lower than for a surface of liquid water. Thus snow crystals grow at the expense of water droplets in a cloud containing both liquid and solid phases of water. To calculate the saturation vapor pressure over water, the latent heat of vaporization $Lv = 2.5$ MJ/kg is substituted for L in eqn. A.2; to calculate the saturation vapor pressure over ice, the latent heat of sublimation $L_d = 2.83$ MJ/kg is entered.

The mixing ratio r is the ratio of the mass of water vapor M_w in an atmospheric volume to the mass of dry air M_d in the volume and can be calculated from the equation

$$r = \frac{M_w}{M_d} = \frac{\varepsilon e}{p - e} \tag{A.3}$$

where $\epsilon = 0.622$ is the ratio of the gas constants for water vapor and dry air and p is pressure. Mixing ratio values are typically between 0.001 and 0.015 kg/kg, although they may reach 0.030 kg/kg in warm, saturated tropical atmospheres.

Specific humidity q is the ratio of the mass of water vapor to the mass of moist air in an atmospheric volume:

$$q = \frac{M_w}{M_d + M_w} = \frac{\varepsilon e}{p} \tag{A.4}$$

Because M_w is usually much smaller than M_d, the numerical values for mixing ratio and specific humidity are often quite close.

Dew-point temperature T_d is the temperature to which air must be cooled to become saturated at constant pressure. Dew-point temperature is always less than or equal to air temperature T, and air becomes saturated when the dew-point temperature equals the air temperature. Thus, the dew-point depression $T - T_d$ is a relative measure of the dryness of the air. Dew-point temperature is determined from eqn. A.2 by replacing e_s by e and T by T_d and solving for T_d to obtain

$$T_d = \left[\frac{1}{T_0} - \frac{R_v}{L} \ln \frac{e}{e_0} \right]^{-1} \quad \text{(A.5)}$$

The dew-point temperature is measured by a dew-point hygrometer, a tiny mirror that is cooled until dew first forms on it. The mirror temperature at the time of dew formation is the dew-point temperature.

Droplet Settling Velocities

Particles in the atmosphere are acted upon by gravity and settle toward the earth's surface with a speed that depends on the balance between gravity and the frictional drag of the atmosphere. Small particles maintain a spherical shape, whereas large droplets become flattened as they fall. This change in particle geometry with particle size affects the atmosphere's drag on the particles and thus their settling velocities or fall speeds. Table A.1 provides typical settling velocities for water droplets of different diameter.

Table A.1 Settling Velocities (cm/s) of Different Diameter Water Droplets in Still Air at Various Altitudes

Diameter			Altitude			
(mm)	micrometers	Droplet type	$z = 0$ km	$z = 2$ km	$z = 4$ km	$z = 6$ km
0.01	10		0.30	0.32	0.33	0.34
0.02	20	typical cloud droplet	1.22	1.26	1.31	1.36
0.05	50		7.60	7.90	8.20	8.52
0.1	100	large cloud droplet	23.77	25.19	26.69	28.29
0.2	200	drizzle	69.00	73.12	77.49	82.12
0.5	500		194.88	211.32	229.15	248.48
1	1000	small raindrop	377.62	409.48	444.03	481.49
2	2000	typical raindrop	637.02	690.76	749.04	812.24
5	5000	large raindrop	897.31	973.01	1055.11	1144.13

Velocities for droplets of intermediate sizes not listed in the table can be calculated directly from Best's (1950) formulas:

$$V = 3040 d^2 \exp(0.0191\, z) \qquad d < 0.05 \text{ mm}$$

$$V = 191 \exp(0.0290\, z) \left\{1 - \exp\left[-(d/0.316)^{1.754}\right]\right\} \qquad 0.05 \le d \le 0.30 \text{ mm} \quad \text{(A.6)}$$

$$V = 932 \exp(0.0405\, z) \left\{1 - \exp\left[-(d/1.77)^{1.147}\right]\right\} \qquad d > 0.30 \text{ mm}$$

where droplet diameter d is in mm, height z is in km, and the settling velocities V are in cm/s.

The Ideal Gas Law: Computing Temperature, Pressure, or Air Density

The *ideal gas law,*

$$p = \rho R^* T \tag{A.7}$$

where p is pressure, ρ is air density, R^* is the gas constant for air, and T is absolute temperature, can be used to calculate pressure, density, or temperature when the other two variables are known. Equation A.7 is solved for the variable of interest after the known values of the other variables are entered into the equation. It should be noted that the so-called gas constant is not really a constant. Air is a mixture of gases, and the gas constant changes with the composition of the mixture, which is a function primarily of the moisture content. This equation is usually solved by assigning the gas constant for dry air ($R_d = 2.87053$ mb K^{-1} m^3 kg^{-1}) to R^* and taking the effects of moisture into account by adjusting the temperature slightly by writing eqn. A.7 as follows:

$$p = \rho R_d T_v \tag{A.8}$$

where T_v, called the virtual temperature, is defined by

$$T_v = T(1 + 0.61\, r) \tag{A.9}$$

where T is in Kelvin and r (see eqn. A.3), the water-vapor mixing ratio, a measure of the air's moisture content, is defined as the number of kilograms of water vapor per kilogram of dry air.

Wind Chill Equivalent Temperature

Several different formulas are used to compute wind chill equivalent temperatures. The formula by Stull (1995) follows:

$$T_{\text{wind chill}} = T_{\text{skin}} - \left(\frac{M + M_0}{M_0}\right)^{0.21} (T_{\text{skin}} - T_{\text{air}}) \tag{A.10}$$

where T_{skin} is 33°C, M is the actual wind speed (m/s), and M_0 is 2 m/s. This definition assumes an effective wind speed of 2 m/s produced by a person walking in calm conditions. Table 4.2 presents wind chill equivalent temperatures calculated using eqn. A.10.

Wind Speed Profiles

An estimate of the wind at one height when the wind has been measured at a different height can be calculated using the equation

$$U_2 = U_1 \frac{\ln(z_2/z_0)}{\ln(z_1/z_0)} \quad \text{(A.11)}$$

where U_1 is the wind speed at height z_1, U_2 is the wind speed at height z_2, z_0 is the roughness as indicated in the table on page 72 and ln is the natural logarithm.

The formula is most appropriate for neutral stability. For example, suppose the measured wind speed at 3 m is 4 m/s. To estimate the speed at 10 m over a flat cultivated area with low crops and occasional obstacles ($z_0 = 0.1$ m) under neutral stability conditions, the formula would be applied as follows:

$$U_2 = 4\frac{\ln(10/0.1)}{\ln(3/0.1)} = 5.4 \text{ m/s}$$

B. Psychrometric Tables

Table B.1 Psychrometric Table for Determining Relative Humidity (%) from Dry Bulb Temperature and Wet Bulb Depression in °F at 850-mb Pressure or 5000-ft Elevation[a]

	Relative Humidity (%)								
	Wet Bulb Depression (T − T_w) (°F)								
Dry Bulb (°F)	1	2	3	4	6	8	10	15	20
0	76	52	27	3					
5	80	60	40	20					
10	84	67	50	33	0				
15	86	72	58	44	16				
20	88	76	64	52	28	5			
25	90	79	69	58	38	18			
30	91	82	73	64	46	29	12		
35	92	84	76	68	52	37	23		
40	93	86	78	71	58	44	31	1	
45	94	87	81	74	62	50	38	11	
50	94	88	82	77	65	55	44	19	
55	95	89	84	78	68	58	49	26	6
60	95	90	85	80	71	62	53	32	13
65	95	91	86	82	73	64	56	37	20
70	96	91	87	83	74	67	59	41	25
75	96	92	88	84	76	69	61	45	30
80	96	92	88	85	77	70	64	48	34
85	96	93	89	85	78	72	65	51	37
90	97	93	90	86	79	73	67	53	40
95	97	93	90	87	80	74	68	55	43
100	97	94	91	87	81	75	70	57	45

[a]Uses Teten's formula for saturation vapor pressure over water at all temperatures.

Table B.2 Psychrometric Table for Determining Relative Humidity (%) from Dry Bulb Temperature and Wet Bulb Depression in °C at 850-mb Pressure or 5000-ft Elevation[a]

	Relative Humidity (%)								
	Wet Bulb Depression ($T-T_w$) (°C)								
Dry Bulb (°C)	0.5	1.0	1.5	2.0	3.0	5.0	7.0	10.0	15.0
−20.0	75	49	23						
−17.5	79	57	36	14					
−15.0	82	64	46	28					
−12.5	85	70	54	39	9				
−10.0	87	74	61	48	22				
−7.5	89	77	66	55	32				
−5.0	90	80	70	60	41	3			
−2.5	91	83	74	65	48	15			
0.0	92	84	77	69	54	25			
2.5	93	86	79	72	59	33	8		
5.0	94	87	81	75	63	39	17		
7.5	94	88	83	77	66	45	25		
10.0	95	89	84	79	69	49	31	6	
12.5	95	90	85	80	71	53	36	13	
15.0	95	91	86	82	73	56	41	20	
17.5	96	91	87	83	75	59	45	25	
20.0	96	92	88	84	76	62	48	30	3
22.5	96	92	89	85	78	64	51	34	8
25.0	97	93	89	86	79	66	54	37	13
27.5	97	93	90	86	80	67	56	40	18
30.0	97	94	90	87	81	69	58	43	21
32.5	97	94	91	88	82	70	60	45	25
35.0	97	94	91	88	82	71	61	47	28
37.5	97	94	91	89	83	72	63	49	30
40.0	98	95	92	89	84	73	64	51	33[a]

Uses Teten's formula for saturation vapor pressure over water at all temperatures.

C. Sources of Information on Weather Monitoring and Instrumentation

Dobson, F., L. Hasse, and R. Davis, 1980. *Air-Sea Interaction: Instruments and Methods*. Plenum Press, New York.

Finklin, A. I., and W. C. Fischer, 1990. *Weather Station Handbook—An Interagency Guide for Wildland Managers*. NFES 2140. National Wildfire Coordinating Group, Boise Interagency Fire Center, Boise, Idaho.

Fritschen, L. J., and L. W. Gay, 1979. *Environmental Instrumentation*. Springer-Verlag, New York.

Lee, R., 1978. *Forest Microclimatology*. Columbia University Press, New York.

Lenschow, D. H., 1986. *Probing the Atmospheric Boundary Layer*. American Meteorological Society, Boston.

Mason, C. J., and H. Moses, 1984. Meteorological Instrumentation. In: D. Randerson (ed.), *Atmospheric Science and Power Production*. DOE/TIC-27601. Technical Information Center, U.S. Department of Energy, Washington, DC, 81–135.

Mazzarella, D. A., 1985. Measurements Today. In: D. A. Houghton (ed.), *Handbook of Applied Meteorology*. John Wiley and Sons, New York, 283–328.

Meteorological Office, 1982. *Observer's Handbook*, 4th ed. Her Majesty's Stationery Office, London.

Middleton, W. E. K., and A. F. Spilhaus, 1953. *Meteorological Instruments*. University of Toronto Press, Toronto.

Munn, R. E., 1985. Observing Networks. In: D. A. Houghton (ed.), *Handbook of Applied Meteorology*. John Wiley and Sons, New York, 473–482.

Neff, W. D., 1990. Remote sensing of atmospheric processes over complex terrain. In:

W. Blumen (ed.), *Atmospheric Processes over Complex Terrain. Meteor. Monogr.*, 23 (no. 45), Amer. Meteor. Soc., Boston, Massachusetts, 173–228.
Rosenberg, N. J., 1974. *Microclimate: The Biological Environment.* John Wiley and Sons, New York.
Sellers, W. D., 1965. *Physical Climatology.* The University of Chicago Press, Chicago.
World Meteorological Organization, 1983. *Guide to Meteorological Instruments and Methods of Observation*, 5th ed. WMO-No. 8. World Meteorological Organization, Geneva.

D. Units, Unit Conversion Factors, and Time Conversions

Système International (SI) Units

The Système International (SI) is an international standard for general and scientific measurement units. The four basic SI units of length, mass, time, and temperature are meters (m), kilograms (kg), seconds (s) and Kelvin (K). Other derived units can be conveniently obtained from these base units (for example, units for velocity, power, momentum, energy). When the value of a quantity in any of these derived or base units is very large or very small, prefixes (table D.1) can be used with the units to eliminate excessive zeros or powers of ten. For example, a power generation rate of 1,000,000 W (i.e., Watts) can be called a 1-MW power generation rate.

Table D.1 Prefixes Used to Describe Multiples or Fractions of Ten

Abbreviation	Prefix	Definition	Scientific Notation	Decimal Notation
T	tera-	one trillion (U.S.)[a]	10^{12}	1,000,000,000,000
G	giga-	one billion (U.S.)[a]	10^{9}	1,000,000,000
M	mega-	one million	10^{6}	1,000,000
k	kilo-	one thousand	10^{3}	1,000
h	hecto-	one hundred	10^{2}	100
da	deka-	ten	10	10
d	deci-	one tenth	10^{-1}	0.1
c	centi-	one hundredth	10^{-2}	0.01
m	milli-	one thousandth	10^{-3}	0.001
μ	micro-	one millionth	10^{-6}	0.000001
n	nano-	one billionth (U.S.)[a]	10^{-9}	0.000000001
p	pico-	one trillionth (U.S.)[a]	10^{-12}	0.000000000001

[a]These definitions of billion and trillion are applicable in the United States and France. Usage of these terms is different in Great Britain and Germany, where a billion is 10^{12} and a trillion is 10^{18}.

Scientific or exponential notation is a second shorthand way of depicting very large or small numbers by eliminating the use of many zeros. The notation 10^n means that the number 10 is multiplied by itself n times (e.g., $10^3 = 10 \times 10 \times 10 = 1000$), where n is called the exponent. The notation 10^{-n} indicates the reciprocal of 10^n, or $1/10^n$ (e.g., $10^{-3} = 1/(10 \times 10 \times 10) = 1/1000 = 0.001$). It is convenient to express large or small numbers in terms of these exponents. Multiplication of numbers in exponential notation is accomplished very easily by simply adding the exponents (e.g., $10^3 \times 10^2 = 10^5$).

A Kelvin unit of temperature (1K) is equivalent to a Celsius unit of temperature (1°C). The Kelvin and Celsius scales, however, differ in respect to their starting points. The Kelvin scale starts at absolute zero, where theory states that molecular motion ceases and a body contains no heat energy. The Celsius scale, on the other hand, has its starting point at the freezing point of water (0°C).

Unit Conversion Factors

The following tables facilitate conversions among units of length, mass, volume, speed, and pressure (table D.2), and units of temperature (table D.3).

Table D.2 Conversion Factors for Units of Length, Mass, Volume, Speed, and Pressure

To convert from	to	multiply by
Length		
feet	meters	0.3048
meters	feet	3.2808
centimeters	inches	0.3937
inches	centimeters	2.54
miles	kilometers	1.6093
nautical miles	kilometers	1.8532
kilometers	miles	0.6214
nautical miles	miles	1.1516
Mass		
kilograms	pounds	2.2046
pounds	kilograms	0.4536
Volume		
gallons	liters	3.7853
liters	gallons	0.2642
Speed (see also table 5.1)		
miles per hour	meters per second	0.4470
miles per hour	knots	0.8684
miles per hour	kilometers per hour	1.6093
knots	meters per second	0.5148
knots	miles per hour	1.1516
knots	kilometers per hour	1.8533
kilometers per hour	meters per second	0.2778
kilometers per hour	miles per hour	0.6214
kilometers per hour	knots	0.5396
meters per second	miles per hour	2.2369
meters per second	knots	1.9425
meters per second	kilometers per hour	3.6
Pressure		
pascals	millibars	0.01
pascals	inches of mercury	0.0002953
pascals	millimeters of mercury	0.007501
pascals	pounds per square inch	0.0001450
inches of mercury	millibars	33.8639
inches of mercury	pascals	3386.39
inches of mercury	millimeters of mercury	25.4
inches of mercury	pounds per square inch	0.4912
millimeters of mercury	millibars	1.3332
millimeters of mercury	pascals	133.32
millimeters of mercury	inches of mercury	0.03937
millimeters of mercury	pounds per square inch	0.01934
pounds per square inch	millibars	68.9476
pounds per square inch	pascals	6894.76
pounds per square inch	inches of mercury	2.0360
pounds per square inch	millimeters of mercury	51.7149
millibar	pascals	100
millibar	inches of mercury	0.02953
millibar	millimeters of mercury	0.7501
millibar	pounds per square inch	0.01450

Table D.3 Formulas for Conversions between Units of Temperature

To convert temperatures from	to	use this formula
Celsius (C)	Fahrenheit (F)	F = 32 + (9/5) C
Fahrenheit (F)	Celsius (C)	C = (5/9) (F − 32)
Celsius (C)	Kelvin (K)	K = C + 273.16
Kelvin (K)	Celsius (C)	C = K − 273.16

Time Conversions

The boundaries of the standard time zones in North America are shown in figure D.1. The abbreviations for these time zones, their standard meridians, and their number of hours in advance of Coordinated Universal Time (UTC) are provided in table D.4. Greenwich mean time (GMT) and Zulu (Z) time, terms that are used on some weather charts, are equivalent to UTC. Table D.5 provides formulas to convert between local standard time (LST), daylight savings time or local daylight time (LDT), and Coordinated Universal Time. Finally, the historical and present definitions of daylight savings time, a politically governed time system, are provided for the United States in the following Point of Interest.

Table D.4 Standard Time Zones in North America

Name of standard time zone	Abbreviation	Standard Meridian	H = Hours Earlier than UTC
Newfoundland	NST	52 1/2° W	3 1/2
Atlantic	AST	60	4
Eastern	EST	75	5
Central	CST	90	6
Mountain	MST	105	7
Pacific	PST	120	8
Alaska	AST	135	9
Aleutian-Hawaii	AHST	150	10
Bering	BST	165	11

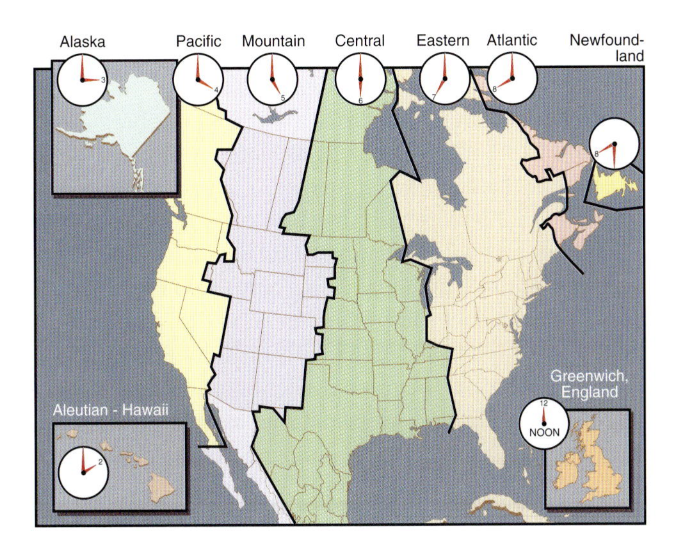

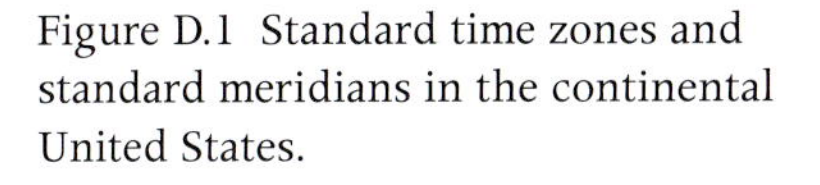
Figure D.1 Standard time zones and standard meridians in the continental United States.

DEFINITION OF DAYLIGHT SAVINGS TIME (I.E., LOCAL DAYLIGHT TIME, LDT) IN THE UNITED STATES (HOFFMAN, 1987)

LDT is observed by advancing the clock one hour. In the United States, under the Uniform Time Act (UTA), which became effective in 1967, clocks were advanced at 2 AM on the last Sunday in April and returned to local standard time (LST) at 2 AM on the last Sunday in October. This act applied to all states, the District of Columbia, and all U.S. possessions. Any state, by law, could exempt itself; Arizona, Indiana, Hawaii, Puerto Rico, the Virgin Islands, and American Samoa have been exempt since a 1972 amendment to the UTA. Local zone boundaries in some states (Kansas, Texas, Florida, Michigan, and Arkansas) were modified by the U.S. Department of Transportation during the years 1986–1988. Congress put most of the nation on LDT during the period 6 January 1974 through 26 October 1975 to conserve energy, but further legislation in October 1974 restored standard time from the last Sunday of that month to the last Sunday in February 1975. At the end of 1975, this temporary legislation was not renewed and the nation returned to the older system of the end of April to the end of October. On 8 July 1986, the start of daylight savings time was changed to the first Sunday of April, leaving the ending time as the last Sunday of October. LDT is, thus, presently observed from 2 AM on the first Sunday of April to 2 AM on the last Sunday of October.

Table D.5 Time Conversions

Conversion	Formula
LST to LDT	LDT = LST + 1
LDT to LST	LST = LDT − 1
LST to UTC	UTC = LST − H
LDT to UTC	UTC = LDT − H − 1
UTC to LST	LST = UTC + H
UTC to LDT	LDT = UTC + H + 1

E. Solar Radiation on Slopes

The intensity of solar radiation received on a unit of surface area in the mountains varies considerably across the landscape because of the differing inclination and azimuth angles of mountain slopes. Because solar radiation receipt is the main driving force for surface energy balance differences in a mountainous region that lead to local wind systems (chapter 11), algorithms for the calculation of solar radiation receipt on slopes assume an important role in mountain meteorology. In this appendix, the differing rates of receipt of solar radiation on variously oriented slopes at different times of year are discussed, and FORTRAN algorithms are provided for making detailed theoretical calculations.

Any sloping plane surface can be defined by specifying its inclination and azimuth angles. The inclination angle is the surface's tilt angle relative to the horizon; the azimuth angle is the direction that the surface faces, as determined from a compass bearing. A vertical, north-facing wall has an inclination angle of 90° and an azimuth angle of 0°. A slope with an azimuth angle of 90° faces east, a 180°-azimuth slope faces south, etc.

On a horizontal surface, sunrise, sunset, and day length, the length of the daily period of direct solar radiation, vary significantly with latitude and day of year (figure 1.1). In mountainous terrain, sunrise, sunset, and day length are also affected by the slope azimuth and inclination angles. This is illustrated for slopes at 40° N latitude in figure E.1, where the calculations assume a transparent atmosphere, a solar constant of 1353 W m^{-2}, and a longitude that is a multiple of 15° (times must be advanced 4 minutes for each degree of longitude west and delayed by 4 minutes for each degree of longitude east of these standard longitudes). The gray shading in the figure indicates nighttime or a shaded slope, and the gray-white boundary at the tops (bottoms) of the subfigures indicates sunrise (sunset) times. The left-hand side of each subfigure indicates insolation on a slope with a 0° inclination angle (a horizontal surface) and is identical for each set of four subfigures in a column. The right-hand side of each subfigure gives the insolation on a vertical wall oriented as stated to the right of its row (e.g., north-facing). The dates at the tops of the columns are approximate dates of the solstices and equinoxes. The subfigures for west- and east-facing slopes are asymmetric in time and slope inclination angle and are mirror images. Maximum insolation occurs in the morning on an east-facing slope, in the afternoon on a west-facing slope, and at noon on a south-facing slope. On north- and south-facing slopes the insolation is symmetric about noon. The day length on a sloping surface can be much shorter than on a horizontal surface at the same date and latitude. Low-angle north-facing slopes are in sunlight for only a short period in midwinter, and higher angle slopes receive no direct sunlight at all, even though an adjacent horizontal surface may be illuminated for hours. For south-facing slopes, peak insolation occurs on slopes of different inclination angles at different times of year. The inclination angle for maximum insolation at 40°N latitude is about 15° on June 21 and 65° on December 21. At the winter solstice, sunrise and sunset occur at the same time on all south-facing slopes because the sun rises in the southeast and sets in the southwest. At the summer solstice, the sun rises in the northeast and sets in the northwest, so sunrise on south-facing slopes is delayed and sunset is advanced. The delay in sunrise time and advance in sunset time increase with steeper slopes. Double sunrises and sunsets can occur on steep north-facing slopes in the summer half-year. The slopes are in direct sunlight early in the morning when the sun rises to the northeast and in the evening when the sun sets to the northwest, but slopes are in shade during much of the day when the sun takes a path through the southern sky during midday.

Computer programs are provided here for the calculation of solar inclination and azimuth angles, sunrise and sunset times, and instantaneous and daily total solar radiation on slopes. The equations used in the calculation of these parameters, as implemented in the following FORTRAN subroutines, were documented by Whiteman and Allwine (1986).

A FORTRAN subroutine called SOLSUB, listed in table E.1, performs all the basic solar radiation computations. Sample main programs that use this subroutine to compute (1) sunrise and sunset times and theoretical values of daily total solar radiation on a slope of any azimuth and incli-

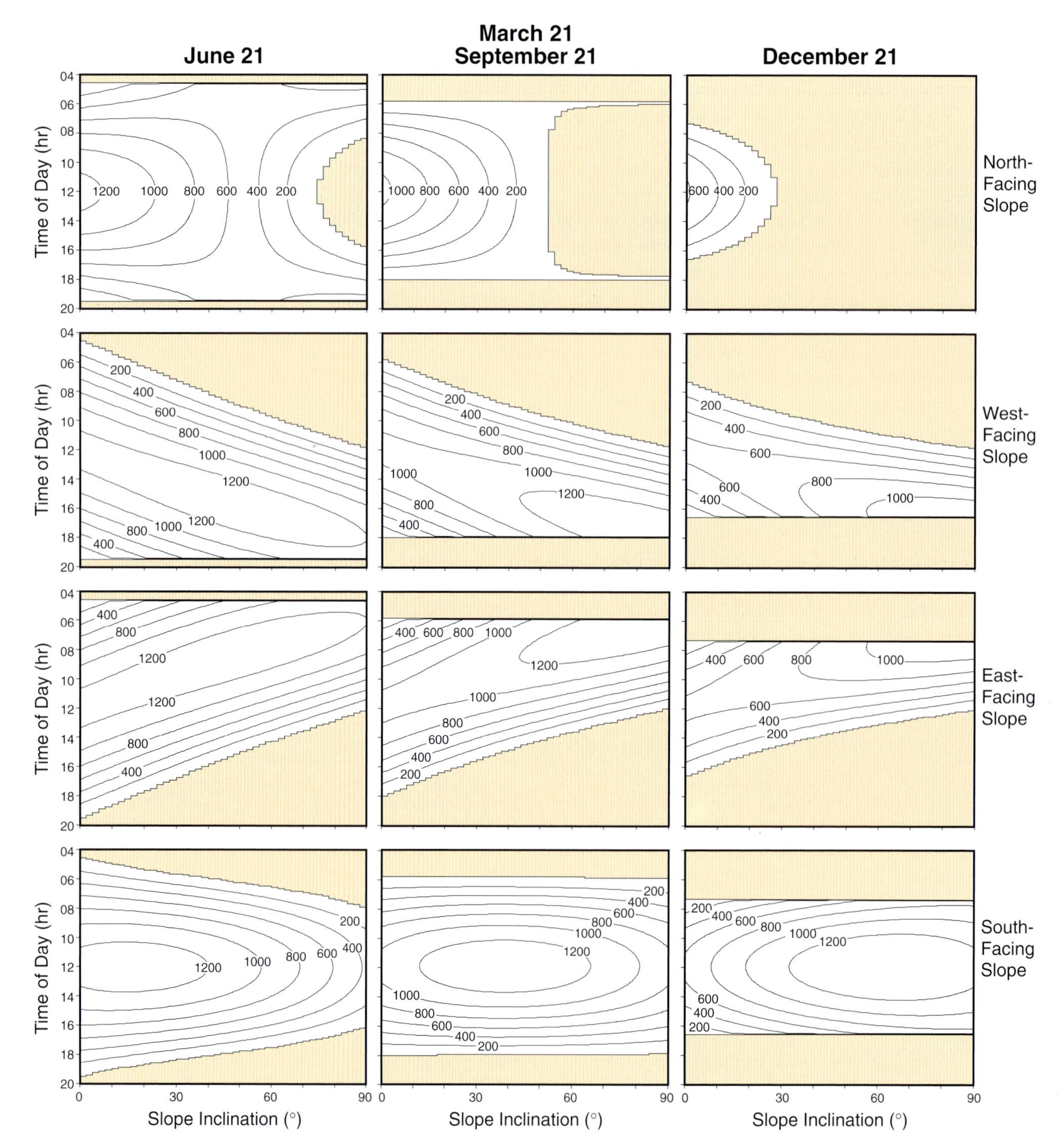

Figure E.1 Theoretical solar radiation (W m^{-2}) at 40°N latitude on slopes of different inclination (0–90°) and azimuth (north-, west-, east-, and south-facing) angles at different times of day (0400–2000 LST) for the approximate dates of the solstices and equinoxes (see section 1.1).

nation angle at any latitude and day of year and (2) sun position angles (solar azimuth and elevation) and instantaneous insolation values on a slope of any azimuth and inclination angle at any latitude, day of year, and time of day are provided in tables E.2 and E.3, respectively.

The programs listed in tables E.1–E.3, with minor modifications, can be used to determine (1) day length and sunrise and sunset times as a function of latitude, (2) the maximum altitude angle of the sun as a function of latitude, (3) the bearing and length of shadows cast by vertical structures, and (4) the radiation differences between opposing valley sidewalls.

Table E.1 Listing of Subroutine SOLSUB

```
      SUBROUTINE SOLSUB (LONG, LAT, AZ, IN, SC, DAILY, MO, IDA,
     +                   IHR, MM, OUT1, OUT2, OUT3, COSBETA)
C+++++++++++++++++++++++++++++++++++++++++++++++++++++++++++++++++++++
C     C. D. WHITEMAN AND K. J. ALLWINE                                |
C     PACIFIC NORTHWEST LABORATORY,   RICHLAND, WA  99352             |
C---------------------------------------------------------------------|
C     USES GARNIER & OHMURA (1968) AND MCCULLOUGH (1968) SCHEMES TO    |
C     CALCULATE EXTRATERRESTRIAL SOLAR RADIATION AT A REQUESTED TIME  |
C     OF DAY ON A SLOPE OF ANY AZIMUTH AND INCLINATION ANGLE GIVEN THE|
C     LATITUDE AND LONGITUDE, DATE, AND SLOPE AZIMUTH AND INCLINATION |
C     ANGLES.  ALSO INTEGRATES INSTANTANEOUS VALUES TO DETERMINE      |
C     THE DAILY TOTAL IF REQUESTED.                                   |
C---------------------------------------------------------------------|
C     INPUTS: LONG = (-180. TO 180. DEG), LAT = (-90. TO 90. DEG),    |
C     AZ = SLOPE AZIMUTH ANGLE (0. TO 359.), IN = SLOPE INCLINATION   |
C     ANGLE (0. TO 90. DEG), SC = SOLAR CONSTANT (E.G. 1353. W/M**2), |
C     DAILY = DAILY TOTAL OR INSTANTANEOUS (.TRUE. OR .FALSE.),       |
C     MO = MONTH (1 TO 12), IDA = DAY (1 TO 31), IHR = HOUR (0 TO 23),|
C     MM = MINUTE (00 TO 59).                                         |
C---------------------------------------------------------------------|
C     OUTPUTS: IF DAILY IS TRUE >> OUT1 = TOTAL RADIATION (MJ/M**2),  |
C     OUT2 = SUNRISE (HOURS LST), OUT3 = SUNSET (HOURS LST).          |
C     IF DAILY IS FALSE >> OUT1 = INSTANTANEOUS RADIATION (W/M**2),   |
C     OUT2 = SUN'S ZENITH ANGL (DEG), OUT3 = SUN'S AZIMUTH ANGL (DEG).|
C+++++++++++++++++++++++++++++++++++++++++++++++++++++++++++++++++++++

      LOGICAL DAILY, FIRST
      INTEGER NDAY(12)
      REAL    LAT, LONG, LONGCOR, LONGSUN, IN, INSLO
      REAL    ACOF(4), BCOF(4)

C *** CONSTANTS.
      DATA ACOF/ 0.00839,-0.05391,-0.00154,-0.00222/
      DATA BCOF/-0.12193,-0.15699,-0.00657,-0.00370/
      DATA NDAY/0,31,59,90,120,151,181,212,243,273,304,334/
      DATA DZERO,ECCENT,PI,CALINT/80.,.0167,3.14159,1./
      RTOD = PI/180.
      DECMAX=(23.+26./60.)*RTOD
      OMEGA=2.*PI/365.
      ONEHR=15.*RTOD

C *** JULIAN DATE.
      D=FLOAT(IDA+NDAY(MO))
C *** RATIO OF RADIUS VECTORS SQUARED.
      OMD=OMEGA*D
      OMDZERO=OMEGA*DZERO
      RDVECSQ=1./(1.-ECCENT*COS(OMD))**2
```

(continued)

Table E.1 Continued

```
C *** DECLINATION OF SUN.
      LONGSUN=OMEGA*(D-DZERO)+2.*ECCENT*(SIN(OMD)-SIN(OMDZERO))
      DECLIN=ASIN(SIN(DECMAX)*SIN(LONGSUN))
      SDEC=SIN(DECLIN)
      CDEC=COS(DECLIN)
C *** CHECK FOR POLAR NIGHT OR DAY.
      ARG=((PI/2.)-ABS(DECLIN))/RTOD
      IF(ABS(LAT).GT.ARG) THEN
        IF((LAT.GT.0..AND.DECLIN.LT.0.) .OR.
     +       (LAT.LT.0..AND.DECLIN.GT.0.)) THEN
          OUT1=0.
          OUT2=0.
          OUT3=0.
          RETURN
        ENDIF
        SR=-PI
      ELSE
C *** SUNRISE HOUR ANGLE.
        SR=-ABS(ACOS(-TAN(LAT*RTOD)*TAN(DECLIN)))
      ENDIF
C *** STANDARD TIME MERIDIAN FOR SITE.
      STDMRDN=NINT(LONG/15.)*15.
      LONGCOR=(LONG-STDMRDN)/15.
C *** COMPUTE TIME CORRECTION FROM EQUATION OF TIME.
      B=2.*PI*(D-.4)/365.
      EM=0.
      DO (I=1,4)
        EM=EM+(BCOF(I)*SIN(I*B)+ACOF(I)*COS(I*B))
      ENDDO
C *** TIME OF SOLAR NOON.
      TIMNOON=12.-EM-LONGCOR

      AZSLO=AZ*RTOD
      INSLO=IN*RTOD
      SLAT=SIN(LAT*RTOD)
      CLAT=COS(LAT*RTOD)
      CAZ=COS(AZSLO)
      SAZ=SIN(AZSLO)
      SINC=SIN(INSLO)
      CINC=COS(INSLO)

      IF (DAILY) THEN
C *** COMPUTE DAILY TOTAL.
        IHR=0
        MM=0
        HINC=CALINT*ONEHR/60.
        IK=(2.*ABS(SR)/HINC)+2.
        FIRST=.TRUE.
        OUT1=0.
        DO (I=1,IK)
          H=SR+HINC*FLOAT(I-1)
          COSZ=SLAT*SDEC+CLAT*CDEC*COS(H)
          COSBETA=CDEC*((SLAT*COS(H))*(-CAZ*SINC)-
     +            SIN(H)*(SAZ*SINC)+(CLAT*COS(H))*CINC)+
     +            SDEC*(CLAT*(CAZ*SINC)+SLAT*CINC)
          EXTRA=SC*RDVECSQ*COSZ
          IF(EXTRA.LT.0.) EXTRA=0.
          EXTSLO=SC*RDVECSQ*COSBETA
          IF(EXTRA.LE.0. .OR. EXTSLO.LT.0.) EXTSLO=0.
          IF(FIRST .AND. EXTSLO.GT.0.) THEN
```

(continued)

Table E.1 Continued

```
            OUT2=(H-HINC)/ONEHR+TIMNOON
            FIRST=.FALSE.
          ENDIF
          IF(.NOT.FIRST .AND. EXTSLO.LE.0.) OUT3=H/ONEHR+TIMNOON
          OUT1=EXTSLO+OUT1
        ENDDO
        OUT1=OUT1*CALINT*60./1000000.
      ELSE
C *** COMPUTE AT ONE TIME.
        T1=FLOAT(IHR)+FLOAT(MM)/60.
        H=ONEHR*(T1-TIMNOON)
        COSZ=SLAT*SDEC+CLAT*CDEC*COS(H)
        COSBETA=CDEC*((SLAT*COS(H))*(-CAZ*SINC)-
     +          SIN(H)*(SAZ*SINC)+(CLAT*COS(H))*CINC)+
     +          SDEC*(CLAT*(CAZ*SINC)+SLAT*CINC)
        EXTRA=SC*RDVECSQ*COSZ
        IF(EXTRA.LT.0.) EXTRA=0.
        EXTSLO=SC*RDVECSQ*COSBETA
        IF(EXTRA.LE.0. .OR. EXTSLO.LT.0.) EXTSLO=0.
        OUT1=EXTSLO
        Z=ACOS(COSZ)
        COSA=(SLAT*COSZ-SDEC)/(CLAT*SIN(Z))
        IF(COSA.LT.-1.) COSA=-1.
        IF(COSA.GT.1.) COSA=1.
        A=ABS(ACOS(COSA))
        IF(H.LT.0.) A=-A
        OUT2=Z/RTOD
        OUT3=A/RTOD+180.
      ENDIF

      RETURN
      END
```

Table E.2 Main Program to Compute Daily Total Solar Radiation Products

```
      PROGRAM SOL1
C+++++++++++++++++++++++++++++++++++++++++++++++++++++++++++++++++++++++
C     CALCULATE DAILY TOTAL EXTRATERRESTRIAL RADIATION, SUNRISE AND
C     SUNSET TIMES, AND DAY LENGTH.
C+++++++++++++++++++++++++++++++++++++++++++++++++++++++++++++++++++++++
      DIMENSION RAD(0:90,4),NDAY(12),EDAY(12)
      LOGICAL DAILY
      REAL LONG, LAT, AZ, IN, SC
      DATA NDAY/0,31,59,90,120,151,181,212,243,273,304,334/
      DATA EDAY/31,28,31,30,31,30,31,31,30,31,30,31/
      DAILY = .TRUE.
      LONG = -105.
      AZ=0.
      IN=0.
      SC = 1353.
      IHR=0
      MM=1

      OPEN(2,FILE='SOLFIG1.OUT',STATUS='UNKNOWN',FORM='FORMATTED')

      DO ILAT=0,90,1
        DO J=1,4
          RAD(ILAT,J)=0.
```

(continued)

Table E.2 Continued

```
      ENDDO
    ENDDO

    DO MO=1,12,1
      DO IDA=1,EDAY(MO)
        DO ILAT=0,90,10
          LAT=ILAT
          CALL SOLSUB(LONG,LAT,AZ,IN,SC,DAILY,MO,IDA,
   +                  IHR,MM,OUT1,OUT2,OUT3,COSBETA)
          IF(OUT1.LE.0.) OUT1=0.
          RAD(ILAT,1)=OUT1
          IF(OUT2.LT.0.) OUT2=0.
          RAD(ILAT,2)=OUT2
          IF(OUT3.GT.24.) OUT3=24.
          RAD(ILAT,3)=OUT3
          RAD(ILAT,4)=OUT3-OUT2
        ENDDO
        WRITE(2,102) MO,IDA,NDAY(MO)+IDA,(RAD(ILAT,1),ILAT=0,90,10),
   +    (RAD(ILAT,2),ILAT=0,90,10),(RAD(ILAT,3),ILAT=0,90,10),
   +    (RAD(ILAT,4),ILAT=0,90,10)
.02     FORMAT(2I3,I4,4(10(1X,F7.2)))
        DO ILAT=0,90,1
          DO J=1,4
            RAD(ILAT,J)=0.
          ENDDO
        ENDDO
      ENDDO
    ENDDO
    CLOSE(2)
    STOP 'NORMAL EXIT'
    END
```

Table E.3 Main Program to Compute Instantaneous Solar Radiation Products

```
      PROGRAM SOL2
C+++++++++++++++++++++++++++++++++++++++++++++++++++++++++++++++++++++
C     CALCULATE INSTANTANEOUS EXTRATERRESTRIAL RADIATION ON SLOPES OF
C     VARIOUS INCLINATION AND AZIMUTH ANGLES FOR CERTAIN LATITUDES,
C     LONGITUDES, AND DAYS OF THE YEAR.
C+++++++++++++++++++++++++++++++++++++++++++++++++++++++++++++++++++++
      DIMENSION IRAD(0:90)
      LOGICAL DAILY
      REAL LONG, LAT, AZ, IN, SC
      DAILY = .FALSE.
      LONG = -105.
      SC = 1353.
      IDA=21

      OPEN(2, FILE='SOLFIG.OUT',STATUS='UNKNOWN', FORM='FORMATTED')

      DO MO=3,12,3
        DO ILAT=30,50,10
          LAT=ILAT
          DO IAZ=0,315,45
            AZ=IAZ
C           IF(IAZ .EQ. 45) STOP 'FORCED EXIT'
            DO IHR=4,20,1
              DO MM=0,55,5
                IF(IHR.EQ.20 .AND. MM.GE. 5) GO TO 1
                IF(IHR.EQ. 4 .AND. MM.EQ. 0) THEN
```

(continued)

Table E.3 Continued

```
                WRITE(2,100) MO,IDA,LAT,LONG,AZ,0,2,4,6,8,
     +          10,12,14,16,18,20,22,24,26,28,30,32,34,36,38,
     +          40,42,44,46,48,50,52,54,56,58,60,62,64,66,68,
     +          70,72,74,76,78,80,82,84,86,88,90
  100           FORMAT(/,'MO=',I2,' IDA=',I2,' LAT=',F5.2,' LONG=',F7.2,
     +          ' AZ=',F5.0,/,4X,46(3X,I3))
                ENDIF
                  DO IIN=0,90,2
                    IN=IIN
                    CALL SOLSUB(LONG,LAT,AZ,IN,SC,DAILY,MO,IDA,
     +                          IHR,MM,OUT1,OUT2,OUT3,COSBETA)
                    IF(OUT1.LE.0.) OUT1=0.
                    IRAD(IIN)=NINT(OUT1*10.)
                  ENDDO
                  WRITE(2,101) IHR*100+MM,(IRAD(IIN),IIN=0,90,2)
  101             FORMAT(I4.4,46I6)
                  DO IIN=0,90,2
                    IRAD(IIN)=0.
                  ENDDO
    1           CONTINUE
              ENDDO
            ENDDO
          ENDDO
        ENDDO
      ENDDO

      CLOSE(2)
      STOP 'NORMAL EXIT'
      END
```

F. Additional Reading

Popular Meteorology Books and Articles

Ahrens, C. D., 1994. *Meteorology Today: An Introduction to Weather, Climate, and the Environment*, 5th ed. West Publishing Co., St. Paul, Minnesota.

Bryant, H. C., and N. Jarmie, 1974. The glory. *Sci. Amer.*, July: 60–71.

Burroughs, W., 1996. *Mountain Weather: A Guide for Skiers and Hillwalkers*. Crowood Press, Ramsbury, United Kingdom.

Carney, T. Q., A. J. Bedard, Jr., J. M. Brown, J. McGinley, T. Lindholm, and M. J. Kraus, 1996. *Hazardous Mountain Winds and Their Visual Indicators*. NOAA Handbook, National Oceanic and Atmospheric Administration, Environmental Research Laboratories, Boulder, Colorado.

Day, J. A., 1966. *The Science of Weather*. Addison-Wesley Publishing Co., Reading, Massachusetts.

deGolia, J., 1993. *Fire: A Force of Nature, The Story Behind the Scenery*. KC Publications, Las Vegas, Nevada.

Doesken, N. J., and A. Judson, 1996. *The Snow Booklet—A Guide to the Science, Climatology and Measurement of Snow in the United States*. Colorado Climate Center, Department of Atmospheric Science, Colorado State University, Fort Collins, Colorado.

Keen, R. A., 1987. *Skywatch: The Western Weather Guide*. Fulcrum Inc., Golden, Colorado.

Keen, R. A., 1992. *Skywatch East: A Weather Guide*. Fulcrum Publishing, Golden, Colorado.

Lehr, P. E., R. W. Burnett, and H. S. Zim, 1957. *Weather: A Guide to Phenomena and Forecasts*. Golden Press, New York.

Ludlam, F. H., and R. S. Scorer, 1957. *Cloud Study—A Pictorial Guide*. John Murray Publishers Ltd, London.
Moran, J. M., and M. D. Morgan, 1994. *Meteorology: The Atmosphere and the Science of Weather*, 4th ed. Macmillan College Publishing Co., New York.
Price, L. W., 1981. *Mountains and Man: A Study of Process and Environment*. University of California Press, Berkeley.
Putnam, W. L., 1991. *The Worst Weather on Earth: A History of the Mount Washington Observatory*. American Alpine Club, New York.
Reifsnyder, W. E., 1980. *Weathering the Wilderness: The Sierra Club Guide to Practical Meteorology*. Sierra Club Books, San Francisco.
Renner, J., 1992. *Northwest Mountain Weather: Understanding and Forecasting for the Backcountry User*. The Mountaineers, Seattle, Washington.
Riehl, H., 1965. *Introduction to the Atmosphere*. McGraw-Hill, New York.
Schaefer, V. J., and J. A. Day, 1981. *A Field Guide to the Atmosphere*. Houghton Mifflin Co., Boston.
Scorer, R. S., 1961. Lee waves in the atmosphere. *Sci. Amer.*, March: 124–134.
Scorer, R. S., 1972. *Clouds of the World: A Complete Color Encyclopedia*. Stackpole Books, Harrisburg, Pennsylvania.
Scorer, R. S., and A. Verkaik, 1989. *Spacious Skies*. David and Charles Publishers, London.
Washburn, B., 1962. Frostbite. *Amer. Alpine J.*, 13: 1–26.
Waugh, A. E., 1973. *Sundials: Their Theory and Construction*. Dover Publications, Inc., New York.
Williams, J., 1992. *The Weather Book: An Easy-to-Understand Guide to the USA's Weather*. Vintage Books, New York.

Nontechnical Meteorological Magazines

Weather, Bracknell, England.
Weatherwise, Washington, DC.

Technical or Specialty Meteorology Books

Angle, R. P., and S. K. Sakiyama, 1991. *Plume Dispersion in Alberta*. Standards and Approvals Division, Alberta Environment, Edmonton, Alberta.
Anonymous, 1980. *Pilot's Handbook of Aeronautical Knowledge*. AC 61-23B. Federal Aviation Administration, Washington, DC.
Barry, R. G., 1992. *Mountain Weather and Climate*, 2nd ed. Routledge, London.
Geiger, R., 1969. Topoclimates. In: H. Flohn, (ed.), *World Survey of Climatology*, vol. 2. Elsevier Publishing Co., New York, 105–138.
Geiger, R., R. H. Aron, and P. Todhunter, 1995. *The Climate Near the Ground*, 5th ed. Friedr. Vieweg & Sohn Verlagsgesellschaft mbH, Braunschweig/Wiesbaden.
Greenler, R., 1989. *Rainbows, Halos, and Glories*. Cambridge University Press, Cambridge.
Hanna, S. R., G. A. Briggs, and R. P. Hosker, Jr., 1982. *Handbook on Atmospheric Diffusion*. DOE/TIC-11223. Published by Technical Information Center, U.S. Department of Energy, Washington, DC.
Houghton, D. D. (ed.), 1985. *Handbook of Applied Meteorology*. John Wiley and Sons, Inc., New York.
Huschke, R. E. (Ed.), 1959. *Glossary of Meteorology*. American Meteorological Society, Boston, Massachusetts.
Kidder, S. Q., and T. H. Vonder Haar, 1995. *Satellite Meteorology: An Introduction*. Academic Press, San Diego.
Lee, R., 1978. *Forest Microclimatology*. Columbia University Press, New York.
Lester, P. F., 1993. *Turbulence: A New Perspective for Pilots*. Jeppesen Sanderson, Inc., Englewood, Colorado.
Lester, P. F., 1995. *Aviation Meteorology*. Jeppesen Sanderson, Inc., Englewood, Colorado.
Lindsay, C. V., and S. J. Lacy, 1976. *Soaring Meteorology for Forecasters*, 2nd ed. Soaring Society of America, Los Angeles.

List, R. J. (Ed.), 1966. *Smithsonian Meteorological Tables*, 6th rev. ed. Smithsonian Institute, Washington, DC.
National Weather Service, 1995. Federal Meteorological Handbook No. 1: Surface Weather Observations and Reports, 5th ed. December 1995. National Oceanic and Atmospheric Administration, Washington, DC.
National Wildfire Coordinating Group Prescribed Fire and Fire Effects Working Team, 1985. Prescribed Fire Smoke Management Guide. NFES No.1279. PMS 420-2. Publ. by National Wildfire Coordinating Group. (For copies contact Boise Interagency Fire Center, BLM Warehouse, 3905 Vista Avenue, Boise, ID 83705.)
Oke, T. R., 1978. *Boundary Layer Climates*. Methuen and Co., Ltd., London.
Orgill, M. M., 1981. Atmospheric Studies in Complex Terrain: A Planning Guide for Future Studies. PNL-3656, ASCOT/80/4, Pacific Northwest Laboratory, Richland, Washington.
Pyne, S. J., P. L. Andrews, and R. D. Laven, 1996. *Introduction to Wildland Fire*, 2nd ed. John Wiley and Sons, Inc., New York.
Quantick, H. R., 1985. *Aviation in Crop Protection, Pollution and Insect Control*. Collins Professional and Technical Books, London.
Randerson, D. (ed.), 1984. *Atmospheric Science and Power Production*. DE84005177 (DOE/TIC-27601). Available from National Technical Information Service, Springfield, Virginia.
Schroeder, M. J., and C. C. Buck, 1970. *Fire Weather, A Guide for Application of Meteorological Information to Forest Fire Control Operations*. Agriculture Handbook 360, U.S. Department of Agriculture Forest Service, Washington, DC.
Scorer, R. S., 1958. *Natural Aerodynamics*. Pergamon Press, Elmsford, New York.
Scorer, R. S., 1997. *Dynamics of Meteorology and Climate*. John Wiley and Sons, Chichester.
Slade, D. (ed.), 1968. *Meteorology and Atomic Energy 1968*. U.S. Atomic Energy Commission, Division of Technical Information, Oak Ridge, Tennessee.
Stull, R. B., 1995. *Meteorology Today for Scientists and Engineers: A Technical Companion Book*. West Publishing Co., Minneapolis-St. Paul.
Tricker, R. A. R., 1970. *Introduction to Meteorological Optics*. American Elsevier Publishing Co., New York.
Turner, D. B., 1969. *Workbook of Atmospheric Dispersion Estimates*. Public Health Service Publication No. 999-AP-26, U.S. Department of Health, Education and Welfare, Cincinnati, Ohio.
Wallace, J. M., and P. V. Hobbs, 1977. *Atmospheric Science: An Introductory Survey*. Academic Press, San Diego.
Wenger, K. F. (ed.), 1984. *Forestry Handbook*, 2nd ed. John Wiley and Sons, New York.
World Meteorological Organization, 1987. *International Cloud Atlas, Vol. II*. World Meteorological Organization, Geneva, Switzerland.

G. METAR and TAF Code Abbreviations

This appendix provides a summary of the abbreviations used in Aviation Routine Weather Reports (METAR) and Aerodrome Forecasts (TAF), as discussed in section 9.3.3. The METAR and TAF codes follow international conventions, so that abbreviations are not necessarily for English words. Table G.1 provides general abbreviations, table G.2 provides weather phenomena abbreviations, table G.3 provides weather descriptor abbreviations, and table G.4 provides cloud type abbreviations.

Table G.1 General Abbreviations Used in METARs and TAFs

Abbreviation	Meaning
AO1	Automated observation without precipitation discriminator (rain/snow)
AO2	Automated observation with precipitation discriminator (rain/snow)
AMD	Amended forecast (TAF)
BECMG	Becoming (expected between 2-digit beginning hour and 2-digit ending hour)
BKN	Broken
CLR	Clear at or below 12,000 ft (AWOS/ASOS report)
COR	Correction to the observation
FEW	1 or 2 octas (eighths) cloud coverage
FM	From (4-digit beginning time in hours and minutes)
LDG	Landing
M	In temperature field, means "minus" or below zero
M	In RVR, indicates visibility less than lowest reportable sensor value
NO	Not available (e.g., SLPNO, RVRNO)
NSW	No significant weather
OVC	Overcast
P	In RVR, indicates visibility greater than highest reportable sensor value
P6SM	Visibility greater than 6 SM (TAF only)
PROB40	Probability 40%
R	Runway (used in RVR measurement)
RMK	Remark
RVR	Runway visual range (visibility)
RY/RWY	Runway
SCT	Scattered
SKC	Sky clear
SLP	Sea level pressure (e.g., 1013 mb reported as 013)
SM	Statute mile(s)
SPECI	Special report
TEMPO	Temporary changes expected (between 2-digit beginning hour and 2-digit ending hour)
TKOF	Takeoff
T01760158, 10142, 20012, and 401120084	In remarks, examples of temperature information
V	Varies (used for wind direction and RVR)
VC	Vicinity
VRB	Variable wind direction when speed is less than or equal to 6 knots
VV	Vertical visibility (indefinite ceiling)
WS	Wind shear (in TAFs, low-level and not associated with convective activity)

Table G.2 Weather Phenomena Abbreviations Used in METARs and TAFs

Abbreviation	Meaning
BR	Mist
DS	Dust storm
DU	Widespread dust
DZ	Drizzle
FC	Funnel cloud
+FC	Tornado/water spout
FG	Fog
FU	Smoke
GR	Hail
GS	Small hail/snow pellets
HZ	Haze
IC	Ice crystals
PE	Ice pellets
PO	Dust/sand whirls
PY	Spray
RA	Rain
SA	Sand
SG	Snow grains
SN	Snow
SQ	Squall
SS	Sandstorm
UP	Unknown precipitation (Automated observations)
VA	Volcanic ash

Table G.3 Weather Descriptor Abbreviations Used in METARs and TAFs

Abbreviation	Meaning
BC	Patches
BL	Blowing
DR	Low drifting
FZ	Supercooled/freezing
MI	Shallow
PR	Partial
SH	Showers
TS	Thunderstorm

Table G.4 Cloud Type Abbreviations Used in METARs and TAFs

Abbreviation	Meaning
CB	Cumulonimbus
TCU	Towering cumulus

Glossary

Absolutely stable an atmospheric stability condition that exists when the **environmental temperature lapse rate** is less than the **moist adiabatic lapse rate.**

Absolutely unstable an atmospheric **stability** condition that exists when the **environmental temperature lapse rate** is greater than the **dry adiabatic lapse rate.**

Absorption the removal of energy or particles from a beam by the medium through which it propagates.

Acid precipitation precipitation in the form of **rain**, **snow**, or **sleet** that contains relatively high concentrations of acid-forming chemicals that have been released into the atmosphere and combined with water vapor; harmful to the environment.

Acid rain See **acid precipitation**.

Acoustic sounder See **sodar**.

Adiabat a line on a thermodynamic chart that relates the **pressure** and temperature of a substance (such as air) that is undergoing a transformation in which no heat is exchanged with its environment.

Adiabatic compression the increase in air **density** that occurs thermodynamically when an air parcel is brought to a lower altitude and thus a higher pressure without gain or loss of heat.

Adiabatic lapse rate the rate of decrease of temperature experienced by a parcel of air when it is lifted in the atmosphere under the restriction that it cannot exchange heat with its environment. When lifted, unsaturated parcels cool at the **dry adiabatic lapse rate**, whereas saturated parcels cool at the **moist adiabatic lapse rate**.

Adiabatic process a process that occurs with no exchange of heat between a system and its environment.

Advect to move by the process of **advection.**

Advection the horizontal transport of atmospheric properties.

Advection fog a fog that forms when warm air flows over a cold surface and cools from below until saturation is reached.

Aeroallergen any of a variety of allergens, such as pollen, grasses, or dust, carried by winds.

Aerosol a system of colloidal particles dispersed in a gas, such as smoke or fog.

Air mass a body of air covering a relatively wide area and exhibiting horizontally uniform properties.

Air mass thunderstorm a **thunderstorm** produced by local **convection** within an unstable air mass, as opposed to thunderstorms associated with **fronts** or instability lines.

Air pollutant harmful substance or product introduced into the atmosphere.

Air pollution potential the meteorological potential for air pollution problems, considered without regard to the presence or absence of actual pollution sources.

Air quality model mathematical or conceptual model used to estimate present or future air quality.

Air toxic toxic **air pollutant**.

Albedo reflectivity; the fraction of **radiation** striking a surface that is reflected by that surface.

Along-valley wind system a closed, thermally driven, **diurnal mountain wind** whose lower branch blows up or down the axis of a valley. The upper branch blows in the opposite direction, thereby closing the circulation.

Alto- a prefix used in naming midlevel clouds.

Altocumulus a type of cloud from a cloud class characterized by globular masses or rolls in layers or patches, the individual elements of which are larger and darker than those of **cirrocumulus** and smaller than those of **stratocumulus**. These clouds are of medium altitude, about 8000–20,000 ft (2400–6100 m).

Altostratus a type of cloud from a cloud class characterized by a generally uniform gray sheet or layer, lighter in color than **nimbostratus** and darker than **cirrostratus**. These clouds are of medium altitude, about 8000–20,000 ft (2400–6100 m).

Ambient of the surrounding area or environment.

Anemometer an instrument for measuring wind speed.

Aneroid barometer an instrument for measuring atmospheric **pressure** in which a needle, attached to the top of an evacuated box, is deflected as changes in atmospheric pressure cause the top of the box to bend in or out.

Anthropogenic caused or produced by humans.

Anticyclogenesis the formation or intensification of an **anticyclone** or high pressure center.

Anticyclone a large-scale circulation of winds around a central region of high atmospheric **pressure**, clockwise in the Northern Hemisphere and counterclockwise in the Southern Hemisphere.

Antiwind the upper or return branch of an **along-valley wind system**, confined within a valley and blowing in a direction opposite to the winds in the lower altitudes of the valley.

Area source an array of pollutant sources that is so widely dispersed and uniform in strength that they can be treated in a **dispersion** model as an aggregate pollutant release from a defined area at a uniform rate. Compare **line source** and **point source**.

Aspect (angle) the cardinal direction or bearing angle toward which a slope faces.

Atmospheric boundary layer See **boundary layer**.

Azimuth angle 1. the direction or bearing toward which a sloping surface faces (e.g., a north-facing slope has an azimuth angle of 360°; a northeast-facing slope, an azimuth angle of 45°); 2. the arc of the horizon measured clockwise from north to the point where a vertical circle through a given heavenly body intersects the horizon (e.g., used for solar azimuth angle or pibal azimuth angle).

Backfire 1. a fire started to stop an advancing fire by creating a burned area in its path; 2. a **firing pattern** or fire ignition pattern for a **prescribed fire**, in which a fire is ignited across the **leeward** edge of an area and burns into the wind. Backfires can be ignited in parallel strips to protect an area from fire escapement.

Banner cloud a cloud plume produced by rising motions in an **eddy** that forms downwind of steep, isolated mountain peaks; may occur even on otherwise cloud-free days.

Barometer an instrument for measuring atmospheric **pressure.**

Barometric pressure the **pressure** of the atmosphere as indicated by a **barometer.**

Barrier jet a **jet** wind current that forms when a stably stratified low-level airflow approaches a mountain barrier and turns to the left to blow parallel to the longitudinal axis of the barrier.

Belt weather kit belt-mounted case with pockets fitted for **anemometer**, compass, **sling psychrometer**, water bottle, and forms. Used to provide on-site weather observations for the fire weather forecaster or fire behavior analyst.

Bernoulli effect See **Venturi effect**.

Billow cloud a cloud consisting of broad parallel bands oriented perpendicular to the wind. The cloud forms when wind shear occurs across a sharp change in temperature in a cloudy atmosphere. When wind shear is strong, the clouds can resemble breaking waves.

Blizzard a severe weather condition characterized by low temperatures and strong winds (32 mph or higher) bearing a great amount of snow (including snow picked up from the ground) that reduces **visibility** to less than 500 ft.

Blocked the state of a flow approaching a mountain barrier at elevations below the barrier height that is too weak or too stable to be carried over the barrier.

Blowdown a tree or stand of timber that has been blown down by the wind; same as **windthrow**

Blowup fire a fire that suddenly increases in intensity or rate of spread. It is usually accompanied by intense **convection** and adversely affects fire control activities or fire suppression plans.

Bora a regional **downslope wind** whose source is so cold that it is experienced as a cold wind, despite **compression heating** as it descends the **lee** slope of a mountain range.

Boundary layer the layer of fluid in the immediate vicinity of a fluid-solid boundary. In the atmosphere, the air layer near the ground affected by diurnal heat, moisture or momentum transfer to or from the surface of the earth.

Box model a computer model used to calculate air pollution concentrations by assuming that pollutants emitted into a box-shaped volume are immediately and uniformly dispersed throughout the volume. The sides and bottom of the box, for example, could be defined by the sidewalls and floor of a valley, and the top could be defined by the **mixing depth**.

Brocken specter an optical effect characterized by concentric rings of color (red outermost and violet innermost) surrounding the shadow of an observer's head when the shadow is cast onto a cloud deck below the observer's elevation; same as **glory**

Build to increase in **pressure**: a building **ridge** or high (**anticyclone**).

Canopy the cover formed by the leafy upper branches of the trees in a forest; height stratum containing the **crowns** of the tallest vegetation present (living or dead).

Canyon wind 1. a **foehn** wind that is channeled through a canyon as it descends the **lee** side of a mountain barrier, for example, **Wasatch wind**. 2. any along-canyon wind.

Cap cloud a stationary lens-shaped cloud that forms over a mountain peak, with a cloud base below the mountaintop.

Capping inversion an elevated **inversion** layer that caps a **convective boundary layer**, thus keeping the convective elements from rising higher into the atmosphere.

Ceilometer a device using a laser or other light source to determine the height of a cloud base. An optical ceilometer uses triangulation to determine the height of a spot of light projected onto the base of the cloud; a laser ceilometer determines the height by measuring the time required for a pulse of light to be scattered back from the cloud base.

Centerfire a **firing pattern** or fire ignition pattern for **prescribed fires**, in which a fire is ignited at a chosen spot and then in concentric circles or spirals around the spot to produce a fire with strong central **convection**, inward spread, and high **fire intensity**. This firing pattern is modified on slopes to account for the faster speed of fire propagation up the slope.

Chemistry model a computer model used in air pollution investigations that simulates chemical and photochemical reactions of pollutants during their transport and **diffusion**.

Chinook the name given to the **foehn** in North America, especially used on the eastern side of the Rocky Mountains and Sierra Nevada.

Chinook arch a **foehn** cloud formation that appears as a bank of **altostratus** clouds east of the Rocky Mountains, heralding the approach of a **chinook**. It forms in the rising portion of standing waves on the **lee** side of the mountains. An observer underneath or east of the cloud sees an arch of clear air between the cloud's leading edge and the mountains below. The cloud appears to converge with the mountains to the north and south because of a perspective effect.

Chinook pause same as **foehn pause.**

Cirro- a prefix used in naming high clouds.

Cirrocumulus a cirriform cloud characterized by thin, white patches, each of which is composed of very small granules or ripples. These clouds are of high altitude (20,000–40,000 ft or 6000–12,000 m).

Cirrostratus a type of cloud from a class of clouds composed of **ice crystals**; the cloud appears as a whitish and usually somewhat fibrous veil, often covering the whole sky and sometimes so thin as to be hardly discernible. **Halos** are often seen in cirrostratus. These clouds are of high altitude (20,000–40,000 ft or 6000–12,000 m).

Cirrus a type of cloud from a cloud class characterized by thin white filaments or narrow bands and composed of **ice crystals**. These clouds are of high altitude (20,000–40,000 ft or 6000–12,000 m).

Class I Area geographic area designated by the Clean Air Act where only a small amount or increment of air quality deterioration is permissible.

Clean Air Act a U.S. law designed to protect and enhance the quality of the nation's air and to protect public health and welfare.

Clear ice a thin coating of ice on terrestrial objects caused by rain that freezes on impact. The ice is relatively transparent, as opposed to **rime**, because of large drop size, rapid accretion of liquid water, or slow dissipation of **latent heat** of fusion.

Climate the composite or generally prevailing weather conditions of a region throughout the year, averaged over a series of years.

Climatology the science that deals with the phenomena of climates or climatic conditions.

Cloud condensation nuclei particles, either liquid or solid, upon which water condenses to form cloud droplets.

Cold air avalanche downslope flow pulsations that occur at more or less regular intervals as cold air builds up on a peak of plateau, reaches a critical mass, and then cascades down the slopes.

Cold air damming a process in which a shallow cold **air mass** is carried up the slope of a mountain barrier, but with insufficient strength to surmount the barrier. The cold air, trapped upwind of the barrier alters the effective terrain configuration of the barrier to larger scale approaching flows.

Cold front a zone separating two **air masses**, of which the cooler, denser mass is advancing and replacing the warmer.

Cold occlusion a **frontal zone** formed when a **cold front** overtakes a **warm front**. When the air behind the cold front is colder than the air ahead of the warm front, the cold air slides under the warm front, lifting it aloft. Compare **warm occlusion**.

Collection efficiency the fraction of droplets approaching a surface that actually deposit on that surface.

Colorado Low a low **pressure** storm system that forms in winter in southeastern Colorado or northeastern New Mexico and tracks northeastward across the central plains of the United States over a period of several days, producing **blizzards** and hazardous winter weather.

Complex terrain mountainous terrain. In general usage, it may also refer to coastal regions and heterogeneous landscapes.

Compression heating the temperature increase produced thermodynamically when an air parcel is compressed by bringing it to a lower altitude and therefore a higher pressure.

Conditionally unstable an atmospheric **stability** condition that exists when the **environmental lapse rate** is less than the **dry adiabatic lapse rate** but greater than the **moist adiabatic lapse rate.**

Conduction flow of heat in response to a temperature **gradient** within an object or between objects that are in physical contact.

Coning pattern of plume **dispersion** in a **neutral** stability atmosphere, in which the plume attains the form of a cone with its vertex at the top of the stack.

Continental Divide the line of summits in the Rocky Mountains that separate streams flowing toward the Gulf of California and Pacific from those flowing toward the Gulf of Mexico, Hudson Bay, and the Arctic Ocean.

Continentality the degree to which the climate of a region typifies that of the interior of a large landmass.

Contrail condensation trail from a jet aircraft.

Convection 1. vertical air circulation in which warm air rises and cool air sinks, resulting in vertical transport and mixing of atmospheric properties; 2. flow of heat by this circulation.

Convection column a rising column of gases, smoke, fly ash, particulates, and other debris produced by a fire.

Convection-dominated fire See **plume-dominated fire**.

Convective boundary layer the boundary layer that forms at the surface and grows upward through the day as the ground is heated by the sun and convective currents transfer heat upward into the atmosphere; same as **unstable boundary layer.**

Convective cloud cloud formed by **convection**, with strong vertical development.

Convective overdevelopment **convection** that covers the sky with clouds, thereby cutting off the sunshine that produces convection.

Convergence a net flow of air into a given region. Compare **divergence**.

Coriolis force a fictitious force used to account for the apparent deflection of a body in motion with respect to the earth, as seen by an observer on the earth. The deflection (to the right in the Northern Hemisphere) is caused by the rotation of the earth.

Corona a white or colored circle or set of concentric circles of light of small radius seen around a luminous body, especially around the sun or moon. The color varies from blue inside to red outside and the phenomenon is attributed to diffraction of light by thin clouds or mist (distinguished from **halo**).

Cross-valley wind system a thermally driven wind that blows during daytime across the longitudinal axis of a valley toward the heated sidewall.

Crown 1. verb, (of a fire) to spread from the understory into the crown of a forest canopy; to spread rapidly across the overstory of a forest; 2. noun, the top or highest part of the forest **canopy.**

Crown fire a fire where flames travel from tree to tree at the level of the trees' **crowns** or tops.

Cumuliform (of cloud elements) having approximately equal vertical and horizontal extent; resembling cotton balls.

Cumulonimbus a cloud indicative of **thunderstorm** conditions characterized by large, dense towers that often reach altitudes of 30,000 ft (9000 m) or more, **cumuliform** except for their tops, which appear fibrous because of the presence of **ice crystals.**

Cumulus a cloud characterized by dense individual elements in the form of puffs, mounds, or towers, with flat bases and tops that often resemble cauliflower. They are found at a lower altitude than **altocumulus**, usually below 8000 ft (2400 m).

Cumulus stage the first phase in the life cycle of a **thunderstorm**, characterized by warm, moist air rising in a buoyant plume or in a series of convective updrafts.

Cyclone a large-scale circulation of winds around a central region of low atmospheric **pressure**, counterclockwise in the Northern Hemisphere, clockwise in the Southern Hemisphere.

Cyclogenesis the formation or intensification of a **cyclone** or low pressure storm system.

Damp haze **haze** in which the individual particles have grown by the absorption of water, often when **relative humidity** is below 100%.

Dart leader a faint, negatively charged **lightning** channel that travels more or less directly and continuously from cloud to ground.

Day length duration of the period from sunrise to sunset.

Dead fuel moisture moisture content of nonliving fuels. For dead fuels, moisture content is governed primarily by exposure to precipitation or soil moisture and by the tendency of the dead fuels to approach equilibrium with the **relative humidity** of the surrounding air.

Decaying stage the third and final phase in the life cycle of a **thunderstorm**, characterized by downdrafts throughout the cloud.

Deepen to decrease in atmospheric **pressure**: a deepening **trough** or low (**cyclone**).

Density mass per unit volume.

Dew-point temperature the temperature to which air must be cooled (at constant **pressure** and constant water vapor content) for saturation to occur.

Diagnostic model a computer model used to calculate air pollution concentrations. A diagnostic model produces a **wind field** over an area by interpolating in space and time from actual wind observations.

Diffusion the spreading of atmospheric constituents or properties by turbulent and molecular motions of the air.

Dispersion transport and **diffusion** of pollutants.

Diurnal daily, especially pertaining to actions that are completed in 24 hours and are repeated every 24 hours.

Diurnal mountain circulation See **diurnal mountain winds**.

Diurnal mountain winds a diurnally reversing closed cellular wind current resulting from horizontal temperature contrasts caused by different rates of heating or cooling over adjacent surfaces; includes **slope**, **cross-valley**, **along-valley**, **mountain–plain**, and **sea breeze winds**, also called **thermally driven circulations.**

Divergence the net flow of air from a given region. Compare **convergence**.

Dividing streamline height in the blocked flow region upwind of a mountain barrier, the height above ground of the **streamline** that separates the blocked flow region near the ground from the air aloft that flows over the barrier.

Domain in numerical models, the geographical area over which a simulation is performed.

Downburst a strong downdraft of air from a **cumulonimbus** cloud, of limited duration and often associated with intense **thunderstorms.**

Downslope wind a thermally driven wind directed down a mountain slope and usually occurring at night; part of the **slope wind system.**

Downslope windstorm a windstorm on the **lee** side of a mountain barrier produced by **foehn** or **chinook** winds.

Down-valley wind a thermally driven wind directed down a valley's axis, usually occurring during nighttime; part of the **along-valley wind system.**

Downwash a deflection of air downward behind a flow obstruction.

Drainer a valley or basin from which air drains continuously during nighttime rather than becoming trapped or pooled.

Drift sprayed or dusted material that does not deposit on the target area.
Drizzle precipitation consisting of numerous minute droplets of water less than 0.5 mm (500 micrometers) in diameter.
Dry adiabatic lapse rate the rate (5.4°F per 1000 ft or 9.8°C per km) at which the **temperature** of a parcel of dry air decreases as the parcel is lifted, under the assumption that no heat is exchanged with its environment.
Dry bulb temperature in a **psychrometer**, the temperature indicated by the **dry bulb thermometer.**
Dry bulb thermometer in a **psychrometer**, the thermometer whose bulb is unmoistened and thus measures the actual air temperature. Compare **wet bulb thermometer**.
Dry haze fine, dry dust or salt particles dispersed in the atmosphere. See **haze**.
Dry line the roughly north–south boundary between moist air in the Mississippi Valley and dry air on the west side of the Great Plains descending from the Mexican Plateau and Southern Rockies. **Thunderstorms** often form along this line, which moves eastward during the morning and westward in the evening.
Duff organic matter in various stages of decomposition on the floor of a forest.
Eddy swirling currents of air at variance with the main current.
Effective topography the topography affecting an approaching flow, which may include not only the actual terrain but also cold **air masses** trapped within or adjacent to the actual topography.
Elevated temperature inversion a **temperature inversion** with its base above the surface of the earth.
El Niño literally, the Christ child; a name given to an extensive ocean warming in the equatorial eastern Pacific along the coast of Peru and Ecuador that often begins around Christmas (hence, the name). The warming brings nutrient-poor tropical water southward along the west coast of South America in major events that recur at intervals of 3–7 years. El Niño is associated with atmospheric circulations that produce wide-ranging effects on global weather and climate.
Emissivity the ability of a surface to emit radiant energy compared to that of a black body at the same **temperature** and with the same area.
Entrainment zone a shallow region at the top of a **convective boundary layer** where fluid is entrained into the growing boundary layer from the overlaying fluid by the collapse of rising convective plumes or bubbles.
Environmental temperature lapse rate the rate of decrease of air temperature with height, usually measured with a **radiosonde.**
Environmental temperature sounding an instantaneous or near-instantaneous **sounding** of temperature as a function of height. This sounding or vertical profile is usually obtained by a balloon-borne instrument, but it can also be measured using remote sensing equipment.
Equinox the time when the sun crosses the earth's equator, making night and day of approximately equal length all over the earth and occurring about March 21 (the spring or vernal equinox) and September 22 (autumnal equinox).
Evaporation-mixing fog a fog that forms when the evaporation of water raises the **dew-point temperature** of the adjacent air.
Evapotranspiration the process of transferring moisture from the earth's surface to the atmosphere by evaporation of water and transpiration from plants.
Extraterrestrial shortwave radiation the theoretically calculated radiation **flux** from the sun at the top of the atmosphere, before radiation traverses the atmosphere.
Extreme fire behavior level of **fire behavior** that ordinarily precludes direct control of the fire. One or more of the following is usually involved: high rates of spread, prolific **crowning** or **spotting**, presence of **fire whirls**, or an intense **convection column**.
Fall line the line of steepest descent of a slope.
Fallstreak same as **virga.**
Fanning a pattern of plume **dispersion** in a **stable** atmosphere, in which the plume fans out in the horizontal and meanders about at a fixed height.
Fill to increase in atmospheric **pressure**: a filling **trough** or low (**cyclone**).
Fine fuels fast-drying dead fuels, generally less than $\frac{1}{4}$ inch in diameter and having a **time lag** of one hour or less. These fuels (e.g., grass, twigs, and needles) ignite readily and are consumed rapidly by fire when dry.
Fire behavior the manner in which a fire reacts to the influences of fuel, weather, and topography. Fire behavior is described in terms of the specific characteristics exhibited by a fire, including its rate of spread, spread direction, flame length, and rate of heat release.
Fire behavior model a computer model that uses a set of mathematical equations to predict certain aspects of **fire behavior** when provided with data on fuel and environmental conditions unique to a site.
Firebrand any source of heat, natural or man-made, capable of igniting wildland fuels; flaming or glowing fuel particles that can be carried naturally by wind, convection currents, or gravity into unburned fuels.
Fire danger the exposure to risk or harm from a fire as determined from factors that affect the start, spread, intensity, and difficulty of suppression of **wildfires** and the damage they cause.
Fire intensity the rate of heat release per unit time and per unit of fire travel distance at the fire front. Numerically, it is the product of the quantity of fuel consumed at the fire front, the heat yield per unit of fuel consumed, and the rate of fire spread.
Fire line a zone along a fire's edge where there is little or no fuel available to the fire.
Fire severity the degree to which a site has been altered or disrupted by fire. Severity is dependent primarily on the product of **fire intensity** and duration.
Firestorm raging fire of great intensity that spreads rapidly.
Fire weather weather conditions that influence fire ignition, behavior, or suppression.

Fire Weather Watch a term used to alert land managers to the potential that weather conditions when coupled with critically dry fuels can lead to dangerous **wildfires.**

Fire whirl a **vortex** similar to a dust devil that forms in the fire area, often on the **lee** edge of the fire. A fire whirl increases **fire intensity** and the frequency of **spot fires**.

Fire wind a wind blowing radially inward toward a fire, produced by horizontal temperature differences (and thus **pressure** differences) between the heated air above the fire and the surrounding cooler **free atmosphere.**

Firing pattern the specific pattern and timing of ignition of a **prescribed fire** to affect the direction and rate of fire spread and **fire intensity.**

Flankfire 1. a fire started to stop an advancing fire by creating a burned area on its flank; 2. a **firing pattern** or fire ignition pattern for **prescribed fires**, in which the **lee** edge of an area to be burned is first protected by a **backfire** and then fires are ignited in parallel lines leading to **windward** from the **leeward** edge.

Flash a sudden, brief illumination of a conductive channel associated with **lightning**, which may contain multiple strokes with their associated **stepped leaders**, **dart leaders**, and **return strokes.**

Flash density the number of **lightning flashes** per unit area in a specified time interval (e.g., one year).

Flash flood a sudden and destructive rush of water down a narrow gully or over a sloping surface, caused by heavy rainfall.

Flow 1. noun, wind; volume of transported fluid; 2. verb, to move along, circulate.

Flow separation the process by which a **separation eddy** forms on the **windward** or **leeward** sides of bluff objects or steeply rising hillsides.

Flow splitting the splitting of a stable airflow around a mountain barrier, with branches going around the left and right edges of the barrier, often at accelerated speeds.

Flux the rate of transfer of fluids, particles, or energy per unit area across a given surface.

Foehn a warm, dry wind on the **lee** side of a mountain range, the warmth and dryness of the air being due to **adiabatic compression** as the air descends the mountain slopes.

Foehn pause a temporary cessation of the **foehn** at the ground because of the formation or intrusion of a cold air layer that lifts the foehn off the ground.

Foehn wall the steep **leeward** boundary of an extensive cloud layer that forms as air is lifted over ridges and mountains during **foehn** conditions.

Foehn wall cloud same as **foehn wall.**

Föhn alternate German spelling of **foehn.**

Fog a cloud with its base at the earth's surface.

Forced channeling channeling of upper winds along a valley's axis when upper winds are diverted by the underlying topography. Compare **pressure-driven channeling**.

Fractocumulus a **cumulus** cloud with a ragged, shredded appearance, as if torn.

Fractostratus a **stratus** cloud with a ragged, shredded appearance, as if torn. It differs from a **fractocumulus** cloud in having a smaller vertical extent and darker color.

Free air wind **synoptic-scale** wind. In fire and land management, it is also called **general wind**.

Free atmosphere the part of the atmosphere that lies above the frictional influence of the earth's surface.

Freezing drizzle **drizzle** that falls as a liquid but freezes into **glaze** or **rime** upon contact with the cold ground or surface structures.

Freezing level the altitude at which the air temperature first drops below freezing.

Freezing rain **rain** that falls as a liquid but freezes into **glaze** upon contact with the ground.

Front an interface or zone of transition between two dissimilar **air masses.**

Frontal inversion an elevated **temperature inversion** that develops above a **frontal zone** when cold air at the surface is overrun by warmer air aloft.

Frontal zone See **Front.**

Frostbite human tissue damage caused by exposure to intense cold.

Fuel loading the amount of fuel present as expressed by the weight of fuel per unit area.

Fugitive dust dust that is not emitted from definable **point sources**, such as industrial smokestacks. Sources include open fields, roadways, storage piles, etc.

Full-physics numerical model a computer model used to calculate air pollution concentrations. A full-physics numerical model uses a full set of equations describing the thermodynamic and dynamic state of the atmosphere and can be used to simulate atmospheric phenomena.

Fumigation a pattern of plume **dispersion** produced when a **convective boundary layer** grows upward into a plume trapped in a **stable** layer. The elevated plume is suddenly brought downward to the ground, producing high surface concentrations.

Gap a major erosional opening through a mountain range.

Gap winds strong winds driven through low passes or major breaks in mountain barriers by cross-gap **pressure** gradients that develop regionally. In North America, the term is used mostly for winds in coastal mountain ranges where major river valleys issue onto seaways.

Gaussian plume model a computer model used to calculate air pollution concentrations. The model assumes that a pollutant plume is carried downwind from its emission source by an average wind. Plume concentrations are obtained by assuming that the highest concentrations occur on the horizontal and vertical midlines of the plume, with the distribution about these midlines characterized by Gaussian or bell-shaped concentration profiles.

Gaussian plume segment model a modification of the **Gaussian plume model** in which wind direction changes during plume transport are handled by breaking the plume into segments.

Gaussian puff model a computer model used to calculate air pollution concentrations. The model assumes that a continuously emitted plume or instantaneous cloud of pollutants can be simulated by the release of a series of puffs that are carried in a time- and space-varying **wind field**. The puffs are assumed to have Gaussian or bell-shaped concentration profiles in their vertical and horizontal planes.

General circulation the continuous circulation of wind and ocean currents that acts to moderate the temperature differences between the poles and the equator.

General wind land management agency term for wind produced by **synoptic-scale pressure** systems on which a smaller scale or **local convective wind** may be superimposed. Also called **free air wind**.

Geostrophic wind 1. the wind that blows parallel to **isobars** with strength proportional to the pressure **gradient** (i.e., spacing of the isobars); 2. the wind obtained theoretically by a balance between **pressure gradient force** and **coriolis force.**

Glaciation 1. the transformation of cloud particles from water drops to **ice crystals**. Thus, a **cumulonimbus** cloud is said to have a "glaciated" upper portion. 2. land surface cover of ice or snow.

Glacier wind a shallow **downslope wind** above the surface of a glacier, caused by the temperature difference between the air in contact with the glacier and the free air at the same altitude. The glacier wind does not reverse diurnally like **slope** and **along-valley wind systems**.

Glaze a smooth, clear coating of ice, which sometimes contains air pockets.

Glory same as **Brocken specter**

Gradient a rate of change with respect to distance of a variable quantity, such as temperature or **pressure**, in the direction of maximum change.

Graupel same as **snow pellets** or **small hail.**

Gravity wave a wave created by the action of gravity on **density** variations in a stratified atmosphere. A generic classification that includes **orographic waves (lee waves** and **mountain waves**) and many other waves that form in the atmosphere.

Greenhouse effect atmospheric heating caused when solar **radiation** is readily transmitted inward through the earth's atmosphere but **longwave radiation** is less readily transmitted outward, as a result of **absorption** by certain gases in the atmosphere.

Greenness amount of viable or living material in vegetation. Measurements of greenness are obtained from **emissivity** data collected from remote sensors on satellites.

Ground heat flux the **flux** of heat from the ground to the earth's surface; a component of the **surface energy budget.**

Ground stroke electrical current propagating along the ground from the point where a direct stroke of **lightning** hits the ground.

Hail showery precipitation in the form of irregular pellets or balls of ice more than 5 mm in diameter, falling from a **cumulonimbus** cloud.

Haines Index See **Lower Atmosphere Stability Index**.

Halo any of a variety of bright circles or arcs centered on the sun or moon, caused by the refraction or reflection of light by **ice crystals** suspended in the earth's atmosphere and exhibiting prismatic coloration ranging from red inside to blue outside.

Haze an aggregation in the atmosphere of very fine, widely dispersed solid and/or liquid particles, which gives the air an opalescent appearance that subdues colors. See also **regional haze**, **layered haze**, **dry haze**, and **damp haze**.

Headfire a **firing pattern** or fire ignition pattern for **prescribed fires**, in which a fire is ignited along a line perpendicular to the wind direction and runs to **leeward**. The headfire is often ignited in strips with the **lee** edges protected by **backfires** to guard against fire escapement.

Heat exhaustion a mild form of **heat stroke**, characterized by faintness, dizziness, and heavy sweating.

Heat Index an index that combines temperature and **relative humidity** to determine an apparent temperature.

Heat low a shallow, **low pressure center** that forms in response to strong surface sensible heat flux.

Heat stroke a condition resulting from excessive exposure to intense heat, characterized by high fever, collapse, and sometimes convulsions or coma.

Hertz an international unit of frequency equal to one cycle per second; named after a German physicist.

High See **anticyclone**.

High pressure center a region of high pressure enclosed by a pressure or height contour.

High pressure system See **anticyclone**.

Humidity a general term referring to the air's water vapor content. See **relative humidity**.

Hydraulic flow atmospheric flow that is similar in character to the flow of water over an obstacle.

Hydraulic jump a steady disturbance in the **lee** of a mountain, where the airflow passing over the mountain suddenly changes from a shallow, high-velocity flow to a deep, low-velocity flow.

Hydrometeor liquid water or ice in the atmosphere in various forms, including **rain**, **ice crystals**, **hail**, **fog**, or **clouds**.

Hygrometer any instrument that measures the water vapor content of the atmosphere.

Hygroscopic absorbing or attracting moisture from the air.

Hygrothermograph an instrument that records, on one record, the variation with time of both atmospheric **humidity** and temperature.

Hypothermia a rapid, progressive mental and physical collapse that accompanies the lowering of body temperature.

Ice crystal precipitation consisting of small, slowly falling crystals of ice.

Ice crystal process the process by which ice crystals are introduced into a supercooled cloud of water droplets and grow at the expense of the water droplets to produce precipitation.

Ice fog a fog that consists of small **ice crystals** rather than water droplets. Ice fog usually forms at temperatures below −20°F (−29°C).

Ice nucleus any particle that serves as a nucleus in the formation of **ice crystals** in the atmosphere.

Ice pellets precipitation consisting of particles of ice less than 5 mm in diameter, occurring either as frozen raindrops or as small hailstones.

Icing formation of a coating of ice on a solid object. See **clear ice** and **rime**.

Ideal gas law the thermodynamic law that applies to perfect gases.

Ignition pattern See **Firing pattern**.

Impingement See **plume impingement**.

Inclination angle the tilt angle of a surface relative to the horizon.

Insolation solar **radiation** received at the earth's surface.

Intertropical convergence zone the dividing line between the northeast **trade winds** of the Northern Hemisphere and the southeast trade winds of the Southern Hemisphere where air converges to produce **convection** and generally rising air.

Inversion See **temperature inversion**.

Iridescence brilliant colored borders or spots of color in clouds, usually red and green, caused by diffraction of light by small cloud particles. The phenomenon is usually observed in thin cirrus clouds within about 30° of the sun and is characterized by bands of color in the cloud that contour the cloud edges.

Isobar a line of equal or constant **pressure**; an isopleth of pressure.

Isotach a line on a weather map or chart connecting points where winds of equal speeds have been recorded.

Isotherm a line of equal or constant temperature; an isopleth of temperature.

Isothermal temperature that remains constant with height or time.

Isotropic of equal physical properties along all axes.

Jet a fast-moving wind current surrounded by slower moving air.

Jet stream strong, generally westerly winds concentrated in a relatively narrow and shallow stream in the upper **troposphere.**

Jet stream cirrus a loose term for filamentous **cirrus** that appears to radiate from a point in the sky and exhibits characteristics associated with strong vertical **wind shear**, such as twisted or curved filaments.

Jet wind speed profile a vertical profile of horizontal wind speeds characterized by a relatively narrow current of strong winds with slower moving air above and below. Large **wind shears** occur above and below the jet speed maximum.

Keetch–Byrum Drought Index a drought index representing the net effect of **evapotranspiration** and precipitation in producing cumulative moisture deficiency in deep **duff** and upper organic soil layers. This index is widely used in the southeastern United States.

Kelvin–Helmholtz waves vertical waves in the atmosphere associated with **wind shear** across stable layers. Can appear as breaking waves and as braided patterns in **radar** images and cloud photos.

Lake breeze a thermally produced wind blowing during the day from the surface of a large lake to the shore, caused by the difference in the rates of heating of water and land.

Lake breeze front the leading edge of a **lake breeze**, whose passage is often accompanied by showers, a wind shift, and/or a sudden drop in temperature.

Lake-effect storm a fall or winter storm that produces heavy but localized snowfall over the **lee** shoreline of a large open-water lake. Air is moistened and warmed as it flows over the lake, and snowfall is produced by **convergence** when the flow encounters increased roughness at the shoreline and is lifted up rising ground.

Land breeze a coastal breeze at night blowing from land to sea, caused by the difference in the rates of cooling of land and water.

Lapse rate the rate of decrease of air temperature with increase of elevation.

Latent heat heat absorbed or released during a change of water phase (from gas, liquid, or solid phases) at constant temperature and **pressure.**

Latent heat flux the **flux** of heat from the earth's surface to the atmosphere that is associated with evaporation or condensation of water vapor at the surface; a component of the **surface energy budget.**

Layered haze **haze** produced when air pollution from multiple **line**, **area**, or **point sources** is transported long distances to form distinguishable layers of discoloration in a stable atmosphere.

Leaf Area Index projected foliage area per unit area of ground surface.

Lee the side or part that is sheltered or turned away from the wind.

Leeward the side away from the wind. Compare **windward.**

Lee wave a wavelike oscillation of a flow that occurs in the **lee** of a mountain range when rapidly moving air is lifted up the steep front of a mountain range and oscillates downwind of the barrier. Compare **mountain wave.**

Lenticular cloud a very smooth, round or oval, lens-shaped cloud that is often seen, singly or stacked in groups, near or in the lee of a mountain ridge.

Lidar a device that is similar in principle and operation to radar but uses a laser to generate pulses of light (rather than radio waves) that are scattered back from aerosols in the atmosphere. The device is used to determine aerosol content and particle movement. From *li*ght *d*etection *a*nd *r*anging.

Lightning a visible electrical discharge produced by a **thunderstorm**. The discharge may occur within or between clouds, between a cloud and the air, or between a cloud and the ground.

Line source an array of pollutant sources along a defined path that can be treated in **dispersion** models as an aggregate uniform release of pollutants along a line. For example, the sum of emissions from individual cars traveling down a highway can be treated as a line source. Compare **area source** and **point source**.

Live fuel moisture the ratio of the amount of water to the amount of dry plant material in living plants.

Local convective wind in fire weather terminology, local, diurnal, thermally driven winds that arise over a comparatively small area and are influenced by local terrain. Examples include **sea** and **land breezes**, **lake breezes**, **diurnal mountain winds**, and convective currents.

Lofting a pattern of plume **dispersion** in a **stable boundary layer** topped by a **neutral** stability layer, in which the upper part of the plume disperses upward while the lower part of the plume undergoes little dispersion.

Longwave radiation a term used to describe the infrared energy emitted by the earth and atmosphere at wavelengths between about 5 and 25 micrometers. Compare **shortwave radiation**.

Long-wave ridge a **ridge** in the hemispheric **Rossby wave** pattern. Compare **short-wave ridge**.

Long-wave trough a **trough** in the hemispheric **Rossby wave** pattern. Compare **short-wave trough**.

Looping a pattern of plume **dispersion** in an **unstable** atmosphere in which the plume undergoes marked vertical oscillations as it is alternately affected by rising convective plumes and the subsiding motions between the plumes.

Loran *Lo*ng *Ra*nge *N*avigation, a system of long range navigation whereby latitude and longitude are determined from the time displacement of radio signals from two or more fixed transmitters.

Low See **cyclone**.

Lower Atmosphere Stability Index an atmospheric index used to indicate the potential for rapid **wildfire** growth. This index, also called the **Haines Index**, contains a **stability** term and a dryness term.

Low-level jet a regular, strong, nighttime, northward flow of maritime tropical air over the sloping Great Plains of the central United States, in which the wind increases to a peak in the lowest kilometer and then decreases above.

Low pressure center a region of low pressure enclosed by a pressure or height contour.

Low pressure system See **cyclone**.

Maloja wind a wind, named after the Maloja Pass between the Engadine and Bergell valleys of Switzerland, that blows down the valley of the Upper Engadine by day and up the valley at night. These wind directions are contrary to **mountain wind system** theory and have been explained as the encroachment of winds from the Bergell into the Upper Engadine.

Marine air intrusion invasion of an **air mass** with marine characteristics into a continental area.

Marine inversion **temperature inversion** produced when cold marine air underlies warmer air.

Massif a compact portion of a mountain range, containing one or more summits.

Mature stage the second phase in the life cycle of a **thunderstorm**, characterized by the presence of both updrafts and downdrafts within the cloud.

Mercury barometer an instrument for measuring atmospheric **pressure**. The instrument contains an evacuated and graduated glass tube in which mercury rises or falls as the pressure of the atmosphere increases or decreases.

Mesopause the top of the **mesosphere**, corresponding to the level of minimum temperature in the atmosphere found at 70–80 km.

Mesoscale pertaining to meteorological phenomena, such as wind circulations or cloud patterns, that are about 2–200 km in horizontal extent.

Mesosphere the atmospheric layer between about 20 km and about 70–80 km above the surface of the earth, extending from the top of the **stratosphere** (the **stratopause**) to the upper temperature minimum that defines the **mesopause** (the base of the **thermosphere**).

Meteorology the science dealing with the atmosphere and its phenomena.

Microburst an intense, localized downdraft of air that spreads on the ground, causing rapid changes in wind direction and speed; a localized **downburst**

Microclimate the climate of a small area, such as a cave, house, or wooded area, that may be different from that in the general region.

Microscale pertaining to meteorological phenomena, such as wind circulations or cloud patterns, that are less than 2 km in horizontal extent.

Midflame wind wind measured at the midpoint of the flames, considered to be most representative of the wind that is affecting **fire behavior.**

Millibar a unit of atmospheric **pressure** equal to $\frac{1}{1000}$ bar or 1000 dynes per square centimeter.

Mixed layer an atmospheric layer, usually the layer immediately above the ground, in which pollutants are well mixed by convective or shear-produced **turbulence.**

Mixing depth vertical distance between the ground and the altitude to which pollutants are mixed by **turbulence** caused by convective currents or vertical shear in the horizontal wind.

Mixing ratio a measure of humidity; the ratio of the mass of water vapor in an atmospheric volume to the mass of dry air in the volume.

Moist adiabatic lapse rate the rate at which the temperature of a parcel of saturated air decreases as the parcel is lifted in the atmosphere. The moist adiabatic lapse rate is not a constant like the **dry adiabatic lapse rate** but is dependent on parcel temperature and **pressure**.

Monsoon a thermally driven wind arising from differential heating between a land mass and the adjacent ocean that reverses its direction seasonally.

Mountainado a vertical-axis **eddy** produced in a **downslope windstorm** by the vertical stretching of horizontal roll **vortices** produced near the ground by vertical shear of the horizontal wind. Mountainadoes, when carried by the prevailing wind, can produce strong horizontal **wind shears** and wind gusts that are much more damaging than the prevailing wind speeds.

Mountain meteorology **meteorology** of a mountainous or topographically complex area.

Mountain–plain wind system a closed, large-scale, diurnal, thermally driven circulation between the mountains and the surrounding plain. The mountain-to-plain flow that makes up the lower branch of the closed circulation usually occurs during nighttime, whereas the plain-to-mountain flow occurs during daytime.

Mountain wave 1. a wavelike oscillation of a flow that occurs above and downwind of a mountain range when rapidly flowing air encounters the mountain range's steep front; 2. generic term for all

gravity waves occurring in the vicinity of or caused by mountains; 3. specific term for waves that form above, rather than downwind of, mountains. Compare **lee wave**.

Mountain wind system the system of diurnal winds that forms in a complex terrain area, consisting of **mountain–plain, along-valley, cross-valley**, and **slope wind systems.**

National Ambient Air Quality Standards in the United States, national standards for the ambient concentrations in air of different air pollutants; designed to protect human health and welfare.

National Fire Danger Rating System a uniform fire danger rating system used in the United States that focuses on the environmental factors that impact the moisture content of fuels. **Fire danger** is rated daily over large administrative areas, such as national forests.

Net all-wave radiation the net or resultant value of the upward and downward **longwave** and **shortwave radiative fluxes** through a plane at the earth–atmosphere interface; a component of the **surface energy budget**.

Neutral an atmospheric **stability** condition that exists in unsaturated (saturated) air when the **environmental temperature lapse rate** equals the **dry (moist) adiabatic lapse rate.**

Nieve penitente a spike or pillar of compacted snow, firn, or glacier ice, caused by differential melting and evaporation. The pillars form most frequently on low-latitude mountains where air temperatures are near freezing, dew points are much below freezing, and **insolation** is strong. Penitentes are oriented individually toward the noonday sun and usually occur in east–west lines.

Nimbostratus a cloud characterized by a formless layer that is almost uniformly dark gray; a rain cloud of the layer type, of low altitude, usually below 8000 ft (2400 m).

Nonattainment Area an area out of compliance with ambient air quality standards.

Obstruction the process by which low-level **air masses** that form on one side or the other of a mountain barrier are prevented from crossing the barrier.

Occluded front a composite of two fronts, formed as a **cold front** overtakes a **warm** or quasi-**stationary front**. Two types of occlusions can form, depending on the relative coldness of the air behind the cold front to the air ahead of the warm or stationary front. A **cold occlusion** results when the coldest air is behind the cold front and a **warm occlusion** results when the coldest air is ahead of the warm front.

Orographic wave a wavelike airflow produced over and in the **lee** of a mountain barrier. Collective term for **lee waves** and **mountain waves**.

Ozone a form of oxygen, O_3; a powerful oxidizing agent that is considered a pollutant in the lower **troposphere** but an essential chemical in the **stratosphere**, where it protects the earth from high-energy ultraviolet **radiation** from the sun.

Palmer Drought Severity Index an index used to gage the severity of drought conditions by using a water balance equation to track water supply and demand. This index is calculated weekly by the National Weather Service.

Particle trajectory model a computer submodel that tracks the trajectories of multiple particles that are released into an atmospheric flow model.

Permafrost a layer of soil at varying depths below the surface in which the temperature has remained below freezing continuously from a few to several thousands of years.

Perturbation model a computer model used to calculate air pollution concentrations. A perturbation model produces a **wind field** from solutions to a simplified set of equations that describe atmospheric motions.

Phenomenological model a computer model used to calculate air pollution concentrations. A phenomenological model focuses on an individual phenomenon, such as **plume impingement** or **fumigation**.

Photochemical smog air pollution containing **ozone** and other reactive chemical compounds formed by the reaction of nitrogen oxides and hydrocarbons in the presence of sunlight.

Pibal abbreviation for **pilot balloon.**

Pilot balloon a small helium-filled meteorological balloon that is tracked as it rises through the atmosphere to determine how wind speed and direction change with altitude; abbreviated as **pibal.**

Plume a continuous flow of air pollutants moving horizontally and/or vertically from an emission source and dispersing at a rate determined by atmospheric conditions.

Plume blight **visibility** impairment caused by air pollution plumes aggregated from individual sources.

Plume-dominated fire a fire whose behavior is governed primarily by the local wind circulation produced in response to the strong **convection** above the fire rather than by the **general wind.**

Plume impingement the collision of an air pollution plume with topography that rises above the plume altitude; often a temporary condition that occurs as the plume sweeps by the face of a hill as the wind shifts.

Polar front the variable **frontal zone** of midlatitudes that separates **air masses** of polar and tropical origin.

Polar front jet a strong, generally westerly **jet stream** wind concentrated in a relatively narrow and shallow current in the upper **troposphere** above the **polar front**; a feature of the **general circulation.**

Point source a pollutant source that can be treated in a **dispersion** model as though pollutants were emitted from a single point that is fixed in space; for example, the mouth of a smokestack. Compare **area source** and **line source**.

Powder snow dry, loose, unconsolidated **snow.**

Precipitation liquid or solid water particles that fall from the atmosphere and reach the ground. See **drizzle**, **hail**, **rain**, **snow**, and **snow pellets**.

Prescribed fire a management-ignited or natural wildland fire that burns under specified conditions where the fire is confined to a predetermined area and produces the **fire behavior** and fire characteristics required to attain planned resource management objectives.

Pressure the exertion of force upon a surface by a fluid (e.g., the atmosphere) in contact with it.
Pressure-driven channeling channeling of wind in a valley by **synoptic-scale pressure gradients** superimposed along the valley's axis. Compare **forced channeling**.
Pressure gradient force the force caused by the change in atmospheric **pressure** per unit of horizontal distance and acting in the direction in which pressure changes most rapidly.
Prevention of significant deterioration a program, specified in the Clean Air Act, whose goal is to prevent air quality from deteriorating significantly in areas of the country that are presently in compliance with ambient air quality standards.
Primary ambient air quality standards air quality standards designed to protect human health. Compare **secondary ambient air quality standards**.
Primary pollutant substances that are pollutants immediately on entering the atmosphere. Compare **secondary pollutant**.
Psychrometer a **hygrometer** whose operation depends on two similar thermometers with the bulb of one being kept wet (the **wet bulb thermometer**) so that the cooling as a result of evaporation causes its temperature to fall lower than that of the **dry bulb thermometer**. The difference between the **wet bulb temperature** and the **dry bulb temperature** is a measure of the dryness of the surrounding air. See also **sling psychrometer**.
Pyro-cumulonimbus **cumulonimbus** formed in the **convection column** of a fire.
Pyro-cumulus **cumulus** formed in the **convection column** of a fire.
Radar a device used to detect and determine the range to distant objects (e.g., **hydrometeors**) or atmospheric discontinuities by measuring the time for the echo of a radio wave to return from it; from *ra*dio *d*etection *a*nd *r*anging.
Radiation energy transport through electromagnetic waves. See **shortwave radiation** and **longwave radiation**.
Radiation fog a fog that forms when outgoing **longwave radiation** cools the near-surface air below its **dew-point temperature**.
Radiosonde an instrument that is carried aloft by a balloon to send back information on atmospheric temperature, **pressure**, and **humidity** by means of a small, expendable radio transmitter. See also **rawinsonde**.
Rain precipitation that falls to earth in droplets with diameters greater than 0.5 mm.
Rain shadow an area of reduced precipitation on the **lee** side of a mountain barrier caused by warming of air and dissipation of cloudiness as air descends the barrier.
Rawinsonde a **radiosonde** that is tracked by **radar**, radio direction finding, or navigation systems (such as the satellite Global Positioning System) to measure winds.
Red Flag Warning a term used by fire weather forecasters to alert land managers to an imminent or ongoing weather event that could cause dangerous fire activity.
Regional haze **haze** that is mixed uniformly between the surface and the top of a **convective boundary layer.**
Regional scale pertaining to meteorological phenomena that are about 500–5000 km in horizontal extent.
Relative humidity the ratio of the actual water **vapor pressure** at a given time to the vapor pressure that would occur if the air were saturated at the same ambient temperature.
Residence time the time, in seconds, required for the flaming front of a fire to pass a stationary point at the surface of the fuel.
Residual layer the elevated portion of a **convective boundary layer** that remains after a **stable boundary layer** develops at the ground (usually in late afternoon or early evening) and cuts off **convection.**
Resonance the state of a system in which an abnormally large vibration is produced in response to an external stimulus, occurring when the frequency of the stimulus is the same, or nearly the same, as the natural vibration frequency of the system.
Retrograding a westward shifting of a Rossby wave pattern that normally propagates to the east.
Return stroke an electrical discharge that propagates upward along a **lightning** channel from the ground to the cloud.
Ridge on a weather chart, a narrow elongated area of relatively high **pressure.**
Ridging the building or intensification of a **ridge** or **high pressure center.**
Rime an opaque coating of tiny, white, granular ice particles caused by the rapid freezing of **supercooled** water droplets on impact with an object. See also **clear ice**.
Ringfire a **firing pattern** or fire ignition pattern for **prescribed fires**, in which the fire is ignited in a ring to produce strong central **convection**, inward spread, and high **fire intensity**; variant of a **centerfire.**
Rossby waves a series of **troughs** and **ridges** on quasi-horizontal surfaces in the major belt of upper tropospheric **westerlies**. The waves are thousands of kilometers long and have significant latitudinal amplitude.
Rotor cloud a turbulent **altocumulus** or **cumulus** cloud formation found in the **lee** of some mountain barriers when winds cross the barrier at high speed. The air in the cloud rotates around an axis parallel to the mountain range.
Sampling frequency the rate at which sensor data are read or sampled.
Santa Ana Wind (in southern California) a strong, hot, dust-bearing **foehn** wind that descends to the Pacific Coast around Los Angeles from inland desert regions.
Sastrugi ridges of **snow** formed by wind on a snowfield.
Saturation vapor pressure the **vapor pressure** of a system at a given temperature, wherein the vapor of a substance is in equilibrium with a plane surface of that substance's pure liquid or solid phase.

Scattering the process in which a beam of light is diffused or deflected by collisions with particles suspended in the atmosphere.

Sea breeze a thermally driven wind that blows during the day from over a cool ocean surface onto the adjoining warm land, caused by the difference in the rates of heating of the ocean and land surfaces.

Sea breeze convergence zone the zone at the leading edge of a **sea breeze** where winds converge. The incoming air rises in this zone, often producing convective clouds.

Sea breeze front the leading edge of a **sea breeze**, whose passage is often accompanied by showers, a wind shift, or a sudden drop in temperature.

Secondary ambient air quality standards air quality standards designed to protect human welfare, including the effects on vegetation and fauna, **visibility**, and structures. Compare **primary ambient air quality standards**.

Secondary pollutant pollutants generated by chemical reactions occurring within the atmosphere. Compare **primary pollutant**.

Sensible heat flux the **flux** of heat from the earth's surface to the atmosphere that is not associated with phase changes of water; a component of the **surface energy budget**.

Separation eddy an **eddy** that forms near the ground on the **windward** or **leeward** side of a bluff object or steeply rising hillside. **Streamlines** above this eddy go over the object.

Severe thunderstorm a **thunderstorm** that produces heavy precipitation, frequent **lightning**, strong, gusty surface winds, or **hail**. A severe thunderstorm can cause **flash floods** and wind and hail damage, and it may spawn tornadoes.

Shortwave radiation a term used to describe the radiant energy emitted by the sun in the visible and near-ultraviolet wavelengths (about 0.1–2 micrometers). Compare **longwave radiation**.

Short-wave ridge a relatively small-scale **ridge** that is superimposed on and propagates through the longer wavelength **Rossby waves.**

Short-wave trough a relatively small-scale **trough** that is superimposed on and propagates through the longer wavelength **Rossby waves.**

Sleet in the United States, a term used to describe tiny **ice pellets** that are formed when **rain** or partially melted **snowflakes** refreeze before reaching the ground. Colloquial usage of the term coincides with British usage, which defines sleet as a mixture of rain and snow.

Sling psychrometer a **psychrometer** that is whirled by hand in the air to evaporate water from the **wet bulb thermometer** until the **wet bulb temperature** reaches a constant value.

Slope wind system a closed, thermally driven **diurnal** mountain wind circulation whose lower branch blows up or down the sloping sidewalls of a valley or mountain. The upper branch blows in the opposite direction, thereby closing the circulation.

Small hail same as **snow pellets** or **graupel.**

Smoke management the use of **meteorology**, fuel moisture, **fuel loading**, fire suppression, and burn techniques to keep smoke impacts from **prescribed fires** within acceptable limits.

Snow precipitation in the form of **ice crystals**, mainly of intricately branched, hexagonal form and often agglomerated into **snowflakes**, formed directly from the freezing of the water vapor in the air.

Snow cornice a mass of **snow** or ice projecting over a mountain ridge.

Snowflake an agglomeration of snow crystals falling as a unit.

Snow grain precipitation consisting of white, opaque ice particles usually less than 1 mm in diameter.

Snow pellets precipitation, usually of brief duration, consisting of crisp, white, opaque ice particles, round or conical in shape and about 2–5 mm in diameter; same as **graupel** or **small hail**

Snow pillow a windrow of **snow** deposited in the immediate **lee** of a snow fence or ridge.

Sodar an instrument similar in principle and operation to radar that uses sound to determine the range to distant objects (e.g., **hydrometeors**) or atmospheric discontinuities in temperature or **humidity** structure that scatter or reflect sound energy; from *so*und *d*etection *a*nd *r*anging.

Solstice either of the two times per year when the sun is at its greatest angular distance from the celestial equator: about June 21 (Northern Hemisphere summer solstice), when the sun reaches its northernmost point on the celestial sphere, or about December 22 (Northern Hemisphere winter solstice), when it reaches its southernmost point.

Sounding a set of data measuring the vertical structure of one or more atmospheric parameters (e.g., temperature, **humidity**, **pressure**, wind) at a given time.

Specific gravity the ratio of the **density** of any substance to the density of water.

Specific humidity the ratio of the mass of water vapor to the mass of moist air in an atmospheric volume.

Spot fire a fire ignited outside the perimeter of the main fire by a **firebrand**

Spotfire a **firing pattern** or fire ignition pattern for **prescribed fires**, in which a fire is ignited by firing equally spaced spots in rapid succession or using simultaneous ignition. The spacing is chosen so that the fires draw together rather than making individual runs.

Spotting outbreak of secondary fires as **firebrands** or other burning materials are carried ahead of the main **fire line** by winds.

Spray block a geographical area to be sprayed.

Stability the degree of resistance of a layer of air to vertical motion.

Stable See **absolutely stable**.

Stable boundary layer the stably stratified layer that forms at the surface and grows upward, usually at night or in winter, as heat is extracted from the atmosphere's base in response to longwave radiative heat loss from the ground. A stable boundary layer can also form when warm air is advected over a cold surface or over melting ice.

Stable core postsunrise, elevated remnant of the

temperature inversion that builds up overnight within a valley.

Standard Atmosphere a hypothetical vertical distribution of temperature, **pressure**, and **density**, which, by international consent, is taken to be representative of the atmosphere for purposes of pressure altimeter calibrations, aircraft performance calculations, aircraft and missile design, ballistics tables, etc.

State Implementation Plan a formal air quality management plan, produced by an individual state, specifying how state air resources will be managed to achieve federal and state standards.

Stationary front a **front** between warm and cold air masses that is moving very slowly or not at all.

Station model a specified pattern for plotting on a weather map the meteorological symbols that represent the state of the weather at a particular observing station.

Steam fog an **evaporation-mixing fog** that develops when a cold air mass flows over a warm body of water.

Stepped leader a faint, negatively charged **lightning** channel that emerges from the base of a **thunderstorm** and propagates toward the ground in a series of steps of about 1 microsecond duration and 50–100 meters in length, initiating a lightning stroke.

Stratiform (of a cloud) having predominantly horizontal development.

Stratocumulus a cloud characterized by large dark, rounded masses, usually in groups, lines, or waves, the individual elements being larger than those in **altocumulus** and the whole being at a lower altitude, usually below 8000 ft (2400 m).

Stratopause the boundary between the **stratosphere** and **mesosphere**

Stratosphere the atmospheric layer above the **troposphere** and below the **mesosphere**. It extends from the **tropopause**, usually 10–25 km high, to a height of approximately 20–25 km, where temperature begins to decrease.

Stratus a cloud characterized by a gray, horizontal layer with a uniform base, found at a lower altitude than **altostratus**, usually below 8000 ft (2400 m).

Streamline the path of an air parcel that flows steadily over or around an obstacle.

Subsidence a descending motion of air in the atmosphere occurring over a rather broad area.

Subsidence inversion a **temperature inversion** that develops aloft as a result of air gradually sinking over a wide area and being warmed by **adiabatic compression**, usually associated with subtropical high **pressure** areas.

Subtropical jet a strong, generally westerly wind concentrated in a relatively narrow and shallow stream in the subtropical upper **troposphere**; a feature of the **general circulation**

Supercooled a liquid cooled below its freezing point without solidification or crystallization.

Surface-based temperature inversion a **temperature inversion** with its base at the surface of the earth.

Surface energy budget the energy or heat budget at the earth's surface, considered in terms of the **fluxes** through a plane at the earth–atmosphere interface. The energy budget includes **net all-wave radiation**, and **sensible**, **latent**, and **ground heat fluxes**.

Surface weather chart an analyzed **synoptic** chart of surface weather observations. A surface chart shows the distribution of sea-level **pressure** (therefore, the position of **highs**, **lows**, **ridges**, and **troughs**) and the location and nature of **fronts** and **air masses**. Often added to this are symbols for occurring weather phenomena.

Swath width the width of the spray swath parallel to the path of the aerial or ground application vehicle over which a desired spray deposition is achieved.

Synoptic relating to or displaying atmospheric and weather conditions as they exist simultaneously over a broad area.

Synoptic scale pertaining to meteorological phenomena occurring on the spatial scale of the migratory high and low **pressure** systems of the lower **troposphere**, with length scales of 1000–2500 km and on time scales exceeding 12 hours.

20-foot wind wind measured or estimated over a 2- to 10-minute period at a standard level 20 feet above the vegetative surface or continuous tree canopy and used as input in fire danger models or for fire suppression activities.

Temperature a measure of the warmth or coldness of an object or substance with respect to some standard scale or value.

Temperature inversion a layer of the atmosphere in which air temperature increases with height.

Terpene any of a class of monocyclic hydrocarbons of the formula $C_{10}H_{16}$, obtained from plants.

Terrain-forced flow an airflow that is modified or channeled as it passes over or around mountains or through **gaps** in a mountain barrier.

Thermal belt a zone of high nighttime temperature and relatively low **humidity** that often occurs within a narrow altitude range on valley sidewalls. The thermal belt is especially evident during clear weather with light winds.

Thermally driven circulation See **diurnal mountain winds**.

Thermistor a resistor whose resistance changes with temperature and can therefore be used as a temperature sensor.

Thermosphere the atmospheric layer extending from the top of the **mesosphere** to outer space. It is a region of more or less steadily increasing temperature with height, starting at 70 or 80 km above the surface of the earth.

Thunder the sound caused by rapidly expanding gases in a **lightning** discharge.

Thunderstorm a local storm produced by a **cumulonimbus** cloud and accompanied by **lightning** and **thunder**

Time constant the time required for a measuring instrument to respond to 63.2% of a stepwise change in a measured quantity.

Time lag 1. same as **time constant**; 2. the time needed under specific atmospheric conditions for a fuel particle to lose about 63.2% of the difference between its initial moisture and its equilibrium moisture contents.

Towering cumulus a tall **cumulus** cloud, extending through low and middle cloud levels.

Trade winds any of the nearly constant easterly winds that dominate most of the tropics and subtropics, blowing mainly from the northeast in the Northern Hemisphere and from the southeast in the Southern Hemisphere.

Transpiration the passage of water vapor into the atmosphere through the vascular system of plants.

Transport wind for **smoke management** computations, the average wind speed and direction through the depth of the **mixed layer**

Trapper a valley or basin in which cold air becomes trapped or pooled.

Tropopause the boundary between the **troposphere** and **stratosphere**, characterized by an abrupt change in temperature **lapse rate** (temperatures decrease with height in the troposphere, but increase or remain constant with height in the stratosphere).

Troposphere the portion of the earth's atmosphere from the surface to the **tropopause**; that is, the lowest 10–20 km of the atmosphere. The troposphere is characterized by decreasing temperature with height and is the layer of the atmosphere containing most clouds and other common weather phenomena.

Trough on a weather chart, a narrow, elongated area of relatively low **pressure**

Troughing the deepening or intensification of a **trough** or **low pressure center**

Turbulence irregular motion of the atmosphere, as indicated by gusts and lulls in the wind.

Undersun an optical effect seen by an observer who is above a cloud deck and is looking toward the sun. Sunlight is reflected upward off the horizontally oriented **ice crystals** in the cloud deck below.

Unstable See **absolutely unstable**.

Unstable boundary layer See **convective boundary layer**.

Upper air weather chart an analyzed synoptic chart of upper-air weather observations at standard levels (850, 700, 500, 300 mb) in the troposphere.

Upslope fog a fog that forms when moist, stable air is cooled as it is lifted up a mountain slope.

Upslope wind a **diurnal** thermally driven flow directed up a mountain slope and usually occurring during daytime; part of the **slope wind system**

Up-valley wind a diurnal thermally driven flow directed up a valley's axis, usually occurring during daytime; part of the **along-valley wind system**

Upwelling the process by which warm, less dense surface water is drawn away from a shoreline by offshore currents and replaced by cold, denser water brought up from the subsurface.

Valley exit jet a strong elevated down-valley air current issuing from a valley above its intersection with the adjacent plain.

Valley volume effect the effect of the reduction in volume of a valley or basin (compared to a volume with a horizontal floor, an equal depth, and an equal area at the top) on temperature change. The temperature change for an equal heat flux is greater in the valley volume than in the flat-floor volume.

Vapor pressure the pressure exerted by the water vapor molecules in a given volume of air.

Ventilation index product of **mixing depth** and **transport wind** speed, a measure of the potential of the atmosphere to disperse airborne pollutants from a stationary source.

Venturi effect the speedup of air through a constriction due to the **pressure** rise on the upwind side of the constriction and the pressure drop on the downwind side as the air diverges to leave the constriction; also called the **Bernoulli effect**

Virga streaks of water droplets or ice particles that fall out of a cloud and evaporate before reaching the ground.

Visibility the distance at which an object can be seen and identified with the naked eye.

Visibility Protection Program the program specified by the **Clean Air Act** to achieve a national goal of remedying existing impairments to **visibility** and preventing future visibility impairment throughout the United States.

Visible spectrum the portion of the electromagnetic spectrum to which the eye is sensitive, that is, light with wavelengths between 0.4 and 0.7 micrometers. Compare **shortwave radiation** and **longwave radiation**.

Volatile fraction mass fraction of a sprayed liquid that could evaporate in the atmosphere before depositing.

Vortex a whirling mass of air in the form of a column or spiral. A vortex can rotate around either a horizontal or a vertical axis.

Wake the region of **turbulence** immediately behind a solid body caused by the flow of air over or around the body.

Warm front a transition zone between a mass of warm air and the colder air it is replacing.

Warm occlusion a **frontal zone** formed when a **cold front** overtakes a **warm front**. When the air behind the cold front is warmer than the air ahead of the warm front, it leaves the ground and rises up and over the denser, colder air. Compare **cold occlusion**.

Warm rain process the collision and coalescence of water droplets in a cloud, producing rain.

Warm sector the region of warm air within a **low pressure center** located between an advancing **cold front** and a retreating **warm front**

Wasatch wind a strong easterly wind blowing out of of the canyons of the Wasatch Range onto the plains of Utah, or down the west slopes of the range.

Weather the state of the atmosphere with respect to wind, temperature, cloudiness, moisture, **pressure**, etc.

Westerlies the prevailing winds that blow from the west in the midlatitudes of both the Northern and Southern Hemispheres.

Wet bulb depression the difference between the **dry** and **wet bulb temperatures** measured by a **psychrometer**, used to determine atmospheric **humidity**

Wet bulb temperature in a **psychrometer**, the temperature indicated by the **wet bulb thermometer**.

Wet bulb thermometer in a **psychrometer**, the thermometer whose bulb is kept moistened as air flows by; used in conjunction with a **dry bulb thermometer** to determine atmospheric **humidity**.

Wildfire an unwanted fire that requires measures of control.

Wind air in natural motion relative to the earth's surface.

Wind chill See **wind chill equivalent temperature**.

Wind chill equivalent temperature the apparent temperature felt on the exposed human body owing to the combination of temperature and wind speed.

Wind-driven fire a fire whose behavior is governed primarily by a strong consistent wind, rather than by the convective circulation produced by the fire itself.

Wind field the three-dimensional spatial pattern of winds.

Wind rose a diagram, for a given locality or area, showing the frequency and strength of the wind from various directions.

Wind shear the rate of wind velocity change with distance in a given direction (e.g., vertically). The shear can be speed shear (where speed but not direction changes between the two points), directional shear (where direction but not speed changes between the two points), or a combination of the two.

Winds of Most Concern to Firefighters winds that dominate the fire environment, rendering fire suppression activities dangerous or ineffective.

Windthrow See **blowdown**.

Windward the side toward the wind. Compare **leeward**.

Abbreviations and Acronyms

A arctic
ABL atmospheric boundary layer
Ac altocumulus
ACSL altocumulus standing lenticularis, a lens-shaped cloud that forms over mountains
AGDISP Agricultural Dispersion Model, a particular atmospheric dispersion model used for treating the transport and **diffusion** of aerially sprayed pest control agents in agricultural applications
As altostratus
AWDN Automated Weather Data Network, a network of automatic weather stations located in the High Plains area of the United States
BKN broken clouds, $\frac{5}{8}$ to $\frac{7}{8}$ of the sky covered
c continental
CAA Clean Air Act, a public law in the United States that regulates air quality
Cb cumulonimbus
CBL convective boundary layer
Cc cirrocumulus
CD-ROM Compact Disk–Read-Only Memory, a diskette on which scientific and music data are stored
Ci cirrus
CI capping inversion
CLR clear, no clouds
Cu cumulus
Cs cirrostratus
DALR dry adiabatic lapse rate
DGPS Differential Global Positioning System, a special form of **GPS** that improves the accuracy of geographical position determinations by applying corrections that are obtained from a separate GPS receiver located at a precisely known location
DOD (U.S.) Department of Defense
EPA (U.S.) Environmental Protection Agency
ETLR environmental temperature lapse rate
ETS environmental temperature sounding
EZ entrainment zone
FA free atmosphere
FEW few clouds, $\frac{1}{8}$ to $\frac{2}{8}$ of the sky covered
FORTRAN Formula Translation, a high-level computer programming language used chiefly for solving problems in science and engineering
FSCBG Forest Service, Cramer, Barry, Grim Model; an aerial spray dispersion model named for its sponsor and developers
GIS Geographical Information System, a computer-based graphics program that allows the superposition of plan-maps of thematic elements, such as roads, rivers, land use patterns, and the like to aid in local or regional planning activities.
GMT Greenwich mean time, same as **UTC**
GOES Geostationary Operational Environmental Satellite, a series of earth satellites launched into an orbit over the equator that keeps them stationary above the same place on earth at all times
GPS Global Positioning System, a navigation system that uses a constellation of artificial earth satellites to make precise determinations of the latitude and longitude of locations on the earth's surface or in the atmosphere.
GypsES Gypsy Moth Decision Support System
HPCC High Plains Climate Center
ITCZ Intertropical convergence zone, a wind **convergence** zone that forms near the equator, frequently producing lines of **thunderstorms** and shifting north and south with the seasons
LAI Leaf Area Index, projected foliage area per unit area of ground surface
LASI Lower Atmosphere Stability Index
LDT Local daylight time
LLJ low-level jet

LST Local standard time
m maritime
MALR **moist adiabatic lapse rate**
METAR Meteorological Aviation Routine Weather Report, a weather report coded in an international format that is used worldwide
ML **mixed layer**
MSL above mean sea level
MST Mountain standard time
NAAQS **National Ambient Air Quality Standards**, (U.S.) national standards for ambient air pollution concentrations designed to protect human health and welfare
NASA (U.S.) National Aeronautics and Space Administration
NDVI Normalized Difference Vegetation Index, an index that measures the **greenness** of vegetation and is evaluated radiometrically from earth satellite data
NEXRAD next generation weather **radars**, a network of **NWS** Doppler radars now being installed and operated throughout the United States
NOAA (U.S.) National Oceanic and Atmospheric Administration
NRCS (U.S.) Natural Resources Conservation Service
Ns nimbostratus
NWS (U.S.) National Weather Service
OVC overcast, sky covered with clouds
P polar
PDSI **Palmer Drought Severity Index**
PM-10 particulate matter with diameter below 10 micrometers
PM-2.5 particulate matter with diameter below 2.5 micrometers
PSD **Prevention of Significant Deterioration**, air quality regulations issued by the **EPA** to prevent the deterioration of air quality in pristine areas
RAWS Remote Automated Weather System, a network of automatic weather stations in the western United States
RL **residual layer**
SBL **stable boundary layer**
Sc stratocumulus
SCT scattered clouds, $\frac{3}{8}$ to $\frac{4}{8}$ of the sky covered
SI Système International, an internationally agreed-upon set of units to be used in scientific work
SIP **State Implementation Plan**
SNOTEL Snowpack Telemetry Network, a network of high-altitude stations in the western United States that provides information on winter snowpack for water supply planning
St stratus
T tropical
TAF Terminal Aerodrome Forecast, an airport weather forecast coded in an international format that is used worldwide
TCU **towering cumulus**
U.S. United States
USDA United States Department of Agriculture
USFS United States Forest Service
UTA Uniform Time Act, a law that governs the measurement of time in the United States
UTC Coordinated universal time; the international time standard kept at Greenwich, England
VMD volume median diameter; a statistical measure of the average droplet size in a spray cloud, such that 50% of the volume of sprayed material is composed of droplets smaller in diameter than the volume median diameter
WIMS Weather Information Management System
WRCC Western Regional Climate Center
WSFO Weather Service Forecast Office
Z Zulu time, same as **UTC**

Index

Figures and tables are indicated by "f," "ff," "t," and "tt" in the locators. Names of mountains are indexed in "inverted" order, for example "Everest, Mount" rather than "Mount Everest."

ABL (atmospheric boundary layer), 42, 46–48, 47ff
absorption, 5
acid precipitation, 206, 223, 224f
ACSLs (altocumulus standing lenticulars), 148
  See also lenticular clouds
Adams, Mount (Washington), 15–16
adiabatic compression, 151, 153
adiabatic processes, 38
Adirondack Mountains (New York), 14
advection fogs, 96
advection of troughs/ridges, 56
aerial spraying, 273–97
  aircraft used for, 273, 274f, 276, 277–78ff, 280–81ff, 280–82, 296
  and atmospheric stability, 284–85
  computer modeling of, 293–95, 295f, 296
  data collection for, 274, 288–91, 290f
  and diurnal mountain flows, 287–88
  drift potential of, 273–74, 282, 284, 287
  equipment for, 276, 279
  evaluation of, 296
  for fire suppression, 273
  forecast support for, 274, 296
  geographic information/positioning technology for, 276–77, 277f
  goals of, 273, 275
  and humidity/temperature, 283–84
  materials sprayed, 275–76
  operational plans for, 274, 295
  for pest control, 273, 274f, 275–76, 284
  phenological/meteorological time constraints on, 275
  and precipitation, 275, 285
  project personnel/training for, 295–96
  and public relations, 297
  for seeding/fertilizing, 273, 275
  spray deposition, 285–86
  and terrain-forced flows, 288
  typical, 275
  and wind fields/droplet trajectories, 279–83, 280t, 281f, 283f
aeroallergens, 205, 213
aerosols, 27, 89, 102, 103, 206
aircraft
  for aerial spraying, 273, 274f, 276, 277–78ff, 280–81ff, 280–82, 296
  vortices from, 280–81ff, 280–82
aircraft sondes, 268
air density
  and altitude, 7
  definition of, 31
  and height, 31, 32f
  and the ideal gas law, 306
air masses
  air pollutants in, 229
  boundaries between (*see* fronts)
  definition of, 73
  mountain ranges as barriers to, 12, 16–17, 50
  source regions and trajectories for, 73–74, 74f
  types of, 73
air mass thunderstorms, 115–17, 116–17ff
air pollutants, 27, 48, 205–36
  acid precipitation, 206, 223, 224f
  in air masses, 229
  and air quality studies/models, 209–12, 265
  and air stability/plume behavior, 213, 217–18, 217f
  assessing potential of, 235–36

air pollutants (*continued*)
and the Clean Air Act, 206–7, 209f
and cold air pools, 190
definition of, 205
dilution/diffusion of, 212–13, 213–14ff
dispersion of, and synoptic weather categories, 218–21, 219ff, 220t
dispersion of, in diurnal mountain flows, 221–22, 229–35, 231f, 233f, 235f
dispersion of, in terrain-forced flows, 221–22, 226–29, 228f
and elevated inversions, 215–16, 216f
and emission standards, 207–8, 209f
factors affecting site concentrations of, 209–10
haze, 207, 215, 222, 224, 225–27ff
and mixing depths, 209, 216, 216f, 224
and National Ambient Air Quality Standards, 206, 207–8, 208t
regional/hemispheric, effects of mountains on, 222–23, 222t
smog, 206, 216, 232
and surface-based inversions, 213–15, 215f
types/sources of, 205–6
in valleys, 230–31, 232
and visibility impairments, 224, 225f, 227f
visibility of, 205
and wind speed, 212–13, 213–14ff
air pressure. *See* atmospheric pressure
air quality. *See* air pollutants
air soundings, 268
air toxics, 208
Alaska Range, 11, 12f, 15
albedo of surfaces, 45, 46f
Aleutian Low, 8f, 9, 17
Allegheny Mountains (Appalachians), 14
along-valley winds, 171, 172f, 174f, 180, 186, 186f, 187–93, 188f, 190–93ff
Alps, 159, 161–62ff
Altamont Pass (California), 165
altimeters, 53
altitude
and air density, 7
and angle of the sun, 5
and atmospheric moisture, 6
and atmospheric pressure, 31, 32–33ff, 32t, 52–54, 55
and boiling point of water, 37t
and friction, 71
and humidity, 244–45
and precipitation, 6, 108f
and solar radiation, 5
and temperature, 4–5, 6, 39, 244–45
and wind speeds, 6–7
altocumulus/altostratus clouds, 82, 84, 104f
altocumulus standing lenticulars (ACSLs), 148
See also lenticular clouds
Anderson Creek Valley (California), 183, 184f
anemometers, 65, 112, 290, 292
Antarctica, 201
anticyclones/anticyclogenesis, 50, 51f
See also high pressure systems; low pressure systems
antiwinds, 188
Appalachian Mountains, 13f
climate in, 11, 13–15
as a climatic divide, 168
cold air damming in, 167
cyclones in, 50
elevations in, 13–14
flow splitting/postfrontal cold air in, 161
forests in, 14
major ridges in, 14
precipitation in, 105, 109
winds in, 151, 159, 161, 172f
Arapahoe Basin Ski Area (Rocky Mountains), 149
Arctic Circle, 19–20
Arizona Monsoon. *See* Mexican Monsoon
atmosphere
aerosols in, 27 (*see also under* precipitation)
boundary layer and surface energy budget, 42–48, 43–47ff
free, 6
gases in, 26–27
humidity of, 28–29, 28f
moisture in, 6
scales of motion of, 25–26, 26f
stability of (*see* stability, atmospheric)
Standard Atmosphere, 32, 32t, 33f, 34, 35f
vertical motions in, 4, 49–50
vertical structure of, 31–33, 32–33ff, 32t
water phase changes in, 29–30, 29f
See also atmospheric pressure; *entries that begin with "air"*; temperature; winds
atmospheric pressure, 49–60
and altitude/height, 31, 32–33ff, 32t, 52–54, 55
analyses of, 54–56, 55f
definition of, 31
hemispheric waves, 25, 56, 57–58ff, 58–59
measurement of, 52
pressure differences over mountain-plain areas, 194–96, 196f, 197, 200f
pressure gradients, 59–60, 59f, 188–89, 188f
pressure sensors, 291
and wind direction, 59–60, 59f
See also entries that begin with "high pressure" or "low pressure"
Automated Weather Data Network (AWDN), 133–34
autumnal equinox, 3
AWDN (Automated Weather Data Network), 133–34
azimuth angle, 180, 312, 315–18tt

Badwater (Death Valley), 21
Baker, Mount (Washington), 15–16
balloons, weather, 70–71, 129, 253, 267–68, 268f, 289, 291
banner clouds, 85, 89, 91, 92f, 145–46
barometers, 52, 53

barrier jets, 159–60ff, 160
basins, 197–98
Beaufort, Sir Francis, 66
Beaufort wind scale, 66, 67t
belt weather kits, 266, 267f
Bergell Valley (Switzerland), 192
Bermuda-Azores High, 7–8, 8f, 9, 14, 19, 216
Bernoulli effect (Venturi effect), 165, 165f, 190
Best's formulas, 305
Bighole Basin (Montana), 197
Big Thompson Flood (Big Thompson Canyon, Colorado), 118
billow clouds, 85, 91–92, 92–93ff, 146
bird migration, 169
Black Hills, 105
Blackwell Fire (Payette National Forest, McCall, Idaho), 237
blizzards, 50
blocking, 165–66, 167f
blowup fires. *See* plume-dominated fires
Blue Mountains (Washington), 108f
Blue Ridge Mountains (Appalachians), 14
boras, 151, 162f
Boulder (Colorado), 153, 156, 156f, 225f
box models, 211
breezes, 199–200, 200f
Brocken specter, 94, 105
Brooks Range (Alaska), 12, 159
Brush Creek Valley (Colorado), 183, 183f, 232, 233f, 234, 235f
Butte (Montana), 113
Buys-Ballot, Christoph Hendrik Diederik, 60
Buys-Ballot Rule, 60, 61f

CAA (Clean Air Act), 206–7, 209f
California Low, 8–9, 8f
Canadian Rockies, 11
Canyonlands National Park (Utah), 227f
canyon winds (Wasatch winds; Utah), 154, 157, 158f
cap clouds, 85, 89, 91f, 146
Cape Denison (Antarctica), 201
capping inversions, 48
Caracena Straits (California), 20, 163–64
carbon dioxide, 26, 27
cars, pollution from, 206, 208
Cascade Range, 12f
  climate in, 11, 15–16
  elevations in, 11, 15–16
  glaciation in, 16
  precipitation in, 17, 105, 106, 108f, 112
  rime accumulations in, 112
  winds in, 152f, 157, 159, 160f, 163
Catskill Mountains (New York), 14
CBL (convective boundary layer), 48
ceilometers, 82, 84
Celsius scale, 38, 38t, 309, 311t
Central Valley (California), 19, 20, 106
CFCs (chlorofluorocarbons), 222–23
channeling, 60, 113–14, 114f
  forced, 165, 166f, 185
  pressure-driven, 161–62, 185, 185f
chemistry models, 212
Chernobyl, 222
chinook arch clouds, 85, 86–87, 87–88ff, 146, 147
chinook pauses, 153
chinook winds (foehn winds; "snoweaters"), 152ff, 156f
  in the Alps, 159
  chinook pauses, 153
  and fires, 249, 249f, 257t
  and forest blowdowns, 119
  in Havre, Montana, 153, 154, 155f
  and health, 159
  and lenticular clouds, 85
  in the Rockies, 151, 154, 155f
  temperature oscillations caused by, 153, 154, 155f
  warmth of, 151, 153, 153f, 155f, 159
chlorofluorocarbons (CFCs), 222–23
circulation cells, 61–62, 62f
cirrocumulus/cirrostratus clouds, 81–82
cirrus clouds, 82, 84
Clausius-Clapeyron equation, 303
Clean Air Act (CAA), 206–7, 209f
Clean Air Act Amendments, 207, 223
clear ice, 111, 112
climate
  in the Alaska Range, 15
  and altitude, 4–7
  in the Appalachians, 13–15
  in the Cascade Range, 15–16
  in the Coast Range, 15
  and continentality, 7
  definition of, xv
  in intermountain regions, 19–21
  and latitude, 3–4
  and regional circulations, 7–9
  in the Rocky Mountains, 18–19
  in the Sierra Nevada, 15, 16
clouds, 81–94
  altocumulus/altostratus, 82, 84, 104f
  banner, 85, 89, 91, 92f, 145–46
  billow, 85, 91–92, 92–93ff, 146
  cap, 85, 89, 91f, 146
  chinook arch, 85, 86–87, 87–88ff, 146, 147
  cirrocumulus/cirrostratus, 81–82
  cirrus, 82, 84
  classification of, 81–82, 82tt, 83f, 84
  cloud condensation nuclei, 89
  cloud vs. rain droplets, 103
  cumulonimbus, 82, 84
  foehn (chinook) wall, 85–86, 86f, 146
  fractocumulus/fractostratus, 85, 92, 93f
  and fronts, 75, 77, 77f, 79, 84
  funnel, 118
  glaciated, 104–5, 104f
  high/middle/low, 81–82, 82t, 137
  jet stream cirrus, 85, 92, 93f, 94

clouds (*continued*)
lenticular (*see* lenticular clouds)
METAR reports on, 137, 137t, 320–22tt
mountain influences on formation of, 84–85, 85f (*see also under* terrain-forced flows)
at night, 82, 84
nimbostratus, 82, 84
optical effects in, 94, 105
precipitation from, 84 (*see also* precipitation)
pyro-cumulus/pyro-cumulonimbus, 258
rotor, 85, 86, 88f, 146, 151f
stratiform vs. cumuliform, 81
stratocumulus, 82, 84
towering cumulus, 82
coastal ranges, flows through, 162–64, 163–64ff
Coast Range, 11, 12f, 15, 16–17, 105
cold air avalanches, 186
cold air damming, 165–66, 167
cold air pools, 190, 191f, 231–32
cold occlusion, 79
collection efficiency, 286
Colorado Lows, 50
Colorado Plateau, 19, 21
Colorado Springs (Colorado), 113
Columbia Basin, 95–96, 163
Columbia Plateau, 19, 21, 165
Columbia River Gorge (Oregon), 163, 164f, 165
Columbia River Valley (British Columbia), 234
communication networks, 266
complex terrain, 7
See also terrain-forced flows
compression heating, 153
computer modeling, 293–95, 295f, 296
continentality, 4–5, 6
convection, 4, 49, 113–14, 114f
convection-dominated fires, 257, 258
convective boundary layer (CBL), 48
convergence, 58–59, 62, 113–14, 114f, 146, 160
conversion tables, 310–11tt
Coordinated Universal Time (UTC), 129, 311
Copper River Valley (Alaska), 163–64
Coriolis force, 59–60, 59f
coronas, in clouds, 94, 105
cross-valley winds, 171, 172f, 186, 186f, 193–94, 195f
Cumberland Mountains (Appalachians), 14
cumulonimbus clouds, 82, 84
cyclones/cyclogenesis, 50, 51f, 75, 76f

DALR (dry adiabatic lapse rate), 39–41, 40–41tt, 102, 103f
data loggers, 290, 292
Davenport-Wieringa roughness length classification, 71t
day length, 3, 4f, 313, 315–17tt
daylight savings time, 311, 312
Death Valley, 11, 21
Denali (Alaska Range), 15, 54, 91
Denver, 13–14, 232
dew-point temperature, 305
DGPS (differential GPS), 278
Dinaric Alps (Eastern Europe), 162f
dispersion of emissions, 209
diurnal winds, viii, 171–202
and aerial spraying, 287–88
air-pollution dispersion in, 221–22, 229–35, 231f, 233f, 235f
along-valley systems, 171, 172f, 174f, 180, 186, 186f, 187–93, 188f, 190–93ff
in basins, 197–98
and convergence zones, 160
coupled period (daytime), 173f, 174, 177–78
cross-valley systems, 171, 172f, 186, 186f, 193–94, 195f
cycle of winds/temperature structure, 172, 173f, 174–79
decoupled period (nighttime), 173f, 174, 175–76, 177f, 288
evening transition period, 172, 173–76ff, 174–75, 288
on forested vs. snow-covered slopes, 181, 182f
formation of, 60–61
and larger scale flows, 182–85, 183–84ff, 197
and local thermally driven systems, 199–202, 200–201ff
morning transition period, 173f, 174, 176–77, 178ff, 287–88
mountain-plain systems, 171–72, 172f, 174f, 186, 194–95, 196f, 197, 200f
over plateaus, 198–99
progression of wind directions on valley sidewalls, 178–79, 179–80ff
slope wind systems, 171, 172f, 174f, 176–77, 177f, 180, 186–87, 187f
stability of, 71
and surface energy budget, 179–82, 181–82ff, 187
See also wind speeds
dividing streamline height, 141–42, 142f, 226–27
DOD (U.S. Department of Defense), 278
Doppler weather radar, 269, 271f
downbursts, 118–19, 119ff
downslope flows, peak, 232–33, 233f
See also diurnal winds
downslope windstorms
boras, 151, 162f
in Boulder, 153, 156, 156f
chinooks (*see* chinook winds)
damage from, 153–54, 156
forecasting of, 150, 157
formation of, 149–50
gap winds, 157–58
and lee waves, 150, 154
and short-wave troughs, 150
sudden onset/cessation of, 153
and topography, 150–51
wind speeds in, 153
downwash, 228–29

drift (aerial spraying), 273–74, 282, 284, 287
droplets
cloud vs. rain, 103
collection efficiency of, 286
evaporation of, 283, 284
and humidity, 283
size/settling velocities of, 279–80, 280t, 283, 286, 294, 305, 305t
trajectories of, 279, 282–83, 283f
droughts, 111, 255–56
dry adiabatic lapse rate. *See* DALR
dry line, 108
Dude Fire (Arizona), 243
duff, 256
dust, 205, 206, 207, 213, 215, 222
dust devils, 145

Eagle Valley (Colorado), 175, 176f, 230
easterly winds (trade winds), 62, 64
East winds (Cascade Range), 152f, 157
eddies (air), 144, 145–46, 146–47ff, 154, 212, 214f
elevated inversions, 215–16, 216f
elevation. *See* altitude
El Niño, 64
Elsinore convergence zone (Santa Ana Mountains, California), 146
emission standards, 207–8, 209f
emissivity, 45–46, 46f
Engadine Valley (Switzerland), 192
entrainment zone, 48
Environmental Protection Agency (EPA), 206, 207
environmental temperature lapse rate. *See* ETLR
environmental temperature sounding (ETS), 33, 40f
EPA (Environmental Protection Agency), 206, 207
equator, 3, 4f, 61
equinoxes, 3
ETLR (environmental temperature lapse rate), 40–41, 40–41ff
ETS (environmental temperature sounding), 33, 40f
evaporation-mixing fogs, 96
evapotranspiration, 241
Everest, Mount, 53, 54
exponential notation, 309

FAA (Federal Aviation Administration), 131, 289
Fahrenheit scale, 38, 38t, 311t
Federal Aviation Administration (FAA), 131, 289
Finkenbach Valley (Germany), 190, 191f
Fire Orders, 240, 240t
fires/wildfires, 157, 157f, 205
backfires for, 238, 238ff
behavior of, 237, 242, 244f, 246–47, 257–59, 258–59ff
convection columns above, 247
crown fires, 237, 256
danger ratings for, 241–42, 248, 256, 271 (*see also* fire weather)
from downslope windstorms, 153, 157, 157f
environment for, 239–40, 240–41tt, 240f
fire lines for, 237–38
and fire weather (*see* fire weather)
fire whirls, 247
and forest composition, 255
and fuels (*see* fuels)
intensity of, 237, 249
lightning-caused, 269
monitoring/forecasting smoke dispersion, 264–71
prescribed, 205, 238–39, 259–65, 260f, 261t, 262f, 264f, 264t
residence time of, 237
and safety, 240–41tt
severity of, 246–47, 246t
shape of, 249–50
and slope angle, 249–50, 250f
spot fires, 238, 239f
winds induced by, 250–51
See also smoke
fire weather, 242–59
atmospheric stability, 251–53, 251f, 252t, 253f
definition of, 242
droughts, 255–56
extreme fire behavior, 257–59, 258–59ff
fire seasons, 254, 254t, 255
Fire Weather Watches/Red Flag warnings, 256, 257tt
humidity/temperature, 243–47, 244–45ff, 246t, 252–53
monitoring/forecasting of, 253, 265–71, 267–71ff
wind, 247–51, 248–50ff
Fire Weather Program, 270–71
fire winds, 202
flagged trees, 66, 69–70, 69–70f
flash density, 111f
flooding, 110, 111, 118, 153, 269
flow separation, 70
flow splitting, 159–60, 160f
foehn pauses, 153
See also chinook winds
foehn (chinook) wall clouds, 85–86, 86f, 146
foehn winds. *See* chinook winds
fogs
definition of, 94
frequency of, 95, 95f
oil, 232
and temperature inversion, 96, 96f
types of, 94–97
in valleys, 177, 178f
visibility in, 94
water content of, 94
Foraker, Mount (Alaska Range), 15
forest blowdowns, 118, 119, 120f, 149
forests
aerial spraying in (*see* aerial spraying)
composition of, and fires, 255
and diurnal winds, 181, 182f
foliage density in, 285–86
pollution from, 206
forest windthrows, 118, 119, 120f, 149

Fort Yukon (Alaska), 20
fractocumulus/fractostratus clouds, 85, 92, 93f
Fraser Plateau (Rocky Mountain Trench), 20, 97
Fraser Valley (British Columbia), 163–64
free air winds, 247
free atmosphere, 6
freezing level, 17, 34–36
friction, 60, 71, 72t
frontal inversions, 216
Front Range (Colorado), 149
fronts, 74–79
  and clouds, 75, 77, 77f, 79, 84
  cold frontal passages, 77, 78t
  and low pressure centers, 74–75
  and mountain barriers, 79, 160–61, 161ff
  occluded, 79
  polar, 75, 76ff
  and precipitation, 79
  sea/lake breeze, 200
  types of, 74, 75f
  warm frontal passages, 77, 78t, 79
frostbite, 35
FSCBG model, 294, 295f
fuels, 243
  definition of, 239
  fine fuels, 245
  and fire severity, 246–47, 246t
  forest fuels, 255
  fuel loading, 239
  moisture content of, 239, 241–42, 246–47, 246t, 255–56, 266–67
  and slope microclimates, 244f
  time lags for, 242, 242t, 243f
fugitive dust, 213
fumigations, 234–35, 235f
funnel clouds, 118

gap winds, 20, 161–65, 165–66ff
gases, 26–27, 206
gas law, 32, 306
gasoline, nonleaded, 208
Gaussian models, 211–12, 293, 294
general circulation/winds, 61–62, 62f, 247
Geographic Information Systems (GIS), 277
Geostationary Operational Environmental Satellite (GOES), 270
geostrophic winds, 60
Geysers area (California), 233
GIS (Geographic Information Systems), 277
glaciated clouds, 104–5, 104f
Glacier Peak (Washington), 15–16
glacier winds, 201–2, 201f
Global Positioning System. *See* GPS
Global Telecommunications System, 130
GOES (Geostationary Operational Environmental Satellite), 270
Gore Valley (Colorado), 190, 230
GPS (Global Positioning System), 70, 277–78, 278f
Granite Fire (Oregon), 245
graupel, 99–100, 101f, 105, 105f
gravity waves, 147, 148f
Great Basin, 19, 21, 106, 165
Great Basin High, 8f, 9
Great Lakes, 143
Great Plains
  climate of, 19
  dry line on, 108
  fogs over, 95
  low-level jets over, 168–69, 168f
  prescribed fires in, 265
Great Smoky Mountains (Appalachians), 14
greenhouse effect, 27
Green Mountains (Vermont), 14
Greenwich mean time (Coordinated Universal Time), 129, 311
Gstettneralm sinkhole (Austria), 197
Gulf Stream (Atlantic Ocean), 9
Gunnison River (Colorado), 191–92
GypsES modeling system, 278f

Hadley cell, 62
Haines Index (Lower Atmosphere Stability Index), 251–52, 252t, 259
halos, in clouds, 94, 105
Handies Peak (Colorado), 146f
Harz Range (Germany), 94
Havre (Montana), 153, 154, 155f
haze, 207, 215, 222, 224, 225–27ff
heat exhaustion, 36
heat flux, 42, 44, 44f, 45–46, 48, 164–65
Heat Index (apparent temperature), 36, 37t
heat lows, 164–65
heatstroke, 36
height. *See* altitude
helical flow trajectories, 186, 186f
high altitudes, measurements at, 12–13
High Plains Climate Center (Lincoln, Nebraska), 133–34
high pressure belts, 4, 6f, 62
high pressure centers, 4, 25, 41–42, 49–50
high pressure systems, 196f
  Bermuda-Azores High, 7–8, 8f, 9, 14, 19, 216
  Great Basin High, 8f, 9
  Pacific High, 7–8, 8f, 17, 216
  and regional winds, 7, 8f
  and tropospheric convergence/divergence, 58–59
  on weather charts, 54, 55f, 127, 128
  and wind direction, 59–60, 59f
  See also anticyclones/anticyclogenesis
Himalayan Monsoon, 109–10
Himalayan Mountains, 109–10
Hood, Mount (Oregon), 15–16
Hudson Bay trading post (Astoria, Oregon), 151
humidity
  and aerial spraying, 283–84
  determination of, 28–29, 307–8t
  and elevation, 244–45

and fires, 243–47, 244f, 246t, 252–53
and fire severity, 246–47, 246t
humidity variables, 303–5, 304f, 305t
sensors for, 291, 292
and surface properties, 245–46, 245f
and temperature, 28–29, 28f, 37t
Hunter, Mount (Alaska Range), 15
hurricanes, 62–63, 145
hydraulic flows, 149, 150, 151, 151f, 154
hydrometeors, 99
hygrometers, 28, 29, 243, 305
hygrothermographs, 37
hygrothermometers, 266
hypothermia, 35

ice crystal process, 103
ice fogs, 97
ice nuclei, 103
ice storms, 35, 36f
icing, 111–12, 112f
ideal gas law, 32, 306
inclination angle, 180, 315–18tt
Inn Valley (Austria), 179, 191, 193f
insect migration, 169
Inside Passage (Alaska and Canada), 20, 164
Intermountain Basin (west of the Rockies), 166
intermountain regions, 19–21, 254t
Internet, weather data on, 135, 136, 266, 271
intertropical convergence zone (ITCZ), 62
iridescence, in clouds, 94, 105
isobars, 54–55
isolation (solar radiation), 19–20
isothermal layers, 34
isotherms, 55f
ITCZ (intertropical convergence zone), 62

Jackson Hole (Wyoming), 106
Japanese Current (Pacific Ocean), 9, 96
Jefferson, Mount (Oregon), 15–16
jet stream cirrus clouds, 85, 92, 93f, 94
jet winds
barrier, 159–60ff, 160
jet streams, 58, 58f, 63–64, 65f
low-level, 107, 168–69, 168f, 249
valley exit, 190–92, 193f

Kauai (Hawaii), 104
Keetch-Byrum Drought Index, 256
Kelvin-Helmholtz waves, 91–92
Kelvin scale, 309, 311t

LAI (Leaf Area Index), 285–86
lake breezes, 199–200, 200f
lake-effect storms, 143
land breezes, 99–200, 200f
lapse rates, 39–41, 39–41tt
LASI (Lower Atmosphere Stability Index; Haines Index), 251–52, 252t, 259
Lassen, Mount (California), 15–16
latent heat, 29–30, 29f
latitude
and angle of the sun, 5f, 315–18tt
and day length, 3, 4f, 313, 315–17tt
and high/low pressure belts, 4
and weather variability, 3
lead, atmospheric, 208
Leaf Area Index (LAI), 285–86
leeside convergence, 113–14, 114f
lee waves, 146, 147, 148ff, 150, 154
length, conversion table for, 310t
lenticular clouds
and chinook winds, 85
as indicators of moisture, 89
as indicators of mountain/lee waves, 146, 147, 148, 148ff
over/downwind of mountains, 63–64
shape/formation of, 87, 90f
lightning
calculating distance from/direction of, 125
cloud-to-ground, 122–23ff, 124
deaths/damage from, 119, 121
definition/formation of, 119, 121–22, 121f
detection of, 268–69
fires started by, 269
flash density of, 111f
forecasting of, 124, 126
frequency of, 119, 123–24ff
ground strokes, 124ff, 125
ground-to-cloud, 122, 122f, 124
and precipitation, 119, 121–22, 124–26
protection from/hazard areas for, 124ff, 125
during severe thunderstorms, 117–18
thunder produced by, 119, 121, 125
Livingston (Montana), 153
LLJs (low-level jets), 168–69, 168f, 249
local convective winds, 247
See also diurnal winds
Logan, Mount (Canada), 54
Logan Sinks (Utah), 197
Loisach Valley (Germany), 179, 180f
Long's Peak (Colorado), 118
longwave radiation, 43, 181
Loran (Long Range Navigation), 70
Los Angeles Basin, 216, 232
Lower Atmosphere Stability Index (LASI; Haines Index), 251–52, 252t, 259
low-level jets (LLJs), 107, 168–69, 168f, 249
low pressure belts, 4, 6f, 62
low pressure centers, 25
and fronts, 74–75
mountain influences on, 84–85, 85f
naming of, 4
and vertical motions, 41–42, 49–50
low pressure systems, 196f
Aleutian Low, 8f, 9, 17
California Low, 8–9, 8f

low pressure systems (*continued*)
and regional winds, 7, 8f
and tropospheric convergence/divergence, 58–59
on weather charts, 54, 55f, 127, 128
and wind direction, 59–60, 59f
See also anticyclones/anticyclogenesis

Mackinnon Pass (Southern Alps, New Zealand), 163f
Maloya wind systems, 192
MALR (moist adiabatic lapse rate), 39–41, 39–41tt, 103, 103f
Mann Gulch Fire (Montana), 243
maps, weather. *See* weather maps/charts
marine inversions/intrusions, 253
maritime influence, 17
Marshall Creek Canyon (Montana), 180–81, 181f
mass, conversion table for, 310t
massifs, temperature within, 6
Matterhorn, 91, 92f
measurements at high altitudes, 12–13
mesopause, 31
mesoscale weather events, 25
mesosphere, 31, 32f
METARs (Meteorological Aviation Routine Weather Reports), 135–37, 135f, 137t, 320–22tt
meteorology, definition of, xv
Mexican Monsoon (Southwest Monsoon), 110–11ff, 197
microbursts, 118, 120f
microclimates, 12–13
microscale meteorology, 25
midflame winds, 248, 248f
Miller-Holzworth method, 264, 265f
minisonde systems, 268
Missoula (Montana), 113
Mitchell, Mount (North Carolina), 14, 224f
mixing depths, 209, 216, 216f, 224, 264, 265f
mixing ratio (water vapor/dry air), 304
Mogollon Rim (Arizona), 110, 111f
moist adiabatic lapse rate. *See* MALR
moisture
and altitude, 6
in fuels, 239, 241–42, 246–47, 246t, 255–56, 266–67
lenticular clouds as indicators of, 89
meters for, 266–67
in soil, 241, 256
Mojave Desert, 165
monsoons, 25, 109–11
monsoon winds, 19
Mossau Valley (Germany), 190, 191f
mountainadoes, 154
mountain barriers
and air masses, 12, 16–17, 50
and fronts, 79, 160–61, 161ff
mountain-plain winds, 171–72, 172f, 174f, 186, 194–95, 196f, 197, 200f
mountains in North America
latitudinal extent of, 11
major ranges and peaks, 12f, 16f
microclimates in, 12–13
mountain waves, 146–49, 148ff, 150f
Mount Washington Observatory (Mt. Washington, New Hampshire), 113f
Mt. Zirkel Wilderness areas (Rocky Mountains), 149

NAAQS (National Ambient Air Quality Standards), 206, 207–8, 208t
Naches Pass (Cascade Range), 163
National Ambient Air Quality Standards (NAAQS), 206, 207–8, 208t
National Climatic Data Center (Asheville, North Carolina), 134
National Fire Danger Rating System, 241–42, 248, 256, 271
National Oceanic and Atmospheric Administration. *See* NOAA
National Water and Climate Center (Portland), 134
National Weather Service (NWS), 127, 134–35, 256, 269, 270–71
National Weather Service Forecast Office (WSFO), 138, 271
Natural Resources Conservation Service (NRCS), 134
Navier-Stokes equations, 294
NDVI (Normalized Difference Vegetation Index), 270
net all-wave radiation (term *R)*, 43–44, 43f, 45–46
NEXRAD Network, 269
nieve penitente, 96f
nimbostratus clouds, 82, 84
NOAA (National Oceanic and Atmospheric Administration), 131, 134, 266, 270
Nonattainment Areas, 206
Normalized Difference Vegetation Index (NDVI), 270
North American Monsoon. *See* Mexican Monsoon
North Park (Colorado), 106
North Pole
angle of the sun at, 5f
day length at, 4f
radiation loss at, 61
weather variability in, 3
wind circulation at, 61–62, 62f
See also Arctic Circle
Northwest Territories, 97
NRCS (Natural Resources Conservation Service), 134
nuclear fallout, 222
NWS. *See* National Weather Service

obstruction of air masses, 165–66, 167–68, 168f
ocean currents, 9
Ohio Valley, 223
oil fog, 232
Oklahoma, 95
Olympic Mountains (Washington), 15, 17, 105, 106, 108f, 160
Olympus, Mount (Washington), 15
orographic lifting, 104, 113–14, 114f
orographic waves. *See* mountain waves

Owens Lake (California), 215
Owens Valley (California), 215
ozone (photochemical smog), 206, 216, 232
ozone layer, 222–23

Pacific High, 7–8, 8f, 17, 216
Palmer Drought Severity Index, 256
Paonia (Colorado), 191
Paradise Ranger Station (Mt. Rainier, Washington), 104
particle trajectory models, 212
passes, flows through, 161–62
Payette National Forest (McCall, Idaho), 237, 238f
permafrost, 20
pest control, 273, 274f, 275–76, 284
phenomenological models, 212
Phoenix, 232
photochemical smog (ozone), 206, 216, 232
Pico de Orizaba (Mexico), 54
pilot balloons, 267–68, 268f
plateaus, 198–99
plume-dominated fires, 257, 258–59, 259f
plumes, air-pollutant, 217–18, 217f, 224, 225f, 226–27, 231, 233–34
polar regions. *See* Arctic Circle; North Pole; South Pole
pollutants. *See* air pollutants
Port Walter (Baranof Island, Alaska), 17
postfrontal cold air, 160–61
precipitation, 99–126
  acid precipitation, 206, 223, 224f
  and activities, 99
  and aerial spraying, 275, 285
  and aerosols, 27, 102, 103
  and altitude/elevation, 6, 108f
  in coastal mountain ranges, 7, 16–17
  day to day/diurnal variations in, 106–7
  and droughts, 111
  and flooding, 110, 111, 118
  formation of, 102–5, 103–4ff
  icing, 111–12
  intensity of, 101, 102t
  and lightning, 119, 121–22, 124–26
  maps of, 106
  and maritime influence, 17
  measurement of, 101–2
  from middle/low clouds, 84
  rain shadows, 17
  record levels of, 104
  seasonal variations in, 108–11, 109f
  during severe thunderstorms, 117–18
  and terrain, 105–6, 107f
  thunderstorms, 107, 112–19
  types of, 99–101, 100t
  and warm fronts, 79
  year to year variations in, 111
  See also snow; thunderstorms
pressure
  conversion table for, 310t
  differences over mountain-plain areas, 194–96, 196f, 197, 200f
  and the ideal gas law, 306
  pressure gradients, 59–60, 59f, 188–89, 188f
  sensors for, 291
Prevention of Significant Deterioration (PSD), 206–7
PRISM model, 107f
Prospect Creek (Alaska), 19–20
PSD (Prevention of Significant Deterioration), 206–7
psychrometric tables, 307–8t
Puget Sound convergence zone, 160
Pyranees, 161
pyro-cumulus/pyro-cumulonimbus clouds, 258

radiation
  longwave, 43, 181
  net all-wave, 43–44, 43f, 45–46
  shortwave (*see* solar radiation)
  shortwave gain vs. longwave loss, 61
radiation fogs, 95–96
radiometers, 291
radiosondes, 33, 34f, 129–30, 134f, 161
rain, 104
  See also precipitation
rainbows, 105
rain droplets, size of, 103
rain gauges, 101–2
rain shadows, 17
Ranier, Mount (Washington), 15, 91, 104, 188, 192
Raton (New Mexico), 113
rawinsondes, 33
RAWS (Remote Automated Weather System), 133–34, 268, 269–70ff
Red Flag Warnings, 256, 257tt
reflectivity (albedo), 45, 46f
regional-scale weather events, 25–26
relative humidity, 28, 28f
Remote Automated Weather System. *See* RAWS
residual layer (of the atmosphere), 48
resonance, of mountain waves, 148
Rhone Valley (Switzerland), 106
Richland (Washington), 17
ridges, 54, 56, 57f, 58
rime, 99, 100, 111–12, 113f
Robson, Mount (Canadian Rockies), 11
Rock Creek Valley (Montana), 120f
Rocky Flats (Colorado), 153
Rocky Mountains
  barrier jets in, 159, 169
  chinook winds in, 151, 154, 155f
  climate in, 11, 18–19
  as a climatic divide, 168
  clouds in, 63–64, 86
  cold air damming in, 167
  cyclones/anticyclones in, 50, 51f
  elevations in, 11, 18
  fire seasons in, 254t
  flow splitting/postfrontal cold air in, 161

Rocky Mountains (*continued*)
forest blowdown in, 149
jet streams in, 63
Northern vs. Southern, 18
precipitation in, 18, 105
pressure systems in, 196f
Rossby waves in, 56, 58
thunderstorms in, 18–19, 107, 110–11, 113, 115f
Rocky Mountain Trench, 19, 20, 50
Rossby waves, 25, 56, 57–58ff, 58, 59
rotor clouds, 85, 86, 88f, 146, 151f
Routt National Forest (Rocky Mountains), 149

Sacramento Valley (California), 20
safety
and fires, 240–41tt
and lightning, 124ff, 125
and smoke, 261
and temperature, 35–36
Salt Lake City, 232
Salt Lake Valley, 95–96
sampling (aerial spraying), 291–93
sand dunes, 66, 70, 144f
San Fernando convergence zone (Santa Monica Mountains, California), 146
San Joaquin Valley (California), 20
Santa Ana Mountains (California), 160, 200
Santa Ana winds (Los Angeles Basin), 152f, 154, 157–58ff
Santa Monica Mountains (California), 160, 200
sastrugi, 66, 68f, 69
satellites, 270, 278
saturation vapor pressure, 103, 303–4, 304f
SBL (stable boundary layer), 47–48, 47f
scattering, 5
Schultz eddy (Caracena Straits, California), 146
scientific notation, 309
sea breezes, 199–200, 200f
sea level, 55
SeaTac International Airport (Washington), 160
Seattle, 105–6, 108f
separation eddies, 142–43, 144f
Sequim (Washington), 17, 106
shadows, 315, 317–18t
Shasta, Mount (California), 15–16
shelter belts, 145
shortwave radiation. *See* solar radiation
SI (Système International), 309
Sierra Nevada, 12f
climate in, 11, 15, 16
cyclones in, 50
elevations in, 11, 16
precipitation in, 17, 105
storms in, 17
winds in, 159, 160f
Sierra Wave Project, 85
SIPs (State Implementation Plans), 206
sleet, 35, 99
sling psychrometers, 28–29, 243
slope winds, 171, 172f, 174f, 176–77, 177f, 180, 186–87, 187f
smog (ozone), 206, 216, 232
smoke
and health/safety, 261
management of, 239, 259–65, 262f, 264f, 264t
monitoring/forecasting dispersion of, 264–71
from multiple fires, 255, 255f
as pollution, 205, 206, 207, 213, 222, 232–33, 233f, 260
smokestacks, 228–29
Snake River Plain (Idado), 21
Snoqualmie Pass (Cascade Range), 163
SNOTEL (Snowpack Telemetry Network), 134
snow, 35
blizzards, 50
density of, 100–101
and diurnal winds, 181, 182f
graupel, 99–100, 101f, 105, 105f
measurement of, 102
nieve penitente, 96f
record levels of, 104
specific gravity of, 100
snow cornices, 67, 68f, 69, 145, 146, 146f
snowdrifts, 66–67, 68f
"snoweaters." *See* chinook winds
snow fences, 67, 68f, 145
Snowpack Telemetry Network (SNOTEL), 134
sodars, 268, 291
soil, 44, 241, 256
solar radiation
and altitude, 5
factors affecting, 45, 46
and heat flux, 48
seasonal/diurnal variations in, 3, 19, 43–44, 43f
and slope angle/aspect, 244, 312–13, 314f, 315–18tt
and surface albedo, 45, 46f
and temperature, 6, 19
in valleys, 195f
solstices, 3, 313
SOLSUB, 313, 315, 315t
sounding systems/devices, 25, 33, 34f, 268
South Boulder Creek (Colorado), 191–92
South Canyon Fire (Colorado), 255, 256
South Park (Colorado), 106
South Pole
angle of the sun at, 5f
radiation loss at, 61
weather variability in, 3
wind circulation at, 61–62, 62f
Southwest Monsoon. *See* Mexican Monsoon
speed, conversion table for, 310t
spray blocks, 277, 278f
St. Helens, Mount (Washington), 15–16
stability, atmospheric
and aerial spraying, 284–85
and air pollutants, 213, 217–18, 217f

determination of, 38–42, 39t, 40–41ff, 251ff, 253
of diurnal winds, 71
and fires, 251–53, 251f, 252t, 253f
index of, 251–52, 252t
and temperature, 33
and terrain-forced flows, 141–42, 142f, 147, 148f
stable boundary layer (SBL), 47–48, 47f
Stampede Pass (Cascade Range), 163
Standard Atmosphere, 32, 32t, 33f, 34, 35f
State Implementation Plans (SIPs), 206
steam fogs, 96
Stikine Valley (Alaska), 163–64
storms
El Niño's effects on, 64
ice, 35, 36f
in the Rocky Mountains, 18–19
and Rossby waves, 56
in the Sierra Nevada, 17
weather conditions associated with, 78t
See also thunderstorms
Strait of Juan de Fuca (Washington), 163–64
stratocumulus clouds, 82, 84
stratopause, 31
stratosphere, 31, 32f, 58, 147
subsidence, 4, 49
subsidence inversions, 216, 253, 253f
summer solstice, 3, 313
Summit Fire (Oregon), 259
Sundance Fire (Idaho), 250
sundogs, 105
sunrise/sunset, 180–81, 181f, 313, 315–17tt
surface energy budget, 42, 43–46, 44–46ff
and diurnal mountain winds, 179–82, 181–82ff, 187
Swiss Plateau (Switzerland), 159
synoptic-scale weather systems, 25, 26
and pollution dispersion, 218–21, 219ff, 220t
Système International (SI), 309

Table Mesa (Boulder), 153
TAFs (Terminal Aerodrome Forecasts), 135–37, 136f, 320–22tt
Taku Straits (Alaska), 163–64
Tehachapi Pass (California), 165
temperature
and aerial spraying, 283–84
and air stability, 33
along-valley differences in, 189–90, 190f
and altitude/height, 4–5, 6, 39, 244–45
atmospheric, 31, 33–38, 34–36ff, 36–37tt
in basins, 197
and chinook winds, 153, 154, 155f
and continentality, 4–5, 6
conversion table for, 311t
and fires, 243–46, 244–45ff, 252–53
freezing level, 17, 34–35
horizontal differences in, 34
and humidity, 28–29, 28f, 37t
and the ideal gas law, 306
inversion of, 6, 34, 96, 96f, 189, 252–53 (*see also* diurnal winds)
over land vs. ocean, 109
within massifs, 6
measurement of, 36–38, 38t, 44–45
pole-equator contrasts in, 61
and safety concerns, 35–36
and snow, 35
of soil, 44
and solar radiation, 6, 19
and surface properties, 245–46, 245f
temporal changes in, 33
and winds, 6
Tennessee Valley, 172f
Terminal Aerodrome Forecasts. *See* TAFs
terpenes, 27, 206
terrain channeling, 60
terrain-forced flows, 141–69
and aerial spraying, 288
air-pollution dispersion in, 221–22, 226–29, 228f
and air speed, 141
and air stability, 141–42, 142f, 147, 148f
around mountains, 158–61, 159–61ff
and atmospheric dispersion, viii
clouds/snow as indicators of, 146, 147–48, 148ff, 158
definition of, viii
eddies/wakes/vortices, 68f, 144–46, 146–47ff, 154
formation of, 60–61
hydraulic, 149, 150, 151, 151f, 154
low-level jets, 168–69, 168f, 249
mountain/lee waves, 146–49, 148ff, 150f
obstructions to, 165–68, 167–68ff
through gaps/channels/passes, 161–65, 163–66ff
and topography, 141, 142–44, 143–45ff, 145t
and wildfires, viii
See also downslope windstorms
Teten's formula, 307–8t
thermal belts, 174, 175f
thermally driven circulations, 141, 199–202
See also diurnal winds
thermistors, 36–37
thermocouples, 36–37
thermometers, 36–37, 290
thermosphere, 31, 32f
Three Sisters (Oregon), 15–16
thunderstorms
air mass thunderstorms, 115–17, 116–17ff
definition/formation of, 112–13
diurnal variations in, 107, 109f
downbursts in, 118–19, 119ff
downdrafts in, 184
and the dry line, 108
forecasting of, 115
frequency of, 113
and the low-level jet, 107
ordinary, 114–15
severe, 114–15, 117–19, 119ff
in summer vs. winter, 113

thunderstorms (*continued*)
and terrain-related/convective mechanisms, 113–14, 114–15ff
transience of, 113
and winds, 249
See also lightning
Tibetan Plateau, 56, 58
time zones/conversions, 311–12tt, 311f
Tincup (Taylor Park, Colorado), 197
tornadoes, 118–19, 145
towering cumulus clouds, 82
towers, 289, 291
trade winds (easterly winds), 62, 64
transport winds, 264
Trinity Alps (California), 160f
tropics, 3, 4f, 61
tropopause, 31
troposphere, 31, 32f, 41, 58–59, 147
troughs, 54, 56, 57f, 58, 59, 150
turbulence, 85, 91–92, 144, 145, 154, 282–83, 283f, 292
Turnagain Arm (Alaska), 163–64
20-foot winds, 248, 248f

undersuns, 105
units/unit conversion, 309–10, 309–11tt
upslope fogs, 94–95
upwelling, 64
U.S. Bureau of Reclamation, 131, 133
U.S. Department of Defense (DOD), 278
U.S. Department of the Interior, 131, 133
U.S. Forest Service, 131, 133
USDA Forest Service, 294
UTC (Coordinated Universal Time), 129, 311

Vail (Colorado), 225f, 230
Vail Ski Resort (Colorado), 190
valleys
air pollutants in, 230–31, 232
basins, 197–98
fogs in, 177, 178f
solar radiation in, 195f
volume of, 189–90, 190f
vegetation greenness, 270
ventilation index, 264–65
Venturi effect (Bernoulli effect), 165, 165f, 190
vernal equinox, 3
virga, 105, 106f
visibility, 27, 94, 207, 224, 225f, 227f
Visibility Protection Program, 207
visible spectrum, 43
volcanic ash/dust, 205, 222
volume, conversion table for, 310t
vortices (air), 68f, 145, 147f

wakes (air), 144, 147f
Walpurgis Night, 94
warm occlusion, 79
warm rain process, 103
Wasatch winds (canyon winds; Utah), 152f, 154, 157, 158f
Washington, Mount (New Hampshire), 14, 113f
Watch Out Situations, 240, 241t
water, boiling point of, 37t
watersheds, 223
waterspouts, 145
water vapor, 26–27
weather, definition of, xv
weather forecasting
aerial spraying supported by, 274, 296
automatic meteorological stations for, 131, 133–34
Buys-Ballot Rule used for, 60, 61f
data collection for, 129–31, 133–34, 134f, 268–70, 269f, 271f
data dissemination for, 134–35
of downslope windstorms, 150, 157
for federal projects, 138
of fires, 253, 265–71, 267–71ff
of gap winds, 164
guidelines for, 128–29, 130, 133t
of lightning, 124, 126
METARs used for, 135–37, 135f, 137t, 320–22tt
of precipitation, difficulties of, 107
of smoke dispersion, 264–71
surface observations for, 130–31
TAFs used for, 135–37, 136f, 320–22tt
of thunderstorms, 115
See also weather maps/charts
Weather Information Management System, 271
weather maps/charts
data used to produce, 127
forecasts made from, 128–29, 133t
fronts on, 127, 128, 128f
high/low pressure systems on, 54, 55f, 127, 128
isobars on, 127, 128f
for precipitation, 106
station models for/symbols on, 127, 128–32ff
surface vs. upper air charts, 54–55, 55f
See also weather forecasting
Weather Radio, 266
westerly winds, 63
Western Regional Climate Center (Reno), 133–34
White Mountains (New Hampshire), 14
White Mountains (California and Nevada), 111f
Whitney, Mount (Sierra Nevada), 16, 21
wildfires, 157, 157f, 205
wildland fires. *See* fires/wildfires
Willamette Valley (Oregon), 17
wind chill equivalent temperature, 35–36, 36t, 306
wind-driven fires, 257–58, 258f
wind-field models, 210–11
wind fields, 279, 282
wind machines, 165
wind rose, 231f
winds, 60–72
antiwinds, 188

circulation of, 61–64, 62–63ff
directions of, 49, 59–60, 59f, 62–63, 64–66
and fires, 247–51, 248–50ff
foehn, 119 (*see also* chinook winds)
frontal, 249
geostrophic, 60
glacier, 201–2, 201f
and height countours, 55f, 56
and high/low pressure systems, 7, 8f, 58
importance of, viii
Maloya, 192
measurement of, 247–48, 248f
observations of, 65–67, 66–68tt, 68–70ff, 69–70, 248
strong break-ins, 183–85
synoptic-scale, 247, 248f
and temperature, 6
and thunderstorms, 117–18, 119f, 249
transport, 264
vertical structure of, 70–71, 71f
whirlwinds, 249
wind shear, 91, 92, 94, 145, 154
Winds of Most Concern to Firefighters, 248–49
See also diurnal winds; terrain-forced flows; wind speeds
wind speeds, 72t
and air pollutants, 212–13, 213–14ff
and altitude, 6–7, 71, 71f
and complex/rough terrain, 7, 72t
day-night changes in, 71
and friction, 60, 72t
and high/low pressure centers, 49
of low-level jets, 168, 168f, 249
observations/measurement of, 65–66, 66–67tt, 70f, 292
profiles of, 306–7
and Rossby waves, 56
over rough/smooth surfaces, 143
of valley exit jet winds, 190–92, 193f
See also terrain-forced flows
windstorms. *See* downslope windstorms
wind surfing, 164f
wind vanes, 112, 290
winter solstice, 3, 313
World Meteorological Organization, 136
WSFO (National Weather Service Forecast Office), 138

Yellowstone area fires (1988), 255
Yukon Plateau (Rocky Mountain Trench), 20, 97
Yukon Territory, 50

Zulu time (Coordinated Universal Time), 129, 311